AF750328

Sustainable Building with Earth

MIX
Papier aus verantwortungsvollen Quellen
Paper from responsible sources
FSC® C105338

If you have any concerns about our products,
you can contact us on
ProductSafety@springernature.com

In case Publisher is established outside the EU,
the EU authorized representative is:
Springer Nature Customer Service Center GmbH
Europaplatz 3, 69115 Heidelberg, Germany

Printed by Libri Plureos GmbH
in Hamburg, Germany

Horst Schroeder

Sustainable Building with Earth

Springer

Horst Schroeder
Weimar, Thüringen, Germany

ISBN 978-3-319-19490-5 ISBN 978-3-319-19491-2 (eBook)
DOI 10.1007/978-3-319-19491-2

Library of Congress Control Number: 2015947254

Springer Cham Heidelberg New York Dordrecht London
© Springer International Publishing Switzerland 2016
This work is subject to copyright. All rights are reserved by the Publisher, whether the whole or part of the material is concerned, specifically the rights of translation, reprinting, reuse of illustrations, recitation, broadcasting, reproduction on microfilms or in any other physical way, and transmission or information storage and retrieval, electronic adaptation, computer software, or by similar or dissimilar methodology now known or hereafter developed.
The use of general descriptive names, registered names, trademarks, service marks, etc. in this publication does not imply, even in the absence of a specific statement, that such names are exempt from the relevant protective laws and regulations and therefore free for general use.
The publisher, the authors and the editors are safe to assume that the advice and information in this book are believed to be true and accurate at the date of publication. Neither the publisher nor the authors or the editors give a warranty, express or implied, with respect to the material contained herein or for any errors or omissions that may have been made.

Printed on acid-free paper

Springer International Publishing AG Switzerland is part of Springer Science+Business Media (www.springer.com)

Translation from German language:
Cornelia Theimer Gardella / Kurt Gardella,
Erfurt, Germany and Santa Fe, NM, USA
Graphic works: Katharina Triebel, Leipzig,
Germany

Foreword

Earth has been used as a building material for millennia. Architecture in the ancient cultures of Egypt, the Middle East, China, Central Asia, and Latin America was closely tied to this material. In Central Europe, as well, there is archaeological evidence of the use of earth as a building material for thousands of years. Within the individual regions, practical experiences with the material and the resulting building rules were passed down for generations leading to construction methods which were affordable and optimally developed for the respective climates. The buildings were constructed from locally available materials which were sourced using environmentally friendly methods. Earthen structures blended well into the landscape and shaped the picture of rural regions and urban settlements over the centuries. "Recycling" of the buildings did not pose any problems: earth building materials could be reused indefinitely or could be returned to the natural cycle without harming the environment.

In modern times, all of these aspects can be more or less summed up under the term "sustainable building." For a long time, building materials and architectural design were mainly assessed in terms of structural design, material technology, and economy. Today, however, ecological criteria, particularly a building's energy consumption and its impact on the environment, have become increasingly important in the interest of sustainable development. Clients are requesting nontoxic, healthy building materials which create a comfortable indoor climate. Other popular aspects are the sensual characteristics of building elements, such as unusual textures as well as pleasant tactile surface qualities and a wide range of colors. These add to the desirability of earth as a building material.

In this context, earth can be seen in a new light after years of being marginalized from conventional construction by industrially mass-produced building materials. Today, private as well as public clients are increasingly opting for the building material earth.

This book describes the planning and execution of earth building projects from a modern perspective: it highlights the preservation of traditions for historic conservation and renovation projects while, at the same time, showing current trends in modern earth building. Special emphasis is placed on aspects of sustainability and on how earth can be combined with other "modern" building materials.

The idea of the life cycle of earth as a building material is the recurring theme of this book: it covers all of the processing steps of the soil including the sourcing and extraction of the material, the soil's preparation and processing into building materials and building elements, the useful life of the finished building and its maintenance, and, finally, the demolition and recycling of the building which completes the cycle. Today, the life cycle model of a building material is a generally accepted methodological approach which is used for the quantitative recording and assessment of the production of building materials and building products with regard to sustainability.

This book sums up my years of experience working with earth as a building material. It reflects the knowledge I have gained through my practical work on building sites, as part of my research and teaching at the Bauhaus University in Weimar, Germany, and as a consultant for various national and international organizations and clients. Above all, it has been my 20 years of serving as the President of the German Association for Building with Earth (Dachverband Lehm e.V.), as well as professional exchanges with members of the association while working on numerous projects, which have contributed to the writing of this book.

The first German edition, published in 2010, was soon out of print. The second edition was published in 2013 and incorporated many changes, especially the newly published DIN standards for earth building materials. Many of my international colleagues have expressed an interest in an English translation of this book. Together with my publisher, I have now decided to respond to this wish and hope that the English translation will be met with the same level of interest as the original German version.

The second, revised German edition, which has been reviewed and slightly updated, serves as the basis for the English translation. I hope that earth builders worldwide benefit from this book.

Weimar, Germany
March 2015

Horst Schroeder

Contents

List of Symbols

Symbol	Unit	Parameter	PG[a]
A	m^2	Area	3, 4, 5
A	Bq	Activity	5
A	kN/cm	Compaction work	3
a	mm	Slump	3
a	Bq/kg	Specific activity	5
A_s	cm^2/g	Specific grain surface area	2
b	$Ws/s^{0.5}\ m^2\ K$	Thermal effusivity coefficient	5
C	$W/m^2\ K^4$, $kcal/hm^2\ K^4$	Radiation coefficient for a "non-black" body	5
C_c	–	Curvature coefficient, grading	1
c_N	g/cm^2, N/mm^2	Cohesive strength according to Niemeyer	3
c_p	Ws/kg K, kJ/kg K	Specific heat capacity	5
C_u	–	Uniformity coefficient	1
d	m	Thickness of the building element	5, 7
d	mm	Grain size	1
d	mm	Grain diameter	1
d	mm	Specimen diameter	1, 4
D	Gy, J/kg	Absorbed dose	5
D_f	–	Deformation ratio	4
D_{Pr}	–	Compaction degree	1
d_x	mm	Grain diameter when x% passes through sieve	1
e	–	Void ratio	1
e	m	Eccentricity of the resultant in the foundation bottom plane	4
E	N/mm^2, MN/m^2	Uniaxial elastic modulus, uniaxial Young's modulus	4
E_S	N/mm^2, MN/m^2	Constrained modulus	4

(continued)

(continued)

Symbol	Unit	Parameter	PG[a]
E_S	W/m^2	Total radiated energy	5
F	kN, N	Force	4
F_s	%	Free swell value	4
G	kN, MN	Dead load	4
G	MN/m^2	Shear modulus	4
H	Sv, J/kg	Dose equivalent	5
h	Sv/a, mSv/a	Dose equivalent rate	5
h	mm	height, specimen ~	1, 4, 7
h_1	mm	Crushed sample height	4
h_o	mm	Initial sample height	1, 4
I	Bq/kg	Activity concentration index	5
I	12-degree scale	Earthquake intensity	4
I_A, AI	–	Activity ratio, ~index	2
I_c, CI	–	Level of consistency	3
I_p, PI	–	Plasticity index	3
I_S	–	Shrinkage index	3
l	mm	Length	7
L_n	dB	Impact sound level	5
$L_{n,w}$	dB	Weighted normalized impact sound pressure level	5
M	Richter scale	Magnitude of an earthquake	4
m_{Ca}	g	Mass percentage of total carbonates, based on m_d	2
m_d	g	Dry mass	1, 3
m_m	g	Moist mass	1, 3
m_s	g	Solid mass	1, 3
m_w	kg/m^2	Mass of water absorbed by capillary action	5
n	–	Porosity	1
Q	J, Ws; 1 J=1 Ws	Heat, quantity of heat	5
q	W/m^2	Heat flow density	5
Q_s	Ws/m^2 K	Heat storage capacity	5
R	m^2 K/W	Heat transfer resistance	5
R	kN	Structural resistance, resultant	4
R	dB	Sound reduction index	5
RH	%	Relative humidity	3, 5
R_{si}, R_{se}	m^2 K/W	Surface heat transfer resistance, interior (i) and exterior (e)	5
R_T	m^2 K/W	Overall heat transfer resistance	5
R_w	dB	Weighted sound reduction index	5
R_w'	dB	Weighted sound reduction index taking adjacent building elements into consideration	5

(continued)

(continued)

Symbol	Unit	Parameter	PG[a]
S	Ws/m³ K	Volumetric heat capacity	5
S	kN	Working load	4
s_d	–	Water vapor diffusion equivalent air layer thickness	5
S_r	–	Degree of saturation	1
t	s, h, a	Time	4, 5
T, Θ	K, °C; 1 K=1 °C	Temperature	5, 6
t_A	h	Cooling behavior	5
U	W/m² K	Overall heat transfer coefficient, U-value	5
V	cm³, m³	Total volume of a sample	1, 3
v_{Ca}	%	Lime content	2
v_{gl}	%	Loss on ignition	2
V_P	cm³, m³	Void volume, pore space	1
V_S	cm³, m³	Solid mass volume	1
w	–	Moisture content	1, 2, 3
w	mm	Width	7
w_a	–	Water absorption capacity	5
w_c	–	Practical/continuous moisture content, equilibrium ~	5
w_{hygr}	–	Hygroscopic moisture content	5
w_L, LL	–	Moisture content at liquid limit	3
w_N	–	Moisture content at standard consistency (NIEMEYER)	3
w_P, PL	–	Moisture content at plastic limit	3
w_{Pr}	–	PROCTOR moisture content, optimal ~	3
w_S, SL	–	Moisture content at shrinkage limit	3
α	–	Shape factor	1
β	N/mm²	Strength parameter	4
β_{AS}, c	N/mm²	Adhesive shear strength	4
β_{AT}	N/mm²	Adhesive (tensile) strength	4
β_C	N/mm²	Compressive strength	4
β_D	N/mm²	Dry compressive strength	4
β_F	N/mm²	Flexural strength	4
β_k	N/mm²	Compressive strength determined in an accelerated test	4
β_S	N/mm²	Shear strength	4
β_{ST}	N/mm²	Splitting tensile strength	4
β_T	N/mm²	Tensile strength	4
β_{TW}	N/mm²	Tensile strength at standard consistency (Niemeyer)	4
γ	kN/m³	Unit weight, specific ~	1

(continued)

(continued)

Symbol	Unit	Parameter	PG[a]
γ_M	–	Partial safety factor	4
γ_{zx}	°	Shear distortion	4
$\Delta\Theta$	K	Temperature amplitude	5
ε'	–	Emissivity	5
ε, ε_x, ε_y, ε_z	%	Strain	4
ε_{bl}	%	Settling	4
ε_c	%	Chemically induced strain	4
ε_{el}	%	Elastic strain +/–	4
ε_f	%	Moisture strain +/–	4
$\varepsilon_{f,l}$	%	Degree of linear shrinkage	4
ε_{fl}	%	Plastic strain, flow	4
ε_T	%	Thermal strain +/–	4
$\varepsilon_{v,el}$	%	Delayed elastic strain	4
Λ	W/m^2 K	Heat transfer coefficient	5
λ (k)	W/m K	Coefficient of thermal conductivity	5
μ	–	Water vapor diffusion resistance factor	5
μ	–	Friction coefficient	4
ν	–	Poisson's ratio	4
ρ	g/cm^3, kg/dm^3	Bulk density, moist bulk density	1
ρ_d	g/cm^3, kg/dm^3	Dry bulk density	1
ρ_{Pr}	g/cm^3, kg/dm^3	PROCTOR density	1
ρ_s	g/cm^3, kg/dm^3	Specific density, solid ~	1
ρ_{sr}	g/cm^3, kg/dm^3	Saturated bulk density	1
τ	N/mm^2	Shear stress	4
Φ	W	Heat flow	5
Φ	–	Creep ratio	4
φ	h	Phase displacement	5
ψ	–	Relaxation	4
σ, σ_x, σ_y, σ_z	N/mm^2	Stress	4
σ_p	N/mm^2	Permissible compressive stress	4
σ_s	N/mm^2	Swelling pressure	3

[a]Parameter group according to Table 1.1

1

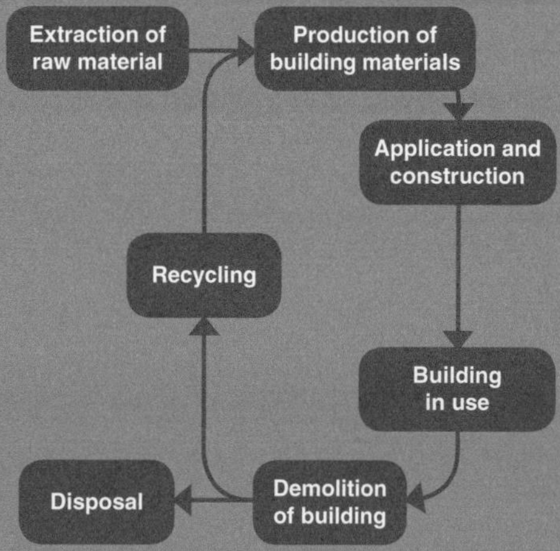

The Development of Earth Building

At around 10,000 B.C. a decisive change took place in the history of humankind: the prevalent mode of food procurement, which until that time was hunting and gathering, was gradually replaced by mixed farming. This new lifestyle was accompanied by the need to build permanent shelters for people and, where necessary, for animals, as well as structures for agricultural storage. Among the building materials used for these structures were natural stone, wood and, above all, earth.

© Springer International Publishing Switzerland 2016
H. Schroeder, *Sustainable Building with Earth*,
DOI 10.1007/978-3-319-19491-2_1

1.1 Historical Roots of Building with Earth

Depending on the prevailing climate and vegetation as well as the geological conditions of a region, different building styles and construction methods developed over the course of the history of humankind: in hot and dry climates without major sources of wood, *massive construction* using load-bearing earthen walls dominates. In addition, these structures function as a "heat buffer" against intense insolation. In transitional climates or mountainous regions with abundant sources of wood, *framed construction* prevails: here a separate wooden skeleton carries the structure's loads. Earth, often in combination with stone, is used as an infill material and has a space-enclosing function. There are also transitional construction types which combine the two systems.

Both building styles can be traced back thousands of years in the different regions of the world.

Based on what is known today, the transition to a sedentary lifestyle started in *Southwest Asia*, in the region of present-day Turkey, Iran, Iraq, Lebanon, Syria, Jordan, and Israel. This is also where archaeological evidence of the first permanent dwellings dating back to 10,000 BC has been found.

Among the oldest permanent earthen houses are those found in the area of present-day's Anatolia in Turkey and in Palestine (Figs. 1.1, 1.2, and 1.3). The approximately 8000-year-old structures of Çatal Höyük, Anatolia, exhibit a surprisingly high standard. The load-bearing exterior walls were constructed of earth blocks with interior wooden supports carrying the roof construction. The roofs were flat, made of poles and grasses or reed and a layer of puddled earth for protection against rainwater. The houses were entered via the roof. The individual structures themselves were grouped together like honeycombs touching each other [1].

Fig. 1.1 Model drawing of an earth block house, Çatal Höyük, Anatolia/Turkey, around 6000 BC [1]

Fig. 1.2 Archeological excavation site, Anatolia/Turkey, around 6000 BC

Fig. 1.3 Earth block structures, Jericho, Palestine/Israel, around 6000 BC

Fig. 1.4 Rammed earth construction in Ancient China, Shang Dynasty, around 1320 BC [2]

Large areas of *China* are covered with clay-rich soils, particularly loess soil. There is evidence of load-bearing structures made of earth as well as framed construction in combination with earthen materials spanning several thousand years.

Figure 1.4 is a historical depiction of the rammed earth building technique which is tied to the following story: Fu Yueh, a minister under a ruler of the Shang Dynasty (around 1320 BC), is said to have been the inventor of this technology and the first "rammed earth master builder." According to legend, Fu Yueh acquired his position in an unusual way: one day, the Emperor had such a vivid dream of a wise and able man that he woke up and had a picture drawn of the person he had seen in his dream. He sent messengers with the picture of this man all over the country to look for him. The messengers encountered Fu Yueh who resembled the person in the picture and, at the time, was busy building a rammed earth house. This is the scene depicted in the illustration. Fu Yueh was called to the court and appointed to a minister position [2].

The production and use of earth blocks has also been known in China for thousands of years. Figure 1.5 shows the production of earth blocks during the time of the Ming Dynasty [3].

The largest and most famous structure in China is the Great Wall of China. It is also the largest structure ever built by humans. It took around 2000 years to build the approx. 50,000 km of total length known today. Depending on local availability,

Fig. 1.5 Production of earth blocks in Ancient China during the time of the Ming Dynasty [3]

Fig. 1.6 The Great Wall of China, section in Gansu Province, built around 220 BC [4]

the materials used included wood, stone, and earth (also in the form of fired bricks) as well as vegetal material for reinforcement. Figure 1.6 shows a section of the wall dating back to the Quin Dynasty built using the rammed earth technique around 2200 years ago [4].

Egypt is another classic example of an earth building country with a building tradition dating back thousands of years. The annual floods of the Nile River brought fertile mud from the Ethiopian highlands which dried in the sun and became solid. When the mud got wet, it became malleable once again. This fundamental knowledge

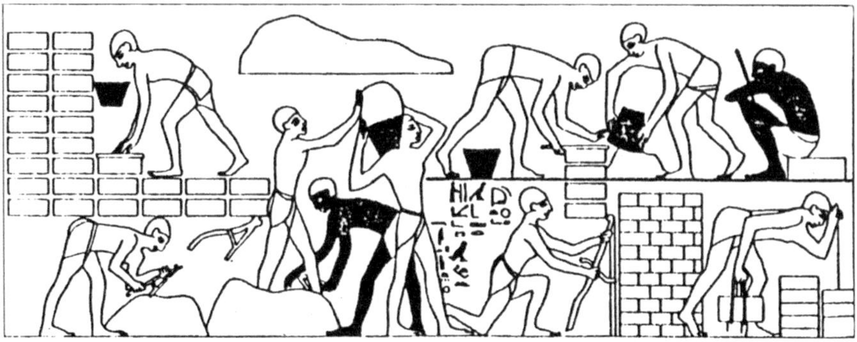

Fig. 1.7 Production of earth blocks in Ancient Egypt, around 1500 BC [5]; depiction in the tomb of Vizier Rekhmire, Theban West

formed the basis of the production of sun-dried mud blocks which could be made stronger and more durable by adding sand or plant fibers and could be improved even further through firing. The Old Testament describes the use of chopped straw in the production of earth blocks [Exodus 5:7f.; 16:18f].

Figure 1.7, an illustration from around 1500 BC, shows the individual technological steps used in the production of earth blocks from soil preparation to using the blocks for building [5]. The symbolic depiction of Hatshepsut, the reigning pharaoh of that time, as a master builder during the production of earth blocks emphasizes the importance of this activity in Fig. 1.8 [6]. The origin of vault construction using sun-dried earth blocks can also be traced back to Egypt. Figure 1.9 shows an earth block vault used as a storage room in the tomb of Ramses II from around 1300 BC.

There is archaeological evidence of an earth building tradition dating back thousands of years in the relatively woodless but clay-rich regions of *Mesopotamia,* located between the *Euphrates* and *Tigris*, in *Afghanistan* and *Iran*. Figure 1.10 shows sun-dried earth blocks from different parts of this region [7]. They exemplify the already highly developed technique of prefabricating building elements.

Earth blocks were also used to construct large religious buildings in this region. Built in the shape of pyramids, these buildings can easily be compared to the ones in Egypt in terms of their size. Figure 1.11 shows the condition of the pyramid (Ziggurat) of Chogha Zanbil after its restoration. It was built around 1500 BC by Elamite rulers in the area of present-day Iran [8]. This category of buildings also includes the Tower of Babel [Old Testament, Genesis 11:3] which was built using sun-dried earth blocks with an exterior cladding of fired bricks.

Furthermore, this region is home to the oldest known written rules for building with earth. Documented on fired clay tablets, they date back to the time of the Babylonian ruler Hammurabi who lived around 1800 BC [9].

To the north are the Central Asian steppes and deserts of *Turkmenistan*, *Uzbekistan*, and *Kazakhstan* whose cultures go back thousands of years and where earth has been used as a building material for more than 5 millennia [10]. Figure 1.12 shows the ruins of the ancient city of Afrasiab, the predecessor to present-day

Fig. 1.8 Pharaoh Hatshepsut during the production of earth blocks made of mud from the Nile River, around 1500 BC [6]

Fig. 1.9 Earth block vault near Luxor/Egypt, around 1300 BC

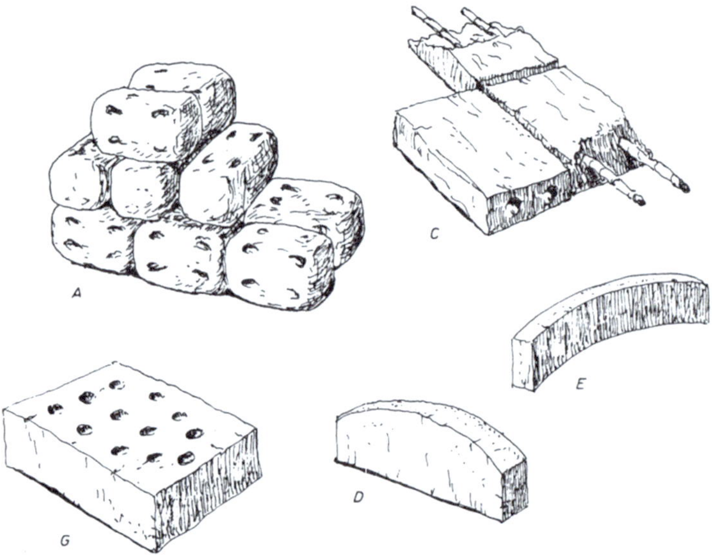

Fig. 1.10 Sun-dried earth blocks and prefabricated building elements in Mesopotamia and Afghanistan, fourth millennium to sixth century BC [7]

Fig. 1.11 Ziggurat Chogha Zanbil, Iran, around 1250 BC [8]

Samarkand in Uzbekistan which was completely destroyed by Genghis Khan during the Mongol invasion in the thirteenth century. Present-day Samarkand, Bukhara, and Chiwa are cities with a 2500-year-old history and earth building tradition.

Along the Indus River, in present-day Pakistan, lies Mohenjo Daro, a city built of earth blocks around the beginning of the third millennium BC [11].

Fig. 1.12 Earth block walls in the city of Afrasiab, present-day Samarkand, Uzbekistan

In the so-called New World as well, in Pre-Columbian *Peru*, people were familiar with different earth building techniques. It is estimated that 130 million sun-dried adobes were used to build the Huaca del Sol pyramid of Moche (around 200–500 AD) whose dimensions are 120×120 m^2. Chan Chan, the largest city in Pre-Columbian America, had approximately 60,000 inhabitants in the fourteenth and fifteenth centuries. Today, the city still covers 25 km^2 and is blanketed by large mounds of adobe block debris. The different city quarters were laid out at right angles and surrounded by high adobe walls. The rammed earth building technique was also known.

Figure 1.13 shows a rammed earth palace wall in Chan Chan decorated with friezes (thirteenth century AD) [8].

There is also a long history of building traditions in North America. Figure 1.14a shows the concept of a Pueblo Indian pit house (Arizona, New Mexico) incorporating a support structure of wooden beams for the flat roof and a layer of puddled adobe as a cover. The houses were entered via a ladder through an opening in the roof (around the second century AD) [12]. This type of house construction is surprisingly similar to the Neolithic earth block homes of Çatal Höyük, Anatolia (Fig. 1.1), which were also entered via the roof.

The small town of Taos along the Rio Grande Valley north of Santa Fe, New Mexico (USA), has preserved a Pueblo Indian settlement with its main core dating back to the thirteenth and fourteenth centuries. The settlement's building style has its roots in the pit house. The buildings as well as their entrances are at ground level and rise up to four floors (Fig. 1.14b). The walls, which are made of adobes, are refinished with earth plaster once a year.

The "Pueblo de Taos" is a UNESCO World Heritage Site.

Fig. 1.13 Ruins of a palace constructed in rammed earth technique, built in Chan Chan in present-day Peru in the thirteenth century AD [8]

Fig. 1.14 Adobe homes of the Pueblo Indians in the Southwestern USA. (a) Pueblo Indian pit house in North America about second century AD [12]. (b) Pueblo de Taos, New Mexico, USA, adobe buildings with flat roofs and earth plaster

1.2 Earth Building as Cultural Heritage

In the course of the centuries, the knowledge of traditional earth building techniques became lost in many parts of the world. Even in the poorest developing countries, "modern" building materials such as concrete and cement have begun to replace or have already replaced earth as a building material. Here, earth or mud often equals poverty. As soon as they can afford it, people use concrete or fired brick for building, particularly in urban areas. In the rural regions of developing countries, however, earth as a building material has survived in the everyday building practice.

It is in large part due to the international activities of the organizations ICOMOS and CRATerre in the field of the preservation of traditional earthen architecture that building with earth has once again become part of peoples' cultural identity in many third-world countries. Within ICOMOS a number of specialized work groups focus on the preservation of historical structures, among them the International Committee for Earthen Architectural Heritage (ISCEAH) which works in the field of earth building (http://isceah.icomos.org).

The inclusion of historical earthen buildings in the UNESCO World Heritage list of architectural monuments [13] has triggered a change in thinking in the countries concerned: presumed poverty is slowly changing to pride in the countries' own historic building traditions and accomplishments. Out of the 759 cultural monuments added to the World Heritage list in 2013, 143 or 19 % are partially or completely built out of earth. Among them are the Great Wall of China; the earth block "tower houses" of Shibam, Yemen; the famous Alhambra in Granada, Spain; or the Potala Palace in Lhasa, Tibet [52].

The status of an architectural monument brings with it the commitment to adhere to the principles of conservation and restoration of historic structures according to the Venice Charter which was agreed upon by the participants of the Second International Congress of Architects and Specialists of Historic Buildings in 1964.

In developing countries, "sustainable" tourism is beginning to develop around these restored earthen architectural monuments leading to desperately needed foreign revenue. The rammed earth houses in Ait Benhaddou in Southern Morocco, shown in Fig. 1.15a, have been designated a UNESCO World Heritage Site and serve as an example of this development. These houses give impressive testament to the engineering skills and accomplishments of their builders. Although this building technique is still known in rural areas today, especially by the older generation, it is at risk of being forgotten. The reason for this can be traced to profound changes within the building practice itself: whereas building used to be an activity mainly conducted by the village community or the extended family, it is now carried out by small businesses for money.

Efforts to preserve special earth building techniques will lead to the creation of new museums similar to the open-air museums found in Germany. In this context, it is essential to document endangered earth building structures and traditional earth building techniques as part of our cultural identity.

In 2011, a "Map of Earthen Heritage in the European Union" was published as part of an EU-funded project. The map was created in cooperation with 50 authors from 27 European countries [14]. It can be found at www.culture-terra-incognita.org. This map shows the geographic areas for the different earth building techniques of half-timber construction with various types of earthen infill, earth block masonry, rammed earth, and cob. The representation of the respective areas in Germany is based on an incomplete data base. Therefore, a map based on geographic information systems (GIS) is currently under development (dev.lehmbau-atlas.de) [15].

Fig. 1.15 Cultural heritage of earth building materials. (**a**) UNESCO World Heritage Site: traditional rammed earth houses in Ait Benhaddou, Southern Morocco. (**b**) "World Monuments in Danger 2008 (WMF)": urban settlements using earth building materials dating back to the Sumerian time period (around 3500 BC) located in the former Iraqi war zone (www.wmf.org)

Such "inventories" have already been taken in many different European countries and regions, such as France [16], Portugal [17, 18], the Czech Republic [19] (http://hlina.info/cs.html), and Italy [20].

For more than 40 years, the activities of the private organization World Monuments Fund (WMF) (www.wmf.org) have been dedicated to keeping at-risk architectural monuments from further deterioration or destruction. Monuments in isolated and difficult-to-access places and war zones are particularly threatened. Every 2 years the WMF publishes a list of the 100 most threatened architectural monuments with the goal of drawing attention to the precarious situation of these monuments and of finding worldwide sponsors for urgently needed stabilization work.

On the WMF 2008 Watch List, the situation of the archaeological excavation sites of the Uruk and Sumerian period (around 3500 BC) in Iraq, which are located in the middle of the Iraqi war zone, is defined as particularly critical. The walls of these urban settlements are made of earthen materials (Fig. 1.15b).

1.3 Historical Development of Earth Building in Germany

Around 8000 years ago, mixed farming slowly made its way to Central Europe and present-day Germany via trade routes from the Southeast. Wood and earth were available almost everywhere as building materials for house construction, but in this region houses had to be planned differently: whereas in the Eastern Mediterranean region the houses needed to protect their occupants, livestock, and supplies from the summer heat, they now had to protect them from precipitation and the cold of winter.

The building designs of that time can be reconstructed today with the help of post holes which stand out from the surrounding building area as dark circular discolorations. The design principle of these houses was based on post constructions with woven branches as the support frame for a coat of straw clay (Fig. 1.16) [1]. Reconstructions of these early wooden post buildings can be seen in a number of open-air museums, such as in Oberdorla or at the State Office for Preservation of Monuments and Archaeology (Thüringischen Landesamt f. Archäologie u. Denkmalpflege) in Weimar in the German State of Thuringia (Fig. 1.17) [21].

Fig. 1.16 Longhouse of Central European woodland farmers around 4000 BC [1]

Fig. 1.17 Model of a Neolithic longhouse in wood post construction with wattle and daub, Thüringisches Landesamt f. Archäologie und Denkmalpflege, Weimar [21]

One of the largest Neolithic settlements in Central Europe dating back to around 4500 BC was recently discovered near Erfurt–Gispersleben during the construction of the new German highway A71. House structures dating back to around the same time period were also found while excavating the foundation for the underground library depot of the Duchess Anna Amalia Library in the town center of Weimar. The historic aspects of earth building of that period are the topic of a student research paper at the Bauhaus University Weimar in cooperation with the Thuringian State Office for Preservation of Monuments and Archaeology [21].

The oldest written record of building with earth in Germany can be found in the work "Germania" by the Roman writer Tacitus from around 100 AD. In his descriptions, the buildings were still very similar to those of the former woodland farmers which are approximately 4000 years older. Their walls consisted of wooden posts which had been rammed or dug into the ground. The openings between the posts were filled with a wattle made of willow branches and covered with a paste-like straw-clay mixture which enclosed the entire surface.

It is assumed that these post-built houses, with their woven lattice of branches and an applied layer of daub, later developed into the load-bearing earth building technique which today is known as *cob* in Central Germany. Probably as a result of continuous repair work during the lifetime of the building, but also for reasons of fire protection, the daub eventually enclosed the load-bearing posts and the wattle with a layer that was several decimeters thick. At some point, the load-bearing function was finally transferred to the earthen building material and the posts and wattle could be omitted.

For the region around Weimar, this gradual transition has been dated to the time after the ninth century AD by Behm-Blancke [22]. The remains of a solid earthen wall dating back to the tenth to eleventh century, which was presumably built using the cob technique, have been found on an early medieval farmstead within the town of Weimar [23]. From that time on there is evidence of the use of limestone slabs for the base of walls. These slabs do not show any depressions for wooden posts. Instead, they were enclosed by "collapsed" earth. This most likely serves as evidence of "load-bearing" earthen walls. Although archaeological literature speaks of "rammed walls" in this context, it is more probable that they are cob walls. Rammed earth using formwork did not establish itself in Germany until the turn of the eighteenth to nineteenth century. The oldest known example of cob building in central Thuringia is a building which combines living quarters and stables in Wülfershausen near Arnstadt. It was built in 1577 and no longer exists [24].

The oldest known example of load-bearing *earth block construction* north of the Alps is the hill-fort Heuneburg an der Donau, southwest of Ulm, dating to the time around 500 BC. It is assumed that it was built under Celtic influence which would indicate a time long before the Roman occupation. Connections to Greek architecture by way of the Danube River are also possible because, at the time, the Greeks had long been familiar with the construction of load-bearing earth block walls.

A second development based on Neolithic walls made of a woven lattice of branches and an applied earthen daub led to half-timber construction. Throughout the centuries and into present times, the *half-timber construction technique* with its regional stylistic differences has defined the architectural appearance of urban settlements and rural areas in Germany and other European countries as well as in Asia. Figure 1.18 shows the development of the load-bearing timber frame: starting with the early post-built house with a ridge post, a woven lattice of branches and an applied daub, continuing on to the center-column timber-frame house with an anchor beam, and finishing with the half-timber house using story framing [25].

The self-supporting wooden posts, which had originally been driven into the building ground, were elevated to protect them from rotting and eventually placed on a foundation of stone slabs. The post-built house thus became a post-and-beam house with the result that the restraining effect of the building ground was lost.

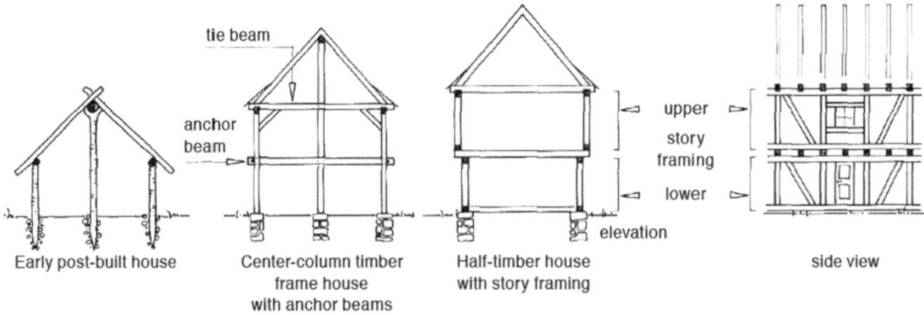

Fig. 1.18 Development of support structures from the post-built house to half-timber construction [25]

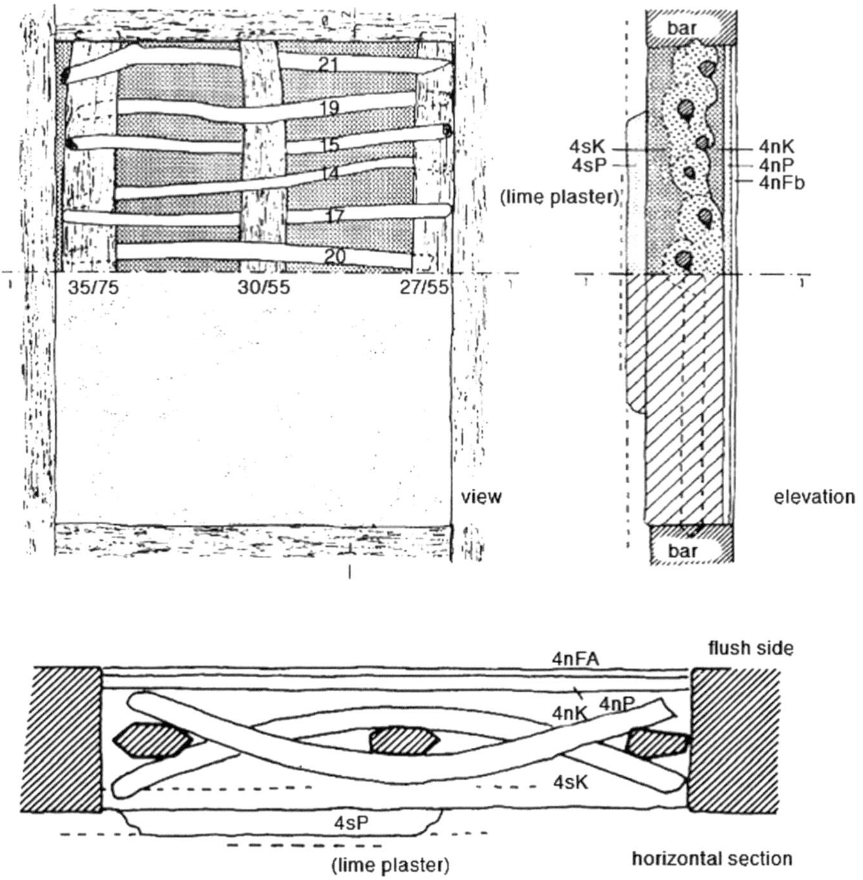

Fig. 1.19 "Gothic House" Limburg/Lahn (1289), lattice work with straw clay [26]

Now the individual construction elements of the longitudinal wall, transverse wall, ceiling, and roof were designed as self-supporting, panel-like systems made up of vertical posts, horizontal sills and noggin pieces, and diagonally positioned braces which were connected with mortise and tenon joints. As with the early post-built houses, the openings between the vertically, horizontally, and diagonally positioned timbers, the so-called *panels*, were filled with a lattice work of stakes and flexible branches and sealed with straw clay.

Today, there is still evidence of a large number of infill techniques. Figure 1.19 shows panel analyses carried out by Volhard [26] on a half-timber house built in 1289, the "Gothic House" (Gotisches Haus) in Limburg a. d. Lahn (see Fig. 4.1). The manner in which the straw-clay mixture was applied can be seen clearly.

The rise of cities in Central Europe starting around the twelfth and thirteenth centuries led to a lack of building land which resulted in the need for constructing second stories. Population growth, ensuing town fires, and wartime destruction led to a shortage of wood which, at the time, was the favored building material. Earth, which was "fire resistant" and readily available, gained in importance. This development also found its expression in the formation of special guilds, comparable to present-day chambers of crafts and trades. Figure 1.20 shows the "Claiber"

Fig. 1.20 Depiction of the earth builder ("Kleiber") Hans Pühler (died in 1608 in Nuremberg), from the house books of the Nuremberg Twelve Brothers Foundations [27, 28] (Stadtbibliothek Nürnberg, Amb. 317b.2°, f. 76r)

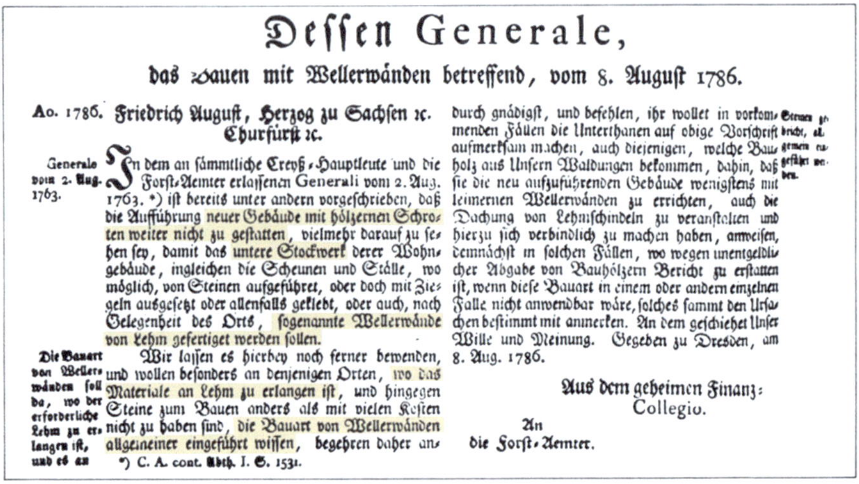

Dessen Generale,

das Bauen mit Wellerwänden betreffend, vom 8. August 1786.

Ao. 1786. Friedrich August, Herzog zu Sachsen rc. Churfürst rc.

Generale vom 2. Aug. 1763.

In dem an sämmtliche Creyß-Hauptleute und die Forst-Aemter erlassenen Generali vom 2. Aug. 1763. *) ist bereits unter andern vorgeschrieben, daß die Aufführung neuer Gebäude mit hölzernen Schroten weiter nicht zu gestatten, vielmehr darauf zu sehen sey, damit das untere Stockwerk derer Wohngebäude, ingleichen die Scheunen und Ställe, wo möglich, von Steinen aufgeführet, oder doch mit Ziegeln ausgesetzt oder allenfalls geklebt, oder auch, nach Gelegenheit des Orts, sogenannte Wellerwände von Lehm gefertiget werden sollen.

Die Bauart von Wellerwänden soll da, wo der erforderliche Lehm zu erlangen ist, und es an

Wir lassen es hierbey noch ferner bewenden, und wollen besonders an denjenigen Orten, wo das Materiale an Lehm zu erlangen ist, und hingegen Steine zum Bauen anders als mit vielen Kosten nicht zu haben sind, die Bauart von Wellerwänden allgemeiner eingeführt wissen, begehren daher andurch gnädigst, und befehlen, ihr wollet in vorkommenden Fällen die Unterthanen auf obige Vorschrift aufmerksam machen, auch diejenigen, welche Bauholz aus Unsern Waldungen bekommen, dahin, daß sie die neu aufzuführenden Gebäude wenigstens mit leimernen Wellerwänden zu errichten, auch die Dachung von Lehmschindeln zu veranstalten und hierzu sich verbindlich zu machen haben, anweisen, demnächst in solchen Fällen, wo wegen unentgeldlicher Abgabe von Bauhölzern Bericht zu erstatten ist, wenn diese Bauart in einem oder andern einzelnen Falle nicht anwendbar wäre, solches sammt den Ursachen bestimmt mit anmerken. An dem geschiehet Unser Wille und Meinung. Gegeben zu Dresden, am 8. Aug. 1786.

Aus dem geheimen Finanz-Collegio.

An die Forst-Aemter.

*) C. A. cont. Abth. I. S. 1531.

Fig. 1.21 Decree by the Saxon Elector Frederick Augustus from 1786, concerning the construction of cob walls [29]

(earth builder) Hans Pühler who died in Nuremberg in 1608, in an illustration from the House Books of the Nuremberg Twelve Brothers Foundations [27, 28]. Today, the German name "Kleiber" has survived from this time and is the old word for the profession of earth builder.

An increasing shortage of wood became the driving force behind the development of earth building in Central Europe. Evidence of this can be found in specific written building regulations. To reduce the use of construction timber, the "Saxon Forestry Regulations" ("Forst-und Holzordnung") of 1560 stipulated that a building's ground floor be constructed out of stone or earth. According to the "General Appointment of Forest Personnel" ("Generalbestallung für die Forstbedienten"), issued in 1575, construction timber was only to be made available if there was no other possibility of constructing the ground floor out of "stones or cob walls." The Ernestine "Territorial Law Code" ("Landesordnung") of 1556 banned solid timber construction in Thuringia and only allowed new construction utilizing half-timber, cob, fired brick, or stone [23]. Figure 1.21 shows a decree by the Saxon Elector Frederick Augustus dating back to 1786 concerning the construction of cob walls [29]. Similar regulations for earthen construction existed in Prussia (1764) and Austria (1753), here in connection with the use of unfired ("Egyptian") bricks [30].

At the end of the eighteenth century, the French master builder and architect François Cointereaux published a series of writings which summarized French experiences with rammed earth (pisé in French) and, by doing so, significantly influenced the development of rammed earth building in Germany [31]. Cointereaux describes the building material, technology, and construction techniques at great length and presents these elements as a unity. Advice on the preparation and processing of earth building materials as well as a detailed description

Fig. 1.22 David Gilly, Prussian state master builder and promoter of earth building, based on an illustration by L.W. Chodowiecki, 1790 [29]

of the tools and equipment needed make these writings the first modern "textbook" for building with earth.

The chief Royal Prussian building officer David Gilly was instrumental in spreading this building technique in Prussia and Silesia. Figure 1.22 shows Gilly in a contemporary engraving by Ludwig Wilhelm Chodowiecki dated 1790 [29]. Influenced by his writings, government lawyer Wimpf constructed multistory rammed earth residential buildings in Hesse. A six-story residential building from 1830 in Weilburg/Lahn is still in use today and is considered Germany's tallest rammed earth building (Fig. 1.23).

Radical technical innovations in firing systems and mechanical engineering in the nineteenth century led to fundamental changes in the construction industry: large-scale mining of hard and soft coal for modern furnaces and kilns and the ensuing transition to gas and oil firing led to the industrialization of brick production. The development of the cement industry and with it the rise of concrete and reinforced concrete as building materials would not have been possible without the transition from wood to coal firing (and later oil and gas).

The goal was to increase the strength of the building materials so that the dimensions of the building elements intended for the same purpose could be decreased. This was particularly well accomplished through the combination of steel and concrete. Earth building materials, however, could not adapt to this development due to their limited strength and the added flaw of low water resistance. Thus, earth as

Fig. 1.23 Six-story residential building using rammed earth technique in Weilburg a. d. Lahn, built by W.J. Wimpf around 1830

a building material became increasingly marginalized until it eventually lost all significance.

In the twentieth century, during and after WWI and WWII, earth as a building material regained some importance, but mainly only because factories for the production of building materials were largely destroyed and transportation was not possible.

"Postwar" earth building was particularly significant for the area of the former GDR: in addition to millions of people who had lost their homes, there were millions of refugees arriving from the eastern territories which had been lost as a consequence of the war. Within a short time, housing had to be created using available building materials, among them earth. It is important to mention Order No. 209 of the Soviet military administration in this regard. This order stipulated the construction of 200,000 homes for so-called new farmers. At least 40 % of the houses were to be built using natural and locally available materials.

Earth as a building material was once again becoming the subject of government regulations, but this time universities, as well, began working with the material. In East Germany, it was mostly projects for farms and settlements for new farmers which were carried out using various earth building techniques. The projects were designed and planned at the Weimar University of Architecture and Fine Arts, the predecessor of today's Bauhaus University [32]. Figures 1.24 and 1.25 show two different types of examples that were built: half-timber construction with earth block infill and massive wall earth block construction [33]. Designs for two-story apartment buildings using rammed earth, among them an 18-family residential

Fig. 1.24 New farmers' homestead using half-timber technique with earth block infill, 1947 [33]

Fig. 1.25 New farmers' homestead using earth block technique, load bearing, 1946 [33]

building built in Gotha in 1951, originated at the same university [34]. Figure 1.26 shows a renovated multifamily apartment building using rammed earth construction built in the 1950s in Mücheln in the Geisel valley near Merseburg. The building's exterior displays a wall frieze depicting the earth building technique used during construction.

The history of earth building in the former Soviet occupation zone and GDR has been studied by Rath [35]. At the time, earth building had reached a high technological standard in the GDR. The new territory movement in the Soviet

Fig. 1.26 Multifamily apartment building using rammed earth technique in Mücheln in the Geisel valley near Merseburg, built in the 1950s with wall frieze

Union, especially in Kazakhstan, also benefited from this knowledge [36]. In addition, a housing project using the rammed earth technique was completed in Hamhung, North Korea, as part of a reconstruction agreement.

Around 1960, earth building ended in the GDR due to political decisions which stipulated an industrialization of the production of building materials and the housing industry.

1.4 Earth Building Today: Ecological and Economic Aspects

After the release of the report "The Limits to Growth," commissioned by the Club of Rome (Meadows, 1972), and the first global oil crisis in 1973, it became apparent that energy consumption could not continue unchecked and keep pace with economic growth. Today, this realization applies to resource consumption as a whole.

In a report by the Intergovernmental Panel on Climate Change in 2007 (IPCC, www.awi.de), it is noted that the CO_2 concentration in the air increased by 35 % from the beginning of the Industrial Revolution at around 1750 until 2005. The rate of increase in the past 10 years has been the highest in 50 years. Today's numbers are the highest in the past 650,000 years. Seventy-eight percent of this increase can be attributed to the use of fossil fuels and 22 % to a change in land use, such as the deforestation of the tropical rainforest. The same time period saw an increase of 148 % in methane concentration in the air.

Although only traces of these two gases can be found in the air, an increase in their concentration is considered to be anthropogenic and one of the causes of the "greenhouse" effect in the atmosphere, leading to global warming. The effects visible so far are listed in the IPPC report in detail. The global surface temperature has risen by +0.74 °C and sea levels by approximately 3 mm per year since 1993, totaling 17 cm in the twentieth century.

1.4.1 Sustainable Building

In the Brundtland Commission report to the United Nations World Commission on Environment and Development entitled "Our Common Future" (1987), the term "sustainability" was first used to describe a lasting development of humankind. Sustainable development ensures that it "meets the needs of the present without compromising the ability of future generations to meet their own needs" [37].

The building process always has a more or less severe impact on natural resources and cycles. To apply the term "sustainability" to construction means that in all phases of the life of a building, and with respect to the users' requirements, the use of existing resources and the environmental impact need to be minimized. Traditionally, during the construction process, the main assessment focused on function and design, structural engineering, material technology, and building practice. Today, the building process is increasingly seen as an optimization task where user demands and environmental requirements by the legislature need to be in harmony. Therefore, the term "sustainable development" comprises three aspects which must be considered as equals over an appropriated period of time [38]:

- Ecology
- Economy
- User demands (sociocultural concerns/functional quality)

Additionally, all buildings constructed according to the requirements of sustainable building must meet predefined technical parameters and corresponding quality levels in terms of planning and construction for each of the three aspects.

In Germany, the general requirements placed on building materials and building elements with regard to their technical quality are regulated by the Model Building Code (MBO) for the states of the Federal Republic of Germany (Musterbauordnung für die Länder der Bundesrepublik Deutschland—MBO). According to the MBO, building products, materials, elements, and systems are produced for permanent installation in structures. "Building products are only to be used if the structures they are used in, along with proper maintenance over a time period which is proportionate to its purpose, meet the requirements of this law or are suitable for their intended use based on this law" (MBO, §3.2).

The MBO lists the following aspects as the main requirements for the building materials' and building elements' *suitability for use*:

- Mechanical strength and stability
- Fire protection
- Hygiene, health, and environmental protection
- Safety in use
- Sound insulation
- Energy conservation and heat insulation

The Regulation of the European Parliament and of the Council for "Laying down harmonized conditions for the marketing of construction products" published in March 2011 [39] introduces an additional requirement: the *sustainable use of natural resources*. According to the regulation, buildings must be designed, built, and demolished after their use in a manner which facilitates the sustainable use of natural resources and guarantees the following:

- The building, its building materials, and building elements need to be recyclable after demolition.
- The building must be durable.
- Environmentally friendly raw materials and secondary building materials must be used in the construction of the building.

This regulation also requires the national governments of the EU to apply principles of sustainable development to building activities in their respective countries. In order to establish sustainable development, it is necessary to explicitly formulate *protection objectives* regarding the environment, the economy, and the users' interests. Examples of general protection objectives are the avoidance of harmful substances, a reduction in the use of energy as well as land and resources, and the prevention of waste through material recovery. Based on the knowledge of cause-and-effect relationships, *action strategies* need to be derived from the protection objectives. These strategies should target three levels: raw and building materials, building construction, and surroundings [40]. The *effects* caused by the building process on these levels need to be described through indicators and by defining assessment standards.

Standards for the assessment of the sustainability of buildings in terms of their environmental, social, and economic qualities are specified in the DIN EN 15643 group of standards. Whereas the ISO 21929-1 international standard defines a framework for the development of indicators and for the compilation of core indicators for buildings, ISO 15392 formulates general principles for sustainable building.

1.4.1.1 Raw Materials and Building Materials

The selection of building materials for sustainable building is of particular importance. Generally, the building materials used in sustainable building projects are labeled as ecological building materials and have a low impact on people's health and the environment for the building's entire lifetime. For the selection of building materials, this means in particular:

The *raw materials* used for the production of the building materials must be sourced in an environmentally friendly and sustainable way. They should be renewable or available long-term and free of harmful substances.

Energy expenditure for extracting the raw materials and production of the building materials should be kept as low as possible. This energy expenditure is also referred to as the primary energy content (Sect. 1.4.3.2). The energy expenditure for subsequent life cycle phases (Sect. 1.4.2) also needs to be taken into consideration.

Energy expenditure for transportation as well as the transit times between the individual steps of the life cycle need to be minimized.

Pollution during the production and processing of the building materials should be largely prevented. This is especially true for the actual lifetime of the building with regard to its occupants (damages, indoor air quality, and health) as well as for the demolition of the building and the disposal of the demolition materials.

The building materials need to be *reusable* or *recyclable.* After the lifetime of a building, the materials need to be reusable or recyclable or, at least, disposed of in an environmentally friendly way using minimal energy. This reduces waste and minimizes landfill space.

The building materials need to be *long lasting* because their lifetime factors into the overall assessment of the building. The building materials used in structures need to fulfill usability requirements for an adequate period of time deemed appropriate for the structure's purpose.

1.4.1.2 Building Construction

Buildings which meet the abovementioned requirements for sustainable construction should be built in a manner which facilitates recycling. This means that the connections between building elements should be detachable, making it easy to separate the individual building elements and thus allowing for the sorting and recovery of the (preferably recyclable) demolition materials.

Buildings should be constructed in a simple and compact manner and should be easy to repair and flexible in terms of their usage. The application of passive solar design in connection with the use of suitable building materials would add to a reduction in heating energy requirements and create a comfortable year-round thermal climate.

Buildings should allow the use of ecological building materials as long as the requirements of the MBO are met. For this, the function and requirements placed on the structure or the building elements need to be precisely defined. Building materials susceptible to moisture (e.g., earth) are thus not allowed, or only to a limited extent, for exterior use. For interior use, appropriate design precautions need to be taken.

When it comes to the protection objective "limitation of energy usage" and, with it, a reduction in CO_2 emissions, legislators have made the requirements for a building's thermal performance more stringent over the past years. Oftentimes, this leads to complicated exterior wall structures, made up of several layers with integrated insulation, forming an inseparable material bond. Leaks within the wall structure often create airflow which can result in moisture damage and mold.

There are special requirements with regard to indoor air quality. Therefore, it is particularly important to use suitable materials for finishing building elements. These building materials should be free of harmful substances which could outgas into the indoor air. In particular, they should be vapor permeable and sorption capable in order to "buffer" rapid changes in indoor humidity, thereby reducing the risk of mold.

DIN EN 15643-2 defines the framework for the assessment of technical aspects of the *environmental* performance of buildings.

In addition to these technical requirements, the assessment of the sustainability of buildings also includes sociocultural factors. “Soft” factors such as health and comfort (Sect. 5.1.3), safety, design quality, and functionality determine the well-being of the building’s occupants. These factors depend on subjective perceptions which result in the assessment of user satisfaction. In connection with the quality of the surfaces of building elements, clients are increasingly looking for an aesthetic design (color, structure/texture, pleasant tactile surface qualities). Basic conditions for the assessment of the *sociocultural* performance of buildings are defined in DIN EN 15643-3.

Buildings which are designed according to the principles of sustainable building also need to be able to hold up to a comparison in terms of economic efficiency. This includes the minimization of life cycle costs and the general improvement of economic efficiency during construction. DIN EN 15643-4 defines the framework for the assessment of the *economic* performance of buildings.

1.4.1.3 Surroundings

Negative impacts on the surroundings of the building caused by the construction process should be kept as low as possible. In this regard, action strategies can be developed in two directions [40]: open space design and urban structure.

Building projects in open spaces should be mindful of conservative and environmentally and socially acceptable land use while being sensitive to existing structures. Users and residents should not be subjected to water and soil pollution or poisoning or to the improper disposal of wastewater, smoke, and solid or liquid waste.

Inner-city green spaces need to be connected to surrounding open spaces within the city or region and, above all, meet the needs for recreation and nature conservation.

The overall urban energy needs should be reduced to an environmentally acceptable level. By implementing the concepts of sustainable mobility, urban life could “slow down,” private transportation could almost entirely transition to public transportation and, as a result, the inner-city street space could be designed as a living space.

1.4.2 The Life Cycle and Material Cycle of a Building

The evaluation of the action strategies described above in terms of their environmental impact in all life cycle phases of a building leads to the key principle of sustainable building: the analysis of the life cycle of the materials used in a building. The goal of this analysis is to reduce waste and keep the environmental impact as low as possible by “closing” the cycle. During an inventory, the entire life cycle is assessed. This includes the sourcing and extracting of the raw material, the use of the material to produce building elements and structures, the use of the finished building including its maintenance, and, finally, the demolition of the building and the recycling of the demolition materials. Transportation between the individual phases as well as production-related material and energy flows are also included in this evaluation.

When passing through each phase of the life cycle, the building material needs to meet the requirements of sustainable building defined in Sect. 1.4.1. These requirements are described with the help of relevant parameters which are determined through standardized testing procedures. For example, a building material needs to have adequate compressive strength to be suitable for load-bearing construction. Meeting the test criteria ensures that the required qualities for a certain life cycle phase have been attained after completing this stage. Only then is the building material or building element suitable for use.

Figure 1.27 shows the life cycle model for earth as a building material [41]. After passing through each life cycle phase, the earthen material attains a new quality: raw soil becomes soil for construction; construction soil is processed into earth building materials; etc. By reusing recycled earthen materials, the life cycle becomes self-sustaining.

Table 1.1 links essential features for assessing the technical, environmental, sociocultural, and economic performance of earth structures and displays them in a matrix. It lists parameter groups and parameters related to the cycle phases of "construction soil," "earth building material" and "earthen structures" according to Fig. 1.27. When planning a specific construction project, the relevant parameters for describing the suitability along with the respective test criteria can be derived from this matrix based on the particular conditions of the project. The last column of Table 1.1 indicates the corresponding chapter numbers in this book.

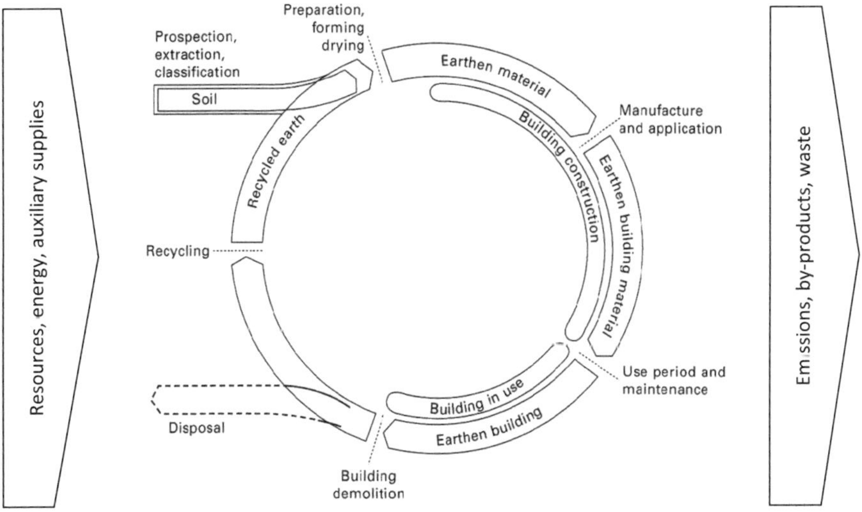

Fig. 1.27 Life cycle of earth as a building material [41, 53]

Table 1.1 Relevant parameters for the processing stages of earth, overview

Parameter group (PG)	Relevant area	Building material parameter	Processing stage Construction soil	 Earth building material	 Earth building element/structures	Section number
1. Physical parameters	Structure parameters	Porosity	●	●		3.6.1.1
	Mass parameters	Bulk density		●	●	3.6.1.2
		Dry bulk density	●	●		3.6.1.3
		Proctor density		●		3.6.1.4
		Specific density	●	●		3.6.1.5
	Grain size parameters	Grain size/grain size distribution	●			2.2.3.1
2. Chemical–mineralogical parameters	Acid–base reaction	pH value			●	1.4.3.3
	Type of clay mineral (class)	Activity	●			2.2.3.2
		Cation exchange capacity	●			2.2.3.4
	Natural additives	Lime	●			2.2.3.4
		Water-soluble salts	●	●	●	2.2.3.4
		Organic additives	●			2.2.3.4
3. Processing parameters	Plasticity	Moisture content	●			2.2.3.2
		Liquid limit/plastic limit	●			2.2.3.2
		Consistency	●			2.2.3.2
		Cohesive strength/standard consistency	●			2.2.3.2

4. Structural parameters	Deformation parameters, load independent	Moisture expansion; shrinkage (−)/ swelling (+)	●	●	○	3.6.2.1 2.2.3.3
		Slump		●		3.6.2.1
	Deformation parameters, load dependent	Modulus of elasticity/Poisson's ratio		●	○	3.6.2.1
	Strength parameters	Dry compressive strength		●	○	3.6.2.2
		Flexural strength		●		3.6.2.2
		Tensile adhesion strength			●	3.6.2.2
		Shear strength			○	3.6.2.2
		Wear resistance			○	3.6.2.2
5. Building physics parameters	Hygric parameters	Capillary water absorption		●	○	3.6.3.1
		Frost test		●		3.6.3.1
		Equilibrium moisture content			●	5.1.2.4
		Water vapor diffusion resistance factor			●	5.1.2.2
		Water vapor sorption			●	5.1.2.5 3.6.3.1
	Thermal parameters	Thermal conductivity		●		3.6.3.2
		Specific heat capacity		●	●	3.6.3.2
		Thermal transmittance coefficient			●	5.1.1.2
		Heat penetration coefficient			●	5.1.1.2
	Sound insulation parameters	Sound reduction index			●	5.1.5
	Fire protection parameters	Flammability (class)		●		3.4.8
		Fire resistance (class)			●	5.1.4.2
	Radiation protection parameters	Activity Concentration Index			●	5.1.6.1

(continued)

Table 1.1 (continued)

Parameter group (PG)	Relevant area	Building material parameter	Processing stage			Section number
			Construction soil	Earth building material	Earth building element/structures	
6. Functional parameters		Erosion resistance			●	5.1.2.6
		Wind resistance			●	5.1.3.3
		Biological durability			○	5.2.3
		Susceptibility to aging			○	5.2.2
7. Parameters describing the construction process	Planning of earth buildings	Fundamental principles of the building trades			●	4.2
	Execution of earth buildings	Foundation			●	4.3.1
		Floors			●	4.3.2
		Wall construction			●	4.3.3
		Ceilings			●	4.3.4
		Roof construction			●	4.3.5
		Plaster			●	4.3.6

8. Building ecology parameters	Consumption of resources, benefits	Energy consumption PEI, CED	●	●	●	1.4.3.1
		Drinking water usage/wastewater	●	●	●	1.4.1.3
		Land use	●	●	●	1.4.1.3
		Recycling potential[a]			●	6.2.2
		Heating value[b]			●	6.2.2
	Environmental impact parameters	Global warming potential, CO_2 equiv.[c]	●	●	●	1.4.3.2
		Ozone depletion potential ODP[d]	●	●	●	1.4.3.2
		Acidification potential, SO_2 equiv.[e]	●	●	●	1.4.3.2
		Overfertilization potential/ eutrophication, PO_4 potential[f]	●	●	●	1.4.3.2
		Photochemical ozone creation potential POCP, C_2H_4 equiv.[g]	●	●	●	1.4.3.2
		Tropospheric ozone precursor equivalent, TOPP equiv.[h]	●	●	●	1.4.3.2
		Risks for the local environment	●		●	2.2.4 6.3
9. Physiological parameters	Limits for harmful substances	Metals/metalloids; TVOC; PAK; AOX; phenol index		●	●	6.2.2.1
10. Architectural and aesthetic parameters	Surface effects	Quality grades Q (plaster)			●	4.3.6.6
	Crack formation	Crack width control			●	4.3.6.6
	Color effects				○	4.3.6.6
	Abrasion	Abrasion dust quantity			●	4.3.6.6
11. Waste technology parameters	Material purity				○	6.1.2
	Reuse/recycling	Levels of harmful substances/ assignment criteria LAGA			●	6.3.2

(continued)

Table 1.1 (continued)

Parameter group (PG)	Relevant area	Building material parameter	Processing stage			Section number
			Construction soil	Earth building material	Earth building element/structures	
12. Building practice parameters	Quantity and mass parameters	Mass and structure parameters	●	●	●	4.2.2.2, 3.6.1
		Unit price	●	●	●	4.2.2.2
		Typical work times		●	●	4.2.2.2
	Building-related costs during use	Repair and renovation cycles[i]			●	5.3.2.2
13. Extraction parameters	Removal, loading, hauling	Extraction class	●			2.2.4.1
		Transport	●	●	●	2.2.4.2, 6
		Risk potential	○			2.2.4.1

● Test method/procedure known, ○ no test method known

[a] Describes how much negative environmental impact can be avoided by "recycling" a material compared to the production of the new material (Sect. 6.2.2). The emissions for the production of a material would have to be reduced by its future recycling potential

[b] The amount of energy which is released when a material undergoes thermal recycling (combustion). One cubic meter of wood has an approximate heating value of 8000–13,000 MJ (=225–365 L of heating oil)

[c] Indicates how much of a certain mass of a "greenhouse gas" contributes to the greenhouse effect. It is expressed as a factor of CO_2 standardized to 1 based on 100 years of residence time in the atmosphere. CO_2 in itself is a major cause of the greenhouse effect and consequently global warming. Ten kilogram of CO_2 emissions is equivalent to the generation and combustion of approximately 3 L of heating oil

[d] The ozone depletion potential combines the effects of various ozone depleting gases, and the reference substance is CFC. The ozone layer in the stratosphere protects from harmful ultraviolet radiation

[e] Reference substance for the acidification potential of an emission into the air resulting in "acid" rain, soils, waters, etc. Secondary effects on buildings are steel corrosion and the degradation of natural stone, concrete, and earth as a building material

[f] Groups substances together in terms of their PO_4 effect. Overfertilization can lead to a concentration of substances toxic to humans in both groundwater and drinking water

[g] Refers to the effects of ethene (C_2H_4). Intense sunlight causes chemical reactions at ground level, producing harmful substances such as ozone which can lead to so-called smog. Higher concentrations of ozone are toxic to humans

[h] Is the quantitative expression of the ground-level ozone creation potential and is formed from the relative ozone creation rate of the air emissions CO, NMVOC (non-methane volatile organic compounds), NO_x, and CH_4. With an increase in TOPP, the risk of summer smog rises

[i] Is the period of time in which a building material can perform its intended function within its assigned use

1.4.3 Environmental Management and Life Cycle Assessment

Environmental management is part of the management system of an organization. It develops action strategies for environmental protection at the company level as well as the official authority level in order to ensure the environmental compatibility of the products and processes developed by the company and its staff performance.

The term "life cycle assessment" describes the systematic, quantitative analysis of the environmental impact of products throughout their lives in the form of ecological assessment results. Here, "environmental impact" refers to the use of resources as well as the environmental effects of emissions at every phase in a product's lifetime. The results of the analysis make it possible to find measures for reducing the environmental impact or for comparing different products.

The life cycle assessment has become a generally accepted methodological approach for the quantitative evaluation of the sustainability of building materials and building products.

On a European level, the following standards for conducting a life cycle assessment are currently available:

DIN EN ISO 14040:2009-11 Environmental management—Life cycle assessment—Principles and framework
DIN EN ISO 14044:2006-10 Environmental management—Life cycle assessment—Requirements and guidelines

According to DIN EN ISO 14040, the life cycle assessment consists of four phases. These phases correspond to each other and cannot be viewed separately:

- Defining goal and scope
- Life cycle inventory analysis
- Impact assessment
- Interpretation

1.4.3.1 Defining Goal and Scope

A determination of the goal and scope must define the use and function of a product and its general life cycle from raw material sourcing to disposal. Figure 1.27 shows this cycle for earthen materials.

In terms of building materials and building products, this phase is used to select and define different material and construction options. To facilitate this process, so-called *functional units* are determined to serve as a reference (such as a quantity unit of a building material or a sample building as a product-specific size). The results of the analysis of the environmental impact can then refer to these functional units. Product units which are to be compared need to match exactly in terms of their functions.

At the beginning, the *system boundaries* need to be determined by deciding which indicators to include in the analysis and which to leave out. "From cradle to factory gate" and "from cradle to grave" are typical examples of system boundaries. The selection of these indicators can influence the result of the life cycle assessment.

1.4.3.2 Life Cycle Inventory Analysis

During the life cycle inventory analysis phase, the defined material and construction variations within the determined system boundaries are established for the relevant material and energy flows. The life cycle inventory analysis contains information on all relevant consumption of raw materials and energy, the kind and quantity of emissions and harmful substances and, if applicable, all quantities of waste generated throughout the entire lifetime of the materials and buildings. This initial information needs to be obtained from the manufacturer. The determined material quantities are linked with their environmental impact during the impact assessment phase. The life cycle inventory analysis itself does not include an evaluation. Collection of the required data can be very time consuming unless existing databases can be used.

Primary Energy Intensity

The energy expenditure needed for the production of building materials including the production and transportation of the source materials is an important indicator for the selection of "ecological" building materials. This energy expenditure is called the "primary energy intensity" (PEI) and is mainly related to the system boundary "from cradle to factory gate."

When it comes to meeting energy needs, the available energy includes renewable (e.g., biomass), inexhaustible (e.g., sun), and nonrenewable (e.g., fossil fuels) sources. The supply of nonrenewable energy sources is limited and they should, therefore, be used sparingly. When determining the PEI, the amount of nonrenewable energy sources is identified. This also makes the PEI a measurement for the environmental impact "use of energy resources" category.

Applied to traditional earth building, the manual processing of suitable excavation material into earth building materials and structures on the building site was and still is the ideal situation as far as the PEI is concerned. Particularly the fact that no transportation of the earth building material was required resulted in a PEI of practically zero.

Modern earth building, however, is largely mechanized and characterized by the physical separation of building material production and product use on the building site. This automatically leads to energy consumption and transportation. Producers of earth building materials in various European countries have begun exporting their products to other countries. Long-distance transportation and high specific energy consumption have a negative impact on an ecological assessment of building materials, e.g., in the case of artificial drying (Sect. 3.3.3). Table 1.2 shows the PEI for common modes of transport according to [40]:

In terms of their PEI, earth building materials are still unrivaled compared to the main conventional building materials. This is even true when additives with a high embodied energy are used. A selection can be seen in Table 1.3 [42]:

Table 1.2 Energy use of common modes of transport

Mode of transport	PEI [kWh/tkm]
Rail	0.43
Passenger car, Western Europe	1.43
Truck, 40 metric tons	0.72
Truck, 28 metric tons	1.00
Truck, 16 metric tons	1.45
Van, <3.5 metric tons	3.10
Cargo ship overseas	0.04
Cargo ship inland waterways	0.27

Table 1.3 Primary energy intensity for selected building materials

Building material	PEI [kWh/m^3]
Earth	0–30
Straw panels	5
Wood, domestic	300
Derived timber products	800–1500
Fired bricks	500–900
Cement	1700
Standard concrete	450–500
Sand-lime bricks	350
Sheet glass	15,000
Steel	63,000
Aluminum	195,000
Polyethylene PE	7600–13,100
PVC	13,000

Cumulative Energy Demand

The *cumulative energy demand* encompasses the energy demand of a building over its entire lifetime (system boundary "from cradle to grave"). According to VDI guideline 4600: 2012-01, it is estimated with the help of specific assumptions and scenarios. The corresponding environmental impacts are compared to this energy demand (Table 1.1, *PG 8*).

Frequently, the PEI of a building material or a building technique is only used for a comparative evaluation. A realistic assessment, however, needs to take all phases in the life of a building into consideration because the actual production of a building material and the construction of a building represent relatively short time periods. Therefore, the advantage of a "low PEI" of an earth building material might come at the expense of the durability of the structure or a higher maintenance requirement. Insulating materials with a higher PEI can offset this "disadvantage" by reducing the required energy for heating, thereby lowering emissions for the entire lifetime of the building. In addition to the positive environmental impact, these materials can therefore also offer financial advantages to the owner. As a result, the determination of how long a building will be in use becomes an additional important system boundary which has an effect on the number of required maintenance cycles for the building element or material layers.

Table 1.4 Selected ecological parameters for stabilized earthen blocks [43]

Option	CED [kWh]	CO_2 equiv. [kg]	SO_2 equiv. [kg]	TOPP equiv. [kg]	Raw material consumption [t]
CSEB (manually produced)	215	94	0.37	0.59	2.53
CSEB (mechanically produced)	655	189	0.49	0.70	3.05
Fired brick	1347	508	1.68	3.03	1.55

1.4.3.3 Impact Assessment

In this phase, the material and energy flows collected during the life cycle inventory analysis are evaluated in terms of their environmental impact on the basis of selected indicators and defined system boundaries: the causes are compared to the impacts. This is when the environmental impacts described in Table 1.1, *PG 8*, are examined. Currently, a number of computer programs containing databases of relevant values can be used for calculating the impact assessment. The data analysis must be carried out according to defined standards. The programs GaBi (www.gabi-software.com), EcoInvent (www.ecoinvent.ch, approx. 4500 data files), and WECOBIS (www.wecobis.de of the German Federal Ministry for the Environment, Nature Conservation, Building and Nuclear Safety) are examples of commonly used environmental databases. The latter has been available since January 2015. It is structured based on the criteria of DIN EN 15804 and currently contains approx. 1300 data files which also cover earth plaster, earth blocks, and rammed earth.

Using the program GEMIS (www.gemis.de), Freudenberg [43] has determined various parameters of a life cycle assessment for compressed stabilized earth blocks (CSEB). Table 1.4 partially compares these parameters for manually and mechanically produced earth blocks, stabilized with 5 % cement, as well as for fired bricks based on 1 metric ton of final product each.

1.4.3.4 Interpretation

Depending on the specific situation, the final interpretation of the calculated results can be carried out in different ways, such as:

- A comparison of suggested design variations (preferred variation)
- Ecological impact assessment (hazards)
- Impact in relation to already existing environmental pollution

Today, life cycle assessments are indispensable for environmental decision making, for example, when trying to determine binding regulations for orders of magnitude for decreasing CO_2 emissions in relevant documents on an international level. Environmental goals can only be achieved if they are defined as guidelines in appropriate standards and regulations. This also includes product standards in the field of building materials. The technical information sheets “Technische Merkblätter 02–04” [44–46] published in 2011 by the German Association for Building with Earth include a procedure for determining the CO_2-equivalent value on the basis of DIN EN ISO 14040. Appropriate procedures have been included in DIN standards for earth blocks and earth mortars (DIN 18945–47) as optional tests (Appendix A).

A life cycle assessment requires a significant effort during the planning stage as well as the willingness to add sustainable building concepts to conventional planning operations. Often, a lack of sufficient data poses a problem. Although the indicators mentioned above allow for a very detailed description of ecological impact categories, there are still harmful environmental effects which are generally known but, so far, have not been able to be measured quantitatively. The reference accuracy of correlations for indicators which have already been defined also remains questionable. Finally, the reliability of the life cycle inventory data collected in the available environmental databases must be examined. This more or less limits the validity of each result.

On the other hand, life cycle assessments are already a suitable tool for checking if seemingly feasible, ecologically founded arguments can hold up in reality. However, the necessary fundamentals and instruments need to be improved further.

1.4.3.5 Environmental Product Declarations and Certification of Buildings

The life cycle analysis according to DIN EN ISO 14040 provides systematic and standardized data for recording energy demands and environmental consumption as well as their environmental impacts over the total life cycle of a building.

In addition, the *environmental performance* of a building according to the principles of sustainable building (Sect. 1.4.1) comprises its technical quality, functional aspects, sociocultural criteria, as well as location, e.g., transportation infrastructure. Finally, costs are an important consideration for the client. Corresponding parameter groups are listed in Table 1.1. These aspects exceed a “pure” life cycle analysis.

Two instruments have been developed for analyzing the environmental performance of a building product:

- Environmental labels/environmental declarations for *manufacturers* of building products
- Environmental building certificates for *owners*/clients

Environmental Product Declarations

Currently, three categories of environmental labeling are available to manufacturers of building products:

- Type I environmental labeling according to DIN EN ISO 14024 consists of symbols or logos which have been awarded by external bodies for outstanding environmental performance. The eco-labels "Blue Angel" and "natureplus" are typical examples. Several earth building products carry the latter [47].

- Type II environmental labeling according to DIN EN ISO 14021 consists of environmental declarations by the producers themselves. This means that the producers are responsible for their own declarations which they can have verified by external bodies.
- Type III environmental labeling according to DIN EN ISO 14025 consists of voluntary standards, commitments, or guarantees for building products. They are provided by producers, organizations, and quality assurance associations in order to establish the "environmental performance" of buildings in the form of a certificate awarded by external bodies. This type of label is known as an *environmental product declaration (EPD)*.

The following standards currently exist for the development of EPDs for building products:

DIN EN 15804 Sustainability of construction works—Environmental product declarations—Core rules for the product category of construction products
DIN EN 15942 Sustainability of construction works—Environmental product declarations—Communication format business-to-business
DIN EN ISO 14025 Environmental labels and declarations—Type III environmental declarations—Principles and procedures
ISO 21930 Sustainability in building construction—Environmental declaration of building products

Such declarations must include all phases of the life cycle of a product by describing the environmental impact during production and use as well as possible health hazards for the users. In order to meet these requirements, a standardized assessment diagram has been developed. It consists of the four life cycle phases (stages) which need to be declared as well as column D which records benefits and loads (consumption) (DIN EN 15804) (Table 1.5).

Environmental Product Declarations (EPDs)

EPDs have become instruments for the selection of products with regard to the environment. They stimulate the use of environmentally friendly products through competition and help to protect the safety and health of consumers by keeping unsafe products off the market.

Earth building materials are inherently environmentally friendly because they do not pose any health risks and have a low PEI compared to other building materials (Table 1.3). Currently, producers of mineral building materials with higher PEIs are providing certified EPDs according to DIN ISO EN 14025 for building materials containing lime, gypsum, and cement based on requirements by their respective industry organizations. When assessing the emissions of greenhouse gases, described by the CO_2 equivalent (GWP) (Table 1.1), producers take advantage of the trade-off, for example, by "consuming" CO_2 during the carbonation of lime or

Table 1.5 Assessment diagram for EPD life cycle phases according to DIN EN 15804

Product stage			Construction process stage		Use stage							End-of-life stage				Benefits and loads
A1	A2	A3	A4	A5	B1	B2	B3	B4	B5	B6	B7	C1	C2	C3	C4	D
Production of raw materials	Transport	Production	Transport	Construction/ installation	Use	Maintenance	Repair	Replacement	Rebuilding/ renovation	Company-related energy consumption	Company-related water consumption	Demolition	Transport	Waste management	Disposal	Reusing, recovering, and recycling potential

by "recovering energy" from waste instead of using fossil fuels. In this manner, producers can reduce the "sustainability gap" between their conventional materials and earth building materials. This emphasizes how environmental product declarations are increasingly assuming the role of a competitive tool on the building material market.

Producers of industrially manufactured earth building materials have to become aware of the fact that the given environmental credibility of earth products will not suffice in the future. To remain successful in an increasingly competitive market, appropriate life cycle assessments (EPDs) for earth building materials need to be drawn up. There are now life cycle assessments for industrially produced, naturally moist earth mortars which, with regard to their manufacturing process (from raw material to delivery ex works), have an energy balance value that is 5–10 times lower than that of building materials made of lime and gypsum [48].

Certification of Buildings

Nowadays, home owners must account for the environmental performance of their houses during the use phase with regard to energy consumption. The Energy Conservation Regulation EnEV 2014 [49] requires owners to present an *energy pass* to anyone interested in renting or purchasing.

However, energy consumption only represents a partial aspect of the environmental performance of a building. Currently, the following standards can be applied in a comprehensive quantitative assessment of the environmental performance of buildings:

DIN EN 15978 Sustainability of construction works—Assessment of environmental performance—Calculation method

DIN EN 16309 Sustainability of construction works—Assessment of social performance—Calculation methodology

DIN EN 16627 Sustainability of construction works—Assessment of economical performance—Calculation method

These standards use the assessment diagram for EPDs according to DIN EN 15804 (Table 1.5).

A number of organizations and associations have developed systems for the certification of the environmental performance of buildings based on criteria catalogs which are more extensive than those found in DIN EN 15804. An example is the certification of buildings issued by the German Sustainable Building Council (DGNB—Deutschen Gesellschaft für Nachhaltiges Bauen) [50]. This system uses a core catalog of six quality categories with additional weighted partial criteria—ecology, economy, sociocultural, and functional aspects—and technical criteria (amounting to 22.5 % each), as well as process quality (amounting to 10 % of the total assessment). The category "location" is included in the total assessment indirectly. For compliance with each quality category, external auditors award combined points leading to the quality seals "bronze," "silver," and "gold."

The German government has decided to make it mandatory to apply the principles of sustainable building to all future federal building projects by using a rating system called "Sustainable Building for Federal Buildings" (Nachhaltiges Bauen für Bundesgebäude—BNB), published by the Federal Ministry of Transport, Building and Urban Development [38, 51]. Federal buildings are thereby intended to serve as role models.

1.4.4 Economic Aspects

Clients often ask if and how much more "expensive" it is to build with earth. Placing earth building components within the overall cost structure of a building can differ considerably from region to region. For example, in typical timber-frame construction with earthen infill, the earth building materials might account for less than 10 % of the total building costs depending on the type and quantity of earth building materials used [41].

With the improved availability of earth building materials, current prices could become more "attractive" to the consumer. When making a decision about using a very "specific" earth building material, clients need to consider additional transportation needs which would have a negative impact on the assessment of the building in terms of ecological criteria.

Conventional cost estimates, calculations, and assessments are carried out according to DIN 276 on the basis of all services performed up to the completion or the final handover of the building. These calculations and assessments do not yet consider the important advantages of earth as a building material with regard to their medium- to long-term "benefits" to society in general and the client in particular. Among the benefits are the undoubtedly low primary energy balance (Sect. 1.4.3.2) and recyclability (Sect. 6.2.2), along with the durability of the materials when properly installed, the architectural aesthetics, and, above all, the health benefits of earth building materials, particularly in terms of indoor air quality (Sect. 5.1.3). These aspects have already been included in comprehensive systems for the certification of buildings (Sect. 1.4.3.5).

Currently, operating costs, including expected maintenance costs during the entire lifetime of the building, can only be assessed in a basic manner. Aspects of sustainability in terms of a quantitative life cycle assessment cannot be recorded at this time. It is, for example, not possible to forecast what the disposal costs for building waste will be in 100 years' time, at the end of the lifetime of a building which is built today. But it is this question in particular which would be of interest when pricing earthen materials and earth building work. A realistic comparison with other building materials would show that building with earth is already cost effective in today's world.

DIN EN 16627 can be applied for assessing the sustainability of buildings with regard to economic aspects.

1.5 Classification of Earth Building as a Field of Science

Figure 1.28 shows an interconnecting model of relevant fields of science which are significant to earth building. Here, the building-material fields are assigned to the fundamental sciences, while the fields dealing with earth building elements and structures are tied to the disciplines of structural planning, engineering, and design. Earth building can therefore be approached in different ways via the various fields of science.

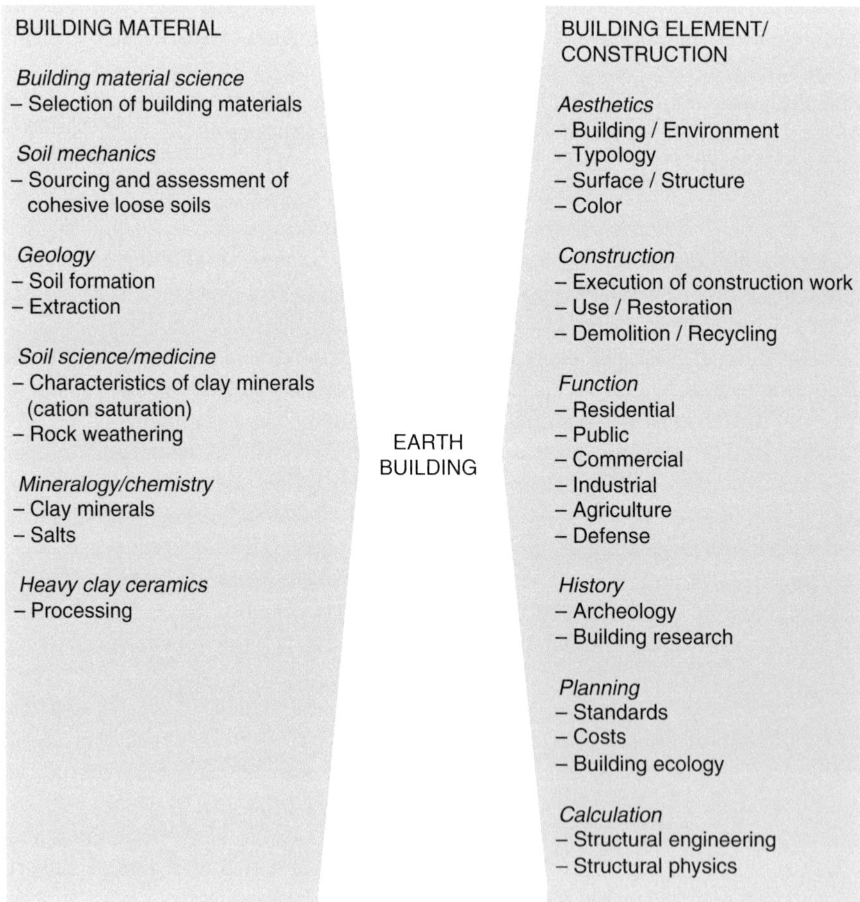

Fig. 1.28 Classification of earth building as a field of science

1.5.1 Terminology

In general terminology, the classification of earth building as part of the construction field is defined by the terms "architecture" and "building" (French, architecture de terre; German, Lehmbau). Architecture is part of human activity with the goal of creating buildings for various human needs.

Depending on whether a structure is built primarily above or below ground, it is classified as above-grade or below-grade construction. Structures above grade are typically also called buildings.

Generally, *earth building* is the processing of earth building materials into building elements and aboveground structures with load-bearing or non-load-bearing structural characteristics. It differs from foundation engineering, road construction, or hydraulic engineering in that earth building structures are generally dry in their finished state (Sect. 3.3).

1.5.2 Building Material and Building Technique

For the conventional building materials, concrete, steel, reinforced concrete and fired brick, and specialized fields of science have developed within the area of civil engineering, particularly over the past 50 years. This development has also led to intensive teaching and research activities at universities. Later, the same became true for timber construction. As described above, for earth as a building material, things have developed differently.

A separate field of science for "earth building" is only in its beginning stages. It will gain an importance and autonomy as the demand for standardized solutions for earth building materials and for their use in the production of functional building elements and structures increases on the building site and as earth building becomes a building engineering technology. Compared to the conventional building materials mentioned above, systematic earth building research has not been conducted for decades due to the fact that earth has not been widely used as a building material. It will take some time to reduce this "research backlog." As part of a research project at the German Federal Institute for Material Research and Testing (BAM—Bundesanstalt für Materialforschung und–prüfung) from 2009 to 2011, selected building material and building element aspects of earth building were systematically examined on a larger scale for the first time. The results of this project were incorporated into the development of DIN standards for earth building.

On the other hand, earth building will always also remain a nonengineering technology because earth as a building material is, unlike concrete and steel, not synthetically produced, but rather a naturally "produced" and non-standardized material. By far the largest part of earth building activities today is carried out by owner-builders in developing countries using traditional methods.

The intertwining of various fields of science is also characteristic for earth building. And so the "raw material" earth has many functions within the various fields, for example, as the *raw material* for the production of ceramic goods in heavy clay ceramics, as *subsoil* for setting up foundations in the field of ground engineering, as *soil* for the cultivation of crops in soil science, and even as for *medicinal purposes* in medicine. Every discipline has developed its own parameters or terminology for specific applications of the "raw material" earth thereby providing the basis for successful communication. Parameters always have content, expressed by a unit of measurement, and a scope which serves as a qualitative statement about a specific property of the value or parameter.

References

1. Burenhult, G.u.a. (Hrsg.): Die Menschen der Steinzeit—Jäger, Sammler und frühe Bauern; Illustrierte Geschichte der Menschheit. Weltbild, Augsburg (2000)
2. Boehling, H.: Chinesische Stampfbauten. Sinologica, Basel. **3**(1), 16–22 (1953)
3. Sung, Y.: Tian gong kai wu, vol. 2, leaf 1. Zhonghua shuzhu, Xinhua shudian zongjing xiao, Beijing. http://depts.washington.edu/chinaciv/home/3intrhme.htm#brik (1959)
4. Luo, Z., Zhao, L.: Chinas Große Mauer—Traditionelle chinesische Kunst u. Kultur. Verlag f. Fremdsprachige Literatur, Beijing (1986)
5. Endruweit, A.: Städtischer Wohnbau in Ägypten. Gebr. Mann, Berlin (1994)
6. Fathy, H.: Architecture for the Poor. University of Chicago Press, Chicago (1973)
7. Galdieri, E.: Tecnología y fantasía en las construcciónes de tierra. In: II. Encuentro de trabajo sobre la tierra como material de construcción, pp. 53–59. Inter-Acción, Navapalos/El Burgo de Osma (1986)
8. Simon, S.: Kulturelles Erbe Erdarchitektur—Materialien, Forschung, Konservierung. In: Naturwissenschaft & Denkmalpflege, pp. 263–273. Innsbruck University Press, Innsbruck (2007)
9. Boysan-Dietrich, N.: Das hethitische Lehmhaus aus der Sicht der Keilschriftquellen. Winter, Heidelberg (1987)
10. Baimatova, N.S.: 5000 Jahre Architektur in Mittelasien—Lehmziegelgewölbe vom 4./3. Jahrtausend. v. Chr. bis zum Ende des 8. Jahrhunderts. n. Chr. Archäologie in Iran und Turan, 7, Deutsches Archäologisches Institut, Eurasien-Abteilung, Außenstelle Teheran. Verlag Philipp v. Zabern, Mainz (2008)
11. Jansen, M.: Mohenjo-daro-Pakistan. The problem of adobe & brick conservation. In: Terra 2003—9th International Conference on the Study and Conservation of Earthen Architecture, pp. 309–318. Yazd (2003)
12. Bardou, P., Arzoumanian, V.: Arquitecturas de adobe. Tecnología y Arquitectura—construcción alternativa, 3rd edn. Ediciones G. Gili, S.A. de C.V., México (1986)
13. ICOMOS (ed.): The World Heritage List. Filling the Gaps—An Action Plan for the Future. Monuments and Sites XII, Paris (2005)
14. Correia, M., Dipasquale, L., Mecca, S. (eds.): Terra Europae. Earthen Architecture in the European Union. Edizioni ETS, Pisa (2011)
15. http://dev.lehmbau-atlas.de
16. CRATerre (ed.), Guillaud, H.: Terra Incognita—Discovering and Preserving European Earthen Architecture. Argumentum/Culture Lab Editions, Lissabon (2008)
17. Correia, M.: Taipa no Alentejo. Lissabon, Argumentum (2007)
18. Fonseca, I.: Architectura de terra em Avis. Argumentum, Lissabon (2007)

19. Syrova-Anýžová, Z., Syrovy, J., Kříž, J.: Inventaire, documentation et méthodologie de conservation de l'architecture en terre en République Tchèque. SOVAMM CD, Brno (2000)
20. Conti, G., Di Chiacchio, A., Cicchitti, M., et al.: Terra cruda—Insediamenti in provincia di Chieti. Edizioni COGECSTRE, Penne (1999)
21. Isensee, B: Materialtechnische Probleme des Hausbaus auf einer jungsteinzeitlichen Siedlung bei Erfurt-Gispersleben. Unpublished student research paper. Bauhaus-Universität, Fak. Bauingenieurwesen, Weimar (2003)
22. Behm-Blancke, G: Die altthüringische und frühmittelalterliche Siedlung Weimar. In: Frühe Burgen und Städte. Beiträge zur Burgen-und Stadtbauforschung, Berlin (1954)
23. Weise, G.: Mineralische Rohstoffe und ihre Nutzung im Weimarer Land, Heft 1. Materialforschungs-und Prüfanstalt an der Bauhaus-Universität, Weimar (1998)
24. Lieberenz, T.: Die ältesten Lehmwellerbauten in Mittelthüringen. In: Häuser aus Lehm und Stroh—vergessene Bauweisen und Materialien, Hohenfeldener Hefte Nr. 4, pp. 27–34. Volkskundliche Kommission für Thüringen e.V./Thüringer Freilichtmuseum Hohenfelden, Hohenfelden (2008)
25. Brändle, E.: Sanierung alter Häuser. BLV Verlagsgesellschaft, München (1991)
26. Volhard, F.: Historische Lehmausfachungen und Putze. In: LEHM '94, pp. 25–27. Internationales Forum für Kunst u. Bauen mit Lehm, Beiträge, Aachen (1994)
27. Hausbücher der Nürnberger Zwölfbrüderstiftungen. Nürnberg 1425-1806. http://www.nuernberger-hausbuecher.de/75-Amb-2-317b-76-r/data
28. Güntzel, J.: Zur Geschichte des Lehmbaus in Deutschland, Bd. 1. Gesamthochschule, Diss., Kassel (1986)
29. Güntzel, J.G.: Zur Propagierung des Lehmbaus in Deutschland im 18. Jahrhundert. In: Bauen mit Lehm, Bd. 1. Ökobuch Verlag, Grebenstein (1984)
30. Hetzl, M.: Der Ingenieur-Lehmbau. Die Entwicklung des ingenieurmäßigen Lehmbaus seit der Französischen Revolution und der Entwurf neuer Lehmbautechnologien. Technische Universität, Fak. Architektur, Diss., Wien (1994)
31. Cointeraux, F: Die Pisé-Baukunst. Leipzig: Zentralantiquariat der DDR. Reprint der Originalausgabe von 1803 (1989)
32. Miller, T., Grigutsch, A., Schultze, K.W.: Lehmbaufibel—Darstellung der reinen Lehmbauweise, Heft 3. Schriftenreihe d. Forschungsgemeinschaften Hochschule, Weimar. Reprint der Originalausgabe von 1947. Verlag Bauhaus-Universität, Weimar (1999)
33. Miller, T.: Landwirtschaftliche Versuchshöfe in Thüringen. Der Baumeister. **44**(9), 282–296 (1947)
34. Grigutsch, A., Keller, B., Fährmann, H.: Erfahrungen beim zweigeschossigen Lehmbau in Gotha. Bauplanung/Bautechnik. **7**(6), 28 (1952)
35. Rath, R: Der Lehmbau in der SBZ/DDR. Unpublished diploma thesis. Technische Universität, Fakultät Architektur, Umwelt, Gesellschaft, Berlin (2004)
36. Архипов, И.И.: Механизированное производство и применение самана в сельском строительстве (Industrial production of straw-mud bricks for agricultural buildings). Научно-иccледовательский институт сельского строительства (НИИСельстрой) РСФСР (Hrsg.). Государственное издательство литературы по строительству, архитектуре и строительным материалам, Москва (1963)
37. United Nations Organization (UNO): Report of the World Commission on Environment and Development: Our Common Future (Brundtland Report). UNO, New York (1987)
38. Bundesministerium für Verkehr, Bau und Stadtentwicklung (BMVBS) (Hrsg.): Leitfaden Nachhaltiges Bauen. BMVBS, Berlin (2011)
39. Europäisches Parlament; rat der Europäischen Union: Verordnung Nr. 305/2011 vom 9.März 2011 zur Festlegung harmonisierter Bedingungen für die Vermarktung von Bauprodukten und zur Aufhebung der Richtlinie 89/106/EWG des Rates. Amtsblatt der Europäischen Union L 88/5 v., Brüssel (2011)
40. Glücklich, D. (Hrsg.): Ökologisches Bauen—von Grundlagen zu Gesamtkonzepten. Deutsche Verlags-Anstalt, München (2005)

41. Dachverband Lehm e.V. (Hrsg.): Lehmbau Verbraucherinformation. Dachverband Lehm e.V., Weimar (2004)
42. Umweltbundesamt (Hrsg.), Krusche, P.u.M., Althaus, D., Gabriel, I.: Ökologisches Bauen. Bauverlag, Wiesbaden (1982)
43. Freudenberg, P.: Ökobilanzierung von gepressten, stabilisierten Lehmsteinen. Unpublished diploma thesis. Fachhochschule, FB Wirtschaftsingenieurwesen, Jena (2007)
44. Dachverband Lehm e.V. (Hrsg.): Lehmsteine—Begriffe, Baustoffe, Anforderungen, Prüfverfahren. Technische Merkblätter Lehmbau, TM 02. Dachverband Lehm e.V., Weimar (2011)
45. Dachverband Lehm e.V. (Hrsg.): Lehm-Mauermörtel—Begriffe, Baustoffe, Anforderungen, Prüfverfahren. Technische Merkblätter Lehmbau, TM 03. Dachverband Lehm e.V., Weimar (2011)
46. Dachverband Lehm e.V. (Hrsg.): Lehm-Putzmörtel—Begriffe, Baustoffe, Anforderungen, Prüfverfahren. Technische Merkblätter Lehmbau, TM 04. Dachverband Lehm e.V., Weimar (2011)
47. Natureplus e.V. (Hrsg.): Richtlinien zur Vergabe des Qualitätszeichens "natureplus"; RL 0607 Lehmanstriche und Lehmdünnlagenbeschichtungen (September 2010); RL 0803 Lehmputzmörtel (September 2010); RL 0804 Stabilisierte Lehmputzmörtel (proposed); RL 1006 Lehmbauplatten (September 2010); RL 1101 Lehmsteine (proposed); RL 0000 Basiskriterien (May 2011); Neckargemünd (2010)
48. Lemke, M.: Nachhaltigkeit von Lehmbaustoffen—Umweltproduktdeklarationen als Wettbewerbsinstrument. In: LEHM 2012 Weimar, Beiträge zur 6. Int. Fachtagung für Lehmbau, pp. 220–229. Dachverband Lehm e.V., Weimar (2012)
49. Zweite Verordnung zur Änderung der Energieeinsparverordnung (EnEV 2009) v. 18.11.2013. Bundesgesetzblatt I, Nr. 61. Berlin (2013)
50. Deutsche Gesellschaft für Nachhaltiges Bauen (Hrsg.): Ausgezeichnet. Nachhaltig bauen mit System. DGNB Systembroschüre, Stuttgart (2012)
51. Bundesministerium für Verkehr, Bau und Stadtentwicklung (BMVBS): Bekanntmachung über die Nutzung und die Anerkennung von Bewertungssystemen für das nachhaltige Bauen. Bundesanzeiger Nr. 70, p. 1642 v. 07. Mai (2010)
52. CRATerre-Ensag/Gandreau, D., Delboy, L.: World Heritage Inventory of Earthen Architecture. World Heritage Earthen Architectural Programme WHEAP. CRATerre-ENSAG, Grenoble (2012)
53. Schroeder, H.: Konstruktion und Ausführung von Mauerwerk aus Lehmsteinen. In: Mauerwerkkalender 2009, pp. 271–290. W. Ernst & Sohn, Berlin (2009)

2 Construction Soil – Sourcing, Extraction and Classification

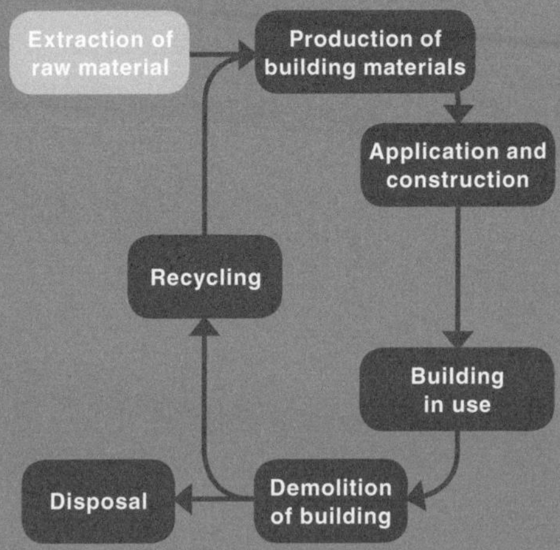

The term "natural materials" refers to all materials which occur naturally and are available in natural deposits or supplies. Naturally occurring clay rich soil can therefore also be termed *natural earth* or *natural soil*. Appropriate test procedures can determine if a natural clay rich soil is generally suitable for earth building. After natural materials have been sourced from their deposits and become part of a technical process they turn into raw materials. They attain a practical value. Natural soil becomes raw soil or *construction soil*.

© Springer International Publishing Switzerland 2016
H. Schroeder, *Sustainable Building with Earth*,
DOI 10.1007/978-3-319-19491-2_2

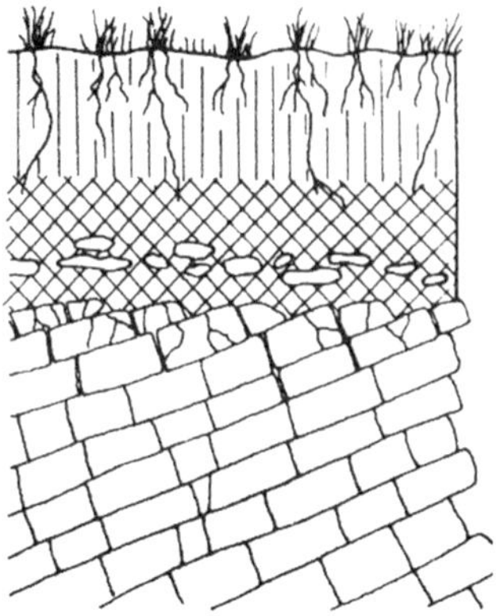

Fig. 2.1 Simplified pedological standard profile [1]

2.1 Natural Soils

2.1.1 The Formation of Natural Soils

Clay-rich soils are part of the top layer of the earth's solid crust which was formed under the influence of weathering, flora, and fauna. This makes them available nearly everywhere. In soil science, this layer of the earth's crust is also termed *soil*, while in engineering geology, it is called *loose rock*.

2.1.1.1 Soil Profile

During soil formation, the decomposition products of inorganic and organic parent material are transformed and structured into new soil components which are characteristic for the soil (clay and humus). These components are then washed off, transported, and mixed with the help of precipitation or groundwater in the soil or by soil tilling and the activities of organisms living in the soil. The result is the differentiation of the original rock into a *soil profile* with an eluviated humus-rich topsoil (A horizon) and an underlying layer of subsoil in which certain materials such as lime are precipitated (B horizon). The unweathered parent rock is called the C horizon (Fig. 2.1 [1]).

While the humus-rich A horizon forms the basis for vegetation and agriculture, soil suitable for construction can be sourced from the lighter-colored, humus-free B horizon. In some cases, horizons A and B are completely missing, for example, on uncovered rocky surfaces.

2.1.1.2 Soil Components

Soil contains solid, liquid, and gaseous components which is why the term *ternary system* can be applied. The components' spatial arrangement and distribution determine their usability for construction purposes (Sect. 3.6.1).

Solid Components

The solid components consist of inorganic and organic matter. The inorganic parts are formed from residual parent rock and parent minerals such as quartz (silicates), feldspar, mica, lime, gypsum, clay minerals, water-soluble salts, oxides of aluminum, and iron.

On the one hand, inorganic or mineral soils can be divided into four main types according to their prevalent *grain size d* (Table 2.2):

Coarse grain: gravel, sand, silt
Ultrafine grain: clay

Generally, soil consists of a blend of these main types. Clay-rich soils are typical examples of mixed-grain soils.

On the other hand, soil types can be differentiated by the amount of clay they contain: *cohesive* soils (high clay content) and *non-cohesive* soils (low or nonexistent clay content). The clay portion contains cohesive parts (clay minerals) and non-cohesive parts (such as quartz and mica). The cohesive clay minerals fulfill the task of holding together the coarse grains of silt, sand, and gravel which make up the soil's "skeleton." Clay-rich soils are cohesive soils.

Depending on the amount of *organic* material, the different soils are furthermore divided into inorganic (no organic material), interspersed with organic material (a low percentage of organic particles), and organic-rich loose rock (a high percentage of organic material). Soils typically contain both inorganic and organic materials.

A classification of soils for building purposes is always based on the soil's solid components (Sect. 2.2.3).

Liquid Components

Liquid soil components are formed by soil water. Soil water can be divided into groundwater that is formed by seepage water which can move freely in the soil voids and bound capillary and sorption water (Fig. 2.33).

Gaseous Components

Gaseous components include air as well as water vapor in the pore spaces.

2.1.1.3 Factors Influencing Soil Formation

The process of soil formation is influenced by a number of factors [2]:

- Climate
- Time
- Vegetation
- Parent material
- Topography
- Human activity

The physical, chemical, and biological processes which are triggered by these forces have a continuous impact on the soil. They create a dynamic balance which, if the ratio between these forces is comparable, leads to analogous profile formation.

Climate and Vegetation

Depending on the degree and fluctuation of air temperature and precipitation, solid rock is loosened mechanically to a greater or lesser extent. This allows precipitation or meltwater containing a number of substances to penetrate the cracks and crevices. Here it transforms minerals with complex structures (silicates, quartz, lime) into simpler chemical compounds (alkali and alkaline earth ions as well as silica) through a weathering process. Clay minerals are then newly formed out of these simpler compounds. Along the water-soluble salts, these clay minerals are transported to lower depths by seeping rainwater (eluviation). Due to their low solubility, Fe and Al ions, on the other hand, remain largely in their original location where they accumulate. The diagram in Fig. 2.2 (according to [2]) shows the process of decomposition and transformation of the base minerals and the new formation of clay.

For example, the corresponding chemical equation for the mineral orthoclase is:

$$\underset{\text{orthoclase}}{2K\left(AlSi_3O_8\right)} + 2H^+ + H_2O \rightarrow 2K^+ + 4SiO_2 + \underset{\text{clay mineral kaolinite}}{Al_2Si_2O_5\left(OH\right)_4}.$$

The newly formed clay minerals are silicates with a layer lattice structure. Due to their small particle size ($d<0.002$ mm) and their large specific surface area, they cause water sorption and are thus responsible for structural characteristics such as shrinkage and swelling. The main clay minerals are kaolinite, illite, and montmorillonite. Their formation is dependent on the intensity of rock weathering and eluviation by seepage: montmorillonite is a mineral with extremely high swelling properties and is formed by inhibited eluviation, while kaolinite is a clay mineral with low swelling properties formed by strong eluviation (Sect. 2.2.3.4).

In the *hot and humid* climate of the tropics, the process of rock weathering happens the fastest and is the most thorough: constant hot temperatures and precipitation as well as the lush vegetation act as catalysts. Simultaneous to the weathering process, a decomposition process takes place transforming organisms, which penetrated into the cracks and later died off, into products with a simpler structure

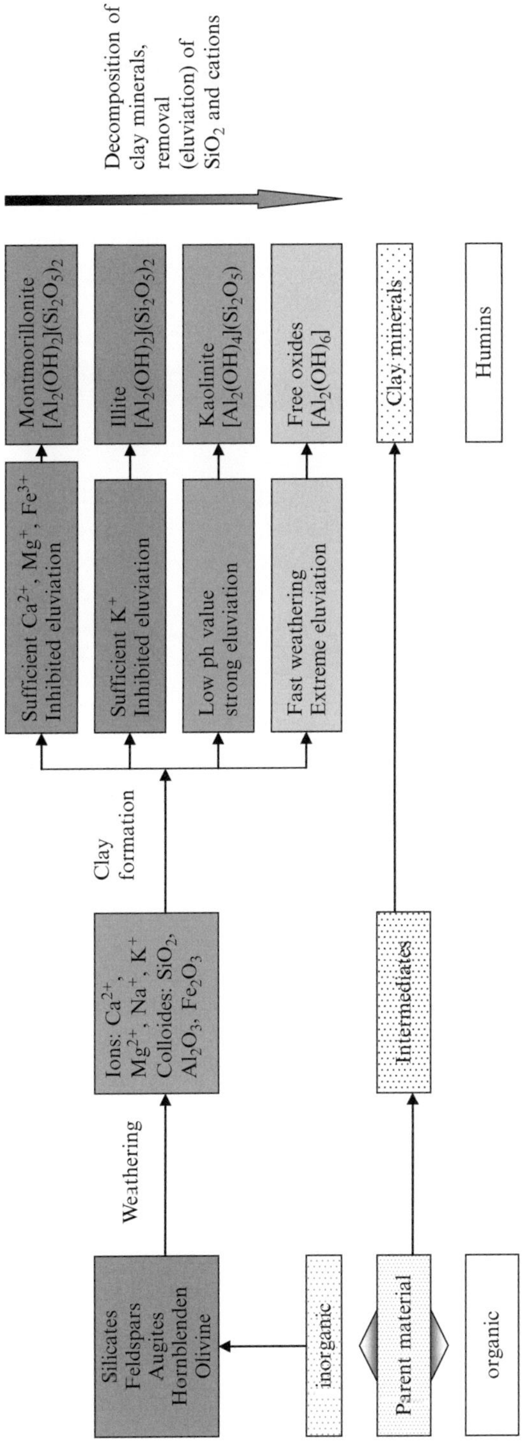

Fig. 2.2 Decomposition, transformation, and composition process of inorganic soil components, according to [2]

(mineralization/humus). Strong eluviation caused by a large volume of seepage water leads to the formation of clay minerals with low swelling properties. The parent rock is *chemically* changed.

In *hot and dry* climates, the parent rock is mainly changed *mechanically* as a result of extreme air temperature fluctuations over the course of the day as well as insufficient or nonexistent precipitation and vegetation. This changes the mineral composition of the rock only slightly. In desert areas, lime or gypsum-rich soils prevail due to a lack of precipitation. In semiarid regions with pronounced dry and rainy seasons, clay minerals with extremely high swelling properties are newly formed as a result of inhibited eluviation.

Cold climates lack the conditions required for rock weathering. Rock and a thin soil cover, if in existence, are permanently and thoroughly frozen (permafrost soils) and only thaw during the summer months for a few weeks. Physical rock weathering due to the freeze–thaw cycle prevails, leading to the creation of rubble zones of loose, angular, and sharp-edged material.

Moderate climates have the same impact on rock weathering as hot and humid climates. However, the intensity of weathering is much lower resulting in a decreased thickness of the weathering zone. Therefore, the newly formed clay is likely to contain clay minerals with high as well as low swelling properties.

Time

Most clay-rich soils in Europe were formed during the youngest geological period, the Quaternary. Within the geological classification of rock, clay soils belong to the group of unhardened, clastic sediments. They were formed by glacial (Holocene) deposits, for example, or can be found as sediments in floodplains. Their formation process started around 1.5 Ma ago and continues into the present.

Over this time period, some clay soils *remained at their place of origin* (residual), while others were *moved* by different transport mechanisms. Different types of transport mechanisms which are determined by the climatic forces of ice (glacial), water (alluvial), and wind (aeolian), and the soil's subsequent deposition (sedimentation), resulted in differences within the soil profile structure. The *unstratified* structure is formed of more or less the same type of loose rock, whereas the *stratified* structure is made up of different types of loose rock. While stratified loose rock in a horizontal formation is referred to as *undisturbed* stratification, all deviations from it are referred to as *disturbed*. When it comes to clay-rich soils, these deviations can be deposits of sapropel, laminated clay, or gravel lenses.

It is also possible that the climatic conditions at the place of origin changed during this time: cold periods were replaced by warm periods and vice versa which affected the weathering conditions of the rock and the characteristics of the soils that were formed. Soils formed at an earlier time are called *fossil*, whereas soils which are presently being formed are referred to as *recent*.

Rock and Topography

Soil formation is furthermore determined by the type of parent rock. It should, however, always be examined in connection with prevailing climatic conditions. For example, in wet–dry tropical climates, pure limestone often produces red soils, while marl produces black soils.

The chemical composition and the mineral structure determine the degree of weathering resistance of the rock and thus also the duration of the soil formation. Depending on the type of formation, solid rock can be classified into three main groups:

Migmatite or solidified rock (depending on the percentage of SiO_2: alkaline (sparse and dark colored) and acidic (abundant and light colored))
Sedimentary rock
Metamorphic rock

In addition, there are subgroups of rock which share the characteristics of rocks from various main groups and can therefore not be clearly assigned to one main group. Laterites are an example of these (Sect. 2.1.2.6).

Topographical landforms such as mountains, plains, valleys, and basins along with their respective flow conditions impact the formation of different soils as well. For example, valleys and basins with insufficient water drainage and high groundwater levels might produce saline soils.

Human Activity

Finally, human activities such as farming, livestock breeding, and construction also lead to changes in soil formation processes. Examples of permanent changes in soil quality are the large-scale deforestation of tropical rainforests and the extensive pasture farming in sub-Sahel Africa.

2.1.2 Designations of Natural Soils

Soils in their natural deposits are classified according to their formation or generation. On geological maps they are depicted as contiguous areas of soils of the same origin and are assigned the same petrographical and lithogenic designations. A general map of the former German Democratic Republic in Fig. 2.3 [3] shows the deposits of different soils differentiated in terms of their formation.

The petrographical designations also indicate the quality of specific group characteristics of loose rock, such as the range of grain distribution and, regarding clay-rich soils, the quantity and quality of clay minerals as "natural" binders. The petrographical designations are therefore included in the system of geotechnical classification of soil types according to DIN 18196 (Sect. 2.2.3.1) by serving as "examples" through the assignment of group symbols.

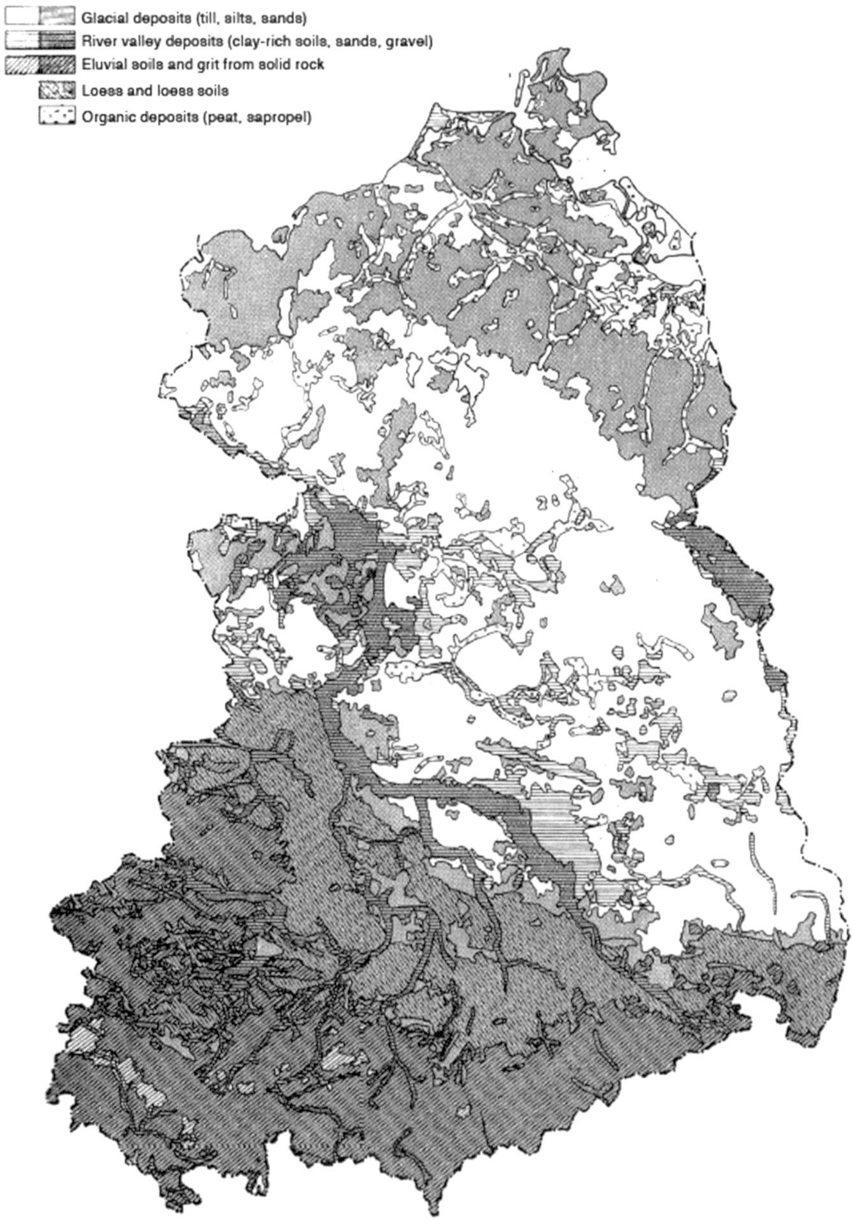

Fig. 2.3 General map of the formation of soil deposits within the former GDR [3]

There are many different types of soil formation. This means that the characteristics of clay-rich soils can differ substantially. Therefore, the petrographical designation of a clay-rich soil can be used immediately on site for making general statements about its suitability for the intended use and processing method and, if necessary, about possible modifications.

2.1.2.1 Loess and Loess Soil

Loess is glacial soil with a high lime content which has been transported by wind. During the weathering process, the lime in the soil is loosened by rainwater, washed out, and deposited in the lower layers of the soil in fine root casts. Through this process, the "decalcified" loess soil obtains a higher cohesive strength compared to loess. The use of the word "soil" in loess soil points to an advanced degree of weathering of the mineral substance which creates a higher clay mineral content than in the source material.

A typical characteristic of loess soil is its steep, narrow grading envelope in the medium to coarse silt region (>75 %) with a low clay fraction (<10 %) (Fig. 2.4, according to [1]). This results in low to medium plasticity as well as the risk of erosion when exposed to water, which is particularly undesirable for building with earth. A certain amount of lime in the soil results in "cementitious properties" after processing and shaping occur and in a comparatively high dry compressive strength.

Main deposits in Germany: northern foothills of upland regions.
Loess soils are widespread in Southeast Europe and Asia. The loess deposits in China, up to several hundred meters high, are particularly famous.
Main minerals: quartz 40–80 %, feldspar 10–20 %, calcite 0–50 %, clay minerals (loess soil).
Color: mostly yellow ocher or grayish when an increased lime content is present.
Application in earth building:
Loess: difficult to process without additives because it is too lean, and when mixed with powdered clay or sand, it is used for plasters and light clay applications.
Loess soil: earth blocks, light clay applications.
Letter symbol: group symbol according to DIN 18196, UL or TM (Sect. 2.2.3.1).

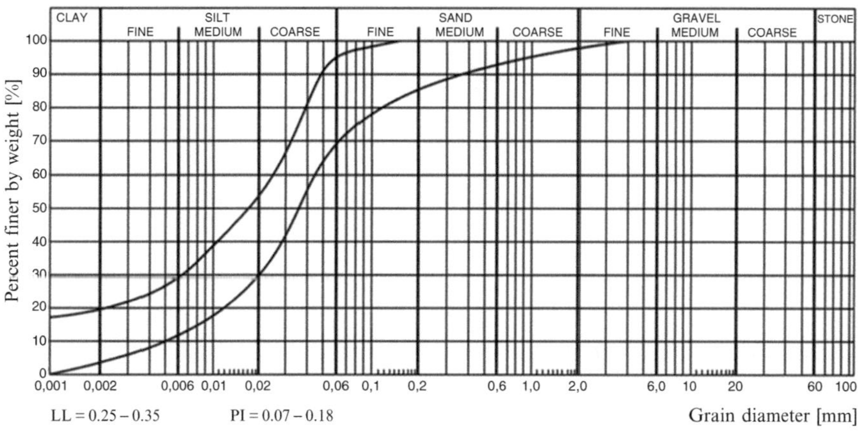

Fig. 2.4 Grain envelope of loess soil, according to [1]

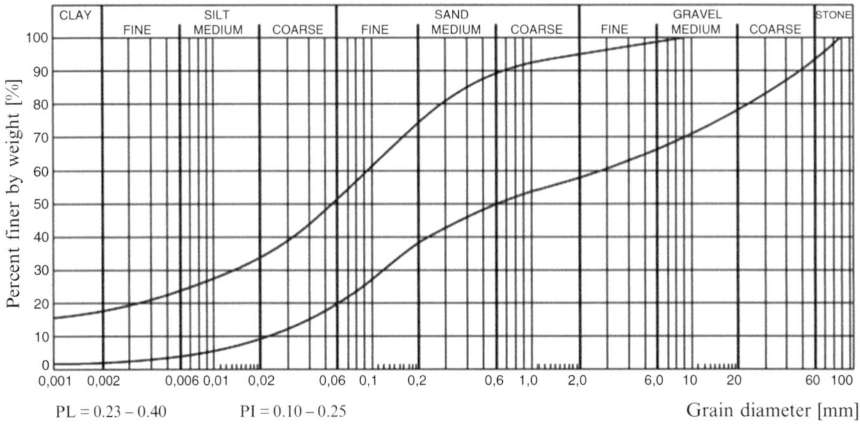

Fig. 2.5 Grain envelope of glacial soil, according to [1]

2.1.2.2 Glacial Marl and Glacial Soil (Till)

Glacial marl refers to soil which was transported by glaciers during the ice age and deposited as unstratified ground moraine. It is a material with a high lime content and a typical broad grain envelope ranging from clay, silt, and sand to gravel and boulders (Fig. 2.5, according to [1]). Similar to loess soil, the soluble lime parts have mostly been washed out of the layers near the surface (glacial soil (till)). A special feature of the structure is the so-called glacial erratic—polished and rounded fragments of magmatic or metamorphic rock from the Scandinavian mountains embedded in a more or less fine-grained matrix. Inclusions of clay, sand, and gravel lenses are common, as well as sections which are completely free of drift material.

Main deposits in Germany: ground and terminal moraine ridges in the North German Plain
Main minerals: quartz 40–50 %, feldspar 5–30 %, clay minerals 5–25 %, lime 5–30 %
Color: ranging from gray to yellowish brown depending on lime content and degree of weathering
Application in earth building: rammed earth
Letter symbol: group symbol according to DIN 18196, TL or ST*

2.1.2.3 Eluvial Soil

Eluvial soil is comparable to glacial soil (till) in terms of its grain size distribution except that it has never been transported and is still found in its primary deposits. This explains why the sand and gravel particles are still angular and sharp edged. The profile shows a gradual transition from eluvial soil to eluvial grit and detritus continuing on to the parent rock material (Fig. 2.6, according to [1]).

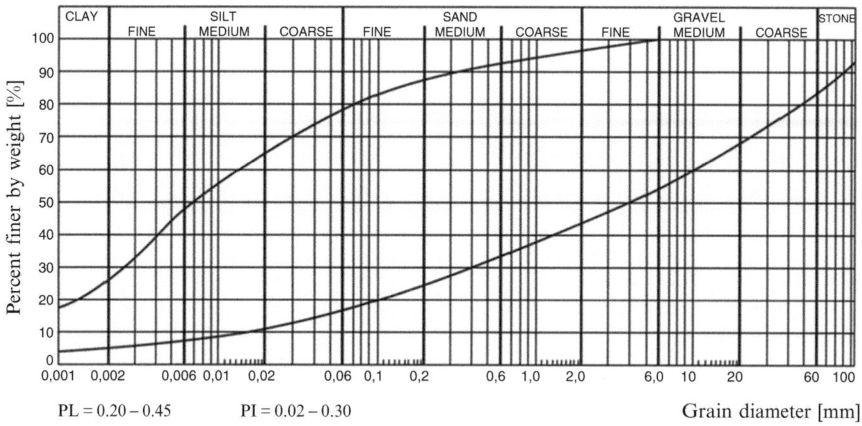

Fig. 2.6 Grain envelope of eluvial soil, according to [1]

Main deposits in Germany: upland regions, Alpine foothills
Main minerals: quartz, feldspar, mica depending on the composition of the parent rock, clay minerals
Color: depending on the parent rock, often brown
Application in earth building: rammed earth
Letter symbol: group symbol according to DIN 18196, GT

2.1.2.4 Fluvial and Slope Wash Soils

Fluvial and slope wash soils consist of medium to strong cohesive sediments left by floods in floodplain areas (fluvial soils). Crumbling slopes form slope wash soils at the base of mountains. Depending on the flow velocity, grain sizes range from clay and silt to sand. Organic deposits are common. Slope wash soils are typically more coarse grained and unsorted (Fig. 2.7, according to [1]).

Main deposits in Germany: floodplains, valley slopes.
Main minerals: see eluvial soils.
Color: yellow to brown.
Application in earth building: light clay applications with organic fibers and lightweight mineral aggregates, earth blocks.
Letter symbol: group symbol according to DIN 18196, SU* or GU*.

2.1.2.5 Clays

From a geotechnical perspective, the term "clay" has two meanings: it refers to a fine-grained soil formed as a result of the weathering process of solid rock and it defines a grain class ($d<0.002$ mm) for construction purposes according to DIN 18196 (Sect. 2.2.3.1).

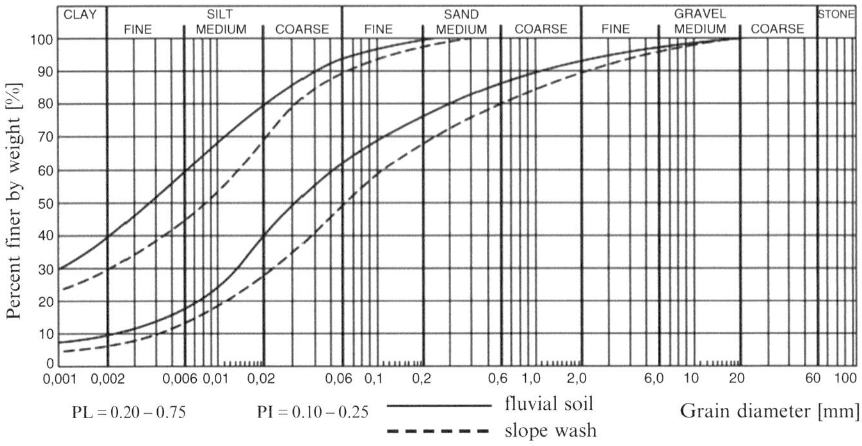

Fig. 2.7 Grain envelope of fluvial and slope wash soils, according to [1]

Naturally occurring clays are mixtures of very fine, mechanically ground *nonplastic* mineral components (such as quartz debris, feldspar remains, and mica) and *plastic* clay minerals newly formed in the final stage of the rock weathering process (Fig. 2.2). In the Tertiary period, in particular, clays were formed through the chemical weathering of acidic migmatites (granite, syenite, and porphyry) as well as alkaline migmatites (basalts). These were deposited and later transported.

The grain sizes of clays lie predominantly in the clay "grain" fraction ($d<2$ µm) at >40 %. The proportion of $d>0.06$ mm (sand) is low at <10 %. The remaining 40 % of grains are in the silt grain fraction using the group symbol T according to DIN 18196.

The color palette of clays is broad: from white and gray to black. Deposits in Germany can be found in Mecklenburg, in the Thuringian Basin, in northern Saxony, and in Lusatia (Lausitz), the area around Meißen.

Looking at the grain size distribution, the difference between pure clay and clay-rich soils becomes clear: compared to clay-rich soils, pure clay lacks the coarse-grain fractions of sand and gravel. However, both soils contain clay minerals which act as *non-hydraulic* binding agents. They harden only when exposed to air and wrap the coarse grains as very fine coatings. They lend the material plasticity and cohesion during the production of earth building materials. After drying they give the material stability and strength while regaining plasticity after renewed wetting. This mechanism can be repeated an infinite number of times, giving earth as a building material, as well as clay, a special ecological quality. Considering that this mechanism is the same for clay-rich soils and pure clay itself, pure clay can be regarded as a "special form" of clay-rich soil from an earth building point of view.

Clays are predominantly used as a raw material in the ceramic industry. They are also processed industrially into powdered clays (Sect. 2.2.1.2) with various applications in construction and earth building.

2.1.2.6 Tropical Residual Soils

Tropical and subtropical climates exert a particular influence on the formation of soils which are used for construction in these regions. The *hot and humid* climate is characterized by constant high air temperatures and high precipitation, whereas the *hot and arid* climate is defined by a high daily air temperature amplitude and low precipitation (Sect. 2.1.1.3). In terms of their construction properties, the resulting soils differ greatly from the soils in moderate climates [4]. Examples of typical tropical residual soils used in earth building are laterites and so-called black cotton soils.

Laterites

Laterites are the typical soils of humid and semi-humid climates. They cover, for example, around one third of Africa and are known for their brick-red to cinnamon color which also explains their name (later: Latin for brick).

Humid climates chemically change the mineral composition of the upper rock layers to such a dramatic extent that they show little resemblance to their parent rock. These newly formed soils are referred to as residual soils. Continuously present seepage dissolves the soluble minerals in the rock and moves them to lower layers. What remains are the insoluble components such as quartz and Fe and Al ions. These metallic oxides cement the residual materials of the parent rock together into new soil aggregates and give the laterite soil its particular color. In addition, clay is formed (Fig. 2.2). The dominant clay mineral in these soils is kaolinite which has low swelling properties (Sect. 2.2.3.4). In terms of grain sizes, all grain fractions are possible.

When laterites harden in the open air, two strength matrices are formed: the water-soluble matrix based on the cohesive strength of the clay minerals which is common in earth building in moderate climates as well as a water-insoluble matrix formed by metallic oxides. The latter is similar to the effects of cement, improving the weathering resistance of the building materials. In this state, laterites form a transition between loose and solid rock (Fig. 3.23).

Black Cotton Soils

The term "black cotton soil" has its roots in the farming of cotton which was extensively cultivated in these soils, particularly in India. The typical color of this group of soils is black or dark gray.

"Black cotton soils" are primarily formed in plains and basin on top of Ca- and Mg-rich alkaline migmatites with insufficient runoff. Their formation is tied to tropical wet–dry climates with major differences in precipitation levels between 200 and 2000 mm per year and a pronounced fluctuation between moisture penetration and drying. During the dry season the soils are very hard and contain deep cracks which, during the rainy season, act as "swallow holes" which help to soften the soils.

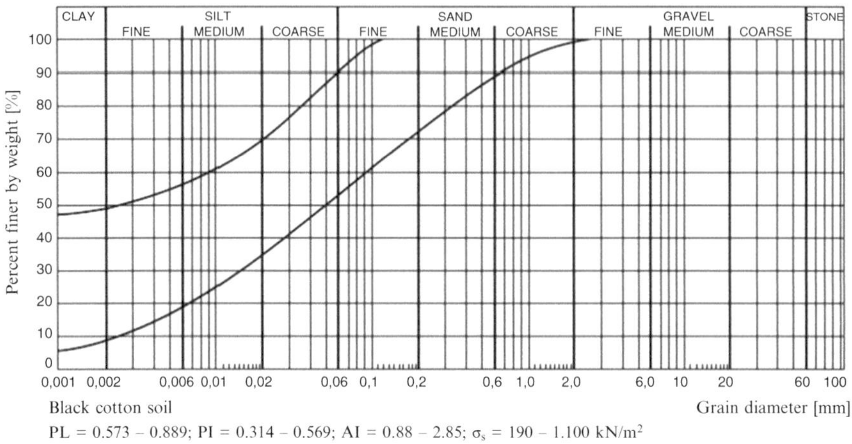

Fig. 2.8 Grain envelope of clay, according to [5]

Most grain sizes lie within the clay and silt fractions. Limited eluviation and a sufficient concentration of Ca and Mg produce clay minerals of the montmorillonite group with extremely high swelling properties (Sect. 2.2.3.4) (Fig. 2.2). These properties manifest themselves in correspondingly high plasticity numbers (Sect. 2.2.3.2).

When used in earth building, these soils therefore need to be extensively leaned through the addition of sand. Figure 2.8 shows a grain envelope of a black cotton soil in Sudan [5].

Desert and Semidesert Soils

In hot and dry regions, soils lack the protective vegetation layer (A horizon). This allows the formation and transport of dust. Water mainly runs in ephemeral or seasonal streams which deposit materials from the surrounding mountains in dry valleys (Arabic, wadis).

Water-soluble components cannot be dissolved and transported from the soils due to a lack of precipitation. This causes lime and salts to accumulate. When the ground is moistened by a higher groundwater level (basins without drains, proximity to the ocean), rising and evaporating groundwater transports dissolved salts to the surface forming crusts of lime and gypsum (Na_2SO_4, $MgSO_4$, NaCl).

2.2 Construction Soil

Construction soil is soil that is suitable for the production of earth building products [6]. A natural soil which is classified as construction soil might be suitable for specific applications without much modification, such as loess soil for earth plaster. In general, however, construction soils have certain "deficits" when it comes to their application which can be largely compensated for by the addition of aggregates and/or additives in carefully planned dosages (Sect. 3.1.2.4).

Industrially produced earth building materials generally consist of different components. Often, "construction soil" as the binding agent is supplemented with dry soils and powdered clays which have also been produced in a plant. This balances the effects of the construction soil's properties as part of the final mixture compared to the characteristic properties which are declared for the "finished" earth building material.

Based on current general earth building practices, it is therefore not very practical to define general fluctuation margins for grain size and plasticity/cohesive strength parameters for testing the suitability of construction soils. Producers of earth building materials can base their selection of a suitable construction soil to be used for their specific building product on the data provided by the supplier (e.g., clay pit). They are subsequently required to verify the declared parameters of their "finished" building product.

2.2.1 Terminology

Construction soil can be extracted from its natural deposits or can be used as a recycling or waste product which has been returned to the life cycle of the material.

2.2.1.1 Open Pit Soil

Open pit soil is construction soil which has been extracted from its deposits in a naturally moist state and is free from humus and root material. Producers primarily offer this product to owner-builders for further preparation and processing into earth building materials.

2.2.1.2 Dry Soil and Powdered Clay

Construction soil to be used as powdered soil and powdered clay is extracted from its natural deposits in the same manner as open pit soil. It is then dried and ground (Sect. 3.1.2.3) and sold in paper bags or big bags (Fig. 3.6) in the form of granulate or powders, primarily to owner-builders for further processing into earth building materials. Granulated products are also called pellets and contain approx. 10–20 % moisture allowing for virtually dust-free processing.

Dry soil is soil without gravel or stones which has been extracted, artificially dried, and usually ground. Dry soil can be used for producing plaster mixes or for specific applications, for example, in oven construction. It can also be directly used as or processed into paints and primers for wood elements and plaster base coats.

Powdered clay or milled clay is characterized by grain sizes primarily in the clay and silt fraction. It is used to increase the plasticity/cohesive strength of lean construction soils or soils with a low plasticity. Powdered clay modified with sand or plant fibers can also be used to produce clay panels.

2.2.1.3 Recycled Earthen Material

Recycled earthen material is obtained from demolished building elements. It is crushed in a dry state, if necessary, and returned to the life cycle of the material (Sect. 6.2.2). Recycled earthen material has found its way into the product catalogs of companies working in the field of the recycling of historic building materials.

The material must be free of impurities, particularly fungus and dry-rot spores. Salts which might have entered and accumulated in the building elements while the structure was in use can reduce the cohesive strength of the recycled earthen material in a similar manner to artificial chemical additives, thereby limiting the possibilities for its reuse. Other limitations might be necessary for hygienic reasons, for example, when odors in demolished stables and farm buildings must be neutralized.

2.2.1.4 Compressed Soil

Compressed soil occurs as a waste product during the gravel washing process in gravel pits. It is first collected in silos and tanks as gravel-wash sludge and mainly contains the finest grains of clay and silt which cannot be used as aggregates by the concrete industry. The filter cake which remains in the silo after the water is drained from the gravel-wash sludge still has a high water content. This can be reduced through the use of a belt filter press. This process decreases the mass of the compressed soil considerably. Several million tons of this material are incurred in Germany per year (several 100,000 tons per year in Saxony alone [7]), but so far a worthwhile use for this material has not been found. One of its current uses is the filling of mining pits. The material's instability, however, largely rules out a subsequent use of this land for construction purposes.

The application of compressed soil in earth building has so far only been experimental. It is conceivable, however, because the material has already undergone intensive processing in the form of dispersion (Sect. 3.1.2.5) during the gravel washing process. To speed up the washing process in the gravel plant, flocculants are sometimes added to the water. The ecological impact of these flocculants on the possible use of the material as construction soil still needs to be examined.

2.2.2 *Sourcing*

The process of sourcing construction soil consists of finding relevant deposits and taking samples for manual and laboratory testing. Sourcing investigations provide information about the strata patterns of the soil, the spatial expanse of the soil deposits, and possible geological disturbances and groundwater levels. The spatial expanse of a deposit can be determined by using suitable geotechnical surveying procedures.

In the past, raw soil was mainly extracted from local pits (clay pits) and the soil was generally available to the owner-builder free of charge. Today, old field or street names still suggest where such deposits are located. Other sources for finding suitable raw soils are geological maps as well as the locations of former brickyards. Freyburg [8] gives an overview of 42 locations in the former GDR together with evaluations of the suitability of the soil as construction soil.

Today, earth building materials are largely industrially produced. This makes it necessary to obtain increasingly larger amounts of construction soil at consistent quality levels. This can only be ensured by continuously monitoring the most important properties of the construction soil.

2.2.2.1 Sourcing Investigation Methods

Producers of earth building materials are increasingly being supplied with construction soil by companies specializing in excavation and construction work. These companies, in turn, save the disposal costs they would otherwise have to pay for the excavated soil.

For sourcing new construction soil, extraction processes common in geotechnical investigations can be applied. Depending on the amount of soil to be extracted, there are various methods which can be used for finding construction soil sources.

Test Excavation

The test excavation pit is an accessible dig above the groundwater level with a maximum depth of 3 m (Fig. 2.9) [15]. If the material is stable, there is no need to use bracing down to a depth of 1.3 m. Down to a depth of 1.8 m, planks must be used for trench support and any excavation deeper than that requires a full trench box. The advantage of a test excavation is the opportunity to both visually assess the strata pattern down to the excavated depth and to take a direct sample for a potential classification as construction soil. The disadvantage is the considerable effort required for manual or machine excavation.

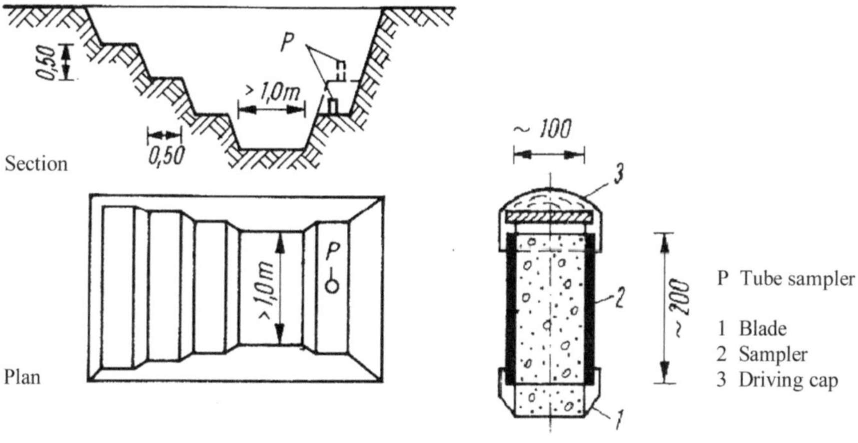

Fig. 2.9 Sourcing investigation method, excavation, trial pit, and tube sampler [15]

Penetration Tests

For penetration tests, steel rods with a groove cut into their lower ends are rammed into the ground down to a depth of 2 m. After the probe is pulled back out of the ground, an approximate assessment of the soil layer that it passed through can be made. Changes in penetration resistance indicate a change in the soil layers. A direct sampling of soil is not possible using this method. The advantage of the penetration test is its ease of application. The disadvantage is that the assessment of the soil quality is only approximate. For this reason, the penetration test only makes sense in combination with test excavation and boring.

For investigating soil in larger areas, penetration and, if necessary, excavation tests can be carried out using a grid with predetermined spacings based on the size of the investigation area.

Potential deposits of construction soil can be more accurately localized with the help of preliminary geophysical investigations using geoelectrical probing and seismic refraction methods.

Boring

Boring can be used to investigate deposits down to virtually unlimited depths. As depths get deeper the effort required to carry out the boring increases. When investigating construction soil, boring is only practical for large excavation sites or for supplementing test excavations and penetration tests.

Boring equipment includes bore rods with extension and attachment pieces and drive mechanisms (manual or engine driven). The type of attachment piece for taking soil samples depends on the consistency of the soil (Fig. 2.10 [9]): spiral augers are suitable for solid or semisolid soil materials. The soil is pulled up inside the auger flutes to collect assessment material or the spiral auger is used to loosen the soil which can then be tested by utilizing other boring tools. For soft or paste-like soils, standard augers are used for sampling. These are steel cylinders with a slit along their entire vertical length and a sharp point at their lower end. The side of the slit which is located in the opposite direction of rotation is curved inward in order to allow the cylinder to collect soil while turning.

2.2.2.2 Sampling

Obtaining Samples

With regard to unaltered obtainable parameters, there are five different quality grades (DIN EN ISO 22475-1) for obtaining soil samples.

Construction soil is tested according to grade 4. This means that the sampling is carried out with unchanged grain composition (Sect. 2.2.3.1). The soil parameters

Tool	Auger (bailer)	Standard auger	Spiral auger	Chisel
Power transmission	ramming	rotating	rotating	ramming
Soil type / consistency	cohesive / pasty, organic / pasty, dry and wet non-cohesive	cohesive /soft –stiff, wet and non-cohesive	cohesive / semisolid - solid	gravel, glacial soils
Drawing	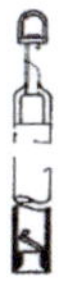			

Borehole after sampling using spiral drill [37]

Fig. 2.10 Sourcing investigation method, boring according to [9]

which can be derived from the test are the processing parameters (Sect. 2.2.3.2) and the amount of natural impurities (Sect. 2.2.3.4).

For testing a soil, a sample of approx. 10 L per sampling site (approx. one bucket) is sufficient. The naturally moist samples are taken from the humus-free B horizon using a spade. The B horizon can be distinguished from the humus-rich top soil (A horizon) by its lighter color (Sect. 2.1.1.1). When construction soil is excavated from a large area, the A horizon is removed and temporarily stored. The minimum depth of the excavation is 0.5 m.

If the soil profile is visibly layered or the soil deposits are disturbed (Sect. 2.1.1.3), several samples from various locations within the sampling site need to be taken and combined into a mixed sample.

Grades 3 and 2 describe how to obtain samples for evaluating further parameters in undisturbed form which can be of importance for testing *earth building materials*: Grade 3, water content (Sect. 2.2.3.2), and Grade 2, wet bulk density (Sect. 3.6.1.2), of the earth building material or the earth building element. These grades can be applied, for example, when testing the attained compaction and the drying process of a finished rammed earth building element.

Manual Test Methods

Manual test methods are used to make a first general assessment of raw soil at the deposit site in order to determine its suitability as construction soil on site. Through hand contact and a few simple tests, these procedures primarily aim at evaluating the grain distribution and plasticity of the soil, as well as the natural moisture content.

Manual test methods can also be used for the qualitative evaluation of consistency and composition of soil mixtures for the processing of construction soil and earth building materials.

A certain amount of experience is necessary to interpret the results of the manual test methods, so the process should be entrusted to professionals. These tests cannot replace required standardized laboratory tests when it comes to industrially produced earth building materials.

Visual Assessment of Grain Composition

In order to visually assess the grain composition of raw soil at the deposit site, the gravel and sand grain size groups can be compared to everyday objects (Table 2.1):

Silt and clay have grain sizes which cannot be seen with the naked eye. Using manual tests, their approximate grain proportions can be assessed by evaluating plasticity and cohesion.

Particles which are bigger than coarse gravel are termed stones. If they are as big as or bigger than a person's head, they are boulders. In addition, the chemical and mechanical stability of the coarse-grain sizes needs to be taken into consideration due to the fact that residual soils on top of limestone also have grain size fractions in the sand and gravel ranges.

Manual Testing at the Soil Sample Site

Sedimentation test (jar test). This test is used for the general assessment of the grain distribution in a soil sample.

Approximately 100 g of the soil to be tested is placed in a jar with high sides. The jar is filled with water and the contents are shaken up for a few minutes until all

Table 2.1 Comparison of the grain size fractions of gravel and sand to everyday objects

Grain size fraction	Comparison
Gravel size fraction	*Smaller than a chicken egg and larger than a match head*
Coarse gravel	Smaller than a chicken egg and larger than a hazelnut
Medium gravel	Smaller than a hazelnut and larger than a pea
Fine gravel	Smaller than a pea and larger than a match head
Sand-size fraction	*Smaller than a match head and down to the limit of what can be perceived with the naked eye*
Coarse sand	Smaller than a match head and larger than a grain of semolina
Medium sand	As large as semolina
Fine sand	Smaller than semolina but every individual grain is still visible with the naked eye

clumps have dissolved. The heavy gravel and sand parts settle to the bottom of the jar first. The finest particles of the clay grains settle last and cloud up the water.

If the water already clears after a few hours, the soil sample contains only a small amount of particles in the finest grain size fraction. With pure clay the test sample remains cloudy even after a few days of settling.

Ball test. This test is used for evaluating the cohesive strength of raw soils. Naturally moist soil is formed by hand into balls with a diameter of 5 cm.

Evaluation

Rich soil sticks to hands while forming (like soft soap).
Lean to semi-rich soil does not stick to hands while forming; the sample keeps its shape after drying.
Overly lean (unsuitable) soil is very difficult to form and disintegrates easily after drying.

Cutting test. A naturally moist sample is cut with a knife.

Evaluation

Rich soil: shiny cutting surface due to high clay content.
Lean soil: dull cutting surface, silt and sand particles dominate, grinding sound while cutting.

Testing of dry compressive strength. A dry sample is crushed and the perceptible resistance indicates the proportion of fine-grain sizes.

Evaluation

Rich soil: sample does not crush under finger pressure.
Lean to semi-rich soil: sample crushes into small pieces under moderate to considerable finger pressure.
Overly lean soil: sample crushes after drying without any pressure or under light finger pressure.

Rub test. A naturally moist soil sample is rubbed between the thumb and index finger.

Evaluation

Rich soil: feels like soft soap and sticks to fingers even after drying.
Lean soil: has a perceptible fine-grain structure; flakes apart after drying and falls off fingers.
Overly lean soil: has a perceptible coarse-grain structure; individual grains fall off fingers after drying.

Smell test. Goal: to rule out soil with a high humus content, recognizable by the typical humus smell of the naturally moist sample. Small amounts of humus are harmless and have positive effects on the natural processing of construction soil through aging (Sect. 3.1.1.3).

Color. The color of the naturally moist soil indicates the chemical composition of the finest grain sizes. After drying is complete, a change of color from dark to light can be observed. Soil colors range from black, gray, beige, ocher, yellow, reddish brown, and cinnamon to red and indicate the presence of the following components:

Light white	Containing Ca and Mg
Dark brown	Mn
Green	Cl
Reddish-yellow brown	Fe
Gray black	Humus, organic content (smell test)

2.2.3 *Testing and Classification*

Construction soil is tested in order to establish its general suitability as an earth building material. The soil's characteristics are determined with the help of parameters from standardized test procedures. These parameters are defined in detail in the German Lehmbau Regeln [6] or need to be derived from the requirements of the individual building project.

Construction soils which are used as a base material or as part of a mixture in the industrial production of earth building materials also need to be tested for fluctuations in their natural composition and with regard to the production process. The test results must be documented to ensure a future traceability. The German Association for Building with Earth has developed Technical Information Sheet 05 [26] in order to offer assistance with this process.

The characteristic properties of construction soil are described as parameter classes which are linked to specific applications. The characteristic properties which determine the suitability of construction soil are "grain" (skeleton), any natural additives, as well as the "binding agent" (clay minerals) (Table 1.1). Exceeding specific limits of these properties can lead to deformations during drying.

The procedures primarily used are tests borrowed from the field of earthworks and foundation engineering for the classification of loose rock according to DIN 18196 (as well as DIN EN ISO 14688-1, 2).

2.2.3.1 Grain Parameters

Grain parameters describe the size, distribution, and shape of mineral particles found in construction soil.

Table 2.2 Grain size fractions of the main soil types for geotechnical work

No.	Soil type	DIN 18196 Letter symbol[a]	DIN 18196 Grain sizes *d* [mm]	USCS Letter symbol	USCS Grain sizes *d* [mm]
1	Cobbles or boulders	X (Bo/Co)	≥63	B	≥76.2
2	Gravel	G (Gr)	$2.0 \le d < 63$	G	$4.75 \le d < 76.2$
3	Sand	S (Sa)	$0.063 \le d < 2.0$	S	$0.075 \le d < 4.75$
4	Silt	U (Si)	$0.002 \le d < 0.063$	M	$0.002 \le d < 0.075$
5	Clay	T (Cl)	<0.002	C	<0.002
6	Organic loose rock	O		O	

[a]Letter symbols according to DIN EN ISO 14688-1

Grain Sizes and Grain Distribution

Terminology

All types of mineral loose rock as well as all loose rock containing organic matter are classified according to the size of their mineral grains in the grain fractions *clay*, *silt*, *sand*, and *gravel* and are assigned a corresponding group symbol (letter symbol in the form of capital letters). For geotechnical tasks, the *grain size ranges* of the fractions are defined in specific regulations. German (European) and American regulations (Unified Soil Classification System, USCS, but also countries using [originally] Imperial measurement systems) differ from each other. This needs to be kept in mind when classifying construction soil (Table 2.2).

Clay-rich soils are mixed-grain materials. They typically contain particles of all grain sizes and can therefore not be assigned to a single-grain fraction. Some clay-rich soils, for example, lack coarse-grain particles (loess soil). However, they all contain a clay fraction.

The "clay" grain fraction with a particle size of $d < 0.002$ mm is formed by the plastic clay minerals and the nonplastic portions of quartz dust, mica flakes, etc. Plastic clay minerals are also possible in the fine silt range with a particle size of $d > 0.002$ mm (Sect. 2.1.2.5). Various sources, such as Russian literature on soil mechanics [23], therefore define the border of the clay–silt grain fractions at $d < 0.005$ mm.

Test Procedures

The grain fractions $d \ge 0.063$ mm (sand, gravel) are determined by sieve analysis. The grain fractions <0.063 mm (silt, clay) are determined by hydrometer analysis (DIN 18123). For mixed-grain soils, a *combined sieve and hydrometer* analysis is used. Accordingly, the clay and silt grain fractions are also called mud grains, whereas the sand and gravel grain fractions can be referred to as pebble (sieve) grains.

Fig. 2.11 Sieve set for the mechanical analysis of sand and gravel grain sizes according to DIN 18123 [10]

Sieve analysis. The sieve analysis uses a set of sieves with standardized wire mesh sizes which represent the main grain sizes of the sample. The sieve set is arranged on a shaker with the largest screen openings on top. Each lower sieve has smaller openings than the one above it. The sieve set is closed with a round receiving pan at the base for grain sizes <0.063 mm and a lid on top (Fig. 2.11) [10]. The sample material which has been dried at +105 °C is poured into the top sieve and screened according to the instructions. The material which remains on each sieve is weighed and entered into a diagram as single-point measurements. The diagram is made up of an orthogonal coordinate system. The proportions of the total dry mass in % are plotted along the y-axis, the corresponding screen openings of the sieves are plotted along the x-axis increasing from left to right as a logarithmic scale. The individual points are combined into a grain size distribution curve, also called grading curve or *grain size distribution*, as a cumulative frequency polygon of the mass percentages (Fig. 2.12) [10].

Hydrometer analysis. The hydrometer analysis is based on the recording of the temporal changes of a suspension density where the grain "diameters" are determined as equivalent values according to STOOKE's Law. In this analysis, approx. 50 g of naturally moist soil is suspended in distilled water in a 1000 mL cylinder with the addition of a chemical which prevents coagulation, such as sodium pyrophosphate. Readings and evaluations are carried out according to a specified chronological sequence (Fig. 2.13) [10]. The hydrometer analysis cannot be applied to particle sizes <0.001 mm.

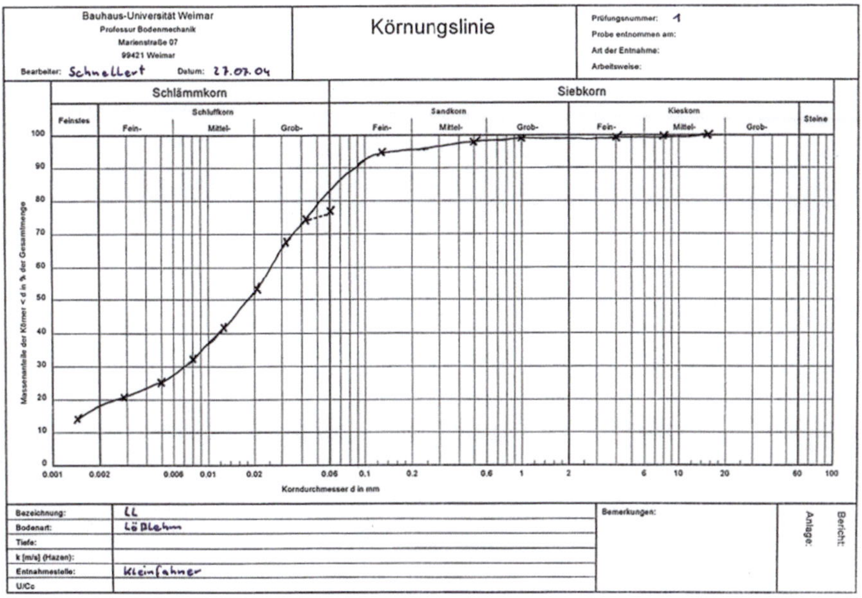

Fig. 2.12 Diagram showing a grain size distribution curve according to DIN 18123 [10]

Fig. 2.13 Hydrometer analysis using a hydrometer for determining silt and clay grain sizes according to DIN 18123 [10]

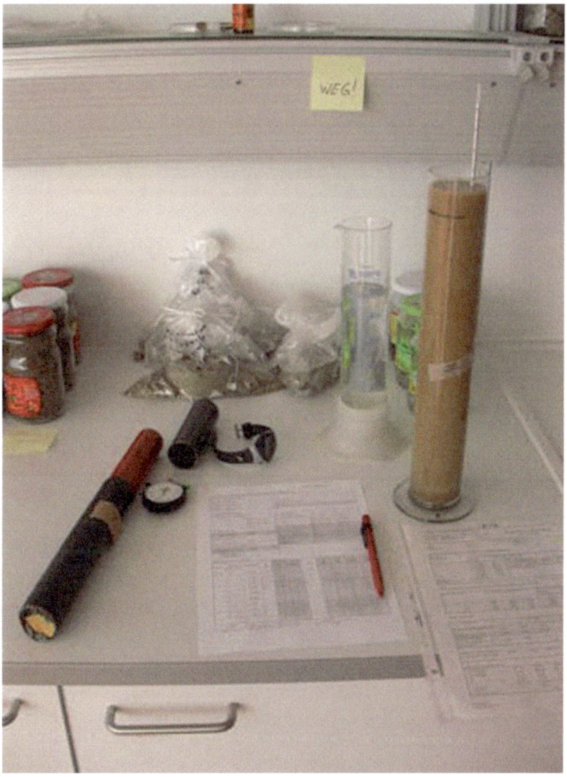

Combined sieve and hydrometer analysis. As the first step in the combined sieve and hydrometer analysis, the soil sample is soaked in water in order to dissolve the "cemented structures" caused by the clay minerals. The subsequent wet screening process breaks down the sample into its different grain fractions and the dried residue is weighed. For this step, the receiving pan is fitted with a rinse water drain from which the fine soil particles <0.063 mm are collected. After drying and weighing the fine soil, the grain size distribution is determined using the hydrometer analysis and the grading curve is adjusted graphically (Fig. 2.12).

Classification

According to grain size fraction/letter symbols. According to Table 2.2 the following grain size fractions are distinguished. The capital letter refers to the corresponding grain fraction that forms the *primary fraction* of the soil or is responsible for its main characteristic properties, namely,

- In coarse-grained soils with a fine-grained portion (silt and/or clay) <5 %
- In mixed-grained soils with a fine-grained portion of 5–40 %
- In fine-grained soils with a fine-grained portion >40 %

Secondary fractions of grain fractions found in a sample do not present any characteristic properties. They are referred to by the corresponding lowercase letters and follow the principal fraction in order of importance. Particularly *high* or *low* portions of secondary fractions are represented by a horizontal line above the letter symbol, an asterisk, or an apostrophe.

Example: *G*, *s**, *u*, *t*′—gravel, high sand proportion, silty, low clay proportion

In order to specify further properties, a second capital letter can be added to the group symbol:

- Nonuniformity of coarse-grained soils: E (close), W (wide), and I (gap graded)
- Plasticity of cohesive soils: L (low), M (intermediate), and A (very high) (Sect. 2.2.3.2)
- Division of mixed-grained soils according to mass percentage of fine grains ≤0.063 mm: U or T low 5–15 % or U* or T* high >15–40 %

Within the grain fractions of silt, sand, and gravel, there is a further division into the subgroups fine, medium, and coarse. The subgroups are referred to by the lowercase letters f, m, and g. These letters are put in front of "pure soil types" which consist only of one grain fraction.

Example: gU—coarse silt ($0.02 > d < 0.06$ mm)

As the clay mineral content increases (and with it the plastic content), the property of "grain size composition" gives way to the question of the processing properties of the soil. This means that, in addition to the grain parameters, processing parameters need also to be identified.

According to the grading curve. Important technological and processing properties (such as compactibility, compressive strength, erosion resistance, deformation, etc.)

can be derived from the grading curve. In this context it is important to find out if the construction soil consists of only a few different grain sizes, which means that the soil is *uniform*, or of many different grain sizes, making the soil *nonuniform*. Accordingly, the respective grain size distribution curves are shallow or steep. Shallow sections within a continuous grading curve indicate a missing grain fraction, whereas steep sections or jumps are a sign of a dominant grain fraction.

For road and dam construction, values have been derived from the grading curve which can be used to evaluate the compactibility of soil types with no or low cohesion.

Generally, nonuniform loose rock compacts more easily than uniform loose rock of the same initial pore space. During the compaction of nonuniform loose rock, the voids formed by the coarser grain sizes are filled by the smaller particles which minimize the pore volume.

The *uniformity coefficient* C_u describes the average gradient of the grading curve: the steeper the gradient the more uniform the soil.

$$C_u = d_{60} / d_{10}$$

d_{60} grain diameter which corresponds to the ordinate 60 % of the grain distribution curve.
d_{30} corresponds to 30 %.
d_{10} corresponds to 10 %.

Loose rock with respective C_u values is identified as follows:

$C_u < 5$	Uniform (e.g., beach sand)
$C_u = 5–15$	Uniformly graded (e.g., sand, loess soil)
$C_u > 15$	Nonuniform (e.g., glacial soil, eluvial soil)

The *curvature coefficient or grading* C_c describes the grading curve between d_{10} and d_{60}: low C_c values indicate that d_{30} is positioned near d_{10}, and high values indicate that d_{30} is positioned near d_{60}.

$$C_c = (d_{30})^2 / d_{60} \cdot d_{10}$$

$C_c < 1$	Poorly graded
$C_c = 1$	Normally graded
$C_c > 1$	Distinctly graded
$C_c = 1–3$	Well graded, e.g., gravelly sand, eluvial soil

The more uniform the loose rock, the lower the C_c.

A second letter added to the group symbols of the grain fractions gravel G and sand S indicates the average increase or gradation of the grain size distribution curve:

Designation	Letter symbol	C_u	C_c
Poorly graded	E	<6	Any
Well graded	W	≥6	1–3
Gap graded	I	≥6	<1 or >3

These values apply to coarse-grained soils (nonbinding gravels and sands) with a proportion of silt and clay fractions at ≤5 % making these types of loose rock generally unsuitable for earth building.

The model of "sphere packing" can basically also be applied to construction soils. According to this model, nonuniform construction soils (such as glacial soils) are, for example, more suitable for rammed earth than uniform soils (loess soils).

Nonuniform soils can reach higher densities and, as a result, a higher compressive strength than uniformly grained soils with the same mineral content. Compaction is counteracted by the clay minerals, however, due to their cohesive strength.

Loose rock ideally graded for purposes of compaction has a grain size composition which follows the *Fuller curve* (Fig. 2.14 [11]):

$$x = 100\left(d_x / d_{100}\right)^n$$

d_x	Mesh size x
d_{100}	Maximum grain diameter
n	Grading coefficient

The grading coefficient $n = 0.5$ applies to "sphere packing" which has no equivalent in natural soils because the shape of a grain typically deviates from the shape of a sphere. Therefore, Houben and Guillaud [12] recommend a coefficient of $n = 0.35$ for sands and gravel and $n = 0.25$ for clay-rich soils.

For soil materials of this kind, the lowest possible pore volume and highest density can be achieved through adequate compaction and an optimal water content. A Fuller curve can be determined for each coarsest grain within the grain material. This provides the densest grain size composition adequate for this particle size after compaction.

This process is used to artificially compose soils for watertight cores of earthen dams. The continuity of the grain size distribution curves alone ensures that the dams are resistant to erosion and watertight. These soils are also referred to as *clay concrete*. For tests with compacted silts, Plehm [13] used a clay concrete containing gravelly sand as its basis (100 parts by weight), as well as Caminau silt (10 pbw) and Guttau powdered clay (15 pbw).

For compaction work in earth building, such as the construction of rammed earth walls, soils can be selected according to this system or can be artificially modified with aggregates. The example described in [38] used a loess soil as construction

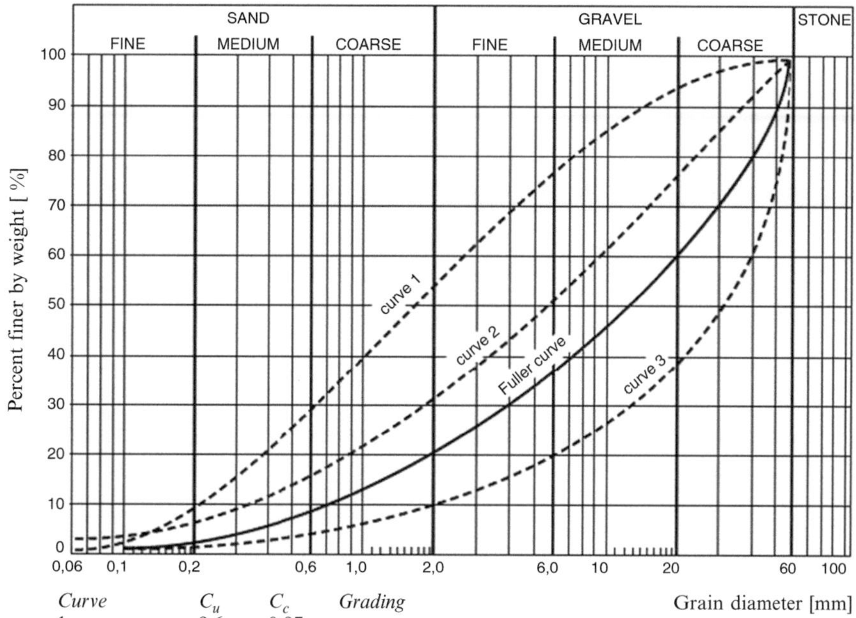

Curve	C_u	C_c	*Grading*
1	9.6	0.97	poor
2	50	2	well
3	18.2	1.6	well
FULLER	35.8	2.3	ideal

Fig. 2.14 Grain size grading for different grain size distributions including Fuller curve with $d_{max} = 60$ mm [38]

soil for the construction of a rammed earth structure. It was modified with coarse aggregates according to the Fuller curve (Fig. 2.15) in order to achieve the desired strength values.

However, there is no general grain size distribution curve which is optimally composed for earth building purposes. Depending on the intended application of the earth building material, the "ideal" grading curve would need to be adjusted according to particular specifications. Rammed earth, for example, requires a gravel portion, whereas plasters need a high sand content. Sun-dried earth blocks, on the other hand, require a higher percentage of clay.

According to plastic components. The grain composition of a soil alone does not allow for a comprehensive assessment of its suitability as construction soil because it fails to evaluate the specific quality of the clay minerals. The clay minerals characterize the processing properties of the construction soil (Table 1.1). This is not addressed by older diagrams based on American road construction research on the different soil types within the clay–silt–sand triangle (Fig. 2.16 [11]). They only classify individual soil samples as a result of a grain composition analysis.

Therefore, when carrying out the geotechnical classification of clay soils according to DIN 18196, processing parameters are always assessed alongside grain size parameters (Sect. 2.2.3.2).

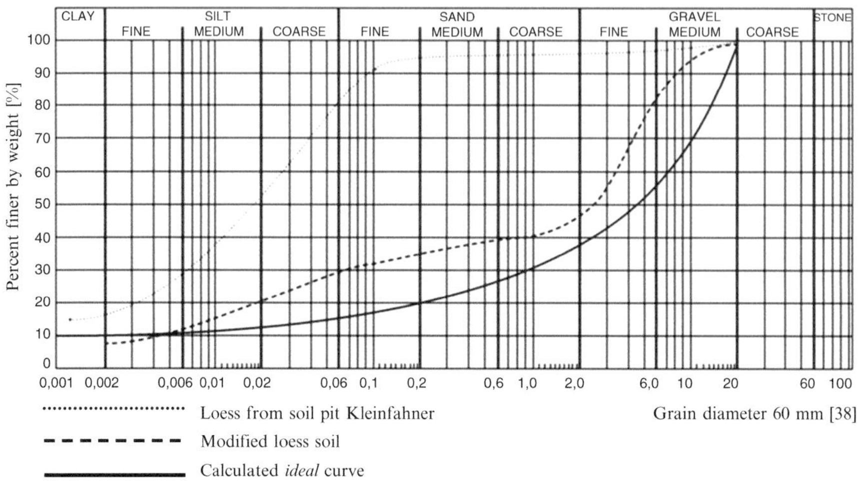

Fig. 2.15 Grain size distribution of a loess soil modified with coarse aggregates to be used as rammed earth [11]

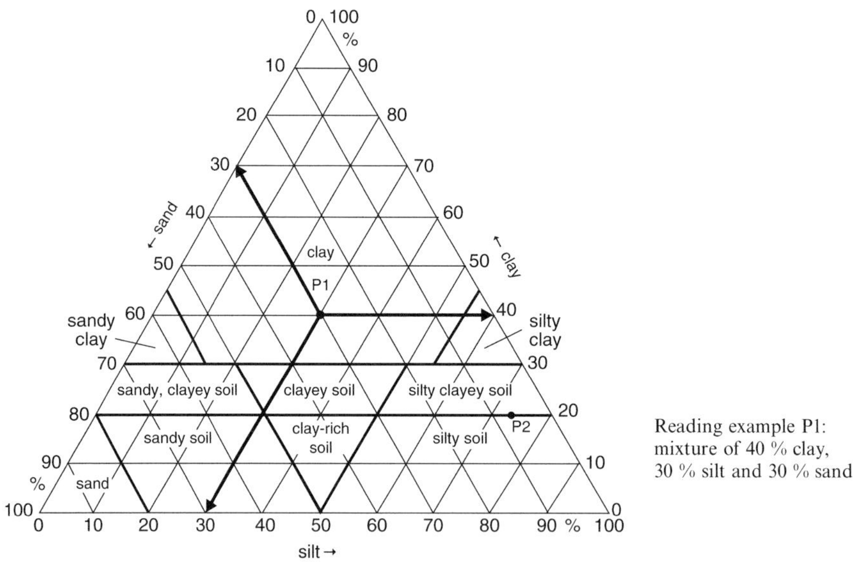

Fig. 2.16 Depiction of different types of soil according to their grain distribution in a triangle [11]

Grain Shape and Grain Angularity

In addition to the grain composition of a soil, the shape and angularity of the individual grains play an important role. The different grain shapes include spherical, prismatic, elongated, and flat, while their angularity can be characterized as angular, rounded, or smooth (Fig. 2.17 [14]).

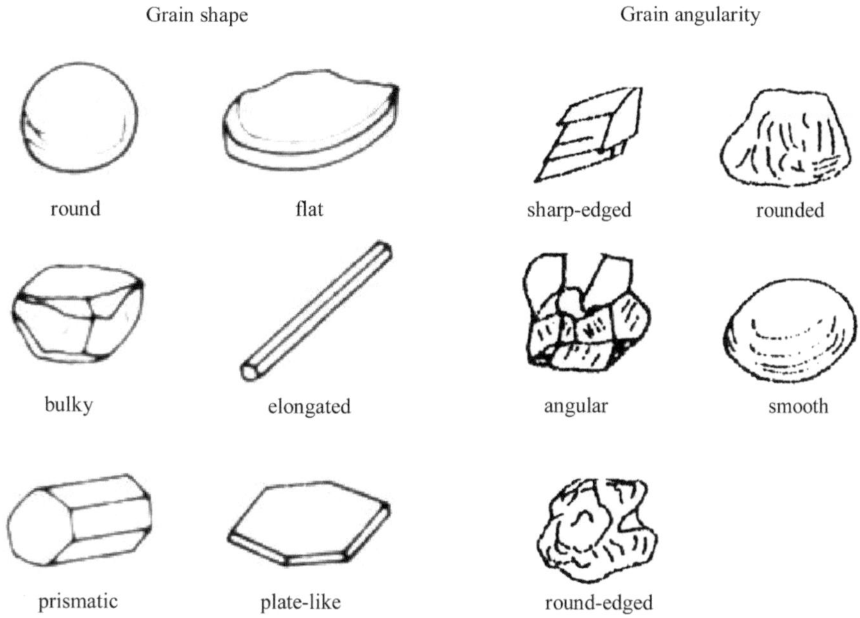

Fig. 2.17 Grain shapes and grain angularity [14]

In coarse-grained soils, particle shape and angularity are influenced by the type of parent rock and its genesis. The grains of transported soils (such as glacial soil) typically range from rounded to smooth, while soils which have not been transported (such as eluvial soil) are sharp edged to angular. The latter exhibit a higher shear strength. In fine-grained soils, grain shape is dependent on the type of mineral: quartz and lime are bulky to prismatic, and clay minerals are generally platelike.

Grain shape and angularity also play a role in the artificial mixture of grain sizes using the Fuller curve.

Grain Surface

A soil's behavior in the presence of water is a characteristic property (Sect. 3.6.3.1). The amount of water which can be adsorptively bound by the soil depends on the soil's *specific grain surface area* A_s, the surface of the grains A of 1 g dry mass m_d. The smaller the grain size, the larger the surface area because the volume increases to the third power of the grain size, while the surface increases to the second power. Grain shape is another influencing factor. The specific grain surface A_s area is determined as follows (Fig. 2.18 [14]):

$$A_s = A / m_d = \alpha / d \cdot \rho_s \left[\mathrm{cm}^2 / \mathrm{g} \right],$$

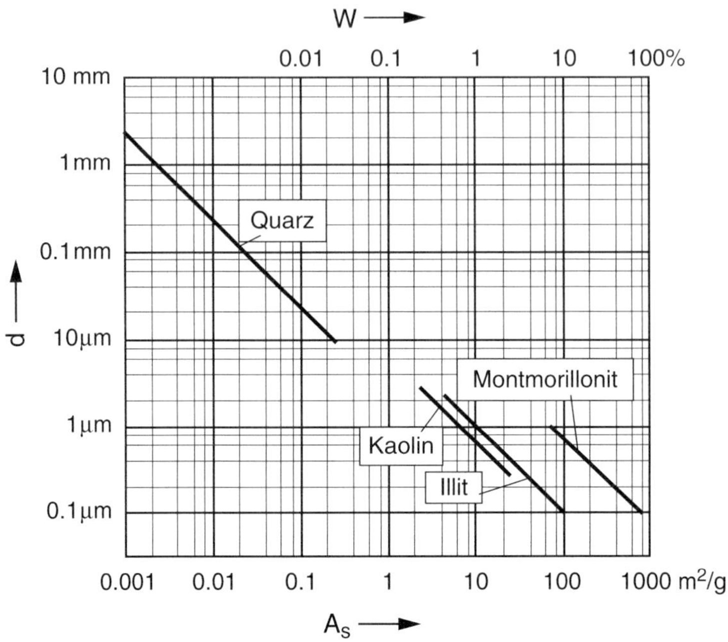

Fig. 2.18 Grain size and specific surface of different clay minerals [14]

d	Grain size
ρ_s	Particle density
α	Shape factor

The shape factors α are:

- For cube and spherical grains (quartz) at $\alpha = 6$
- For plates with a thickness of $0.1d$ (kaolinite, illite) at $\alpha = 24$
- For plates with a thickness of $0.01d$ (montmorillonite) at $\alpha = 204$

The amount of water adsorptively bound on the grain surface is proportional to the grain surface area and increases with dispersity. The figure also shows the moisture content w for an adsorptively bound water layer of 1×10^{-6} mm = 10 Å thickness. The molecular attractive forces of the grains exert extremely high pressure on the adsorptively bound water leading to an increase in density and viscosity (Sect. 2.2.3.4).

2.2.3.2 Processing Parameters

Processing parameters describe the resistance of the soil sample to (plastic) shaping and compaction depending on time and the type of processing. This resistance, also called *cohesion*, is formed by the surface forces of the fine-grained mineral components of the construction soil. Their strength depends on the grain diameter, the structure of the clay minerals, and the moisture content.

Moisture Content

Terminology

The *moisture content* w of a soil sample is the ratio of the weight of the pore water m_w to the dry weight of the soil sample m_d:

$$w = m_w / m_d \, [-].$$

The *degree of saturation* S_r is the ratio of the voids containing water compared to the total voids (Sect. 3.6.1.2).

Test Methods

To calculate the moisture content w according to DIN 18121-1, the weight of the pore water m_w is determined as the difference in weight before and after the drying of the soil sample at 105 °C.

According to DIN 18132, the *capillary water absorption* of construction soils can be calculated with the help of the testing device described by Enslin-Neff (Fig. 2.19 [9, 15]). In this procedure, the water absorption m_w of 1 g of dried soil m_d with $d \leq 2$ mm is determined under predefined conditions dependent on time.

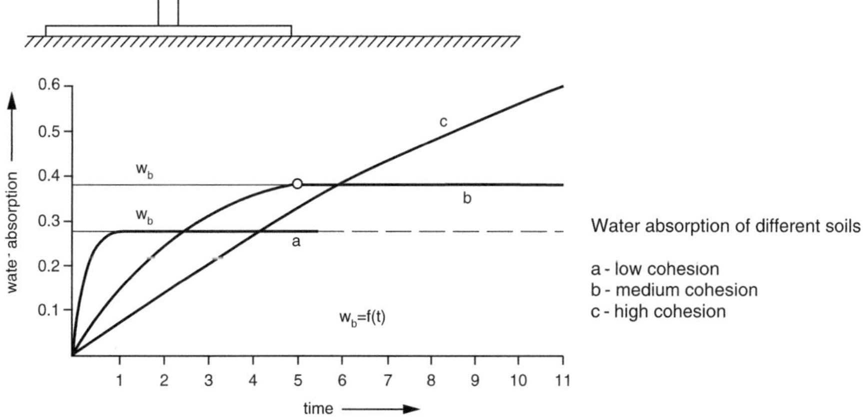

Fig. 2.19 Determining water absorption according to Enslin-Neff [9, 15]

The maximum value which is attained after 4 min is referred to as the *water absorption capacity* w_b:

$$w_b = m_w / m_d [-].$$

Lab and Calculation Values

The state of a construction soil sample in terms of the water content in its pores can be defined with the help of the *degree of saturation* S_r as follows:

S_r (–)	State
0	Dry
0–0.25	Moist
0.25–0.50	Very moist
0.50–0.75	Wet
0.75–1.0	Very wet
>1.0	Saturated

In terms of a soil's water absorption capacity w_b, it can generally be said that lean construction soils ($w_b \leq 0.45$) tend to absorb a relatively small amount of water very quickly. In contrast, clay-rich soils or pure clays ($w_b = 0.6–1.5$) absorb a large amount of water but over a long period of time [15].

Water resistance can also be assessed with the help of the coefficient of permeability *k* which is used in earthworks and dam construction. A coefficient of *k* 1×10^{-5} mm/s describes a soil with good water resistance.

Plasticity Index PI, Consistency Index CI

Terminology

The plasticity index PI (also known as the plastic region) is the general geotechnical classification property for defining the level of plasticity of a soil. It is determined by calculating the difference between two standardized moisture content levels which are independent of the natural moisture found in the soil: the water content at the *liquid limit LL* and the water content at the *plastic limit PL*.

$$\mathrm{PI} = \mathrm{LL} - \mathrm{PL} [-].$$

In a soil sample, the moisture content LL describes the transition from *paste-like* to *liquid* consistency. The moisture content PL describes the transition from *rigid* to *semisolid*, nonplastic consistency. Soil can be shaped within the limits of the plastic region. The critical moisture content levels are depicted in the "consistency diagram" according to Atterberg (Fig. 2.20).

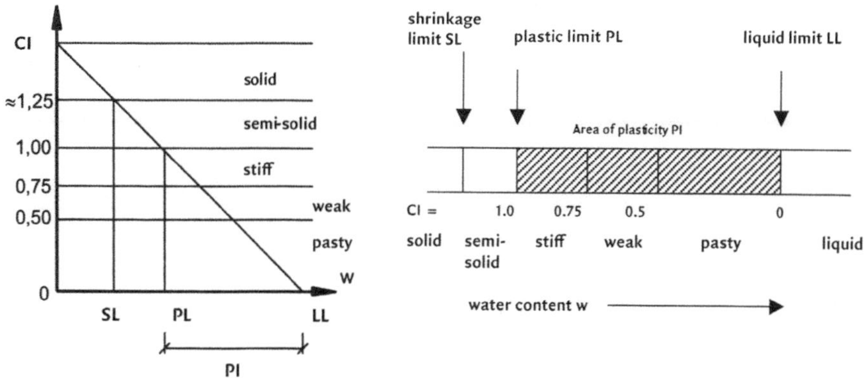

Fig. 2.20 Representation of the critical moisture content levels in the consistency diagram according to DIN 18122

The moisture content at the *shrinkage limit SL* describes the transition of a sample's consistency from *semisolid* to *solid*. Here, soil can no longer be shaped using standard compaction.

The consistency or state of a soil sample, expressed by the *consistency index CI*, compares the current moisture content w of the sample with the specific moisture content levels LL and PL which have been calculated for the same construction soil according to DIN 18122:

$$\mathrm{CI} = (\mathrm{LL} - w) / \mathrm{PI}[-].$$

In geotechnical terminology, loose rock soils with a clay mineral content are referred to as *cohesive/plastic* (clay-rich soils and pure clays), whereas those without a clay mineral content (sands, gravel) are called *non-cohesive/nonplastic* soils. The empirically obtained A-line on the plasticity chart

$$\mathrm{PI} = 0.73(\mathrm{LL} - 20)[-]$$

separates the clays from the silts (Fig. 2.21). There are three plasticity levels: slightly plastic, medium plastic, and highly plastic. For a classification of soil types according to DIN 18196, these levels are added to the letter symbol as a second letter L, M, or A following the group symbol (Sect. 2.2.3.1):

Example: UL, slightly plastic silt (e.g., loess)

Test Methods

Liquid limit LL. The moisture content at the liquid limit LL is determined using the Casagrande method (DIN 18122-1). The standardized testing apparatus includes a brass dish which the prepared soil sample is spread into. The dish is then suspended from a device equipped with a camshaft (Fig. 2.22 [10]). A smooth groove is made

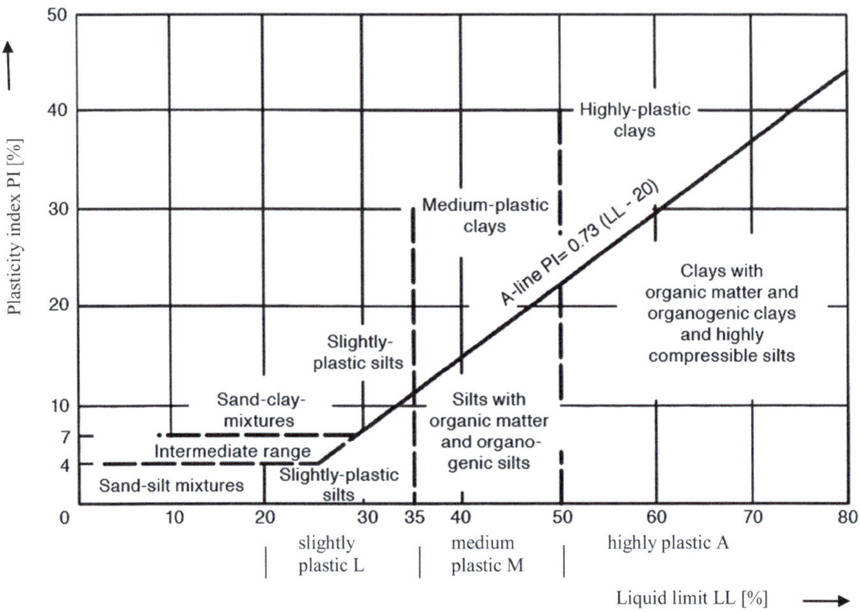

Fig. 2.21 Classification of construction soils in terms of their plasticity according to DIN 18123

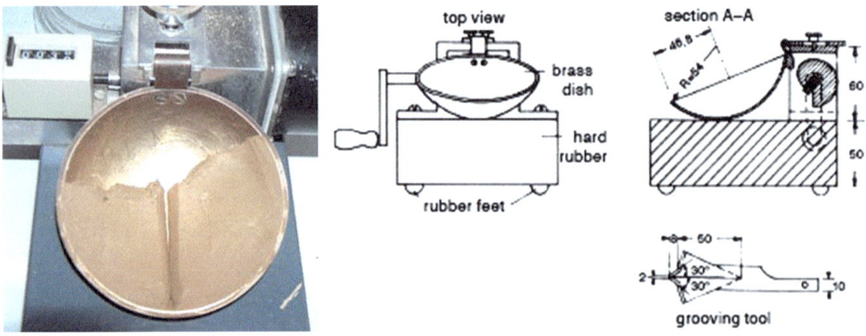

Fig. 2.22 Apparatus for determining the liquid limit LL according to Casagrande [10, 14]

perpendicular to the camshaft using a grooving tool. The dish is lifted to a height of 10 mm by turning the camshaft. The repeated dropping of the dish onto a hard rubber base causes the groove to close. The liquid limit is reached when the groove has closed over a distance of 10 mm after 25 drops.

Plastic limit PL. The moisture content at the plastic limit PL is determined by rolling out a prepared sample on an absorbent mat. The plastic limit PL is reached when a rolled sample with a thickness of 3 mm starts cracking and breaking down into smaller rolls with a length of 10–15 mm (Fig. 2.23 [10]).

Fig. 2.23 Determination of the plastic limit PL [10]

Shrinkage limit SL. The shrinkage limit SL is reached upon the completion of the volume decrease of the soil sample by drying out the physically bound water in its pores (DIN 18122-2). For this procedure, soil is spread into a ring and dried until the weight of the sample remains constant. This state marks the shrinkage limit. A visual sign is the soil's color change from dark to light.

Lab and Calculation Values

Liquid limit SL. The liquid limit SL, as a function of dispersity and mineral constituents, is a measure of the water-binding capacity or hydration of the soil material: the higher the SL the higher the plasticity or cohesive strength of the soil. The proportion of active clay minerals increases the liquid limit.

Examples of SL [9]:

Loess soil: 0.25–0.35
Fluvial soil: 0.30–0.75
Eluvial soil: 0.20–0.45
Black cotton soils [4]: 0.66 (average value with $n = 627$; $s = 16.96$; $\nu = 25.7$ %)

Plasticity index PI. Compared to soils with a high plasticity index PI, soils with a low PI react much faster to the addition of the same amount of water, making them easier to work with.

Examples of PI [9]:

Loess soil: 0.07–0.18
Fluvial soil: 0.12–0.45
Eluvial soil: 0.02–0.30
Black cotton soils [4]: 0.36 (average value with $n = 627$; $s = 12.95$; $\nu = 35.7$ %)

The parameters PI and LL correlate with many geotechnical properties. For example, if the PI and LL of a soil are known, qualitative statements can be made about the clay content and the properties of the clay minerals without extensive testing.

Shrinkage limit SL. According to Muhs [11], the shrinkage limit SL of soils with low cohesion is approx. 5–15 %. For soils with high cohesion, it is approx. 15–40 %. The shrinkage limit is dependent on the initial moisture content. (Krabbe [16]) further states for the SL:

$$\mathrm{SL} \sim \mathrm{LL} - 1.25\mathrm{PI}.$$

Using the shrinkage limit SL, the swelling capacity of a construction soil can be assessed (Sect. 2.2.3.3) [17] with the help of the *shrinkage index SI*:

$$\mathrm{SI} = \mathrm{LL} - \mathrm{SL}\left[-\right].$$

Swelling capacity SI [%]

Low	0–20
Medium	20–30
High	30–60
Very high	>60

Activity ratio I_A. The activity ratio I_A according to Skempton allows for a qualitative assessment of water absorption capabilities as well as a detection of prevailing clay minerals:

$I_A = \mathrm{PI}/(m_{dT}/m_d)$ [–]
m_{dT}—dry mass of the clay fraction $d < 2$ µm
m_d—dry mass of the total sample

Assessment:

$I_A < 0.75$	Inactive (e.g., kaolinite)
$0.75 < I_A < 1.25$	Normal
$I_A > 1.25$	Active (e.g., montmorillonite)

Table 2.3 shows average I_A values for certain clay minerals [14, 15].

The hydration of the clay minerals can be increased further through intensive processing as well as suitable additives (Sect. 3.4.2). Soil samples which have been treated in this manner have higher values for LL than untreated samples of the same soil. Conversely, the hydration of the clay minerals can be decreased by adding synthetic binders (lime, cement).

Consistency index CI. Earth building materials are processed at different levels of consistency depending on their intended application. It is therefore important for the

Table 2.3 Average values of the activity ratio I_A for selected clay minerals

Mineral	LL [%]	I_A [–]	w_a [%]
Kaolinite	60	0.40	80
Illite	100	0.90	
Ca montmorillonite	500	1.50	300
Na montmorillonite	700	7.00	700
For comparison: quartz powder	0	0	30

Table 2.4 Consistency levels of earth building materials, processing properties

Consistency value CI	Designation of consistency [short form]	Consistency of earth building materials during processing, examples	Characteristics	Consistency classes according to DIN 18319
0	Liquid [fl]	Clay slurry for light clay	Watery mixture	
0–0.50	Paste-like [br]	Earth masonry mortar	Squirts through the fingers of a clenched fist	LBM 1[a]
0.50–0.75	Soft [we]	Straw clay	Easily workable	LBM 1
0.75–1.00	Rigid [st]	Rammed earth	Workable	LBM 2
1.00–1.25	Semisolid [hf]	Rammed earth for dry compressing	Can be rolled out, crumbles and tears but cannot be formed into clumps	LBM 2
>1.25	Solid [fe]	Earth block	Dry and light colored, can only be broken apart, separated parts cannot be combined again	LBM 3

[a]LBM—cohesive loose rock, grain size $d \leq 63$ mm, mineral constituents

earth building field to develop an understanding of the different levels of consistency. This understanding can be gained through practical experience (Table 2.4).

Earth building materials are shapeable or plastic at paste-like, soft, and rigid ($0 \leq CI \leq 1$) consistencies.

Plasticity According to Pfefferkorn

The ceramics industry used to apply the Pfefferkorn method for assessing the plasticity of ceramic materials [28]. For this method, a cylindrical soil sample is deformed by a free-falling disk. The ratio of the initial height h_o of the sample to the height of the crushed sample h_1 is determined as the deformation ratio $D_f = h_o/h_1$. This crush test supplies information about the material's consistency. $D_f = 2.5$–4 describes a consistency of clay mixtures which can be easily shaped by hand (e.g., material for a potter's wheel). $D_f = 1.25$ characterizes the ideal consistency for extruded products.

Standard Consistency and Cohesive Strength

Terminology

To assess the suitability of construction soils, DIN 18952-2 (which has been withdrawn) applied the cohesive strength test according to Niemeyer [18]: in this test, clay mineral and grain composition are not determined individually but rather as an "external" effect of both parameters in the form of resistance. This resistance is measured as the cohesive strength c_N (cohesion in geotechnical terms) of a soil specimen at standard consistency in a tensile "break" test. Here, the term *standard consistency* describes the test consistency of the soil sample which has been defined for determining the cohesive strength.

Test Method

The standard consistency is determined empirically as follows: 200 g of a uniformly prepared soil sample is compressed by pounding it several times on a hard, nonabsorbent surface. Immediately afterward it is formed into a ball. The ball is dropped onto a smooth, hard, and firmly mounted nonabsorbent surface from a height of 2 m. The standard consistency of the soil sample has been attained when the ball flattens to a width of 50 mm upon impact.

In order to determine the cohesive strength c_N, the test sample at standard consistency is formed into a figure-eight shape with the help of a wooden mold and then placed between two metal brackets. The upper bracket is suspended from a boom. A load-applying device is attached to the lower bracket. This device consists of a box which is filled with sand or water at a maximum flow rate of 750 g/min until the specimen breaks. The load is applied to the thinnest section of the figure-eight specimen (5 cm) and is converted into a "cohesive strength" (pull) value per cm^2 (Fig. 2.24 [10]).

Lab and Calculation Values

According to their determined cohesive strength c_N, construction soils are classified into various categories ranging from "lean" to "very rich" (Table 2.5 [18]):

According to DIN 18952-2, soils with a cohesive strength c_N of <50 g/cm^2 (0.005 N/mm^2) are deemed unsuitable for earth building purposes. They can, however, still be used for a number of applications, such as ceiling loose fill material. For these extremely lean soils, the cohesive strength test does not supply reliable results.

Developed in the 1940s, the classification of construction soils based on their cohesive strength as a processing parameter has remained confined to German-speaking countries. Internationally, a classification according to the geotechnical parameters PI and LL is used. The growing importance of earth as a building material, also internationally, raises the question of a "translation" of both systems. Research conducted at the Bauhaus University Weimar has examined this question. To start, 16 local soils were tested which represented the classes "slightly plastic,"

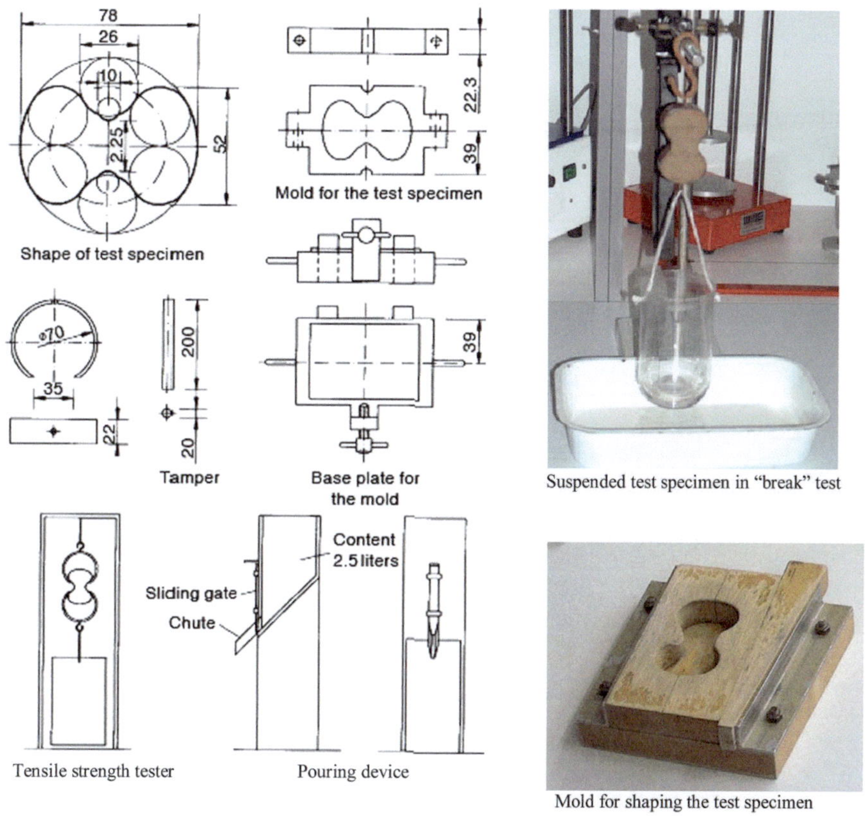

Fig. 2.24 Determination of the plastic limit PL according to Niemeyer [10]/DIN V 18952-2

Table 2.5 Classification of construction soils based on their cohesive strength according to Niemeyer

Cohesive strength c_N [g/cm²] or (N/mm²)	Construction soil classification	Degree of linear shrinkage $\varepsilon_{f,l}$ [%]	Moisture content at standard consistency w_N [%]
50–80 (0.005–0.008)	Very lean	0.9–2.3	
81–110 (0.0081–0.011)	Lean	0.9–2.3	9.5–12
111–200 (0.0111–0.02)	Semi-rich	1.8–3.2	11–15
201–280 (0.0201–0.028)	Rich	2.7–4.5	12–20
281–360 (0.0281–0.036)	Very rich	3.6–9.1	
>360 (0.036)	Pure clay	>9.1	15–23

"medium plastic," and "highly plastic" in terms of plasticity and "lean" to "very rich" in terms of cohesive strength (Table 2.6 [19]).

In these tests, it could generally be confirmed that an increase in cohesive strength is connected to an increase in the soils' plasticity. While there was a distinct correlation

Table 2.6 Parameters of tested construction soils

No.	Designation of sample	Moisture content at standard consistency w_N [%]	w_{pr} according to proctor [%]	Degree of linear shrinkage $\varepsilon_{f,l}$ [mm/20 cm]	Cohesive strength c_N [g/cm^2]	Plastic limit PL [%]	Liquid limit LL [%]	Plasticity index PI [%]
1.	Kromsdorf	19.86	16.22	8.0	60.6	19.27	35.86	16.53
2.	Weimar-Umgehstr	27.59	26.25	15.0	493.2	27.11	67.20	40.09
3.	Leuben	17.52	16.46	7.0	169.0	17.18	41.45	24.27
4.	Hochstedt 1/1	17.57	18.40	7.0	94.9	18.97	31.30	12.33
5.	Hochstedt 2/1	20.18	20.50	7.7	134.3	21.77	39.16	17.05
6	Hochstedt 3/1	17.52	20.91	10.3	137.3	20.62	35.70	15.08
7.	Hochstedt 4/1	18.45	19.00	8.7	110.3	20.47	33.79	13.32
8.	Hochstedt 5/1	21.04	20.30	10.7	259.6	16.14	46.00	29.85
9.	Weimar-Klinik 1/1	21.45	21.74	9.6	155.7	20.23	46.91	26.68
10.	Weimar-Klinik 2/1	21.91	21.91	10.3	164.5	21.16	53.90	32.74
11.	Weimar-Klinik 3/1	29.54	27.21	19.5	350.5	25.30	64.30	39.00
12.	Nohra 1/1	24.90		12.7	513.4	22.94	50.50	27.56
13.	Mörsdorf 1/1	18.00	19.69	8.7	91.3	18.27	33.75	15.48
14.	Weimar-Klinik 1/1a	21.64	20.40	8.0	130.3	21.86	45.50	23.64
15.	Nordhausen 1	20.36	17.29	6.0	127.5	21.93	40.75	18.82
16.	Erdmannsdorf 1/1	29.27		6.0	366.9	26.31	70.25	43.94

Table 2.7 Correlative references between geotechnical and earth building parameters

	Moisture content w_N at standard consistency	Cohesive strength c_N
LL	$W_N = 0.32LL + 7.21; r_{xy} = +0.94$	$c_N = 7.88LL - 155.94; r_{xy} = +0.72$
PL	$W_N = 1.19PL - 3.37; r_{xy} = +0.79$	
w_{pr}	$W_N = 1.10w_{pr} - 1.84; r_{xy} = +0.79$	
PI		$c_N = 9.06PI - 17.01; r_{xy} = +0.70$
Terminology		
	Plasticity	*Cohesive strength c_N*
	Slightly plastic	Lean
	Medium plastic	Semi-rich to rich
	Highly plastic	Very rich, pure clay

between the earth building and the geotechnical classification systems used to describe the moisture content of the consistency of the samples, the correlations between cohesive strength c_N and liquid limit LL as well as for the plasticity index PI were less clear for the tested samples (Table 2.7 [19]).

Influencing Variables

Due to slow water distribution in clays, very rich soils and pure clays require much more intensive processing than lean soils in order to attain the "standard consistency" for testing. Standard consistency defines the same degree of receptiveness of the clay minerals, and whereas lean soils need little water, very rich soils and pure clay require much more. This is also the reason why the moisture content w_N at standard consistency is much higher for very rich soils and pure clays than for lean soils. Table 2.6 illustrates this tendency for selected soil samples by showing the parameters plasticity PI, cohesion c_N and moisture content w_N at standard consistency.

2.2.3.3 Deformation Parameters

Deformation in construction soils through the absorption and the release of water is called swelling and shrinkage. Compared to other mineral building materials, they can reach a considerable scale.

Assessing the quality of these deformations in construction soils can determine if stabilizing measures need to be taken during the production of earth building materials. This is particularly true if soils or clays which are known to be expansive due to their formation genesis (e.g., "black cotton soils," Sect. 2.1.2.6) are to be processed into earth building materials.

The overview in Sect. 3.6.2.1 systematically lists the deformations of earth building materials.

Shrinkage

Terminology

Construction soils experience a decrease in volume during drying as a result of the evaporation of pore water. This causes three-dimensional deformations which are referred to as *shrinkage*. They occur independent of loads and are reversible.

When testing the suitability of a soil for construction, its deformation is generally only examined in one direction: the ratio of change in linear dimension compared to the initial length of a specimen, also known as the *linear degree of shrinkage* $\varepsilon_{f,l} = \Delta l/l$ [%] or linear shrinkage.

The ceramic industry differentiates between "drying shrinkage" and "firing shrinkage." Drying shrinkage describes the transition from shaped body to "green" body before firing, typically at temperatures <200 °C. Firing shrinkage is the deformation caused by the sintering process during ceramic firing at temperatures >800 °C. This is not part of the production of earth building materials which is why the term "shrinkage" is sufficient.

Test Methods

According to DIN 18952-2, the shrinkage test of construction soils can be carried out by determining the reduction in length of a standardized specimen with the dimensions 220 × 40 × 25 mm (Fig. 2.25 [10]). The soil needs to be prepared to standard consistency (Sect. 2.2.3.2). It is then placed into a mold. The mold is removed and the test specimen is air-dried until its length remains stable. In this state, the moisture content of the soil correlates with the moisture content at the shrinkage limit SL (Sect. 2.2.3.2). While the soil is still moist, two reference marks are

Fig. 2.25 Testing the linear degree of shrinkage according to DIN V 18952-2 [10]

scratched into the sample 200 mm apart from each other. The reduction in length is measured between these two lines as the linear degree of shrinkage $\varepsilon_{f,l}$. It should not exceed 2 %. The final result is the average value determined by three tests.

Lab and Calculation Values

According to Niemeyer [18], the correlations between cohesive strength c_N and linear degree of shrinkage shown in Table 2.5 exist when using the test specimen dimensions mentioned above.

The cohesive strength of the construction soil provides information about the extent of the linear degree of shrinkage and if measures need to be taken in order to reduce shrinkage. Therefore, the Lehmbau Regeln [6] do not require a shrinkage test for construction soils. At a later point in time, this test is carried out on the actual earth building material which is ready for processing. This material consists of the construction soil, aggregates, and water added to reach the required processing consistency. Only the actual soil mixture can provide a realistic picture of the extent of shrinkage deformations to be expected. Depending on the aggregates used, the dimensions of the test specimen differ from one earth building material to another (Sect. 3.6.2.1).

Influencing Variables

Shrinkage of a soil sample is influenced by the proportion and structure of the clay minerals, based on the total mass: with an overall unchanged proportion of clay minerals, a high degree of deformation can be expected in soils with a dominance of three-layer minerals (e.g., montmorillonite), whereas prevailing two-layer minerals (e.g., kaolinite) result in low degrees of deformation (Sect. 2.2.3.4). High degrees of deformation generally manifest themselves as cracks.

The risk of cracking decreases with a lower initial moisture content of the soil sample and a better-graded grain distribution curve (Sect. 2.2.3.1).

Figure 2.26 [27] shows the difference in shrinkage behavior of smectite and kaolinite with the same initial consistency: the smectite sample has crumbled into individual chunks, while the kaolinite sample exhibits "normal" crack formation consisting of individual chunks attached to the ground.

Swelling

Terminology

Whereas soils shrink while drying, they increase their volume through water absorption. These deformations are also three dimensional and are called *swelling*. They occur independent of loads and are reversible. Another use of the term "swelling" applies to elastic deformations which develop temporarily in the form of compression of the specimen after a load has been applied and which reverse immediately after the load has been removed (Sect. 3.6.2.1).

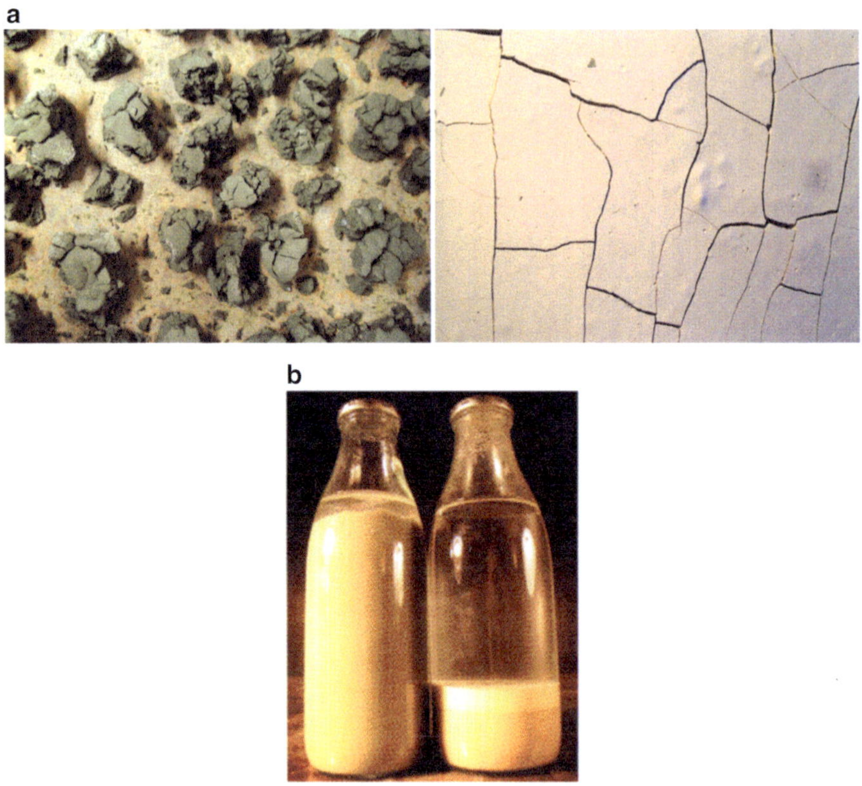

Fig. 2.26 Shrinkage and swelling of clay minerals [27]. (**a**) Shrinkage based on the same initial consistency: *left*, smectite; *right*, kaolinite. (**b**) Swelling based on equal amounts of a mixture of clay minerals and sand: *left*, smectite; *right*, kaolinite

Test Methods

The *free-swell value* F_s ("free-swell test" [17]) can be used to estimate the expansion capabilities of a soil sample in a rapid test.

For the test, 10 cm³ of a soil sample are passed through a 0.4-mm sieve and mixed with 100 cm³ of distilled water in a graduated cylinder. After settling has occurred, the difference between the final and initial volumes based on the initial volume of the sample in % is used to determine the free-swell value F_s.

A more exact method of calculating the swelling potential of construction soils is to determine the *vertical swell* h' with the help of an oedometer, a device which is commonly used in soil mechanics (Sect. 3.6.2.2). This test examines the volume increase of a waterlogged soil sample which is not prevented from expanding laterally and is under a device-specific imposed load tension of 1.6 kN/m² (0.0016 N/mm²).

The measured increase in height of the sample Δh based on the initial height h_o is the vertical swell h'

$$h' = \Delta h / h_o \, [\%].$$

Lab and Calculation Values

Free-swell value F_s [%] and swelling potential according to [17]:

<50	Low
50–100	Moderate
>100	Very high

Bentonite clays reach free-swell values in the range of $F_s = 1200–2000$ %.

Influencing Variables

The absorbed water is integrated into the structure of the clay minerals: three-layer minerals (e.g., montmorillonite) have the capacity to absorb a large amount of water, whereas two-layer minerals (e.g., kaolinite) absorb relatively little (Sect. 2.2.3.4).

Figure 2.26 [27] shows two bottles filled with water and the same amount of a sand/clay mineral mixture: smectite on the left, kaolinite on the right. Whereas the sand in the bottle filled with kaolinite settles to the bottom, the sand in the bottle filled with smectite is "lifted" and evenly distributed within the suspension due to the high swelling potential of smectite.

With an increase in plasticity PI, as well as clay content and water absorption capacity w_a, the vertical swell h' also increases as a result of swelling as does the free-swell value F_s.

2.2.3.4 Chemical–Mineralogical Parameters

The chemical–mineralogical compositions of construction soils, particularly the quantity of clay minerals and their structure, are decisive factors in regard to a soil's processing properties and its deformation parameters. In addition to relevant geotechnical processing parameters, qualitative statements can be made about the processing behavior of construction soils based on the results of suitable tests.

Large quantities of extracted soil and the industrial production of earth building materials require continuous monitoring of the quality of the construction soils used, especially after changing from one extraction site to another. In addition to testing the physical–mechanical parameters, chemical–mineralogical analyses also need to be carried out.

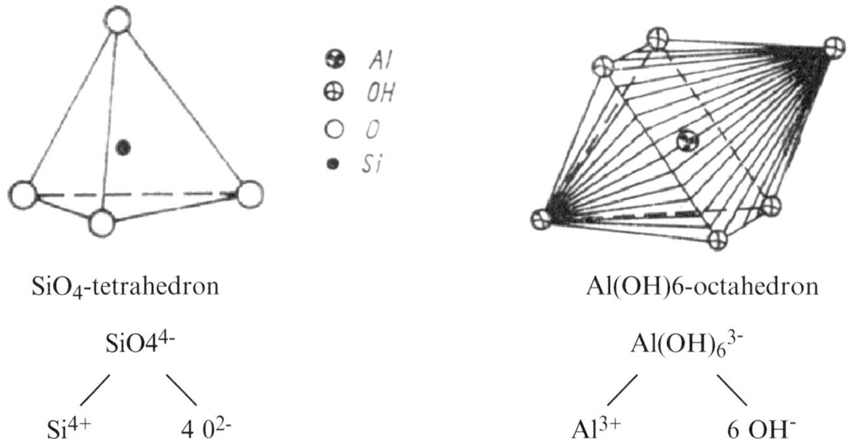

Fig. 2.27 Structural basic units of clay minerals [9]

Clay Minerals

Terminology

In chemical terms, clay minerals are aluminum silicates. They mainly contain the elements Si, Al, oxygen, and hydrogen. Other elements include Fe and different elements of the alkali and alkaline earth groups, particularly Mg, Ca, and K.

Clay minerals are classified based on their internal structure. The structural building blocks of all clay minerals consist of a *Si-O-tetrahedron* formed by Si and oxygen and an *Al-OH-octahedron* formed by Al, oxygen, and hydrogen (Fig. 2.27 [9]). Both building blocks have an excess of negative charge because for every central ion with a positive charge, there are several accompanying ions with negative charges. The tetrahedron SiO_4, containing a tetravalent Si atom with a positive charge, has a tetravalent excess of negative charge resulting from the eight negative charges of the four oxygen ions. In the octahedron $Al(OH)_6$, there are six negative charges from the six OH ions for every Al atom with a trivalent positive charge causing a trivalent excess of negative charge.

These excess charges are balanced by cross-linking with other tetrahedra and octahedra. First, the tetrahedra and octahedra form stable configurations of hexagonal rings (Fig. 2.28 [9]) which build netlike sheets through the addition of further hexagonal rings. In addition, bridge bonding formed by specific oxygen ions causes cross-linking of tetrahedra and octahedra. This succession of ion sheets is also called *packaged sheets or layers*. Several layers are called *layer stacks*. Several layer stacks form a clay mineral crystal lamella which is visible as a single structure under a scanning electron microscope (Fig. 2.31).

The cross-linking between tetrahedron and octahedron sheets not only occurs through one tetrahedron sheet attaching to one octahedron sheet but also through

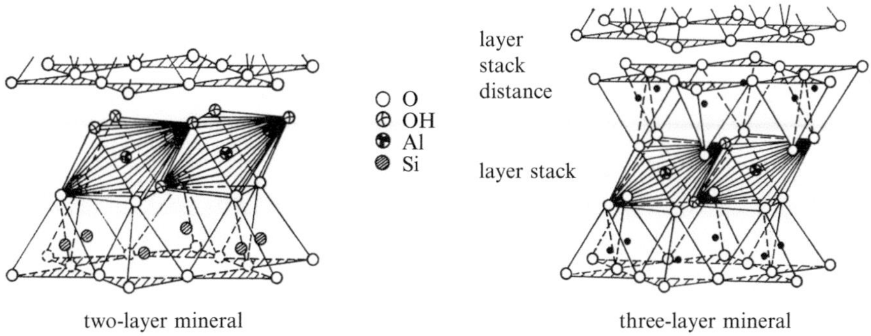

Fig. 2.28 Two- and three-layer structure of clay minerals [9]

two tetrahedron sheets attaching to either side of a octahedron sheet. Figure 2.28 [9] shows hexagonal ring sections of a two-layer and a three-layer stack. Depending on the structure of the crystal (consisting of two- or three-layer stacks), the minerals are called *two-layer* or *three-layer minerals*.

Kaolinite is the most common two-layer mineral and is mainly formed from acidic migmatites during intensive (tropical) weathering (Sect. 2.1.2.6).

The three-layer mineral *illite* is typically formed from mica under temperate and humid climate conditions. It can be found, in different proportions, in nearly all types of cohesive loose rock. The three-layer mineral *montmorillonite* develops primarily through the weathering of alkaline rock under semiarid climate conditions (Sect. 2.1.2.6).

Test Methods

Analyses of a soil's mineral content provide information about its processing behavior. The overall percentage and the type of clay minerals are examined in these tests. The test results make it possible to assess plasticity and cohesive strength as well as drying behavior (shrinkage). Nonplastic minerals (such as quartz, feldspar, calcite, dolomite) act as "tempering agents." The chemical analysis (Al_2O_3/Fe_2O_3) offers further insight into water-binding capabilities.

Due to the effort they require and their limited validity, *wet chemical* analysis procedures are rarely carried out. They are used to obtain quantitative information about the chemical composition of the soil material.

Differential thermoanalysis (DTA) is a *thermal* analyzing method suitable for soil testing. For this analysis, the sample is heated, and the resulting endothermic and exothermic effects provide reliable information about the type of clay minerals present and their concentration. The relevant processes occur in the temperature range of 100–750 °C.

Among the test methods based on the application of X-rays, *X-ray diffraction* can be used to provide significant soil testing results. With this method it is possible to obtain reliable information about the soil's mineralogical composition by

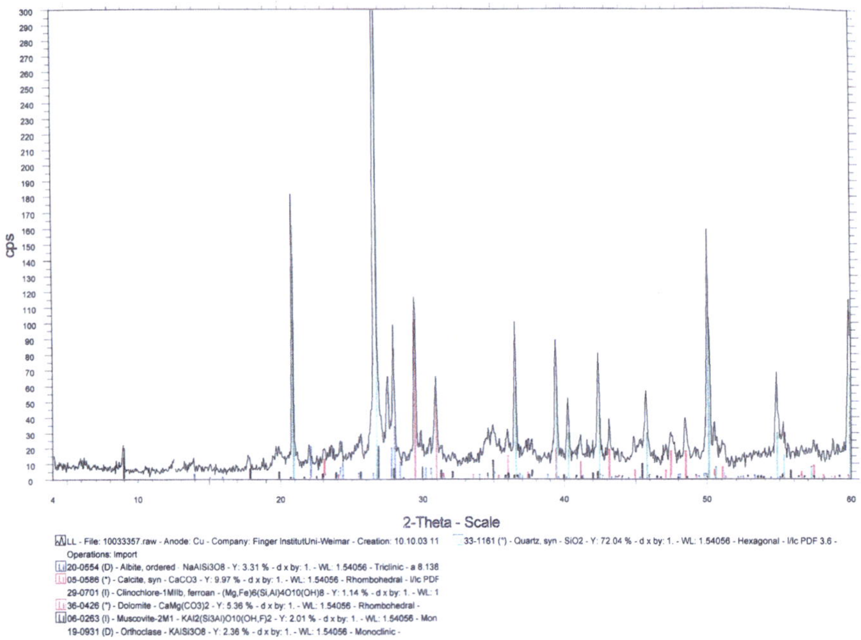

Fig. 2.29 X-ray diffraction analysis of a loess soil [10]

determining intensity values, glancing angles, and layer thicknesses of the crystal lattices (*d*-values). The Thuringian loess soil shown in Figs. 2.15 and 2.29 [10] exhibits a relatively large proportion of the nonplastic minerals quartzite and calcite.

In addition to qualitative assessments, quantitative determinations can be made in order to obtain information about the chemical and mineralogical composition of the soil.

Table 2.8 [29] lists results for Gotha loess soil.

Influencing Variables

Mineral structure. The stability of the mineral structures described above greatly influences the soils' and clays' characteristic properties such as plasticity, shrinkage, and swelling. In this context, clear differences can be seen between two-layer and three-layer minerals:

The crystal lattice of *two-layer minerals* is rigid and electrically neutral on the surface due to the complete assignment of all charges. This reduces the capability of these minerals to store water or ions dissolved in pore water as well as their tendency to shrink or expand. Only between the layer stacks and the individual crystal lamella, in other words at the outer edges of the layers, do free electric charges become available through breaks in the hexagonal rings. This is therefore the only place where water molecules can be stored and where shrinkage and swelling can occur.

Table 2.8 Gotha loess soil: mineral content and chemical analysis

Mineral content	
Quartz	35 %
Plagioclase (feldspars)	8–10 %
Calcite	10–12 %
Dolomite	2–3 %
Kaolinite	8 %
Illite or mica minerals (with a prevailing portion of mica)	20 %
Intercalated minerals	8 %
Swellable minerals of the montmorillonite group	8 %
Chemical analysis	
Loss on drying at 100 °C	1.4 %
Loss on ignition at 1000 °C	7.5 %
SiO_2	65.2 %
Al_2O_3	10.9 %
Fe_2O_3	3.5 %
CaO	5.6 %
MgO	2.1 %
K_2O	2.55 %
Na_2O	0.91 %

The crystal lattice of *three-layer minerals* is unstable. In addition to its shape described above, the tetravalent central Si ion can be replaced by a trivalent Al ion. Furthermore, in the octahedra, the trivalent Fe, the divalent Mg, and even the monovalent Li can take the place of the trivalent Al. Sometimes the central atom is missing altogether. However, if the higher-valent positively charged central atoms are replaced by lower-valent atoms and the number of negative charges of the surrounding ions remains unchanged, the result is an excess of negative charges in the entire lattice. This excess needs to be neutralized with positively charged pore water cations or with water molecules.

Another result of the unstable crystal lattice of the three-layer minerals is that the distances between the layer stacks are not fixed. Instead they are dilatable. This makes it possible for additional water molecules to attach between the layer stacks increasing plasticity as well as shrinking and swelling capabilities. These specific properties are often undesirable in earth building and can be influenced by chemical additives, typically by the binding agents lime and cement.

Often, minerals do not fit into the described pattern of two-layer and three-layer minerals. Such minerals are known as transition minerals or minerals with a mixed layer structure. They do not consist of a succession of uniform layer stacks but are made up of different stacks with a regular or irregular mixed layer structure. Figure 2.30 [9] gives an overview of the most important minerals in the respective groups along with a structural diagram.

Structure type	*Mineral group /* **Mineral name**	**Formation through weathering of**	**Deposits in**
Two-layer minerals 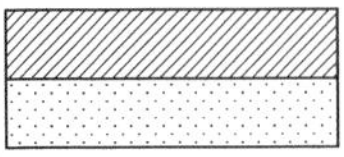 O T Basic formula $Al_2Si_2O_5(OH)_4$	*Kaolin minerals* **Kaolinite** Halloysite Meta-Halloysite	Acidic rock containing feldspar	Kaolin (impure kaolins), ceramic clays
Three-layer minerals 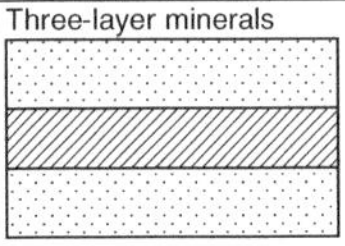 T O T Basic formula $Al_2Si_4O_{10}(OH)_{12}$	*Montmorin minerals* **Montmorillonite** Beidellite Nontronite *Micaceous clay minerals* **Illite** Vermiculite	Volcanic ash, Alkaline rock (basalt, gabbro) Mica	Bentonite (highly-plastic clay) Cohesive loose rock (clay-rich soils)

T Tetrahedron sheet, O Oktahedron sheet

Fig. 2.30 Structural diagram and main clay minerals [9]

Figure 2.31 gives an idea of the actual size ratio between the clay mineral crystal lamella and the nonplastic quartz fragments in the "clay grain" fraction $d<2$ µm. The differences in size illustrate the role of the clay minerals as binding agents between the coarser grains [11].

The shape of the clay mineral crystal lamellae and the type of contact between them greatly influence how construction soils are further processed and dried and which mechanical properties the finished building products possess.

Figure 2.32 [27] shows two different clay mineral crystal lamellae: (a) *needles*, single and cemented (illite/montmorillonite), and (b) *plates* (chlorite) (also Fig. 2.31b (kaolinite)).

According to Fig. 2.32 [20], different types of contact between the individual clay mineral crystal lamellae can also be distinguished: *dot-like* (c, Fig. 2.32b: chlorite), *linear* (d, Fig. 2.30b: illite), and *planar* (e, Fig. 2.31b (kaolinite)).

Planar contact between the crystal lamellae, also known as *band structure*, is the most stable and, therefore, shows more resistance to processing and compaction. Dot-like contact is an open structure which is also called the *house-of-cards structure*. When drying, this structure shows low resistance to water on its way to the evaporation surface thereby reducing the required drying time. Shrinkage and swelling can be largely "buffered" by the open structure. This makes any measurable deformation insignificant.

Cation-exchange capacity. The type of *cation saturation* greatly influences the plasticity of soils and clays because any additional excess electric charges enable them to bind more water. With a qualitative change in sorbed cations toward $H^+ \Rightarrow Al^{3+} \Rightarrow Ba^{2+} \Rightarrow Ca^{2+} \Rightarrow Mg^{2+} \Rightarrow K^+ \Rightarrow Na^+$, the ability to bind water, called hydration, is improved. This in turn increases the plasticity of the soil (see Fig. 2.31, Na- and Ca-bentonite). Following this order ("Hofmeister series"), the ions on the left are more easily replaced by ions which are further to the right and vice versa.

The capacity to exchange such ions is called the cation-exchange capacity and is expressed in milliequivalent per g or per 100 g of soil [mequ]. Kaolinite minerals

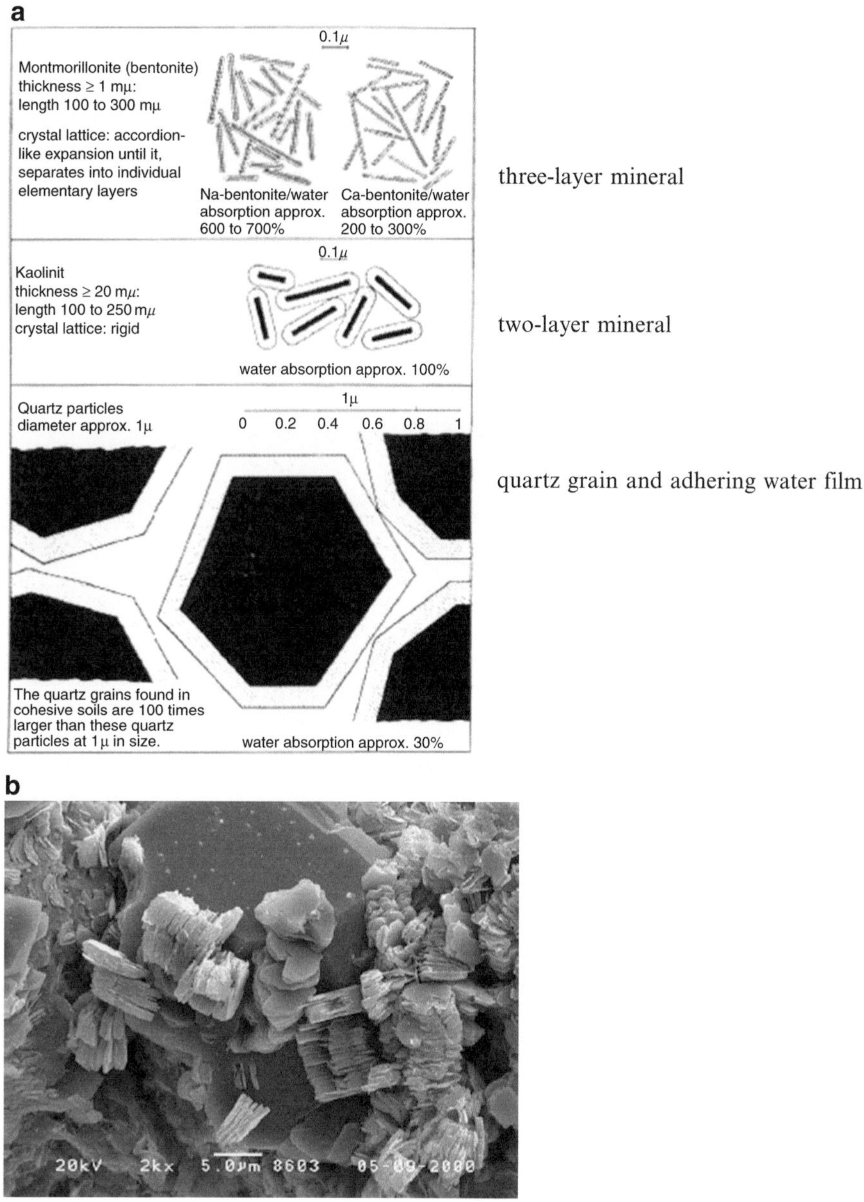

Fig. 2.31 Size ratio of clay minerals and quartz particles with adhering water films [11]. (**a**) Size ratio of three-layer/two-layer clay minerals and quartz grains with adhering water films [11]. (**b**) Quartz grain and adhering clay mineral lamellae in SEM [Wikipedia]

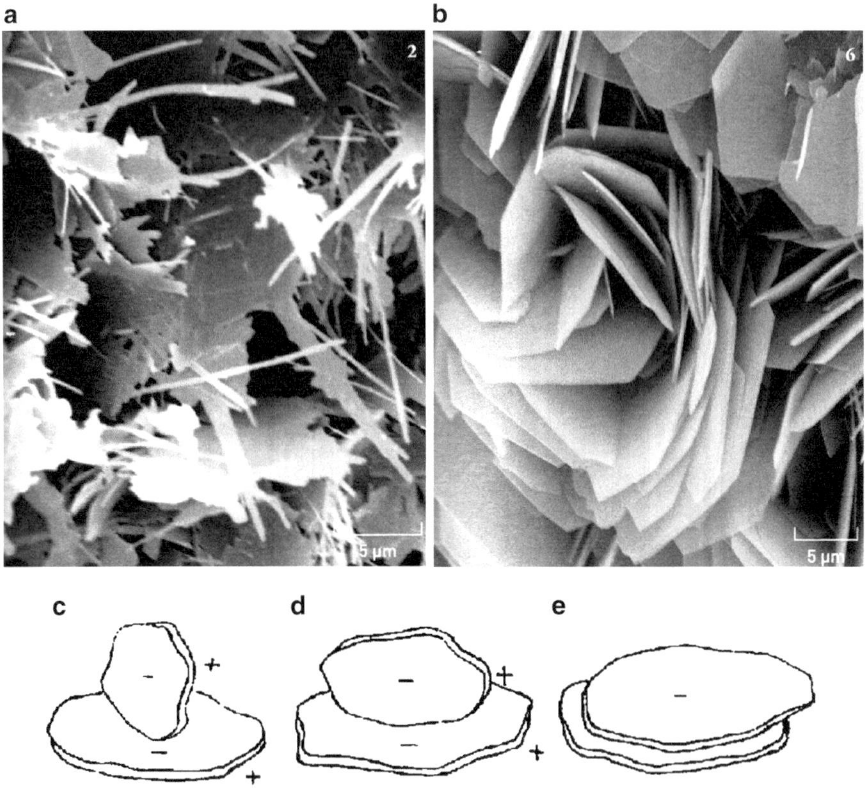

Fig. 2.32 Structure and types of contact between clay minerals. (**a**) Needles/acicular or cemented, illite; structure, "house of cards," SEM [27]. (**b**) Platy, chlorite; structure, "house of cards," SEM [27]. Types of contact between clay mineral lamellae [20]. (**c**) Dot-like. (**d**) Linear. (**e**) Planar

exhibit a comparatively low cation-exchange capacity, whereas montmorillonite minerals have a high cation-exchange capacity (Table 2.9).

Surface tensions and bound water. The *surface tensions* working on the extremely small particles of the clay mineral lamellae, with their interaction mechanisms between the solid and liquid phases of water, are the cause for the cohesion of these small particles and, therefore, for the "cohesive" properties of all loose rock containing clay minerals.

The surface tensions are of electrical nature and lead to the formation of a force field surrounding each particle causing the sorption of water molecules (dipoles) and of ions dissolved in the groundwater. The water molecules arrange themselves into continuous *water films* around the solid particles. Depending on their distance from the solid mineral core, these water films have different properties (Fig. 2.33 [3]): located directly on the surface of the solid substance, the adherent water acts like a solid body (sorption water) due to the extremely high surface tension. As the distance increases, the water properties become similar to those of viscous asphalt

Table 2.9 Cation-exchange capacity and swelling behavior of selected clay minerals, according to [36]

No.	Clay mineral	Exchange capacity [mequ/100 g]	Evaluation	Swelling
1	Kaolinite	0–15	Moderate	Moderate
2	Montmorillonite	60–150	Very strong	Very strong
3	Illite	3–40	Moderate	Medium
4	Vermiculite	100–150	Very strong	Strong
5	Halloysite	5–50	Medium	Medium
6	Chlorite	3–40	Medium	Medium

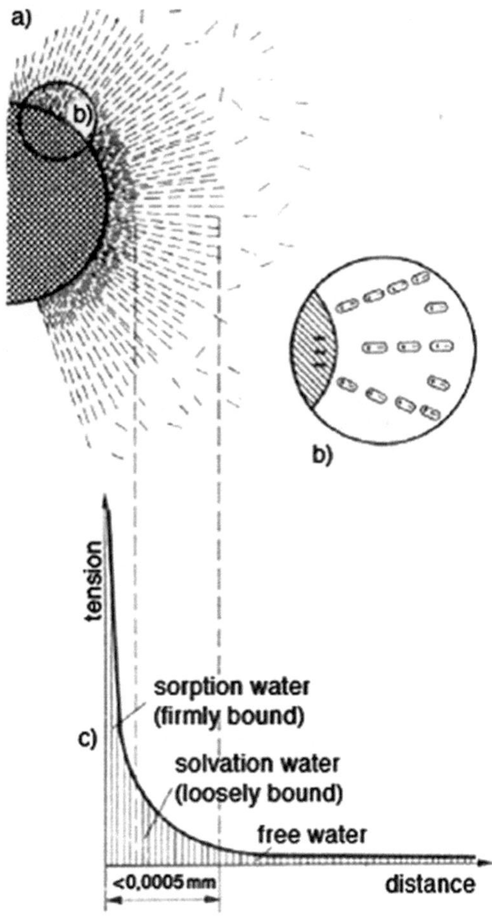

Fig. 2.33 Principle of the interaction between clay mineral and water molecule [3, 23]. (**a**) Depiction of the water bond. (**b**) Alignment of water dipoles on the grain surface. (**c**) Forces of attraction depending on the distance between mineral particle and water

(solvation water) and only at a distance of >0.5 μm do they regain their "liquid" properties. In addition to this "liquid" water, water vapor can be found within the fine pore spaces. This vapor moves independent of gravity under the influence of molecular forces.

The surface tensions and the water film thickness are influenced by the specific surface area (particle size) and the type of mineral substance, by the availability of free cations and by the temperature: the smaller the particle size the higher the cohesive strength. The more unstable the crystal structure of the particle the thicker the water films, under the condition that enough water is available. In the case of an increased thickness of the water sheaths, the attraction between the individual particles decreases and the grain skeleton becomes more unstable. This, however, improves the capacity of the particles to move past each other (workability).

The small particles stick to the surfaces of the coarser grains with the help of hydrogen bonds. The thinner the water layer film the more stable the bonds. The cation sheath of the clay lamellae adhering in this manner binds additional particles. With the continued drying of the grain mixture, the clay mineral lamellae which are suspended in the pore water find themselves in an increasingly smaller space. They crowd together into the corners of the pores which are formed where the coarser grains touch each other (Fig. 2.34 [20]). Finally, a stable bridge is formed between the coarse grains giving the grain skeleton adhesive strength. In soil mechanics, this phenomenon is called cohesion (Sect. 2.2.3.2).

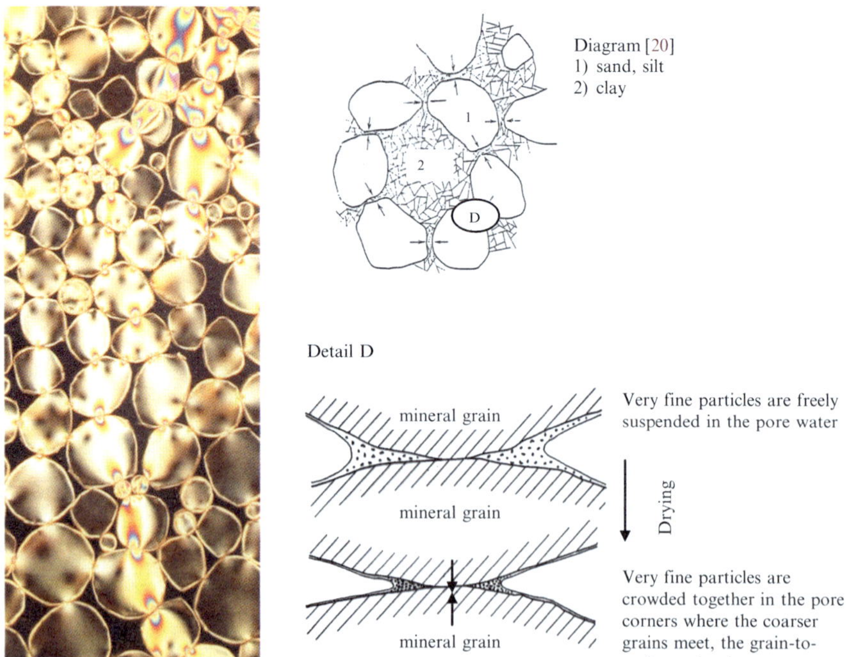

in polarized light [27]

Fig. 2.34 Principle of load transmission in earth building elements with "grain-to-grain"—tension, diagram/polarized light in polarized light [27]

Further drying increases the attractive forces within the pore spaces where the grains meet. This forms a stable strength matrix within the entire grain structure similar to concrete. This strength matrix enables load transmission through "grain-to-grain" pressure and "load-bearing" earth building. However, the strength of this bond between the clay minerals as "binding agents" and the coarse grains is always lower than the strength of the individual coarse grain. An "overload" therefore always results in a break along the surfaces of the coarse grains or the pores. In contrast to concrete, this bond is dissolvable by water.

Color. Construction soils exhibit a wide range of colors which serve as an indication of the prevalence of specific chemical elements within the clay mineral structure (Sect. 2.2.2.2).

Natural Additives

In addition to the mineral soil components which form the basis of a classification for construction purposes (Sect. 2.1.1.2), construction soils can also contain natural additions. These include water-soluble salts and organic matter. These natural additions can influence the construction properties of the soil, such as plasticity and strength properties.

Lime Content

Terminology. Lime components are the most common natural addition in construction soil. They are caused by weathering and eluviation of soluble rock and top soil products.

Test methods. A *qualitative* assessment of the lime content in construction soil can be carried out on location using hydrochloric acid. For this test, diluted hydrochloric acid is added to the soil sample and the foaming quality is assessed: the more intensive the reaction the higher the lime concentration (Table 2.10).

A *quantitative* analysis is carried out using the loss-on-ignition method described in DIN 18129. In this test, approx. 20 g of dried construction soil is weighed out in a porcelain crucible and heated at approx. 900 °C for 2 h. After the sample has

Table 2.10 Qualitative assessment of the lime content of soils according to DIN EN ISO 14688-1

Lime content v_{Ca} [%]	Reaction of the sample after adding HCl	Evaluation
<1	No foaming	Lime-free
1–5	Weak to pronounced, short foaming	Some lime content
>5	Strong, long-lasting foaming	High lime content

$v_{Ca} = m_{ca}/m_d$
m_{Ca} mass percentage of total carbonates, based on m_d
m_d dry mass of the sample

cooled in the desiccator, it is weighed to determine the weight loss and the lime content v_{Ca} is evaluated.

Quantitative testing can also be carried out by X-ray diffraction (Sect. 2.2.3.4).

Influencing variables. The presence of lime in a soil limits the activities of the clay minerals. This leads to a decrease in the water absorption capacity and plasticity of the construction soil. A certain amount of lime forms a stable lime matrix between the coarser grains after drying which can improve the strength properties of earth building structures.

Lime can be added to a soil mix in order to purposefully change the properties of available construction soils for specific applications (Sect. 3.1.2.4).

In desert and semidesert soils (Sect. 2.1.2.6), the natural lime content can be quite significant. Bazara [21] has detected proportions of more than 20 % in construction soils from the wadi Hadramaut/Yemen. At comparatively low dry bulk densities (dominant silt grains), these soils can at times reach high degrees of dry compressive strength. More than 8 N/mm^2 (Shibam) have been measured in specimen cubes with an edge length of 8 cm [22].

Lime-rich soils have also been used for earth building purposes in southern England [30].

Water-Soluble Salts

Terminology. In addition to lime, other water-soluble salts can occur naturally in soils, particularly sulfates (gypsum), chlorides, and sodium and calcium nitrates. Moisture moves these salts through the building element where they crystallize on the surface when the water evaporates.

Test methods. These salts are damaging to building elements and can be detected in laboratory tests using the chemical substances of nitrate of silver $AgNO_3$ and barium chloride BaCl.

Influencing variables. Crystallization causes a softening of the structure on building element surfaces which leads to the different types of damage shown in Sect. 5.2.1.2, Fig. 5.13. Depending on their water solubility, they form different horizons (Fig. 5.12).

The characteristic property of "harmful concentration" generally refers to soluble anions in the individual salts and is expressed by different levels of contamination. The following classification in Table 2.11 [25] refers to plasters.

The permissible concentration of damaging salts for industrially produced earth building materials is given as follows according to DIN 18945–47:

Nitrate	<0.02 mass in %
Sulfate	<0.10 mass in %
Chloride	<0.08 mass in %

The total percentage of damaging salts should not exceed 0.12 mass in %.

Table 2.11 Contamination levels of plasters with damaging salts

No.	Sulfate [mass in %]	Chloride [mass in %]	Nitrate [mass in %]	Concentration [mmol/kg]	Evaluation
1	Up to 0.024	Up to 0.009	Up to 0.016	Up to 2.5	Level 0—no contamination
2	Up to 0.077	Up to 0.028	Up to 0.05	Up to 8.0	Level I—low contamination
3	Up to 0.24	Up to 0.09	Up to 0.16	Up to 25.0	Level II—medium contamination
4	Up to 0.77	Up to 0.28	Up to 0.50	Up to 80.0	Level III—high contamination
5	Above 0.77	Above 0.28	Above 0.50	Above 80.0	Level IV—extremely high contamination

Table 2.12 Classification of soils with organic matter according to DIN 18128

Loss on ignition v_{gl} [%]	Designation
<5	Inorganic soils
5–30	Organogenic soils or soils with organic matter
>30	Organic soils (e.g., peat)

$v_{gl} = \Delta m_{gl}/m_d$

m_d dry mass of the sample before ignition

Δm_{gl} mass loss during ignition based on m_d

Different earth building regulations in other countries, however, specify the permissible content of water-soluble (damaging) salts found in construction soils in the range of 1–2 mass in % (e.g., [31, 32]). This is more than ten times the value given in [33]. These regulations mainly evaluate the possible effects of salts on chemically acting additives, particularly cement. It seems that possible damaging effects of the salts on the structure itself played only a minor role when these values were determined.

Organic Matter

Terminology. Organic matter in the soil consists of living soil organisms and dead plant and animal material as well as soil-specific conversion products (humic matter) (Sect. 2.1.1.2). These decomposition and intermediate products are commonly referred to as “humus.”

Test methods. The proportion of organic matter in a soil is determined and evaluated with the help of the loss-on-ignition method according to DIN 18128 (see also loss on ignition, lime content, Sect. 2.2.3.4) (Table 2.12).

Influencing variables. Organic matter increases the water absorption capability of construction soils considerably and, with this, their plasticity. It also reduces the dry compressive strength of earth building materials.

Organogenic and organic soils in the form of cut sod were traditionally used in Scandinavia, Great Britain, and especially Ireland for the construction of houses [24]. With the roots facing up, the sod was stacked like masonry units into one-story load-bearing wall structures (sod houses). These houses became increasingly stronger as the sod dried (Sect. 4.3.3.1).

2.2.4 *Extraction, Transport, and Quality Monitoring*

2.2.4.1 Extraction

Before extracting construction soil the organic topsoil needs to be removed and deposited separately from the earth building soil. Organic matter, tree roots, gravel lenses, etc. also need to be screened out in order to avoid mixing them with the construction soil.

In traditional earth building, construction soil was extracted manually with the help of hoes, spades, and shovels, a method which was very labor intensive and not very effective. The amount which could be extracted per day depended on the deposit density of the construction soil: a group of 10 workers with 15 wheelbarrows could extract 3.2 m^3 of loosely compacted and around 0.8 m^3 of very densely compacted soil [34].

Today, construction soil is extracted mechanically with the help of suitable equipment such as bulldozers with blades, front-end loaders, and scraper dozers (Fig. 2.35a). With these machines one person can extract approx. 100 m^3 per hour (of densely compacted construction soil) [34, 35].

Thin layers extracted from the naturally formed soil can be processed more easily than thick clods (d>20 cm, Fig. 2.35b). This means that the quality of extraction can influence and shorten subsequent processing in a positive manner.

The extraction of construction soil includes the individual processes of removal, loading, and hauling. According to DIN 18300, the work is divided into classes, based on the soil's state during extraction. Suitable construction soils are mainly class 3 soils (easy to extract soil types) and class 4 (moderately difficult to extract soil types). Class 2 (liquid soil types) and class 5 (difficult to extract soil types) are only marginally suitable as construction soil.

2.2.4.2 Transport

Before transport, the weight of the excavation material needs to be determined. These calculations can be carried out on the basis of the empirical values for bulk density γ stated in DIN 1055-2 (Table 3.5).

As a general guideline, transit between the individual technological processes should be kept as short as possible. In traditional earth building, the ideal case was the direct processing of the excavated soil from the foundation pit into earth building materials. These were transported manually by carrying or with the help

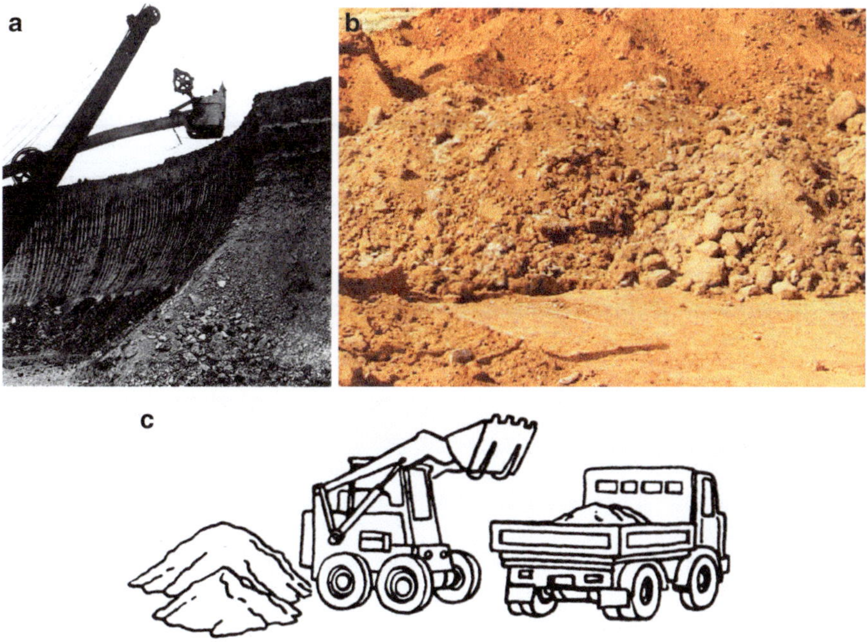

Fig. 2.35 Soil extraction and transport [34, 35]. (**a**) Extraction with the help of an excavator. (**b**) Extracted soil with agglomerate sizes $d > 200$ mm. (**c**) Transport via wheel loader and truck

of wheelbarrows. For shorter distances <5 km draft animals were also used. Using a wheelbarrow, one person can transport approximately 11 m^3 of construction soil per day [34].

Modern earth building can only be economical if manufacturing is centralized and the production volumes are sufficiently high in order to guarantee adequate inventories. This inevitably leads to longer transit routes >5 km requiring transport by trucks and tractor trailers (Table 1.2) (Fig. 2.35c).

2.2.4.3 Quality Monitoring

Once a construction soil is extracted from its deposits, it becomes a building product with defined properties which are subject to the respective inspection process. This process mainly involves suitability testing and quality monitoring.

Suitability Test

The suitability test is the initial assessment of the soil (in its deposit) in order to determine if it might be suitable for use as construction soil (Sect. 2.2.3). This test needs to be carried out if a previously untested soil is intended to be used as construction soil. In Germany, the suitability test is based on the Lehmbau Regeln [6].

Quality Monitoring

Quality monitoring is a continuous inspection of the characteristic properties of the construction soil which is used as (one) raw material for the industrial production of earth building materials. In 2011, the German Association for Building with Earth published a guideline for quality monitoring of construction soil as a raw material for industrially produced earth building products [26]. This guideline is used on a voluntary basis, but its application can also be required by the responsible certification or inspection body if there are considerable fluctuations in the quality of earth building products. Using this guideline makes it possible to identify and track fluctuations in soil composition during extraction.

Characteristic Properties

In Germany, the "producers" of construction soils (e.g., brickyards, clay pits) are currently not required to issue a declaration of characteristic properties as defined by the Construction Products Regulation. Relevant information is typically published in product data sheets but with a focus on the ceramic industry (chemical analysis) as the principal customer.

The properties of construction soil relevant for quality monitoring at the time of extraction are according to [6]:

- Plastic/cohesive properties (Sect. 2.2.3.2)
- Grain size distribution (Sect. 2.2.3.1)
- Humus content by smell test and visual test (Sect. 2.2.2.2)
- Damaging salts (Sect. 2.2.3.4)

For very lean construction soils, plasticity/cohesion tests are only suitable to a limited extent. The same is true for testing the linear degree of shrinkage (Sect. 2.2.3.3) and the dry compressive strength (Sect. 3.6.2.2) of soils which are at least semi-rich. In these cases, other test methods may also be used as long as these tests are able to show fluctuations in the soil composition with sufficient precision.

Testing for damaging salts contained in a soil at a concentration high enough to affect the quality of the earth building materials only needs to be conducted if there is suspicion of the existence of these salts in the given soil deposit.

Inspection Intervals

The producer, or a third party nominated and contracted by the producer, is responsible for the plant's internal monitoring of the relevant properties of the construction soil.

Testing needs to be carried out semiannually or no less than once per 1000 tons of extracted soil. The test results need to be documented and saved for a minimum of 5 years. If the producer has long-term experience in terms of the uniformity of the deposit, the frequency of the tests can be reduced to once per 3000 tons of soil. However, testing needs to be increased if the deposit exhibits frequent geological disturbances.

References

1. Klengel, K.J., Wagenbreth, O.: Ingenieurgeologie für Bauingenieure. VEB Verlag f. Bauwesen, Berlin (1981)
2. Gellert, J.F., et al.: Die Erde—Sphären, Zonen und Regionen, Territorien. Urania-Verlag, Leipzig (1982)
3. Klengel, K.J.: Frost und Baugrund. VEB Verlag f. Bauwesen, Berlin (1977)
4. Schroeder, H.: Problems of the classification of tropical soils. In: Planning and building in the tropics, Heft 84, pp. 103–133. Schriften der Hochschule f. Arch. u. Bauwesen Weimar (1990)
5. Horn, A., Schweitzer, F.: Tropische Böden als Baugrund und Baustoff. In: Vorträge Baugrundtagung, pp. 9–28. Dt. Ges. f. Erd-u. Grundbau, Berlin (1978)
6. Dachverband Lehm e.V. (Hrsg.): Lehmbau Regeln—Begriffe, Baustoffe, Bauteile, 3 überarbeitete Aufl. Vieweg+Teubner/GWV Fachverlage, Wiesbaden (2009)
7. Dietzschkau, A.: Zielgruppen für mögliche Lehmprodukte aus der Kiesgewinnung—eine praktische Marketingkonzeption. Unpublished diploma thesis. FB Wirtschaftswissenschaften, Westsächs Hochschule (FH), Zwickau (1998)
8. Freyburg, S.: Recherche zur Eignung von bau-und grobkeramischen Lehmen und Tonen für die Lehmbauweise. Unpublished study. Institut f. Bau-und Grobkeramik, Weimar (1983)
9. Fuchs, E., Klengel, K.J.: Baugrund und Bodenmechanik. VEB Verlag f. Bauwesen, Berlin (1977)
10. Schnellert, T.: Untersuchung von Transportprozessen der Einbaufeuchte in Baukonstruktionen aus Stampflehm während der Austrocknung. Unpublished diploma thesis. Bauhaus-Universität, Fak. Bauingenieurwesen, Weimar (2004)
11. Muhs, H.: Die Prüfung des Baugrundes und der Böden. In: Mitt. Degebo TU Berlin, H. 11. Springer, Berlin (1957)
12. Houben, H., Guillaud, H.: Earth Construction: A Comprehensive Guide. Intermediate Technology Publications, London (1994)
13. Plehm, H.: Beitrag zur Frage des Einflusses der Verdichtung auf Berechnungswerte von bindigen und schluffigen Erdstoffen. In: Mitteilungen der Forschungsanstalt für Schifffahrt, Heft 31. Wasser und Grundbau, Berlin (1973)
14. Smoltczyk H. (Hrsg.): Grundbau-Taschenbuch, Bd. 1 Grundlagen, 3. Aufl. Ernst & Sohn, Berlin (1980)
15. Bobe, R., Hubaček, H.: Bodenmechanik, 2. Aufl. VEB Verlag f. Bauwesen, Berlin (1986)
16. Krabbe, W.: Über die Schrumpfung bindiger Böden. In: Mitt. der Hannoverschen Versuchsanstalt für Grundbau und Wasserbau, Franzius-Institut der TH Hannover, Heft 13, pp. 256–342. (1958)
17. Horn, A.: Schwellversuche an expansiven Böden. In: Mitt. Degebo TU Berlin, H. 32, Beiträge zur Bodenmechanik und zum Grundbau aus dem In-und Ausland, pp. 81–87
18. Niemeyer, R.: Der Lehmbau und seine praktische Anwendung. Reprint der Originalausg. von 1946. Ökobuch Verlag, Grebenstein (1982/1990)
19. Schroeder, H.: Klassifikation von Baulehmen. In: LEHM 2000 Berlin; Beiträge zur 3. Int. Fachtagung für Lehmbau des Dachverbandes Lehm e.V., pp. 57–63. Overall-Verlag, Berlin (2000)
20. Kezdi, A.: Handbuch der Bodenmechanik, Bd. 1 Bodenphysik. VEB Verlag f. Bauwesen, Berlin (1969)
21. Bazara, A.: Erarbeitung einer Konzeption zur konstruktiven und funktionellen Rekonstruktion der Stockwerk-Lehmkonstruktionen in der VDR Jemen. Unpublished diploma thesis. Hochschule f. Architektur und Bauwesen, Sektion Bauingenieurwesen, Weimar (1986)
22. Bazara, A.: Bautechnische Grundlagen zum Lehmgeschossbau im Jemen. Berichte aus dem Konstruktiven Ingenieurbau, Heft 29. Fraunhofer IRB, Stuttgart (1997)
23. Цытович, Н.А.: Механика грунтов—краткий курс (Soil mechanics—a brief summary), 3-е изд. Высшая школа, Москва (1979)
24. McDonald, F., Doyle, P.H.: Ireland's Earthen Houses. A. & A. Farmar, Dublin (1997)

25. Scholz, J.; Zier, H.-W.: Schäden an den Fresken Moritz von Schwinds im Palas der Wartburg und Möglichkeiten der Restaurierung. In: Europäischer Sanierungskalender 2009 (4. Jahrgang), pp. 343–359. Beuth Verlag, Berlin (2009)
26. Dachverband Lehm e.V. (Hrsg.): Qualitätsüberwachung von Baulehm als Ausgangsstoff für industriell hergestellte Lehmbaustoffe—Richtlinie. Technische Merkblätter Lehmbau, TM 05. Dachverband Lehm e.V., Weimar (2011)
27. Fontaine, L., Anger, R.: Bâtir en terre—Du grain de sable à l'architecture. Citè des sciences et de l'industrie, Édition Belin, Paris (2009)
28. Krause, E., Berger, I., Nehlert, J., Wiegmann, J.: Technologie der Keramik, Bd. 1: Verfahren—Rohstoffe—Erzeugnisse; Krause, E., Berger, I., Plaul, T., Schulle, W.: Technologie der Keramik, Bd. 2: Mechanische Prozesse; Krause, E., Berger, I., Kröckel, O., Maier, P.: Technologie der Keramik, Bd. 3: Thermische Prozesse. VEB Verlag f. Bauwesen, Berlin (1981–1983)
29. Mönnig, H.U., Fischer, F., Schroeder, H., Wagner, B., Mücke, F.: Spezifische Eigenschaftsuntersuchungen für Erdbaustoffe einschließlich der Eigenschaftsverbesserung durch Modifikation und Möglichkeiten des Feuchtigkeitsschutzes für in Erdstoff errichtete Gebäude. Unpublished research report. Hochschule f. Arch. u. Bauwesen, WBI, WB Tropen-u. Auslandsbau, Weimar (1988)
30. Pearson, G.T.: Conservation of Clay & Chalk Buildings. Donhead, Shaftesbury (1997)
31. Regulation & Licensing Dept., Construction Industries Div., General Constr. Bureau: 2006 New Mexico Earthen Building Materials Code. CID-GCB-NMBC-14.7.4. Santa Fe, NM (2006)
32. Bureau of Indian Standards: Indian Standard 2110-1980: Code of practice for in situ construction of walls in buildings with soil-cement (1991 1st rev., 1998 reaffirmed). New Dehli (1981)
33. Dachverband Lehm e.V. (Hrsg.): Lehmsteine—Begriffe, Baustoffe, Anforderungen, Prüfverfahren. Technische Merkblätter Lehmbau, TM 02. Dachverband Lehm e.V., Weimar (2011)
34. International Labour Organisation ILO (Ed.): Techn. Series TM No. 6. Small scale brickmaking Techn. Series TM No. 12. Small-scale manufacture of stabilized soil blocks. ILO, Geneva (1984–1987)
35. Rigassi, V.: Compressed Earth Blocks, vol. 1: Manual of Production; Guillaud, H.u.a.: Compressed Earth Blocks, vol. 2: Manual of Design and Construction. Vieweg-Verlag, Gate/CRATerre-EAG, Braunschweig (1995)
36. Schneider, U., Schwimann, M., Bruckner, H.: Lehmbau für Architekten und Ingenieure—Konstruktion, Baustoffe und Bauverfahren, Prüfungen und Normen, Rechenwerte. Werner-Verlag, Düsseldorf (1996)
37. Dachverband Lehm e.V/Abu Dhabi Authority for Culture & Heritage (Ed.): Planning and Building with Earth—course handbook for architects, engineers, conservators & archaeologists. Unpublished. Weimar (2009)
38. Schroeder, H.; Bieber, A.: Neue Stampflehmprojekte in Thüringen. In: LEHM 2004 Leipzig, Beiträge zur 4. Int. Fachtagung für Lehmbau, S. 190–201, Weimar: Dachverband Lehm e. V

3

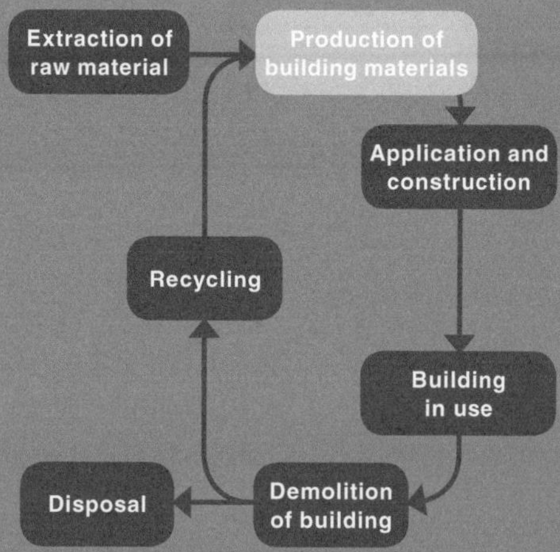

Earth Building Materials – Production, Requirements and Testing

After sourcing the construction soil from its natural deposits, earth building materials are prepared. This process includes various methods of preparation, shaping and drying, and turns construction soil into an *earth building material.*

Earth building materials are unshaped or shaped building materials made of unfired construction soil with or without aggregates or additives. Their suitability for a specific project must be confirmed through appropriate testing.

© Springer International Publishing Switzerland 2016
H. Schroeder, *Sustainable Building with Earth*,
DOI 10.1007/978-3-319-19491-2_3

3.1 Processing of Construction Soil

Construction soil is processed in order to turn the extracted soil into a homogeneous, workable material which can be further prepared and shaped. The addition of appropriate aggregates and additives improves the soil's properties. In this process, the construction soil is also mixed with water in order to attain the required consistency.

When processing the excavated soil, it is important to break down and blend the existing structure of the soil with its local disturbances. This includes natural geological layers and layers created during excavation. During this process, the clay minerals contained within the very fine grain particles in the form of agglomerates are allowed to absorb additional water molecules. This loosens their bond to the coarser grains and makes the soil easier to process. The quality of processing determines the properties which can be attained in the ensuing production of earth building materials.

The general goals of processing raw clay mineral materials also apply to the production of fired bricks. This means that many of the preparation, shaping, and drying methods are similar, except for one difference: at the end of the technological chain there is a fired brick which, in addition to special requirements placed on the raw material, also requires more energy for firing. The development of industrial production methods for fired bricks in the second half of the nineteenth century also improved processing procedures which guaranteed a high-quality product. Earth building, however, did not benefit from this development.

There are natural and mechanical processing methods. While natural processing typically consists of a wet process, mechanical processing methods can be both wet and dry.

3.1.1 Natural Processing

During natural processing the construction soil is exposed to the prevailing weather conditions. Here, time is a key factor. The physical and chemical processes change ("break up") the structure of the construction soil. They are caused by the effects of exposure to sun and frost and by the rotting and decomposition of organic particles contained within the soil.

In the cultures of Central Asia, China, and Japan, natural processing methods were among the building tasks with the highest level of responsibility. They were time intensive, required considerable care, and could, at times, take several years [1].

3.1.1.1 Winter and Summer Weathering

During the winter weathering process, freezing water inside the pores of the soil increases in volume, causing the natural structure of the soil to break apart. In preparation for winter weathering, the material is piled into heaps of approx. 1 m in the fall. Further mechanical processing of the soil after the winter is generally not necessary. This method of processing requires sufficient and suitable storage facilities and adequate time in the construction schedule for at least one winter season.

Similarly, for weathering in summer, the heaped construction soil is exposed to fluctuations in temperature and moisture which lead to corresponding deformations caused by swelling and shrinkage. This also results in a loosening of the clay particles which are attached to the coarser-grained mineral particles.

3.1.1.2 Soaking

The process of soaking involves mixing the construction soil with water and allowing it to rest for a period of time. The resulting swelling breaks apart the structures formed by the clay minerals, making the material easier to work with. In the ceramics industry, extracted soil is filled into soaking pools or bins made of reinforced concrete, mixed with water and mechanically processed. This method can be used to mix various soils, clays, and sands together. After soaking, the soil blend can be sliced out of the mix for further processing.

3.1.1.3 Aging

In contrast to soaking, aging is a biological process of decomposition which involves fermentation. Algae or bacteria grow in the soil or clay material, causing an increase in plasticity. This process can be enhanced by adding suitable additives (Sect. 3.1.2.4). Humus particles in the soil have the same effect.

3.1.2 Mechanical Processing

In the past, soil was manually crushed using human and animal muscle power and the help of simple tools. Today, a full range of machine systems is available, some of which originated in unrelated industries (horticulture and agriculture, meat processing and the food industry). These machine systems break the construction soil down into the required agglomerate sizes using different mechanical operating principles and both wet and dry methods.

3.1.2.1 Crushing, Chopping, and Kneading

Coarse crushing of the soil breaks down or crushes clods in the soil ($d > 20$ cm), which have resulted from the excavation process, into agglomerate sizes <2 cm. Historical methods of coarse crushing include stomping (by humans and animals)

Fig. 3.1 Traditional processing of construction soil: kneading and crushing in a pug mill [3]

with the addition of water, as well as breaking up clumps of dry soil using hoes, drop weights, or with the help of animal-operated pug mills reaching a daily output of approx. 10 m^3 [2] (Fig. 3.1 [3]). Hand-operated clay shredders were also used (daily output approx. 3 m^3) (Fig. 3.2 [2]).

Starting in 1850, machine-operated processing systems became more common in the fired brick manufacturing field. Simplified versions of these systems were also introduced into soil processing and are still in use today. Examples of widely used systems are (Fig. 3.3 [4]):

- *Pan grinders/jaw crushers*: Clumps of soil are crushed by counter-rotating rollers or by one fixed and one rotating roller (daily output approx. 7 m^3).
- *Impact hammers*: A horizontally positioned disc with attached steel brackets rotates around its vertical axis at high speed breaking up clumps of soil and solidified loose rock (daily output 15–40 m^3 depending on the specific machine system).

3.1.2.2 Sieving

During sieving, coarsely broken down soil is sorted according to grain or agglomerate size. In this process, unusable rocks and coarse-grained particles as well as organic matter, such as tree roots, are separated from the soil. Soil clumps which are left on the sieve can be broken down mechanically and added to the sieving process again. The screening device used needs to be able to support the coarse material and allow the fine material to pass through its openings.

Depending on the final grain size, common mesh sizes are between approx. 2 and 7 mm. Stationary sieves and screens such as an inclined upright screen or a hand sieve are used for smaller amounts of material which can be moved manually (Fig. 3.4 [5]). The openings of the screen might become clogged if the soil is too wet or contains too much clay.

For larger amounts of material, sieve sorting is carried out by machine-operated sieving systems such as rotating or vibrating sieves (Fig. 3.5 [74]).

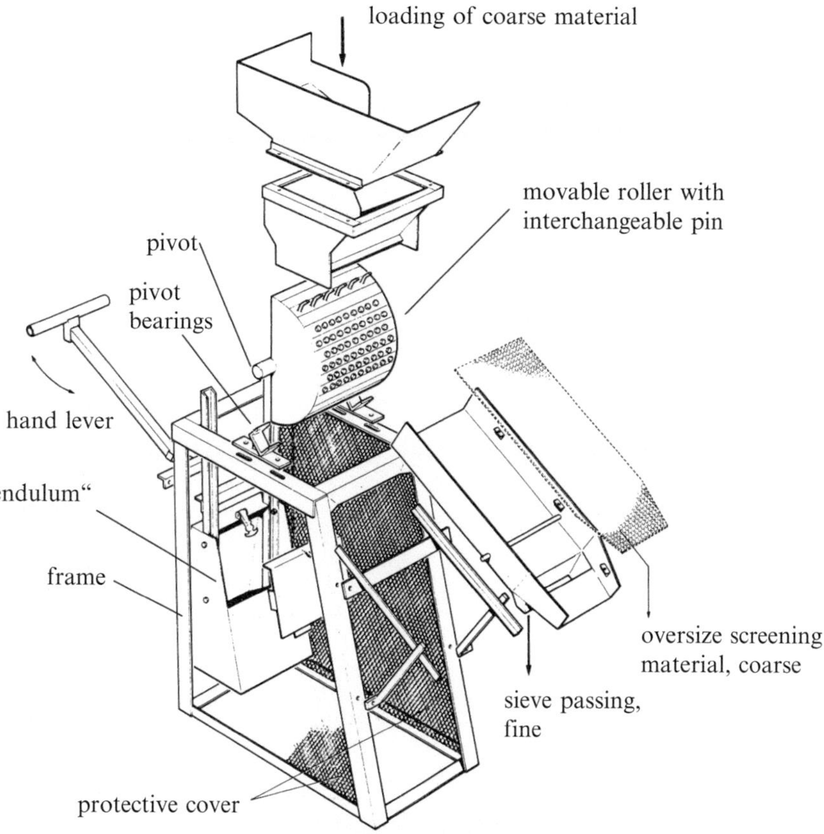

Fig. 3.2 Traditional processing of construction soil: chopping in a clay shredder [2]

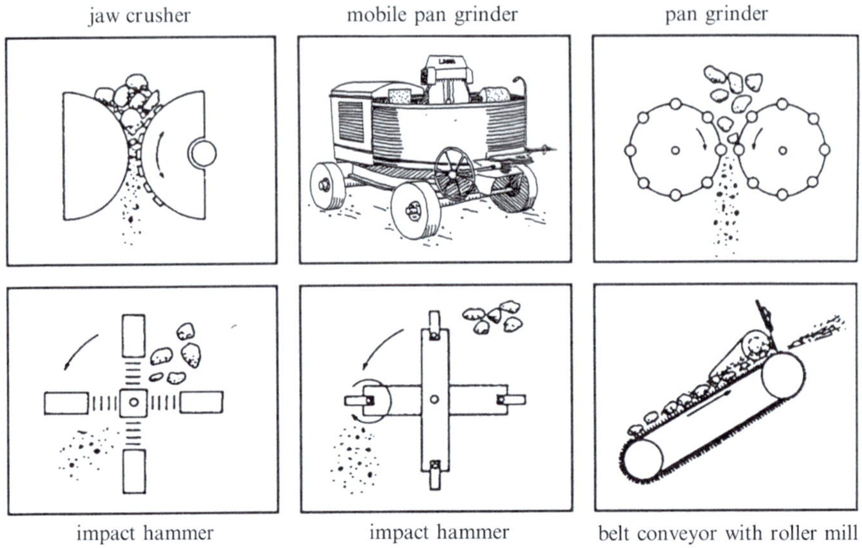

Fig. 3.3 Mechanisms for the processing of construction soil: coarse crushing, breaking, kneading [4]

Fig. 3.4 Processing of construction soil: sorting by manual sieving [5]

3.1.2.3 Grinding and Granulating

A process of fine grinding is used to turn fine-grained wet or artificially dried construction soils into powdered earth building materials with grain sizes $d < 0.063$ mm.

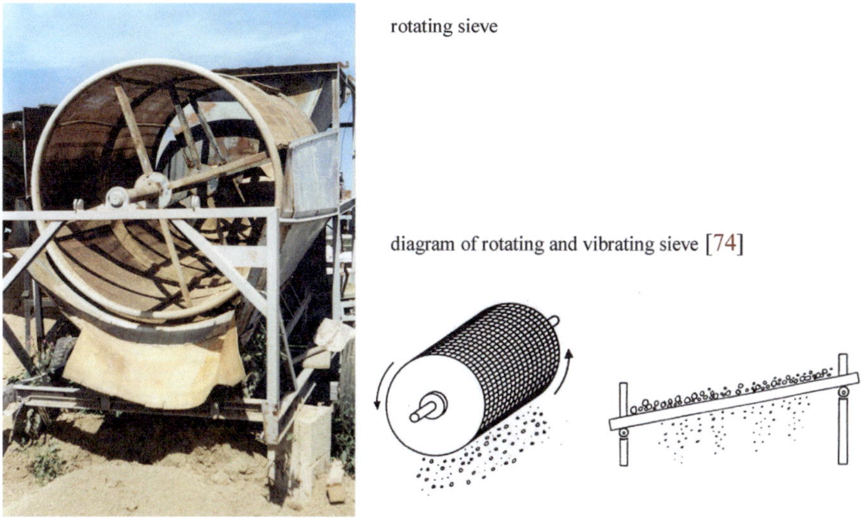

Fig. 3.5 Processing of construction soil: sorting by machine sieving

Fig. 3.6 Powdered clay, forms of supply [7]. (**a**) Granulated powdered clay, (**b**) powdered clay, "bagged," (**c**) silo

Common trade names for this material are "powdered soil and powdered clay" or "milled clay" (Sect. 2.2.1.2). These products can be purchased "bagged" or in silos (Fig. 3.6b, c).

In the ceramics industry, different types of mills are used for grinding raw soil, for example:

Tumbling—or ball mills (Fig. 3.7 [6], *Technologie der Keramik, Bd. 2: Mechanische Prozesse*): A rotating steel cylinder (drum) contains grinding balls made of flint stones (from river beds). The grinding effect takes place when the drum turns and

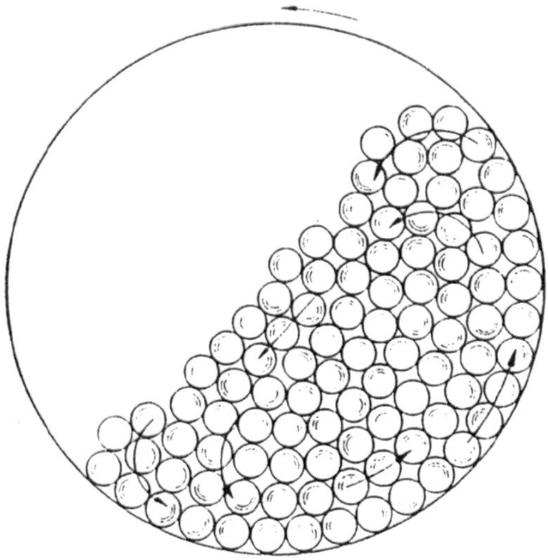

Fig. 3.7 Processing of construction soil: fine grinding in a ball mill, schematic diagram [6]

the balls, which move to the top of the ball cluster, roll down the incline of the material pile. The grinding balls hit each other, grinding down the material which finds its way between the balls. A further grinding effect is caused by the shifting of the balls within the cluster.

Roller mills: Two rollers of equal diameter rotate quickly in opposing directions on parallel shafts. This causes the material to be pulled up into the gap between the rollers set at a width of <1 mm.

The final step in the processing of raw powder can be *granulating*. For this, a semi-wet process is used to agglomerate the heated raw powder to a pellet size of approx. 1–30 mm with the addition of atomized water (Fig. 3.6a [7]).

3.1.2.4 Batching, Combining, and Mixing

Construction soils often do not possess the properties required for their intended use. A number of aggregates and additives can be used specifically for the purpose of improving these properties.

Batching

During the batching process, the construction soil, along with any aggregates and additives (Sect. 3.4.2), is picked up by machines operating in either a volumetric or gravimetric manner and fed onto a downstream conveyor. *Volumetric* batching means that a predetermined volume of solid material per time unit is extracted from the storage pile and fed onto the conveyor. In *gravimetric* batching, on the other

Fig. 3.8 Automatic batching machine for the production of earth mortar (Company Claytec) [8]

hand, the solid material is weighed using appropriate equipment before the speed or extraction surface of the batching is regulated. Figure 3.8 shows a fully automatic batching and mixing plant for the production of earth plaster (Company Claytec) [8].

Batching can also be done manually with the help of simple volumetric measuring devices (buckets, portable boxes).

Combining

During the combining process, different material streams—construction soil, aggregates and additives as well as any required water—are combined according to a predefined formula. Different streams can also be homogenized by pouring them into layers which can then be removed and processed immediately, or after some delay.

Mixing

The goal of mixing is to create a homogeneous and plastic material which retains its composition over a long period of time. To achieve this, the construction soil, the aggregates and additives as well as any required water are mixed via a kneading and

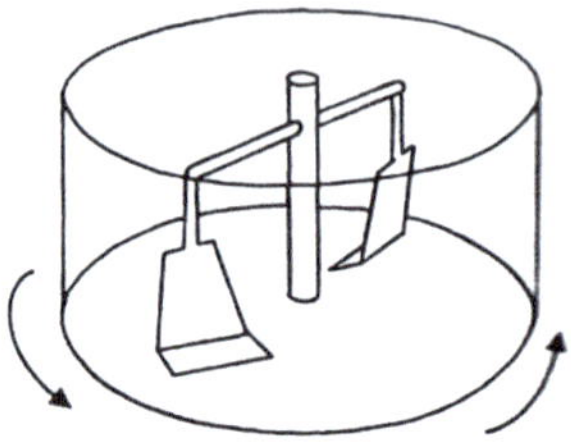

mixing in a pan mixer [9], diagram [74]

mixing with a paddle mixer [9]

traditional mixing method [10]

Fig. 3.9 Processing of construction soil: different mixing methods with aggregates

cutting process. Here, the use of the term "plastic consistency" describes the ability of the material to react to external forces by changing its shape without losing cohesion between the individual components. This ability is created by the adhesive strength of the clay minerals.

Figure 3.9 shows modern and historical techniques for mixing construction soil with aggregates [9, 10].

3.1.2.5 Slurrying

The terms slurrying or dispersing refer to converting construction soil into a liquid material through wet processing by using, for example, an electric-powered paddle mixer. This method disperses the capillary binding forces between the different grains of the construction soil and separates the clay mineral agglomerate coating from the coarse grains. Impurities such as lumps of lime or gypsum are also dissolved or sorted out in this manner.

Fig. 3.10 Pouring clay slurry straw layer [9]

Pouring a clay slurry over lightweight aggregates or submerging these aggregates into the slurry applies a clay mineral coating which, after drying, acts as a binding agent and ensures the dimensional stability of the shaped building material or building element (Fig. 3.10 [9]).

Gravel quarries apply the principle of wet separation of clay mineral coatings from coarse grains in connection with sieve classification with the help of dispersing agents (surfactants). The resulting clay mineral waste product, known as compressed soil, can also be used for the production of earth building materials (Sect. 2.2.1.4).

There is one big disadvantage to applying the slurrying method to break the capillary binding forces between the different grain sizes of construction soil: it uses a lot of water which is typically of drinking quality.

The use of hot water steam is one water-saving method of processing construction soil. Exposure to hot steam heats the material to approx. 90 °C while using a relatively small amount of additional water. This process increases the material's plasticity and, with it, its malleability. Accordingly, the shrinkage deformations of the shaped products during the drying phase are also low. A similar effect is produced by the addition of hot water which heats the material to approx. 30 °C.

Exposing clay or soil material to hot water or steam is a process used in the ceramics industry which is also applied in the production of shaped earth building materials (earth blocks, clay panels). However, such methods are much more energy intensive.

3.2 Shaping

The goal of the shaping process is to produce a shaped earth building material or earth building element with a defined cohesion using a processed and unshaped material mix of plastic consistency. The shaping procedures need to ensure that a shaped product is created which is uniform in terms of its material and structure and can be used as a building material or building element after drying. During the shaping process, inhomogeneities in the shaped product need to be avoided. Examples of inhomogeneities are separation during the mixing of components made of different materials or grain sizes, differences in compaction during the manufacturing of the shaped product (particularly during compression shaping) as well as in the orientation and alignment of anisometric particles, especially of clay minerals (plates and needles).

During the manufacturing of shaped products, various compaction methods should be used in order to remove most of the air or water which is trapped in the pores of the shapeless and typically plastic material. This guarantees the required strength of the building element or structural component after drying. During compaction, not only the frictional resistance of the non-binding sands and gravels needs to be overcome but also the cohesive strength of clay minerals attached to the coarse grains. In this process, the particles move past each other and fill the pores of the loose grain material with fine and very fine mineral particles. This is only possible if the soil is sufficiently wet or if the applied level of compaction is high enough.

3.2.1 *Aspects of Shaping*

The shaping methods for earth building materials can generally be grouped together based on two aspects:

- According to the *format design* of the shaped products as modular building materials intended for further processing or as complete building elements.
- According to the *consistency* or the moisture content of the material mix.

Modular format design consists of the production of earth building materials as elements, blocks, panels, or clumps made of unshaped, generally plastic earth building material. Several or many of these individual building elements are assembled in a wet or dry state with or without masonry mortar to form a finished earth building element according to the rules of masonry construction.

Building element format design is characterized by the production of complete building elements out of unshaped, plastic earth building materials by direct shaping. This area can be divided into direct manual shaping without formwork and shaping with formwork, which involves the placement and compaction of the earth building material in layers.

The ceramics industry differentiates between the following types of shaping based on the consistency or moisture content of the material mix ([6], *Technologie der Keramik, Bd. 2: Mechanische Prozesse*):

- *Compression shaping*: The materials form a relatively dry and shapeless mass with a grainy, powdery consistency without any noticeable cohesion and a moisture content of <15 mass in %
- *Plastic shaping*: The material is plastic and malleable and characterized by obvious cohesion of the construction soil and its aggregates, moisture content is in the range of approximately 15–25 mass in %
- *Shaping by casting*: The material is prepared as a viscous-stable, pourable suspension (slurry) with a moisture content of approx. 25–40 mass in %

These classifications can generally also be applied to the shaping processes used in earth building.

3.2.2 Technological Procedures

Table 3.1 provides a general overview of the technological shaping procedures used in earth building with regard to the aspects of format design, consistency, and moisture content of the materials, as well as the required compaction pressure when using machines ([6], *Technologie der Keramik, Bd. 2: Mechanische Prozesse*). As there are many compaction devices on the market, this table only shows a selection of typical examples.

The technological shaping procedure used in the production of the individual earth building material becomes part of the building material's properties and needs to be declared in its designation (e.g., for earth blocks according to DIN 18945).

3.2.2.1 Compression Shaping

The compression shaping method is used in modular format design as well as building element format design. The consistency of the material is solid to semisolid. Figure 3.11 [4, 11] gives an overview of the technological procedures of compression shaping with the help of compaction devices.

Table 3.1 Shaping methods for earth building materials, according to ([6], *Technologie der Keramik, Bd. 2: Mechanische Prozesse*])

No.	Type of shaping	Technological process	Format design: Building material	Format design: Building element	Consistency CI	Moisture content [mass in %]	Compaction pressure [MPa]	Machine	Figure/source
1	**Compression**				*Solid—semisolid*	*<8–15*			
1.1		Dry compression	x		Solid	<8	<40	Tile press	[12], [13]
1.2		Moist/wet compression	x		Semisolid	8–15	1–2	Lever press	Fig. 3.12
							4–6	"Lightweight" press	Fig. 3.13a
							<20	"Heavy-weight" press	Fig. 3.13b
1.3		Superimposed load		x	Semisolid	<10		Flat roller	Fig. 3.11
1.4		Tamping/vibration	x x	x x	Semisolid	<10		Hand tamper Electric/pneumatic tamper	Figs. 3.14a, b, 3.16
				x x				Vibrating table, vibrating plate Sheepsfoot vibration roller	Fig. 3.11 Fig. 3.15
2	**Plastic**				*Soft–stiff*	*15–25*			
2.1		Manual shaping	x	x	Soft	15–25		Manual	Figs. 4.35, 3.22
2.2		Hand-throwing, manual	x		Soft	15–25	Impulse	Wooden form, molding table	Fig. 3.24a, b
2.3		Extrusion shaping	x x		Soft—stiff		0.5–5.0 10.3	Extrusion press	Fig. 3.25a Fig. 3.25c
2.4		Spraying		x	Soft	~18	15–20 m/s	Spray gun	Fig. 4.32
3	**Casting**				*Paste-like*	*25–40*			
3.1		Grid casting	x		Paste-like	25–40		"Egg layer"-machine	Fig. 3.27b
3.2		Strip casting	x		Paste-like	25–40		Strip casting plant	Fig. 3.27c

Compaction method	Compaction tool	Diagram
static compaction by superimposed load / compression compaction	press, smooth-wheel roller	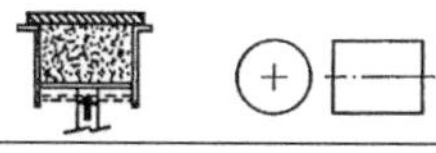
impulse compaction or ramming compaction	pneumatic and electric rammer, hand tamper	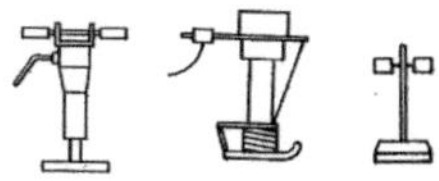
compaction by vibration	vibrating plate compactor	
combination of static compaction / compaction by vibration	vibrating sheepsfoot and grid rollers	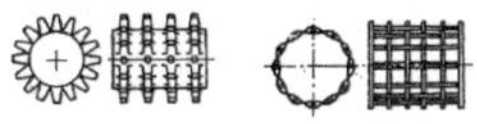

Fig. 3.11 Compaction tools and methods for compression shaping of earth building materials, overview [11], [4]

Modular Format Design

This type of format design is used in the production of earth blocks and clay panels. The material is compacted in a sturdy, metal shaping chamber under the application of a uniaxial static compressive load. It is also possible to apply pressure from two sides using uniform compression movements in a rigid/cushioned shaping chamber.

Compression shaping is listed as the characteristic property "compression-molded (p)" in the declaration of earth blocks according to DIN 18945.

The various compression shaping methods can be divided into "dry compression" and "wet or moist compression" depending on the applied level of compression and the moisture content of the material.

Dry Compression

A free-flowing and relatively dry material is compacted at a moisture content of <8 % and a compression level of up to 40 MPa. Under these conditions, the prepared soil material is no longer plastic. The solid soil components slide against each other and shift into a more densely packed mass while most of the pores become filled. With an increase in compression, the solid particles are additionally plasticized or broken apart. This increases the density of the particle packing further. Some specific earth building products are manufactured using this method, such as clay panels used for interiors [12, 13].

The advantages of this shaping method are the high mechanical strength of the earth building materials, the negligible shrinkage deformations as well as the cost- and time-saving aspects in terms of drying. Dry compression requires high compaction which can only be achieved by using suitable, cost-intensive systems.

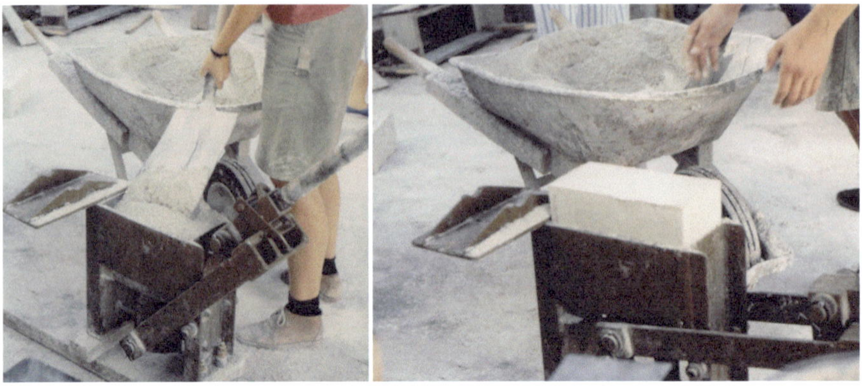

Fig. 3.12 Shaping with a lever press [5]

Wet or Moist Compression

Wet or moist compression is carried out at a moisture content of 8–15 % and a compression level of up to 20 MPa. At this level of compression the material becomes free flowing and plastic. Compaction is limited by the incompressible water content.

Nowadays, many different press types are available for wet or moist compression which can be classified according to various criteria, such as power, compaction level, daily output, and mobility.

Hand and Lever Presses: The simplest presses are manually powered punch presses which contain a compaction chamber for shaping one or two earth blocks. Crumbly material is first poured into a sturdy shaping chamber. By applying compression, the earth building material achieves its final shape. During this static compaction process, the platelike clay mineral particles align in the direction of the tensile forces applied through compaction, in other words, perpendicular to the applied compressive force. The building material "remembers" this process which means that the load placed on the dried building material during its use should only be applied in the direction of the initial compression.

A very popular press in developing countries is the CINVA Ram (Fig. 3.12 [5]), developed by the Colombian engineer Pablo Ramirez and patented in 1957. The advantages of this lever press are numerous: its ease of operation and transportation, also in rough terrains, its independence from the power grid, its relatively low initial cost, and, most of all, the superior quality of the earth blocks it produces. By today's standards, the productivity of this shaping method is very limited. It requires 3–5 people for optimal production in order to achieve a daily output of 300 blocks or more. The applied compression level is in the range of 1–2 MPa.

a

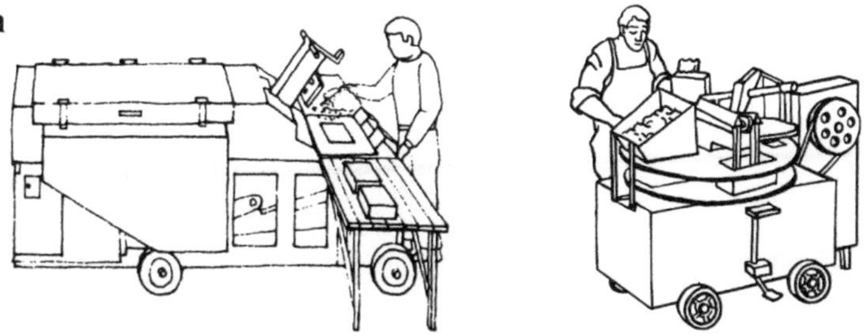

mobile "light-weight" earth block presses, electric/diesel power [14]:
fixed table with a single shaping chamber rotating table with several shaping chambers

b

Terrablock press: mobile "heavy-duty" production unit, motor / hydraulic power [2], [14]

Fig. 3.13 "Lightweight" and "heavy-duty" earth block presses, selected examples

Motorized "Lightweight" Presses: One way to classify these presses is according to the design of the work table (Fig. 3.13a [14]):

- A fixed table with a single shaping chamber
- A rotating table with several shaping chambers

These presses are mobile, powered by electricity or diesel, with a daily output of approx. 800–3000 earth blocks. Hydraulic presses with comparable output numbers are also available. The applied compression level is in the range of 4–6 MPa.

Hydraulic "Heavy-Duty" Production Units: The "Terrablock" press shown in Fig. 3.13b is a hydraulically powered "heavy-duty" mobile production unit. It has a daily output of approx. 7500 earth blocks. Production is fully automatic and computerized. The hopper holds construction soil for approx. 10 min of continuous production. Integrated sieves filter out large clumps of soil which did not get broken up during the coarse grinding phase, as well as rocks and organic matter. A vibrating system facilitates the continuous transport of the construction soil from the lower

a manuel

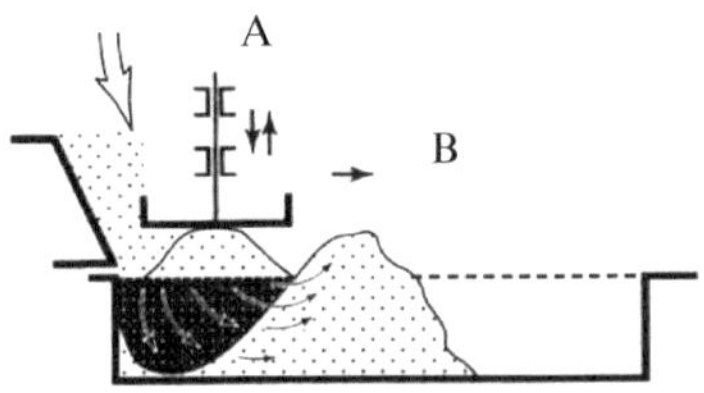

b mechanized

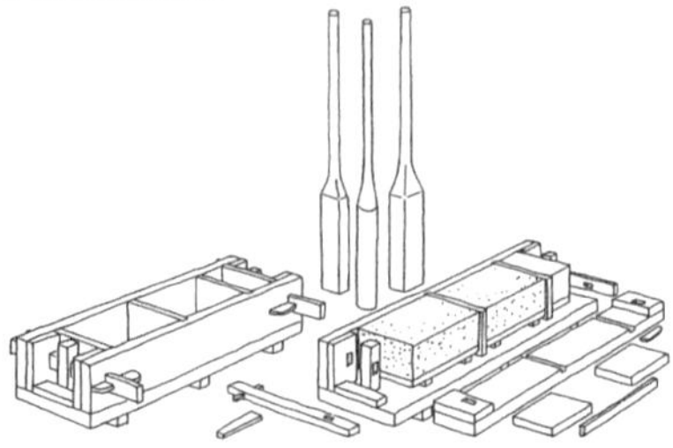

Start position
A rammer motion during compaction
B direction of rammer motion during compaction in relation to steel formwork

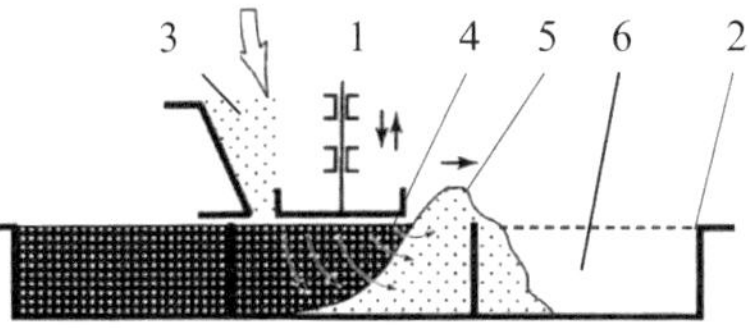

1 rammer
2 steel formwork
3 loading of soil mixture
4 compacted wedge of soil mixture
5 uncompacted soil mixture
6 unfilled part of steel formwork

Ongoing compaction process

Fig. 3.14 Shaping of earth blocks: ram compaction [10, 15]

part of the hopper to the shaping chamber. A hydraulic press compacts the prepared construction soil inside the shaping chamber. When compressing is finished the shaped product is automatically ejected from the shaping chamber onto a conveyor belt. The applied compression level can reach up to 20 MPa [2].

Ram Compactors: They represent a special technology for producing "compacted" earth blocks: the method is part of the category "modular format design/wet and moist compaction" but is classified as "ram compaction" in the following paragraph. "Compacted" earth blocks can be produced manually or mechanically.

Figure 3.14a [10] shows molds and compaction tools for the manual production of "rammed" earth blocks. This method is suitable for the production of large-format earth blocks.

Figure 3.14b shows the principle of a mechanized ramming system for the production of earth blocks [15]. The steel formwork is continuously filled with earth building material during the compaction process. An expanding wedge of compacted earthen material is built up in the direction of ramming in front of the rammer. This wedge of compacted earth pushes the uncompressed earthen material forward until it is reached by the rammer and also compacted. Such a system can produce approx. 250 earth blocks per hour at the dimensions of 390 × 190 × 90 mm (~1.7 m^3) (http://ruskachely.ru).

Format Design of Building Elements

For the format design of building elements using compression, the building element is shaped by adding layers of the earth building material to a formwork (for wall structures) or to a space with lateral boundaries (for floors and ceilings) which are subsequently compacted. The interior dimensions of the formwork represent the final dimensions of the finished building element. Suitable compaction methods are "static or compression compaction" as well as "rammed compaction or compaction by vibration."

Static or Compression Compaction

For static or compression compaction, the earth building material is subjected to a pulsating load and compacted by static rollers inside the formwork or the lateral boundaries.

Smooth-Wheel Rollers produce compaction in the effective range of depth by applying a load. At the same time, however, they introduce a horizontal shear stress which can lead to wavy deformations of the material layer and lateral cracks in the roller's direction of motion.

Grid and Sheepsfoot Rollers combine the static load of the roller's weight with a kneading effect. This results in effective compaction, particularly in connection with cohesive earth building materials. The kneading effect is created by the special design of the roller.

The roller drum of a *grid roller* is surrounded by a steel grid. The grid-like design of the roller drum makes the roller surface three-dimensional thereby preventing the shear stresses which typically occur with smooth-wheel rollers.

Sheepsfoot rollers are also called spiked or bristle rollers. The roller drum is covered with rectangular or truncated feet with oval or angular soles. The compaction effect is caused by the application of pinpoint pressure and horizontal kneading. This method pushes air and water out of the coarse pores within the uncompressed earthen material. The holes created by the sheep's feet through pressure (and possibly also vibration) increase the surface area of the earthen material considerably. This allows more water to evaporate before the next layer of soil is added. Sheepsfoot rollers compact the material from bottom to top causing the penetration depth of the feet to decrease and the feet to lift as the number of passes increases.

The rollers described above can also be designed as *vibration rollers*. In this case, static compaction is combined with dynamic compaction by means of exciter systems.

In recent years, grid rollers and sheepsfoot vibration rollers have also been successfully employed in various rammed earth construction projects (Fig. 3.15 [8]).

Fig. 3.15 Shaping of building elements by compacting rammed earth with a sheepsfoot roller [8]

Ramming or Impulse Compaction/Compaction by Vibration

The "ramming" method of shaping is used both in modular format design for the production of earth blocks (as described above) and in the production of complete building elements (building element format design).

Construction soil prepared to a crumbly, free-flowing consistency with a moisture content of <12 mass in % is loaded into formwork boxes, formwork frames, or formwork for specific building elements. After the soil is added in layers, manually or using a special pouring device, it is compacted with the help of ramming tools and a specific ramming frequency. These ramming tools can be manually or mechanically operated. After the formwork is removed, the shaped products and building elements are allowed to air-dry.

Hand rammers have been traditionally employed in rammed earth construction and are still in use today. Their net weight is between 5 and 8 kg with a contact area of 100–200 cm^2.

Pneumatic and electric rammers move in "bounce" and "impact" phases creating a combination of ramming and vibration between the upward bounce and the landing of the press. This results in the compaction of the earth building material within the formwork or lateral boundaries.

Today, pneumatic and electric rammers are used for processing rammed earth material. Their net mass is limited to a maximum of 15 kg because of the compaction pressure they apply to the formwork. The rate of frequency of these rammers is

Fig. 3.16 Compaction of rammed earth using a pneumatic rammer

700/min (Fig. 3.16). Electric rammers do not require a compressor for their use on the construction site. Their tamping plates are made of steel or hard rubber with a round or square contact area.

Vibrating Rollers and Vibrating Plates: The compaction effect caused by *vibrating rollers* is determined by a number of technical parameters with regard to the compaction tool and the properties of the rammed earth material. The impulses which are applied to the rammed earth material in rapid succession (>1200/min) have the ability to momentarily reduce adhesion within the coarse-grained range. This allows larger pores to be filled by finer grains and causes excess pore water pressure or increased air pressure in the finest grain range. Through this, the molecular cohesive forces are partially neutralized and pore water is brought to the surface. Sufficient superimposed loads and amplitudes can result in a denser particle structure. However, it is important to pay attention to the compaction pressure on the formwork.

In addition to vibrating rollers, *vibrating plates* can attain the same compaction effect. They consist of a wear-resistant base plate with a reinforced rim which the motor and vibrator are securely attached to. Their exciter power is higher than their net mass which causes them to lift off of the surface. Similar to smooth-wheel rollers, this can cause wavy deformations of the material layer and lateral cracks in the direction of motion of the vibrating plate.

Formwork Systems

Formwork Systems for the format design of earth building elements can be divided into temporary and permanent formwork systems. Depending on the earth building materials and the type of compaction, different aspects of the formwork system need to be considered.

Temporary formwork can be removed immediately after the earth building material has been poured in and compacted as long as the newly shaped earth building element is strong enough. In contrast, *permanent* formwork remains within the wall structure and typically serves as the substrate for (earth) plaster. It should be permeable and should not significantly impede drying.

Rammed earth construction is related to monolithic concrete construction. It requires sufficiently rigid formwork which consists of side panels made of wood or wooden composites serving as temporary movable or climbing formwork.

The formwork panels in traditional rammed earth construction are strengthened with vertical braces. Wooden or steel ties or anchors are attached horizontally to the top end of the braces. The braces keep the formwork panels together and absorb the soil compaction pressure. The ties guarantee the correct spacing for the required wall thickness. At the bottom end of the braces, the anchors and the formwork panels rest on the rammed wall, and have to be removed after completion of the rammed wall section (Fig. 3.17).

traditional shaping, Ait Benhaddou, Marokko

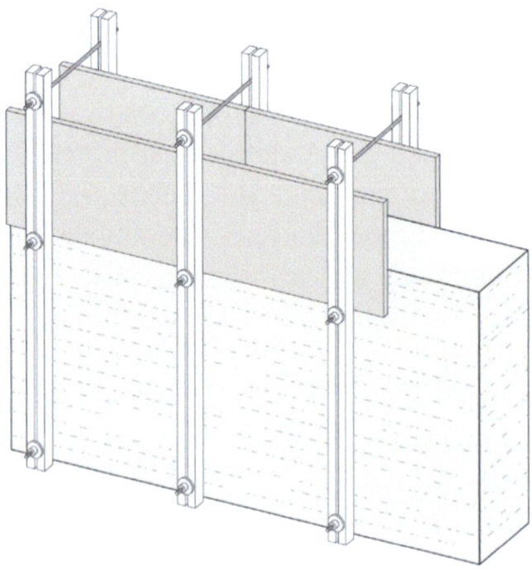

basic arrangement of formwork system [75]

Fig. 3.17 Traditional shaping in rammed earth construction using formwork panels

Fig. 3.18 Prefabricated rammed earth wall elements

When using formwork systems in rammed earth construction the following aspects need to be considered:

- The formwork panels must not be allowed to bend outward during compaction.
- The formwork must be easily adjustable.
- The individual parts of the formwork must be easy to transport.

Modern rammed earth construction uses formwork systems common to concrete construction. They are designed for a formwork pressure of approx. 60 kN/m^2. Before the rammed earth material is poured in, the interior surfaces of the formwork are treated with linseed oil which has proven to be an effective release agent. The removal of the formwork completes the shaping of the rammed earth building element.

Figure 3.16 shows the formwork for a rammed earth wall section which is made up of two layers of fiberboard in a curved line. The second, exterior board easily overlaps the joints of the interior formwork panels.

A special application of this technique is the industrial prefabrication of wall elements made of rammed earth ($d \geq 250$ mm). Using special installation technology, these wall elements can be assembled into ceiling-high, load-bearing and non-load-bearing wall structures (Fig. 3.18) (Sect. 3.5.8). The formwork and compaction techniques are the same as those used for producing rammed earth walls.

Straw-Clay and Light-Clay Construction: Straw-clay and light-clay construction use both temporary and permanent formwork. Even though the level of compaction for these earth building materials is less compared to rammed earth, the formwork

Fig. 3.19 Shaping during the installation of light straw clay: formwork panels as temporary formwork

needs to be similarly strong. Stability is attained by ensuring that the vertical posts of the support frame as well as the non-load-bearing fill frame are positioned correctly (Sect. 4.2.3.2).

Temporary formwork is used for light clay with organic fiber additives and allows for the removal of the formwork immediately after installation. The formwork can easily be attached to the framing members (for example, using clamps) (Fig. 3.19).

Permanent formwork should be used for "lighter" light-clay systems because they typically do not have the necessary stability for the formwork to be removed immediately. Reed mats are suitable for this type of formwork (70 stalks/m). As the light-clay construction work progresses, the mats are continuously unrolled and attached to the timber frame (Figs. 3.20 and 4.45). Other suitable formwork can be made of lime-bound lightweight panels with organic fiber material. For exposed exterior timber framing, a combination of temporary exterior formwork and permanent interior formwork is possible. Lightweight panels should be diffusion-open and should not significantly impede drying.

Cob: The traditional cob building technique was also used to produce building elements, but mostly without the help of formwork. Cob building elements were shaped by using a pointed spade or shovel to apply a perpendicular or flush cut to the protruding and still moist earthen material of the erected wall. Figure 3.21 [16] shows this type of shaping in the Uzbek version of cob building called "pakhsa."

Fig. 3.20 Shaping of walls made of wood-chip light clay using permanent formwork [9]

Fig. 3.21 Traditional shaping of cob walls (Uzbek: Pakhsah) by cutting [16]

3.2.2.2 Plastic Shaping

Plastic shaping methods are primarily used in modular format design. For these methods, the consistency of the material ranges from soft to stiff and the applied compaction pressures are lower than those used during compression shaping.

a

Duenner mud loaf method: production and stacking of mud loafs (Germany) into wall structures [3]

b

"pottery technique": production and stacking of earth clumps (Ghana) [76]

Fig. 3.22 Production and use of clump-shaped earth building elements

In addition to the specific compaction method used, building material declarations for industrially produced earth blocks must identify the dimensions of the blocks with a letter symbol (Sect. 3.5.7, in Germany generally the common DIN formats). Furthermore, solid blocks and perforated blocks (in DIN 18945 identified by the letter symbol "g") need to be differentiated between.

Manual Shaping

The oldest and most primeval shaping method in earth building is the production and use of manually shaped earth clumps without defined dimensions. These earth clumps are made out of construction soil with a plastic consistency and are found in a range of shapes: strips, balls, stone-like, and slab-like (Figs. 1.1–1.3). The clumps were used in load-bearing walls or as earthen infill in a wet (without mortar, Fig. 3.22a) or dry state (with masonry mortar). Formwork was not required for these applications.

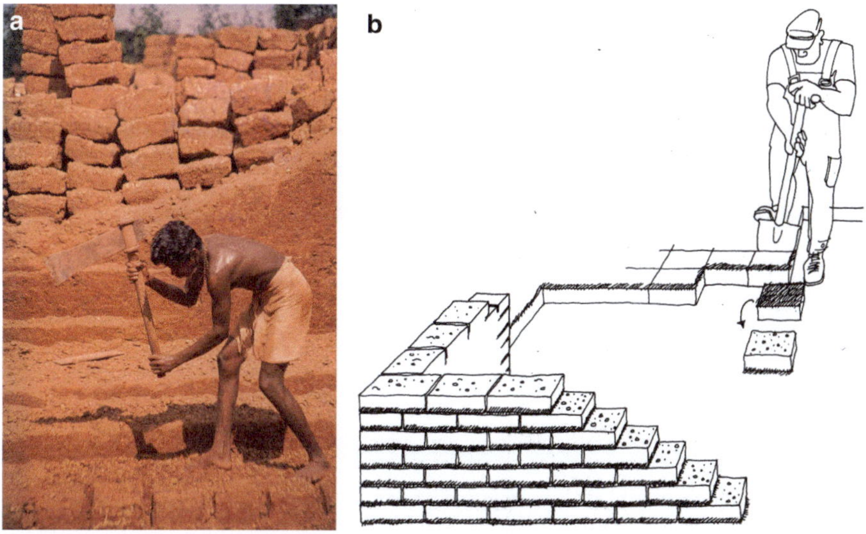

cutting of air-hardened laterite blocks [Wikipedia]

cutting of sod blocks and wall construction by stacking with the roots facing up [17]

Fig. 3.23 Manual shaping of block and slab-like building elements by cutting

Today, direct manual shaping is still widespread in traditional earth building, particularly in Africa. Here, earth clumps with a plastic consistency are stacked in layers to form walls. The “joints” which form between the clumps are evened out in order to form a smooth wall surface (Fig. 3.22b).

In the wet grassland regions of Northern Europe and North America, square chunks of sod and turf were cut out of the ground with spades to be used in the construction of houses. They were stacked into load-bearing walls with their roots facing up. These blocks have different local names which include “turfs” in Ireland, “sods” in Great Britain, and “terrones” in the USA (Fig. 3.23b [17]).

Laterites are a fringe earth building material (Sect. 2.1.2.6). With a suitable clay mineral content, they can be shaped in their naturally moist state. When exposed to air, however, they harden irreversibly and are quarried manually using a pick ax or machines and cut to the desired dimensions, similar to easily removable solid rock (Fig. 3.23a).

Hand-Throwing

The “hand-throwing” shaping method represents the transition to earth blocks with defined dimensions due to its use of formwork. This method is labeled as “hand-thrown (f)” as a characteristic property in the designation of earth blocks in DIN 18945.

b

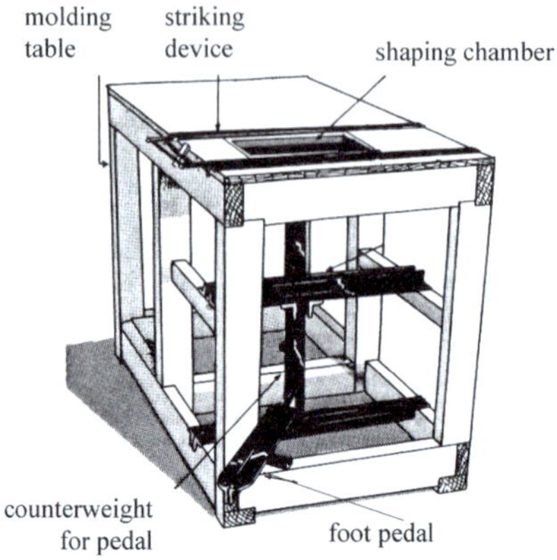

hand-throwing using molding table [2]

throwing of earth into the shaping chamber [18]

a

hand-throwing into wooden forms [16]

Fig. 3.24 Shaping of earth blocks: hand-throwing

For *traditional* hand-throwing, the earth building material is prepared to a soft consistency of 15–25 % by mass, manually thrown with force into a shaping chamber or frame which is then struck with a board without any further compaction ("hand-molded block," Fig. 3.24a [16]). The momentum applied during hand-throwing aligns the clay mineral platelets perpendicular to the direction of the motion. The formwork can usually be removed immediately and as soon as the earth blocks are strong enough they can be turned up on edge to air dry.

Molding tables represent an advancement over simple formwork frames, particularly in regard to occupational physiology. Here, workers do not have to bend down to work on the ground but can instead stand at a table (Fig. 3.24b).

Examples of technological improvements in molding tables are

- A foot pedal for removing the block from the form.
- Trimmers in slide rails for cutting off excess material.

Molding tables were also equipped with a closing plate and a lever for applying pressure to the plastic material inside the shaping chamber. This completes the transition to the hand lever press which allows for the application of higher compression [2, 18].

During the process of *mechanical* throwing, the plastic material is discharged from a hopper onto a conveyor belt in volumetrically metered "batches." The conveyor belt then accelerates and throws the "batch" into a steel formwork which gets completely filled with the earthen material. Excess material is struck off by a wire. No additional compaction occurs.

Extrusion Shaping

Earth blocks can also be shaped in extruders: the prepared soil mix is added, compacted with a mixing auger, and vented in a vacuum chamber until it is finally pushed through an extrusion nozzle forming a continuous block which can be cut to the desired dimensions. This method is labeled "extruded (s)" as a characteristic property in the designation of earth blocks in DIN 18945.

Extruders can be manually or mechanically filled and operated, and are designed as stationary or mobile units (Fig. 3.25). Historical extruders were driven manually

a

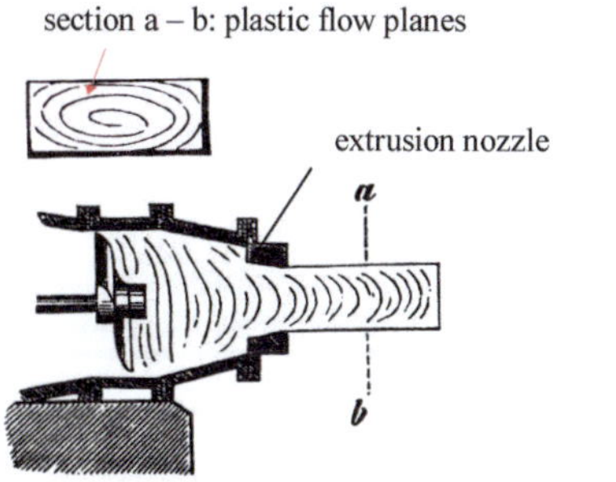

standard extruder with spiral drive shaft [19]

b

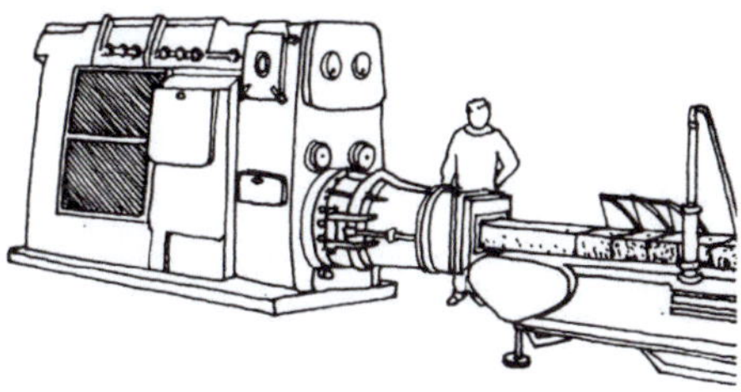

stationary extruder with worm gear drive [14]

c

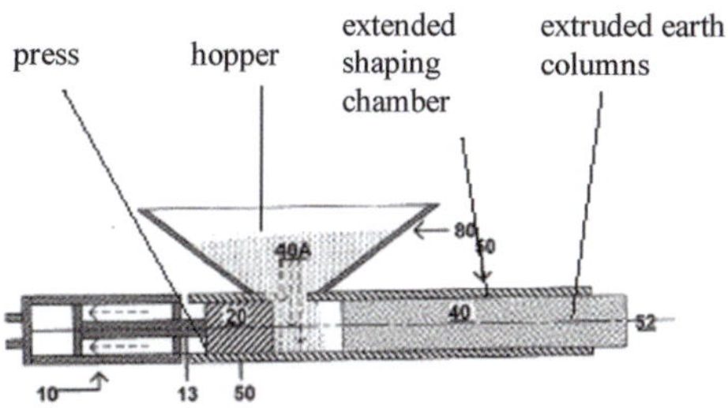

extruder with hydraulic press, EarthCo Megablock system with earth block element compared to standard earth block format [20]

Fig. 3.25 Shaping of earth blocks: extruding. Section *a–b*: plastic flow planes or by draft animals. In traditional fired brick extrusion, the brick is advanced by a spiral drive shaft (Fig. 3.25a), in modern facilities, hydraulic presses with worm gear drives are used (Fig. 3.25b).

Extrusion is the common shaping method used by the fired brick industry. Here, the material consists of pure clays. After the shaping process, the "green" bricks are fired at temperatures between 800 and 1000 °C which make them water resistant. The amount of pressure applied depends on the consistency of the material (soft to stiff) and ranges from 0.5 to 5.0 MPa [6]. Stiffer materials result in more dimensionally stable products with better drying characteristics but also require higher compression forces.

During the shaping process in the extruder, slip surfaces with an increased moisture content form between the more solid areas of the material. This is particularly true for pure clays. These slip surfaces lead to the formation of shell-like structures within the brick along the plastic flow planes (Fig. 3.25a [19]). These structures blend together during the firing of the bricks.

Unfired "clay" bricks or "green" bricks form shrinkage cracks when drying. This leads to spalling in the case of renewed water penetration, e.g., in exterior walls which are exposed to precipitation. Examples of such damage after World War II are documented in [18] (Fig. 5.20).

In the USA, earth blocks up to 3 m in length at defined wall thicknesses are produced in addition to common earth block formats. These extruded columns are used as finished wall elements. Figure 3.25c shows the EarthCo Megablock System with a block cross section of 30 × 46 cm compared to a standard earth block format [20]. For this technique, the material is loaded into a hopper and slides into the pressing chamber which contains an extended shaping chamber. The stiff earthen material is compressed into a continuous block by the alternating piston strokes of the press and pushed into the extended shaping chamber. The frictional resistance of the walls of the shaping chamber (approx. 150 cm in length) allow for the application of up to 10.3 MPa of compression force resulting in high compaction of the earth "megablocks" produced.

Spraying

In earth building, the "spraying" shaping method is particularly popular for the production of building elements including wall and ceiling infill, wall linings as well as reinforced load-bearing wall construction (with steel) (Fig. 4.32) [21]. Today, the "plaster" building element is also applied through spraying, using earth plaster mortar and sprayed earth plaster [22].

For this technique, the material is prepared from a mixture of construction soil and aggregates, possible additives and water. It is mixed to a consistency suitable for pumping. As with the spraying method used for the application of plaster mortars, the material is then sprayed under high pressure into a formwork for building elements (or onto a plaster base) in single or multiple layers. The required compaction of the earth building material is achieved by the mechanical impact of the soil mix

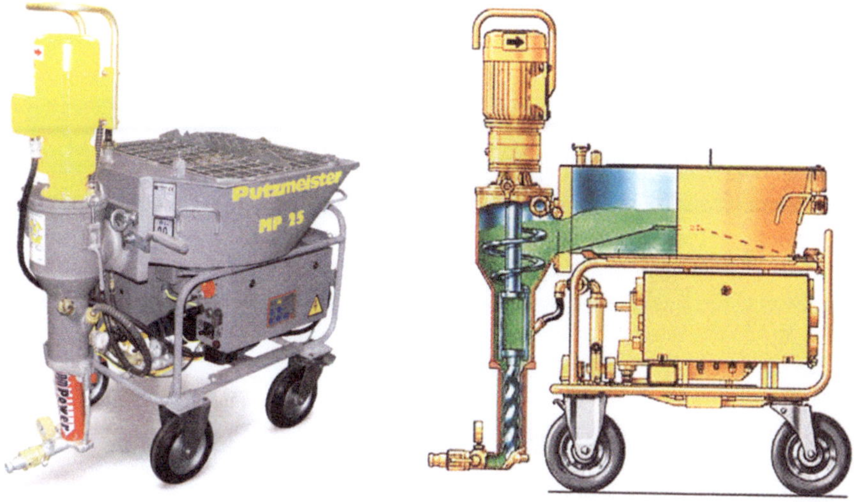

Fig. 3.26 Plastering machine (Putzmeister system) consisting of a continuous mixer with horizontal star wheel drive and mixing helix

onto the building element formwork or the base. The machine systems used for this technique include a pan or continuous mixer (Sect. 3.1.2.4) and a mortar pump. The spraying pressure as well as the composition and consistency of the soil mix can be regulated [23, 24] (Sect. 4.3.3.2).

Figure 3.26 shows a plastering machine with a horizontal star wheel which transports the initial dry material to a mixing chamber with the help of a mixing helix. The soil is then mixed with water inside the mixing chamber. In order to achieve better homogenization of the material additional post-mixer aggregates can be added before the mix is transferred to the first mortar hose. The cohesive strength of the clay minerals ensures the adhesion of the material to the substrate during application as well as the dimensional stability of the building element while drying.

3.2.2.3 Casting

Casting is used in earth building as a shaping method for building elements. For casting, the material has a paste-like consistency and a moisture content between 25 and 40 % by mass. No additional static compaction takes place, but vibration compaction through agitation of the grid molds is possible.

Grid Casting

The technology of grid casting for the production of earth blocks was developed by the German-American Hans Sumpf in the Southwestern USA in the 1930s. This technology is still popular today in New Mexico and Arizona.

The process can be compared to slipform construction of concrete pavement: an earth building material of viscous consistency is manually or mechanically poured into a steel grid mold which is placed on a firm surface. The material is then struck off and compacted by vibration (Fig. 3.27a, b [14, 25]). The mechanical version of this method produces up to 20,000 earth blocks daily or up to 1.5 million per season. The machine systems are also referred to as "laydown" or "egg layer" machines.

The arid climate in this region speeds up the drying process of the earth blocks. Depending on the weather, the required drying time is 1–3 weeks.

Instead of the grid-shaped molds, a simple frame mold can also be used. Here, the "earth slab" is laid down on a firm surface and cut into blocks with a plate cutter while still in a plastic state. The earth blocks are allowed to dry on site until they are strong enough to be transported and stacked.

Strip Casting

The "casting" shaping method can also be applied to produce thin clay panels. During the shaping process, belt presses are used to remove water from the viscous earthen material. Reinforcement can be integrated into the clay panels to increase their flexural strength and their stability during transport. Figure 3.27c shows a plant for the production of clay panels (Claytec-clay panel, Company Muhr).

3.3 Drying of Earth Building Materials and Earth Building Elements

Immediately after shaping, earth building materials and earth building elements are not dimensionally stable and display low levels of strength. They only achieve their intended structural and physical properties in a dry state which means that the mixing water necessary for the processing and shaping of the earth building materials needs to dry out again. Shaping and subsequent drying thus complete the production process of earth building materials.

3.3.1 Drying Process

The drying process of an earth building material or an earth building element can generally be divided into three phases (Fig. 3.28 [6], *Bd. 3: Thermische Prozesse*):

Phase 1: The pore water flows to the surface of the building element and the drying process begins when the water changes from a liquid to a gaseous state and is released into the surrounding air by means of convection.

a)

b)

loading the formwork manually (a) and using an "egg layer" – machine (b) [25] with prepared viscous earth building material

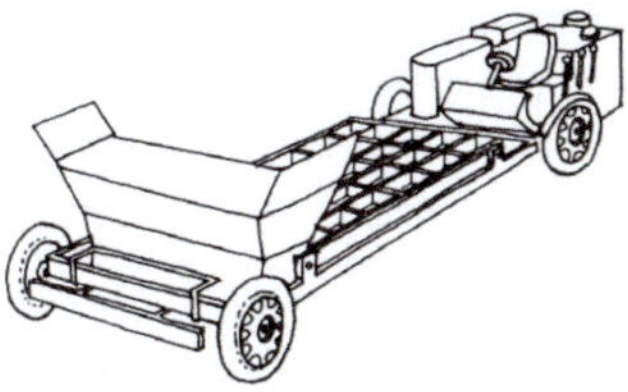

Diagram "egg layer" – machine [14]

c) plant for the production of clay panels (CLAYTEC clay panel, Company Muhr).

Fig. 3.27 Shaping of earth blocks: casting

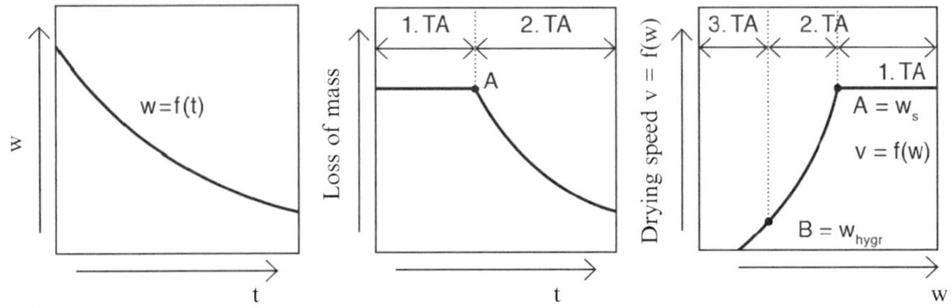

Fig. 3.28 Drying process in earth building materials and earth building elements (according to [6])

Phase 2: As the capillary menisci break down, the evaporation surface moves into the interior of the building element. Capillary moisture migration extends from the inner core to the "evaporation level" (Point A). The water vapor which is formed there diffuses to the surface of the building element through the drier and lighter-colored layer which becomes increasingly thicker. At the surface, it is absorbed by the air and transported by convection. In this manner, the evaporation level eventually moves all the way to the core of the building element.

In the second phase of drying, a loss in weight begins to take place which becomes visible "on the outside" as shrinkage deformation (Sect. 2.2.3.3) and cracking. In soil mechanics, this state is described as the water content reaching the shrinkage limit SL (Sect. 2.2.3.2).

Phase 3: When the maximum attainable *hygroscopic moisture content* w_{hygr} (completely dry soil) is reached, the drying level (Point B) disappears. This moisture content refers to moisture which is absorbed directly from the air by the building material or building element through capillary pores. It levels off depending on the prevailing humidity and temperature. When buildings are in use the *practical moisture content* w_c (also known as the continuous moisture content) refers to the moisture content which eventually sets in as the average value within the building element (Sect. 5.1.2.4).

3.3.2 Drying Speed

The drying speed of a wet earth building element depends on a number of factors: the initial moisture content, the thickness of the building element, the prevailing weather and local conditions for natural drying as well as the way water is bound to the mineral substance of the earth building material with its respective clay mineral structure. Therefore, average drying times for earth building materials and earth building elements can only be given as rough time frame estimates ([22], 1st and 2nd editions).

The drying process is considered ideal when the pore water at the evaporation level or the water vapor on the surface of the building element evaporate at the same speed at which moisture is being transported from the interior of the building element. The same is true for shrinkage deformations: the decrease in volume of a sample which shrinks uniformly and on all sides corresponds to the volume of the evaporated water quantity.

In strong sunlight during the summer, the evaporation speed on the surface is typically higher than the speed of the moisture transfer in the interior of the building element. In most cases, sun exposure is uneven due to different prevailing conditions (such as the geometry of the building elements or external shading). The inhomogeneity of an earth building material can also lead to uneven drying conditions. The formation of cracks on the surface of the building element is a visible result of this process. In the worst case, these cracks can be several centimeters wide and run the complete length of the building element (e.g., very rich soils with a high moisture content at the time of installation (Fig. 4.24)). It is therefore very important to protect the surfaces of building elements from direct sunlight immediately after the formwork has been removed, for example, by covering them with tarps.

The opposite can be true in conditions of high humidity and poor air flow. Here, convection, particularly on the interior building element surfaces, is not high enough to initiate a moisture transfer through evaporation from the interior to the surface of the building element. In such situations, earth building materials containing organic aggregates, in particular, can be subject to mold. Similar scenarios might develop if drying can only occur in one direction (e.g., a light-clay panel lining >30 cm attached to an old building) or if the building elements are particularly thick. It is therefore essential to provide sufficient ventilation, especially for interior building elements.

At the Bauhaus University in Weimar, Germany, the drying time and drying process of seven specimens of different soil mixtures as well as a rammed earth wall section were examined [26]. The specimens were dried naturally. The samples for determining the moisture content were taken from the center of each specimen. The tests also identified the influence of the initial moisture content on the strength development of the specimens (Fig. 3.47, Sect. 3.6.2.2).

Figure 3.29a shows the drying process of the specimens at different initial moisture content levels over the course of 90 days and the respective measured values (Table 3.2). The loess soil specimens I and II (without aggregates) display little change in moisture content during the first phase of drying. After 2 weeks, the moisture content starts continuously decreasing before the residual water content level is reached after 90 days. This corresponds to drying phase A in Fig. 3.28.

The rammed earth specimens with straw fibers V–VII show a continuous decrease in moisture content from the beginning of the drying time, even in the center of the specimens. It appears that the capillary water transfer from the center of the specimens to the surface is aided by additional transport channels along the fibers.

After 90 days, the results for the rammed earth mixtures without (III and IV) and with straw fibers (V–VII) do not vary considerably, showing a residual water con-

a drying process / changes in moisture content over the course of 90 days in specimens of different earth building materials

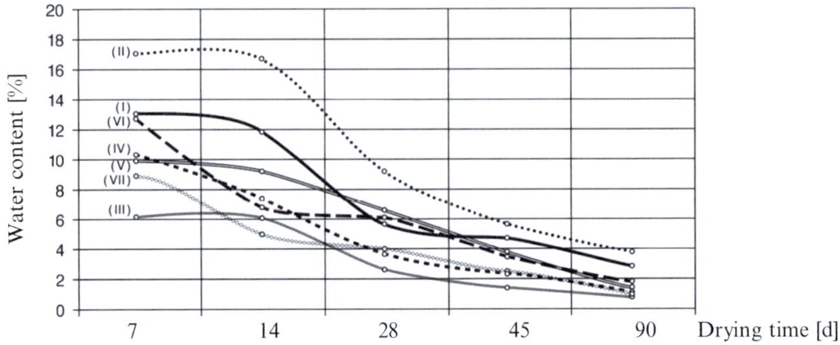

(I) loess soil, natural, $w \sim w_{pr}$
(II) loess soil, natural, $w > w_{pr}$
(III) rammed earth with coarse aggregate, $w \sim w_{pr}$
(IV) rammed earth with coarse aggregate, $w > w_{pr}$
(V) rammed earth with coarse aggregate+ straw fibers, $w \sim w_{pr}$
(VI) rammed earth with coarse aggregate+ straw fibers, $w > w_{pr}$
(VII) rammed earth with coarse aggregate+ straw fibers, $w < w_{pr}$

b moisture profile in the rammed earth test wall during the drying phase [26]

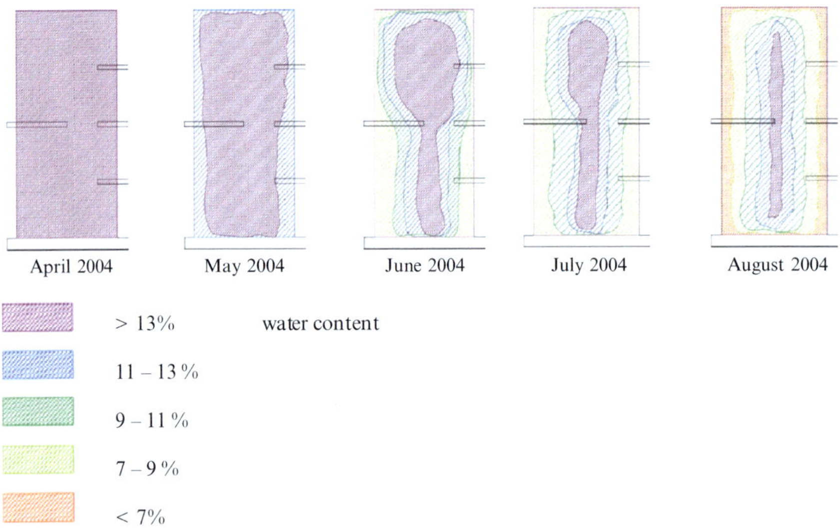

Fig. 3.29 Drying process in a rammed earth test wall

tent of <1.7 %, independent of their moisture content at the time of installation. Both loess soil samples I ($w \sim w_{Pr}$) and II ($w > w_{Pr}$) are much higher, with a residual water content of 2.8 and 3.8 %. The results for w_{Pr} for loess soil and rammed earth with and without fiber aggregates vary accordingly.

Table 3.2 Drying process and changes in water content in different earth building material specimens over the course of 90 days, measured values [26]

No.	Mix	Water saturation	Moisture content at time of installation	Moisture content after 7 days	Moisture content after 90 days
I	Loess soil, natural	$w \sim w_{Pr}$	0.147	0.131	0.028
II	Loess soil, natural	$w > w_{Pr}$	0.203	0.172	0.038
III	Rammed earth with coarse aggregate	$w \sim w_{Pr}$	0.099	0.061	0.007
IV	Rammed earth with coarse aggregate	$w > w_{Pr}$	0.124	0.106	0.013
V	Rammed earth with coarse aggregate + straw fibers	$w \sim w_{Pr}$	0.123	0.101	0.014
VI	Rammed earth with coarse aggregate + straw fibers	$w > w_{Pr}$	0.155	0.128	0.017
VII	Rammed earth with coarse aggregate + straw fibers	$w < w_{Pr}$	0.093	0.088	0.009

Figure 3.29b shows the development of the moisture profile in the rammed earth test wall while drying over the course of 90 days. Even after 6 months of natural drying, the moisture content at the core of the 50 cm thick rammed earth wall was still the same as the initial moisture content at the time of installation.

3.3.3 Drying Methods

In terms of energy economy, *natural* drying (air drying without additional artificial heating) of earth building materials which have been processed in a wet state is the most energy efficient, but also the most time-consuming drying method (Fig. 3.30). In this context, producers of earth building materials face the following challenges: depending on the prevailing climate, long drying times are required as well as large amounts of space in order to ensure the continuous operation of the production machines.

Artificial drying of earth building materials (e.g., in tunnel or chamber dryers) is generally more cost-effective, particularly for large and continuous production volumes, because it shortens the drying time. This results in a more efficient use of machines and a reduction in the required space for drying. However, it has a negative impact on the energy balance for the production of the earth building materials (Sect. 1.4.3.2). Brick yards for fired bricks might have an advantage here because they generally have suitable drying systems at their disposal.

When working with earth building materials which are used in a wet state (plaster, walls, floors), artificial drying with sufficient ventilation is often used, independent of seasonal conditions. In winter, artificial drying is always required because of

Fig. 3.30 Natural drying of earth blocks

the risk of frost, but it is also particularly effective because the outside air is often very dry. In summer, certain weather conditions with high humidity levels make it almost impossible to naturally dry indoor spaces to prevent mold. Therefore, the drying process should be monitored and documented (Sect. 4.3.6.3).

The method of drying also plays a role in testing: to prepare specimens used for determining dry compressive strength, DIN 18952-2 recommended storing the specimens for 5 days under standard atmospheric conditions followed by artificial drying at 80 °C until the residual moisture level is reached. According to [27], however, artificial drying reduces the compressive strength by up to 30 %. It remains to be seen if the method of drying also affects the strength properties of naturally and artificially dried earth blocks.

3.4 Designation, Certification, and Production Control

The Lehmbau Regeln [22], developed by the Dachverband Lehm e.V. and officially approved by the building authorities, define consistent designations and corresponding letter symbols for earth building materials in Germany according to Table 3.3. The material name is part of the earth building material declaration.

In addition, the designation system also describes the characteristic properties of the respective earth building material which are given as classes and descriptions,

Table 3.3 Designation of earth building materials according to the Lehmbau Regeln [22]

No.	Name of building material	Letter symbol
1	Rammed earth	STL
2	Cob	WL
3	Straw clay, clay mixed with fiber	SL, FL
4	Light clay	LL
4.1	– Light wood-chip clay	HLL
4.2	– Light straw clay	SLL
4.3	– Light clay mixed with fiber	FLL
4.4	– Light mineral clay	MLL
5	Earthen loose fill	LT
6	Earth blocks	LS
7	Clay panels	LP
8	Earth mortar	LM
8.1	– Earth masonry mortar	LMM
8.2	– Earth plaster mortar	LPM
8.2.1	– Clay thin layer finishes	LDB [34]
8.3	– Sprayed earth mortar	LSM

Table 3.4 Characteristic properties for the designation of earth building materials, overview

No.	Characteristic properties	Description/Class
1	Place of production	On site or factory mix
2	Aggregates/additives	Mineral and organic
3	Consistency/processing	Wet (semisolid, stiff, soft, paste-like, liquid) and dry (solid)
4	Degree of prefabrication/ format design	Unshaped (earthen mortar, ready-to-use mixtures) and shaped (earth blocks and panels)
5	Shaping procedure	Ramming, compressing, hand-shaping, spraying, casting
6	Dry bulk density	Low ($\rho_d < 1.2$ kg/dm^3) Medium ($1.2 \leq \rho_d \leq 1.7$ kg/dm^3) High ($\rho_d > 1.7$ kg/dm^3)
7	Type of application/ application class	Load-bearing (carrying loads from building elements, e.g., ceiling, roof, live loads) and non-load-bearing (e.g., infills in framed structures)
8	Fire performance	According to DIN 4102-1: Building Material Classes A1, A2, B1, B2

e.g., the physical-mechanical properties, certain aggregates (and possible additives), their intended use as well as aspects related to production and function. Table 3.4 provides an overview of the characteristic properties of earth building materials.

The *designation* according to DIN 18945-47 is carried out in a predefined order of characteristic properties of the respective earth building material. Presently, such designations only exist for earth blocks and earth mortar (Sects. 3.5.7, 3.5.6.1, and 3.5.6.2). For all other earth building materials according to Table 3.3 a specific form of designation has not (yet) been defined.

The *certification* procedure verifies the conformity of the declared data with the results of tests and inspections of the production process of the earth building materials. This certification is supplied by external testing authorities according to a predefined pattern in the form of a conformity certificate (Ü mark in Germany, CE mark on a European level) as well as by the manufacturers in the form of a conformity declaration. By establishing test and inspection intervals of the characteristic properties the *constancy of performance* of an earth building material is proven beyond its production period.

3.4.1 Place of Production

Earth building materials can be produced traditionally "on site" or "industrially" using modern methods (e.g., factory mortar mix). For earth building materials produced industrially or in a plant, it must be proven that their characteristic properties conform to the requirements of the relevant technical regulations (DIN standards/ Lehmbau Regeln). It should be pointed out that the Lehmbau Regeln also officially allow the traditional processing of earth building materials "on site."

3.4.2 Aggregates and Additives

To improve certain characteristics of construction soil, aggregates and additives can be added during the production of earth building materials. These added materials then become part of the designation of the building material.

The earth building materials defined in the Lehmbau Regeln [22] and in DIN 18945-47 are characterized by the properties of their clay minerals. They are the sole binding agents for the soil mixture during processing and ensure dimensional stability and strength during use (Sect. 2.2.3.4). Chemically stabilized earth building materials and clay products containing synthetic binders are therefore not covered by the Lehmbau Regeln. This does not mean, however, that they should be excluded. In many countries, these building materials are used on a daily basis and are subject to local building regulations (Sect. 4.2.1.3). When producing earth building materials according to DIN 18945-47 all aggregates added to the construction soil must be declared.

Aggregates mainly change the physical properties of earth building materials. They decrease shrinkage during drying and increase tensile strength and resistance to erosion. Lightweight aggregates improve the insulation properties of the building elements produced from these earth building materials.

Aggregates can be of mineral or organic origin (Fig. 3.31a [28]). Examples of mineral aggregates are sand, gravel, and lightweight aggregates in the form of thermally modified products such as expanded clay, expanded glass, or expanded slate. However, the production of thermally expanded products requires more energy.

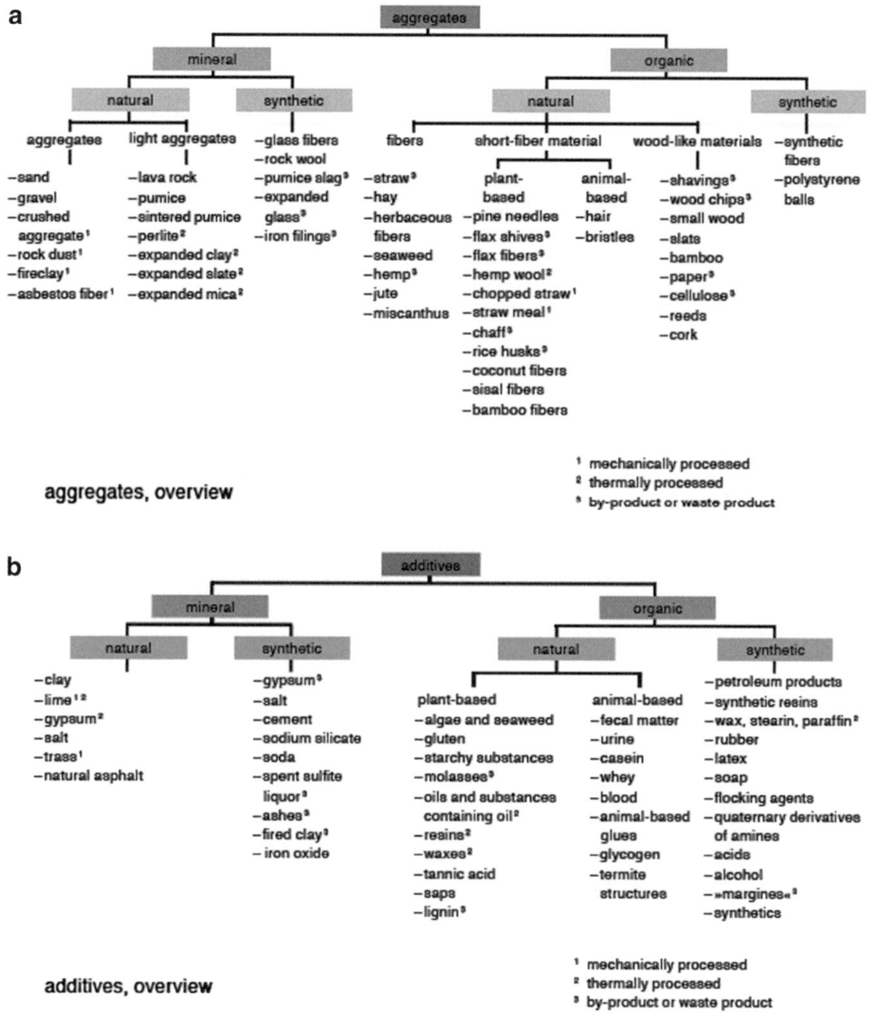

Fig. 3.31 Earth building materials, aggregates and additives [28]

Recycled glass used for the production of expanded glass must not have higher levels of contaminants than the primary raw material and needs to be tested for its environmental impacts according to DIN EN 13055 (D).

Organic aggregates are mainly plant fibers such as chopped straw, wood chips, and hemp shavings. Animal hair, for example, calf hair and pig bristles, can also be used.

Additives can also be of mineral or organic origin (Fig. 3.31b [28]). They alter the chemical structure of the clay minerals in the soil thereby reducing unfavorable properties such as shrinking and swelling. They furthermore increase the compressive strength as well as the earth building elements' resistance to abrasion and weathering. Organic additives include, but are not limited to, sap and animal excrement.

For earth building, the most significant mineral additives are represented by the large group of binders. *Binders* can be divided into hydraulic and non-hydraulic. Some of the most common hydraulic binders are cement and hydraulic lime $CaCO_3$. They completely harden through a chemical process when exposed to air or under water (after some initial air hardening). In the production of earth building materials they are often referred to as "chemical stabilizers."

The chemical stabilization of construction soil with additives restricts an important and special ecological quality of earth as a building material which needs to be preserved: by adding water, dry construction soil can be made plastic again (replasticized), further processed (Sect. 6.2.2.2) or easily returned to the natural cycle without significant energy expenditures.

Chemical stabilization of earth building materials, especially by adding binders such as lime and cement to rammed earth mixtures and earth blocks, is most notably common in developing tropical countries as well as in Australia and the USA. Particularly in developing countries, different local substitutes are added to the construction soil in place of expensive cement. These substitutes, which include ashes of plant by-products (rice husks), can also alter the chemical structure of the clay minerals.

The use of additives based on synthetic hydrocarbons always needs to be considered very carefully. Such additives could pollute the indoor air during the building's useful life. Possible adverse health effects in connection with the use of bitumen (asphalt emulsion) have not been examined. Synthetic additives are also problematic in terms of their ability to break down in the natural cycle when soil from demolished structures is disposed of.

Clay minerals (Sect. 2.2.3.4) are non-hydraulic binding agents. This means that they harden physically through air drying (air-setting mortars) while retaining their plastic characteristics. Renewed water absorption results in a "replasticization." This makes it possible to return the earth building materials to the shaping process and keep them in the life cycle without any added energy. The life cycle principle is an essential component of ecological and sustainable building. Clay minerals can also serve as additives for construction soil (e.g., as clay powder) if the soil is too lean for a specific application.

Examples of other non-hydraulic binders are building limes (non-hydraulic lime $Ca(OH)_2$), gypsum ($CaSO_4 \bullet 2H_2O$), anhydride ($CaSO_4$), and magnesium oxychloride cement ($MgCO_3$). In contrast to clay minerals, these binders harden chemically and can therefore not be replasticized.

3.4.3 Consistency and Processing

The earth building field differentiates between "wet" and "dry" consistency states for the processing of earth building materials. Accordingly, all earth building materials that are processed in a wet state, such as rammed earth, cob, straw clay, light straw clay, and earth mortars (including earth plasters) are classified as *wet earth construction*. Their application requires adherence to specific consistency states which, in certain cases, is tested using defined test methods (e.g., for earth plasters).

Dry earth construction mainly covers interior work using clay panels. Earth masonry construction is a combination of dry (earth blocks) and wet earth construction (earth masonry mortar, earth plaster).

3.4.4 Degree of Prefabrication and Format Design

Pre-processed, shapeless mixtures can be used for different shaping methods to produce modular earth building materials, such as earth blocks and earth panels, or entire building elements, such as rammed earth walls (Sect. 3.2.2).

Formwork is used to shape the building elements. In traditional earth building, the surface of the building element is also shaped by manual manipulation, such as cutting in cob construction.

The actual dimensions of shaped earth building materials need to conform to the manufacturer's declared data. In regard to earth blocks, it is important to know if the blocks are solid or perforated. Perforated blocks need to meet specific requirements concerning the permitted percentage of holes and the minimum web thickness.

In Germany, the dimensions of earth blocks (with or without perforations) largely correspond to the DIN formats for masonry construction with rectangular shapes. Special size variations are also permissible.

3.4.5 Shaping Methods

According to Table 3.1 and Sect. 3.2.2.1, the following shaping methods for earth building materials can generally be differentiated between:

- Compression (compression molding and extrusion shaping)
- Hand-throwing (hand-shaping)
- Ramming
- Spraying
- Casting
- Manual shaping (earth clumps)

The shaping methods influence the mechanical properties of the earth building materials used in the production of the building element or structure.

3.4.6 Bulk Density Class

According to DIN 18945-47 the dry bulk density of earth building materials is given in classes at intervals of 0.1 kg/dm^3. “Light” can be added to the name of the building material for bulk density classes <1.2 kg/dm^3, e.g., “light clay” or “light-clay block.”

3.4.7 Type of Application

The mechanical properties of earth building materials determine their suitability for certain applications. For example, they stipulate if a material can be used in load-bearing or non-load-bearing building elements, or if it is for interior or exterior use. As an example, the respective usage classes for earth blocks are defined in DIN 18945 (Sect. 3.5.7).

3.4.8 Fire Performance

According to DIN 4102-1 and DIN EN 13501-1, building materials are classified into *nonflammable* (Building Material Class A) and *flammable* (Building Material Class B) in terms of their fire performance. The fire performance of a building material is influenced not only by the actual type of material but also by its shape, specific surface area and mass as well as its combination with other materials, the connectors used and the type of installation.

The fire performance of earth building materials according to DIN 18945-47 must be tested according to DIN 4102-1 and DIN EN 13501-1. Soil and mineral aggregates (sand, gravel, etc.) are classified as nonflammable (A1) according to DIN 4102-4.

The fire performance of earth building materials is therefore of particular relevance when natural and artificial organic aggregates are used in the production of light clays. Common organic aggregates which could affect the fire performance of earth building materials are chopped straw, wood chips, wood shavings, sawdust, and crushed cork. To be classified as Building Material Class A1 (nonflammable) earth building materials are allowed to have a homogeneously distributed organic aggregate content ≤1 % of the mass or volume, whichever is greater.

Chapter “5.5 Fire Performance” of the Lehmbau Regeln [22] has been invalidated by the German Institute of Construction Technology (DIBt). This means that the respective Table 5.6 can no longer be applied. For earth building materials not defined in DIN 18945-47, the determination of their building material class and their fire performance is currently not regulated. In order to be classified as Class A1 “nonflammable,” the abovementioned percentage of “homogeneously distributed organic aggregate content ≤1 % of the mass or volume" can also serve as a guideline for all remaining earth building materials in Table 3.3.

For the time being, DIN 4102-1 is in effect as a national standard in Germany, together with DIN EN 13501-1 which was introduced in 2010. Its application is only mandatory for harmonized European permits carrying the CE mark (e.g., ETA, Fig. 4.18).

DIN EN 13501-1 differentiates between Classes A (nonflammable) and F (highly flammable). In addition to fire performance, it also includes and classifies smoke emissions (s1–s3) and flaming droplets (d0–d2). It is therefore not easy to make a direct comparison to conventional building material classes according to DIN 4102-1. However, the "nonflammable" building material classes are identical in both norms.

3.4.9 Certification and Monitoring of the Constancy of Performance

The procedure defined in DIN 18200 serves as the basis for the conformity certification used for assessing a building product. It consists of in-factory production control and regular external monitoring which includes an initial type testing of the building product.

The *performance of a building product* describes its performance in terms of its declared characteristic properties. It can be expressed through levels, classes, or a short description (see overview in Table 3.4). Inspection of the characteristic properties is carried out in predefined cycles throughout the entire production process of the (earth) building material and is referred to as *monitoring the constancy of performance*.

Figure 3.32 shows the *systems* used for assessing and verifying the constancy of performance of building materials according to Regulation (EU) No. 305/2011, Appendix V, which consist of different inspections initiated by the manufacturer or by an authorized body. A national Technical Assessment Body (in Germany: DIBt) determines which characteristic properties need to be verified for each material and which inspection system must be used.

3.4.9.1 Inspections Initiated by the Manufacturer

Before manufacturers can establish in-factory production control they need to carry out an *initial type testing* to determine if the requirements applicable to the building products (= characteristic properties) are met. In-factory production control can only be started after an initial type testing has been successfully completed.

In-factory production control consists of production monitoring which must be carried out regularly and documented by the manufacturer in order to guarantee that production of the earth building materials is in accordance with technical

Inspection systems	1+	1	2+	3	4
Initial type testing					
Testing of samples taken from the factory					
In-factory production control					
initial monitoring					
Random sample inspection					
Initial inspection of the factory and the in-factory production control					
Continuous monitoring, assessment and evaluation of in-factory production control					
Type of conformity	Z	Z	Z	E	E

Inspection initiated by:

Manufacturer — E declaration of conformity

Authorized body (external monitoring) — Z certificate of conformity

S 2+: for earth masonry mortar M2 to M4, earth blocks strength class ≥ 2

S 4: for earth masonry mortarM0 and earth plaster mortar, earth blocks strength class 0

Fig. 3.32 Systems for assessing and verifying the constancy of performance of building products according to Regulation EU No. 305/2011, Appendix V (BauPVO)

regulations and declared parameters are attained. Manufacturers are responsible for this monitoring procedure, and can implement it themselves or commission an external lab.

If the results of the inspections initiated by the manufacturer conform with the characteristic properties which are required by DIN 18945-47 a *Declaration of Conformity* is issued for the respective earth building material (Fig. 3.32) in the form of a conformity mark. The "Ü mark" contains the following information:

- Manufacturer's name
- Proof of conformity certification, such as technical regulations, building registration number or number of the respective test certificate, or "individual approval" by an authorized body
- Symbol or name of certifying body, if required

3.4.9.2 Inspections Initiated by an Authorized Body

External inspections are production controls which are carried out at fixed intervals in the production facilities of the manufacturer by independent testing institutes on behalf of the certification and monitoring body. They consist of initial type testing and compliance monitoring.

For *initial type testing*, an authorized body checks if the product meets all demands in terms of its declared properties and if the declarations on its packaging, and the enclosed information are correct. This test also verifies that all requirements for continuous and adequate production as well as for in-factory product control are met. Compliance monitoring can only begin after successfully passing initial type testing.

Compliance monitoring covers in-factory production control in terms of personnel and technical requirements to guarantee proper production methods and correct designation of the products. It is carried out at least twice a year at reasonable intervals and without prior notice.

If the results of the inspections initiated by the authorized body conform with the characteristic properties required by DIN 18945-47 a *Certificate of Conformity* is issued for the earth building material (Fig. 3.32).

3.5 Requirements and Characteristic Properties

Earth building materials according to Table 3.3 can be used if their characteristic properties and their constancy of performance meet the requirements of the Lehmbau Regeln or DIN 18945-47.

3.5.1 Rammed Earth

Terminology

Rammed earth is a shapeless mixture of construction soil, possible aggregates (and additives), and water. The dry bulk density ρ_d of rammed earth ranges from 1.7 to 2.2 (and up to 2.4) kg/dm^3 depending on the soil and aggregates used. In recent years, the use of thermally expanded lightweight mineral aggregates has produced dry bulk densities of <1.7 kg/dm^3.

Construction Soil

For load-bearing rammed earth wall construction, a well-graded (even and consistent) grain size distribution curve is particularly important. Well-suited construction soils are soils with mixed grain sizes including coarse grains (eluvial soils or glacial till, Chaps. 2.1.2.2 and 2.1.2.3), which can be classified as lean to rich with low to medium cohesion in terms of their cohesive strength and plasticity. Houben and Guillaud [14] identify a wide range of fluctuation in plasticity values (PI=0.03–0.30 and LL=0.24–0.46) and point to the difficulties of clearly defining this range.

Table 3.5 Recommended grain size composition for rammed earth [29]

No.	Grain fraction	Minimum [%]	Maximum [%]
1	Clay + silt	20–25	30–35
2	Sand + gravel	50–55	70–75

Maniatidis and Walker [29] recommend the following upper and lower limits for the individual grain fractions (Table 3.5).

Similar recommendations are given in [14] in the form of grading envelopes of the various grain size fractions for the composition of rammed earth.

Processing, Aggregates and Additives

Possible shortcomings of an available construction soil in terms of particle size (degree of shrinkage, erosion risk) can be balanced through the addition of the missing grain sizes (coarse sand, gravel or pea gravel, also thermally expanded lightweight aggregates). The Fuller curve model can serve as a guideline (Sect. 2.2.3.1). Often, small amounts of organic aggregates (straw or other suitable plant fibers) are added as well.

The grain parameters (Sect. 2.2.3.1) should, however, never be isolated from the processing parameters (Sect. 2.2.3.2).

In most developing countries, but also in the USA and Australia, the addition of lime and cement binders to rammed earth is common practice (Sect. 3.4.2).

Today, a rammed earth mix is typically prepared in pan mixers, but manual mixing is also possible. Mixing is complete when the soil mix has a homogeneous, fine and crumbly structure with evenly distributed moisture. The recommended consistency is stiff to semisolid.

Application

Rammed earth construction with formwork can be used for producing load-bearing and non-load-bearing building elements. It is also suitable for the production of rammed or compressed earth blocks which, after drying, can be used in the construction of load-bearing or non-load-bearing structures in the same manner as standard masonry. In addition, rammed earth is used for floors and to prefabricate large-format to ceiling-high wall elements.

Characteristic Properties/Requirements

- Dry bulk density according to Sect. 3.6.1.3
- Dry compressive strength according to Sect. 3.6.2.2
- Maximum grain size according to Sect. 2.2.3.1
- Linear shrinkage according to Sect. 3.6.2.1
- Fire performance/fiber additives according to Sect. 3.4.8

The consistency of the mix for delivery to the building site should not be wetter than "naturally moist." The mix should have a homogeneous, crumbly structure with evenly distributed moisture. During transport of the rammed earth material to the building site it is important to prevent demixing.

Designation

Based on DIN 18945-47 the following designation is proposed for rammed earth:

Rammed earth—load-bearing/non-load-bearing—Lehmbau Regeln—letter symbol with maximum grain size and fiber aggregate—strength class—bulk density class.

Example Designation of a rammed earth mix for load-bearing applications according to the Lehmbau Regeln with a maximum grain size of 20 mm without fiber aggregates, compressive strength class 2 N/mm^2, and bulk density class 2.0 kg/dm^3.

Rammed earth—load-bearing—LR—STL 20–2–2.0.

3.5.2 Cob

Terminology

Cob is a shapeless mixture of construction soil, straw and other suitable fibers, and water. Depending on the proportion of fiber in the mixture, cob has a dry bulk density ρ_d of 1.4–1.7 kg/dm^3.

Construction Soil

In terms of cohesive strength and plasticity, suitable construction soil is a fine-grained soil, lean to rich, with low to medium cohesion (loess soil). Due to their high degree of shrinkage very rich or highly cohesive soils (clays) are difficult to process and can cause cracking despite the reinforcement effect of the straw fibers.

Rocky construction soils complicate the preparation process as well. However, they were used regionally for the production of similar building materials in areas where suitable fine-grained soils where not available.

Processing, Aggregates and Additives

For the manual preparation of cob, straw or other available and suitable plant fibers with a fiber length of approximately 30–50 cm are spread out in 4–5 layers at a height of approximately 5 cm. These fiber layers are alternated with 10-cm-high layers of paste-like or viscous construction soil. The total height should not exceed 60 cm. Niemeyer [30] recommends 25 kg of straw per m^3 of soil material. The mix can also be prepared mechanically in mixers.

Application

Cob is used for repairing existing structures. Generally speaking, new construction of load-bearing and non-load-bearing walls is also possible, but this is rarely carried out due to the considerable manual effort it requires. In contrast to rammed earth construction, cob walls are built without formwork. Cob can also be used to produce earth blocks.

Characteristic Properties/Requirements

- Dry bulk density according to Sect. 3.6.1.3
- Dry compressive strength according to Sect. 3.6.2.2
- Shrinkage of building element specimen
- Fire performance according to Sect. 3.4.8

If the cob mix is prepared manually, the mixing water in the layered material will be evenly distributed after 1 day of resting. It should therefore be possible to mix the materials with only a small amount of additional water (Fig. 3.33 [10]). The goal is to achieve a homogeneous, plastic material with uniform processing characteristics. The fiber material needs to be evenly coated with the clay mix. The recommended consistency of the soil portion is stiff.

Fig. 3.33 Processing of cob [10]

Designation

Based on DIN 18945-47 the following designation is proposed for cob:
Cob—load-bearing/non-load-bearing—Lehmbau Regeln—letter symbol—strength class—bulk density class.

Example Designation of a cob mix for load-bearing application according to the Lehmbau Regeln, compressive strength class 1 N/mm^2, and bulk density class 1.8 kg/dm^3:
Cob—load-bearing—LR—WL—1–1.8.

3.5.3 Straw Clay and Clay with Added Fibers

Terminology

Straw clay and clay with added fibers is a shapeless mix of construction soil, sand (if needed), organic fibers, and water. Depending on the fiber content, the dry bulk density ρ_d is between 1.2 and 1.7 kg/dm^3.

Construction Soil

The recommended construction soil is lean to semi-rich, silty and with low cohesion (e.g., loess soil, Sect. 2.1.2.1).

Processing, Aggregates and Additives

Suitable organic aggregates are soft straw, hay, or other soft plant fibers with a length up to 25 cm. The Lehmbau Regeln [22] recommend 40–60 kg of fibers/straw per m^3 of soil material. The material is prepared in the same manner as cob. It is also possible to process old infill material which might be amended with sand and/or straw to make it leaner. The recommended consistency of the soil portion is stiff to soft and depends on the intended use.

Application

Straw clay or clay with added fibers has many uses as a building material: as earthen infill material for half-timber or timber-frame construction, as earthen infill for timber beam ceilings, as plaster-like coats, or for producing shaped earth blocks and panels, typically for non-load-bearing applications.

Characteristic Properties/Requirements

- Dry bulk density according to Sect. 3.6.1.3
- Shrinkage of building element specimen
- Fire performance according to Sect. 3.4.8

The goal is to achieve a homogeneous, plastic material with uniform processing characteristics. The fiber material needs to be evenly coated with the clay mix. The recommended consistency of the soil portion is stiff to soft and depends on the intended use.

Designation

Based on DIN 18945-47 the following designation is proposed for straw clay/clay with added fibers:

Straw clay/clay with added fibers—non-load-bearing—Lehmbau Regeln—letter symbol—bulk density class.

Example Designation of a mix of straw clay/clay with added fibers for non-load-bearing applications according to the Lehmbau Regeln with a bulk density class of 1.4 kg/dm^3:

Straw clay/clay with added fibers—non-load-bearing—LR—SL—1.4.

3.5.4 Light Clay

Terminology

Light clay (LL) is a shapeless mix of construction soil, organic and/or lightweight mineral aggregates, and water. The dry bulk density ρ_d is between 0.3 and 1.2 kg/dm^3. In terms of bulk density classes, light clay is grouped into:

- Light mixes: $\rho_d = 0.3–0.8$ kg/dm^3
- Heavy mixes: $\rho_d > 0.8–1.2$ kg/dm^3

The predominant aggregate can be integrated into the name of the building material, e.g.:

Light straw clay: Straw fibers as an organic aggregate (Fig. 3.19)
Light wood-chip clay: Wood chips as an organic aggregate (Fig. 3.20)
Light mineral clay: E.g., expanded clay as a lightweight mineral aggregate (Fig. 3.34)

Fig. 3.34 Light clay, expanded clay used as a lightweight mineral aggregate for infill [77]

Construction Soil

With an increase in the proportion of aggregates, the cohesive strength of the construction soil also needs to increase. This means that suitable soils for heavy mixes are lean and have low cohesive strength (loess soils) while soils for light mixes are rich to very rich or fine grained and have medium to high cohesive strength (fluvial soils, with some clay powder added).

Processing, Aggregates and Additives

Aggregates which can be used are organic fiber material (all types of straw, wood chips) and/or lightweight mineral aggregates (thermally expanded materials, pumice, perlite). Mixtures of both types of aggregates are also possible. Organic fibers should not be longer than the shortest length of the finished building element or building material. For lightweight aggregates, the Lehmbau Regeln [22] recommend the following amounts as aggregates per m^3 of building element:

- Straw from bales: Approx. 60–90 kg/m^3
- Wood chips: Approx. 300 kg/m^3
- Lightweight mineral aggregates: Approx. 300–600 kg/m^3

The preparation of *heavy mixes* is done manually in the same manner as cob or can be mechanized (Sect. 3.1.2.4). In *light mixes*, the clay minerals need to coat all aggregates with a very fine layer. To achieve this, the construction soil is prepared manually as a slurry or in suitable mixers. The aggregates can then be blended in a pan mixer (Fig. 3.9). For organic aggregates with long fibers (straw), the slurry is either poured over the straw (Fig. 3.10) or the straw is immersed into the slurry (Sect. 3.1.2.5). The straw-clay mixture should rest on a level surface for a period of time, preferably overnight. This helps distribute the water evenly within the mixture. In this state, the consistency of the soil portion is somewhat stiff.

Application

Light clay is used for non-load-bearing building elements. It has many different uses as a building material: as earthen infill material for half-timber or timber-frame construction, particularly for exterior walls and as a lining, as earthen infill of timber beam ceilings, as plaster-like coats, or for producing shaped earth blocks and panels.

Characteristic Properties/Requirements

- Dry bulk density according to Sect. 3.6.1.3
- Dry compressive strength and degree of linear shrinkage tested on a building element specimen
- Slump of the clay slurry according to Sect. 3.6.2.1
- Fire performance according to Sect. 3.4.8

The goal is to achieve a homogeneous, plastic material with uniform processing characteristics. The fiber material needs to be evenly coated with the soil mix. The recommended consistency of the soil portion is stiff to soft and depends on the intended use.

Designation

Based on DIN 18945-47 the following designation is proposed for light clay:

Light clay—Lehmbau Regeln—letter symbol with mineral/organic fiber aggregate—bulk density class.

Example Designation of a light-clay mix for non-load-bearing applications according to the Lehmbau Regeln with a fiber aggregate and a bulk density class of 0.9 kg/dm^3:

Light clay—LR—LL f—0.9.

3.5.5 *Earthen Loose Fill*

Terminology

Earthen loose fill is loose, free-flowing earth building material made of construction soil (with or without aggregates) used as fill material for horizontal building elements (e.g., in ceilings). Water is added if needed. In terms of dry bulk density, the following types can be distinguished between:

- Earthen loose fill: $\rho_d > 1.2$ kg/dm^3
- Light-clay loose fill: $\rho_d = 0.3–1.2$ kg/dm^3

The names can also be formed based on the predominant aggregate, e.g., light wood-chip clay loose fill; straw-clay loose fill; and construction soil loose fill (without aggregate).

Construction Soil

There are no specific requirements in terms of cohesive strength/plasticity and particle size of the construction soil. Preference can be given to recycled soil under consideration of the points given in Sect. 2.2.1.3.

Processing, Aggregates and Additives

The construction soil, possible aggregates and water are mixed manually or mechanically and processed into a loose material. Water is mainly added to prevent dusting during installation. Possible aggregates are lightweight minerals and organic fibers.

Application

Earthen loose fill is used for bulk filling of ceilings and building voids. Soils and clays which have been prepared as powders and then granulated are particularly suitable for this application (Fig. 3.6a).

Characteristic properties/requirements

- Fire performance according to Sect. 3.4.8

In order to check assumed loads the bulk density of the building material can be tested according to Sect. 3.6.1.2.

Designation

Based on DIN 18945-47 the following designation is proposed for earthen loose fill:

Earthen loose fill—Lehmbau Regeln—letter symbol with mineral/organic fiber aggregate—bulk density class.

Example Designation of an earthen loose fill mix according to the Lehmbau Regeln with an organic fiber aggregate, bulk density class 1.4 kg/dm^3:

Earthen loose fill—LR—LT f—1.4.

3.5.6 Earth Mortars

Terminology

Earth mortars are mixes of suitable fine-grained construction soils, mineral and/or fine-fibered organic aggregates and water. Earth mortars with a dry bulk density of $\rho_d < 1.2$ kg/dm^3 are referred to as light earth mortars.

Based on DIN EN 998, earth mortars are distinguished by the following names which have also been incorporated into DIN 18946 "Earth Masonry Mortar" and 18947 "Earth Plaster Mortar":

Fresh earth mortar is a completely mixed, ready-to-use earth mortar. *Hardened earth mortar* is earth mortar that has hardened. *Stabilized earth mortars* are mortars with the addition of chemically setting binders which have irreversibly changed the mortars' strength and ability to replasticize.

Oversize is the screen size of the test sieve on which no sieve residue is left after sieving according to DIN EN 1015-1. The *maximum grain size* refers to the screen size of the upper test sieve D of the grain size fraction for which no sieve residue or only single oversized particles can be detected. The particle size group of the aggregate is described by the lower (*d*) and upper (*D*) screen size as *d*/*D*.

Construction Soil

Suitable construction soils are fine-grained, silty to sandy and lean to semi-rich soils, and soils with low cohesion (e.g., loess soil).

Processing, Aggregates and Additives

DIN EN 998 defines general terms for mortars according to their place and method of preparation. These terms are correspondingly applied to earth mortars in the Lehmbau Regeln [22] and DIN 18946-47.

Earth mortar can be mixed out of different materials on the building site as *site-sourced mortar*. Earth mortar made from open pit soil first needs to be prepared according to Sect. 3.1, and all grain sizes >5 mm must be screened out. Suitable mortar mix formulas are generally based on experience at the specific location.

For *industrially produced mortars*, the cohesive binding materials (such as dry soil, Sect. 2.2.1.2) are supplied by the manufacturer and mixed with additional aggregates (e.g., sand) on site following a predefined mixing formula. By adding water, the plasters then reach their required consistency. Site-sourced mortars and industrially produced mortars are subject to the Lehmbau Regeln [22].

Ready-to-use bagged mortars are finished blends of construction soil and aggregates. They are delivered to the building site dry (in paper bags or silos) or naturally moist (in so called big bags, Fig. 3.30). After water is added (according to the manufacturer's instructions), these mortars can be used immediately or after a predefined period of preparation and tempering. Ready-to-use bagged mortars are subject to DIN 18946-47.

Reused mortar (recycled earthen material, Sect. 2.2.1.3) is earth masonry or plaster mortar which has been salvaged from demolished building elements and prepared with water to attain a workable consistency. It should not contain any chemical or biological impurities and can be amended with sand and/or straw as needed (Lehmbau Regeln [22]).

The properties of earth mortar can be influenced according to their intended use by specifically adding mineral and/or organic aggregates and additives (Sect. 3.4.2).

Characteristic Properties/Requirements

The suitability test for *site-sourced mortar* can be carried out on sample surfaces or building element specimens.

For *ready-to-use bagged mortars*, the characteristic properties for the intended use must be verified and declared for initial type testing and as part of in-factory production control.

Deformations

Fresh Earth Mortar Consistency. In order to attain the required performance characteristics, the consistency of fresh earth mortars can be tested before they are further processed. Testing determines the *flow consistency a* of the test earth mortar according to DIN EN 1015-3 (Fig. 3.40, Sect. 3.6.2.1). The specimens used for shrinkage testing are prepared using fresh earth mortar with a flow consistency of 140 mm according to the Lehmbau Regeln.

Degree of Linear Shrinkage. The degree of linear shrinkage of earth mortar is tested according to Sect. 3.6.2.1. The test specimens are prepared according to the Lehmbau Regeln with fresh earth mortar with a flow consistency of 140 mm.

Forms of Supply

The manufacturer needs to indicate the consistency at which the earth mortar is supplied (dry/wet). The most common form of supply is dry mortar in paper bags or silos (Fig. 3.6). The moisture content of dry mortar with organic fibers packaged airtight should not exceed the equilibrium moisture content of the mortar under standard climatic conditions (23 °C/65 % RH). A moisture content higher than "dry" needs to be declared on the packaging by the manufacturer in addition to information about the amount of water needed to achieve the required processing consistency.

The manufacturer's product information also needs to include recommended storage conditions and the expected shelf life when storing the product with the building material supplier and on the building site. Mortar can also be supplied as ready-to-use fresh mortar.

Concentration of Harmful Salts

The concentration of harmful salts, which could enter the mortar mix through the addition of aggregates and additives, should not exceed certain limits. Generally, the property "harmful concentration of salts" refers to soluble anions of the individual salts and is given in different levels of contamination. The following classification in Table 3.6 [31] is for plasters.

The permissible concentration of harmful salts in ready-to-use earth mortar is defined in DIN 18946 and 18947 or [32, 33] as follows:

Nitrates	<0.02 mass in %
Sulfates	<0.10 mass in %
Chlorides	<0.08 mass in %

The total concentration of harmful salts should not exceed 0.12 mass in %.

Fire Performance

Earth mortars according to DIN 18946 and 18947 continue to be classified as building material class A1 without additional testing according to DIN 4102-4 as long as their homogeneously distributed organic aggregate content is ≤1 % of their mass or volume, whichever is higher. Building material classes A1, A2, and B1 must be certified once a year as part of the compliance monitoring process conducted by an authorized body. Building material class B2 must be verified as part of the in-factory production control by the manufacturer (Tables 3.9 and 3.11 and Sect. 3.4.8).

Table 3.6 Contamination levels of harmful salts in plasters

No.	Sulfates [mass in %]	Chlorides [mass in %]	Nitrates [mass in %]	Concentration [mmol/kg]	Assessment
1	Up to 0.024	up to 0.009	up to 0.016	Up to 2.5	Level 0—no contamination
2	Up to 0.077	up to 0.028	up to 0.05	Up to 8.0	Level I—low contamination
3	Up to 0.24	up to 0.09	up to 0.16	Up to 25.0	Level II—medium contamination
4	Up to 0.77	up to 0.28	up to 0.50	Up to 80.0	Level III—high contamination
5	0.77 and higher	0.28 and higher	0.50 and higher	80.0 and higher	Level IV—extremely high contamination

Application

Earth mortars are divided into the following groups based on their application:

- Earth masonry mortar
- Earth plaster
- Sprayed earth mortar

The application is part of the name of the building material. Mortars which have been specifically declared for a particular application are only intended for use in this particular field. For example, an earth masonry mortar is unsuitable as an earth plaster.

3.5.6.1 Earth Masonry Mortar

Terminology

Earth masonry mortar is earth mortar for masonry construction according to DIN 18946.

Construction Soil

Suitable soils for earth masonry mortar are lean to semi-rich and soils with low cohesion.

Processing, Aggregates and Additives

The construction soil is typically amended with medium-grained to coarse-grained sand ($d < 2$ mm) and organic fibers, if needed. It is designated accordingly with "m" (mineral) or "f" (fiber reinforced).

According to DIN 18946, the following aggregates are allowed in earth masonry mortars:

Mineral (m): Natural aggregates according to DIN EN 12620, crushed brick made from mortar-free bricks, expanded perlite/expanded clay/expanded slate/natural pumice according to DIN EN 13055-1,

Organic (f): Plant parts and fibers, animal hair, chemically untreated chopped wood (wood chips).

According to DIN EN 12878, the addition of inorganic pigments is also permissible. All aggregates which have been added to the soil mix must be fully declared.

The individual components need to be blended into a homogeneous material using suitable mixers in the manufacturer's plant. Demixing must be largely prevented during bagging and transport.

Application

Earth masonry mortar is used in load-bearing and non-load-bearing earth block masonry construction (DIN 18946) and other masonry construction using bricks, concrete, or natural stone (Lehmbau Regeln).

Characteristic Properties/Requirements

Due to their function as connections between earth blocks or clay panels as well as between bricks or natural stones, earth masonry mortars need to be able to transmit compressive and shear stresses within the building element. This requires sufficient adhesion between the earth masonry mortar and the (earth) blocks. Earth masonry mortars also need to balance out dimensional deviations in the building materials within permissible limits and seal joints within the masonry structure to make them windtight.

With regard to the characteristic properties which need to be declared by the manufacturer, the following requirements are placed on earth masonry mortar according to DIN 18946.

Maximum Grain Size/Oversized Grain

The oversized grain of earth masonry mortars must be smaller than 8 mm. Tests should be carried out according to Sect. 2.2.3.1.

Dry Bulk Density

The bulk density of hardened earth masonry mortar is divided into classes according to Table 3.7.

Standard earth masonry mortars which are used in load-bearing masonry construction have a dry bulk density ρ_d of approx. 1.8 kg/dm^3.

Table 3.7 Bulk density classes of hardened earth masonry mortar

No.	Bulk density class	Average value of dry bulk density [kg/dm^3]
1	0.9	0.81–0.90
2	1.0	0.91–1.00
3	1.2	1.01–1.20
4	1.4	1.21–1.40
5	1.6	1.41–1.60
6	1.8	1.61–1.80
7	2.0	1.81–2.00
8	2.2	2.01–2.20

Table 3.8 Strength classes for earth masonry mortar

No.	Strength class	Compressive strength [N/mm^2]	Adhesive shear strength [N/mm^2]
1	M0	–	–
2	M2	≥2.0	≥0.02
3	M3	≥3.0	≥0.03
4	M4	≥4.0	≥0.04

Strength

Based on the specific stress earth mortars are exposed to, their strength properties are classified as compressive strength and adhesive shear strength. They are classified into strength classes (Table 3.8).

The minimum *compressive strength* for earth masonry mortar in load-bearing masonry (strength classes ≥ M2) is 2.0 N/mm^2. Earth masonry mortar of the strength class M0 should have a minimum compressive strength of 1.0 N/mm^2. The respective minimum for adhesive shear strength is 0.02 N/mm^2. Tests should be carried out according to Sect. 3.6.2.2.

Degree of Linear Shrinkage

The degree of linear shrinkage should not exceed 2.5 %. For fiber-reinforced earth masonry mortars it should not be more than 4 %. Tests should be carried out according to Sect. 3.6.2.1.

Resistance to Water Vapor Diffusion

It can be assumed as $\mu = 5/10$ without testing or can be determined using the test method according to Sect. 3.6.3.4.

Thermal Conductivity

It should be determined according to Sect. 3.6.3.4 and after rounding up the dry bulk density values ρ_d to the next 0.1 kg/dm^3 according to DIN 4108-4.

Designation

According to DIN 18946:

Earth masonry mortar—main DIN number—letter symbol with maximum grain size and organic fiber/mineral aggregate—strength class—bulk density class.

Example Designation of an earth masonry mortar according to DIN 18946 with a maximum grain size of 4 mm and an organic fiber aggregate, strength class M2, bulk density class 1.6 kg/dm^3:

Earth masonry mortar—DIN 18946—LMM 04 f—M2—1.6.

Monitoring of the Constancy of Performance and Certification

The constancy of performance during the production of earth masonry mortar according to DIN 18946 is monitored according to the systems defined by the DIBt which are shown in Fig. 3.32.

Earth masonry mortar of strength class M0:

System S4: Initial type testing and in-factory production control by the manufacturer; Declaration of Conformity by manufacturer after successful testing.

Earth masonry mortar of strength classes M2–M4:

System 2+: Initial monitoring of factory and in-factory production control, continuous monitoring of in-factory production control (compliance monitoring) by the authorized certifying body; after successful testing, Certificate of Conformity issued by authorized certifying body; initial monitoring and in-factory production control are carried out by the manufacturer; after successful testing, Declaration of Conformity by the manufacturer.

Before in-factory production control takes place, the manufacturer must conduct an initial type testing which needs to correspond to the listed requirements specific to earth masonry mortar. Testing must include the characteristic properties listed in Table 3.9. In-factory production control cannot commence until the initial type testing has been successfully completed.

Compliance monitoring of in-factory production control and of personnel and technical requirements to ensure adequate production and correct labeling of the earth masonry mortar must be carried out at least twice a year by the authorized certifying body. Compliance monitoring cannot commence until the initial monitoring has been successfully completed.

The Product Data Sheet (= Declaration of Performance by the manufacturer) includes all characteristic properties which have been declared in the building material designation and in the material requirements. It can also include a voluntary declaration of the CO_2 value.

The packing slip must include the following information:

- Manufacturer and plant with manufacturer's logo
- Designation, quantity, and form of supply of the delivered earth masonry mortar
- Recipient and date of delivery
- Conformity mark (which also needs to be displayed on the enclosed information sheet/packaging if applicable)

3.5.6.2 Earth Plaster Mortar

Terminology

Earth plaster mortar is an earth mortar used to cover interior wall and ceiling surfaces as well as exterior surfaces which are protected from the weather. The plaster can be applied in one or multiple layers. A *plaster coat* is applied in one or more

Table 3.9 Test system for monitoring the constancy of performance of earth masonry mortar according to DIN 18946

No.	Characteristic property of earth masonry mortar	Initial type testing	In-factory production control/ compliance monitoring	Scope of testing of in-factory production control	Declaration building material name/ product data sheet[d]	Chapter
1	Oversized grain	○	○	Every 400 tons	Product data sheet	2.2.3.1
2	Bulk density class of the hardened mortar	○	○/●	Every 400 tons	Building material name, Product data sheet	3.6.1.3
3	Linear degree of shrinkage	○	○	Every 400 tons	Product data sheet	3.6.2.1
4	Strength class according to Table 3.8 M0[a] M2–M4[b]				Building material name Product data sheet	
5	Dry compressive strength M2–M4	○	○/●	Every 400 tons	Product data sheet	3.6.2.2
6	Adhesive shear strength M2–M4	○	○		Product data sheet	3.6.2.2
7	Fire performance, building material class B2[a] A1, A2, B1[b]	○	○ ○/●	Once per year	Product data sheet	3.4.8
8	Harmful salts (suspicion)	○				2.2.3.4
9	Grain size fraction (upper sieve size D)				Building material name Product data sheet	2.2.3.1
10	Aggregates mineral/ organic				Building material name, Product data sheet	3.4.2
11	Type of earth mortar in terms of form of supply				Product data sheet	3.5.6
12	Thermal conductivity				Product data sheet	3.6.3.2
13	Water vapor diffusion resistance factor				Product data sheet	5.1.2.2
14	CO_2 equivalent[c]				Product data sheet	1.4.3.1
15	Activity concentration index[c]				Product data sheet	5.1.6.1

Inspection initiated by:
Manufacturer ○: initial type testing; in-factory production control
Authorized body ●: initial monitoring; compliance monitoring
[a]Declaration of Conformity by manufacturer
[b]Certificate of Conformity by authorized body
[c]Voluntary
[d]Integrated into: building material name, product data sheet

steps “wet on wet.” A *plaster system* consists of one or more plaster coat(s). The *base coat plaster* refers to the bottom layer(s) of a multilayer plaster system while the *top coat plaster* is the top layer of a plaster system (DIN 18947).

Earth plaster mortars of the bulk density class ≤1.2 according to Table 3.7 can be referred to as light-clay plaster mortars.

In terms of individual requirements, earth plaster mortar is considered a building material while earth plaster is considered a building element (Sect. 4.3.6).

Clay thin-layer finishes can be classified as waterborne coating materials according to DIN EN 13300. They include (mostly colored) earth plaster mortars which are applied at a maximum thickness of 3 mm, clay “putty knife” masses, and earth paint coats. Clay thin-layer finishes are regulated in Technical Information Sheet 06 published by the DVL [34]. In contrast to [22] and DIN 18945-47, this sheet also includes stabilized earth building materials.

Construction Soil

Suitable soils for earth plaster mortar are silty-sandy soils (loess soil) with a sufficient portion of coarse-grained silt to medium-grained sand to reduce shrinkage. In addition, the clay minerals contained in the soil need to have enough cohesion to bind the silt and sand grains to the plaster surface and to keep abrasion to a minimum after the plaster has dried. For very lean soils this can be achieved by adding clay powder.

Processing, Aggregates and Additives

The construction soil is typically amended with medium- to coarse-grained sand ($d<2$ mm) and, if applicable, organic fiber material, and designated accordingly with “m” (mineral) or “f” (fiber reinforced). Fiber materials in the plaster act as reinforcement to prevent cracking during the drying process. They increase the earth plaster’s mechanical resistance to abrasion and impact and improve thermal insulating properties in the plaster’s finished state. Sharp-edged sands display higher interlocking resistance within the soil skeleton and should therefore be given preference over rounded sands for use as aggregates.

According to DIN 18947 the following aggregates are allowed in earth plaster mortars:

Mineral (m): Natural aggregates according to DIN EN 12620, crushed brick made from mortar-free bricks, expanded perlite/expanded clay/expanded glass (see Sect. 3.4.2)/expanded slate/natural pumice according to DIN EN 13055.

Organic (f): Plant parts and fibers, animal hair, chemically untreated chopped wood (no wood composites).

According to DIN EN 12878 the addition of inorganic pigments is also permissible. All aggregates which have been added to the soil mix must be fully declared.

The individual components need to be blended into a homogeneous material using suitable mixers in the manufacturer's plant. Demixing should be prevented during bagging and transport.

In the earth building traditions of Central Asia, North Africa, and Arabia, the use of earth plasters with different chemically modifying additives and aggregates is widespread. In interior spaces they mainly have decorative functions and are an integral part of the local building culture. The application of earth plasters also has a very long tradition in Japan where it requires great skill.

In rural areas of Germany, earth plasters were traditionally also used for exterior walls for economic reasons. In addition to lime, available local waste products were added as aggregates and additives to improve the plasters' weather resistance. These additives included fresh cow dung, whey, and animal blood (Fig. 3.31) which chemically altered the clay mineral structure. These traditional mixes have demonstrated their suitability and can also play a role in today's earth building practice, particularly in reconstruction and preservation work.

Currently, the lack of adequate criteria does not allow for a differentiated ecological assessment of chemically modifying additives in earth plasters in particular, and earth building materials in general. Possible health hazards for the user and questions of biodegradability of the additives after their return to the natural cycle are of particular concern.

The organization natureplus e.V. has developed the guidelines called "Earth Plaster" and "Clay Paints and Clay Thin-Layer Finishes" for their quality seal [35]. The section on "Composition, Banned and Restricted Materials" states the following:

The product "earth plaster" must contain 100 % mineral and renewable resources. Clay minerals are the only permissible binders, synthetic binders and chemically modifying additives are banned. Earth plaster mortars are specifically not allowed to contain the following materials:

- Biocides
- Organohalogen materials
- Synthetic additives and fibers (e.g., acrylates, polyvinyl acetates)
- Lime, gypsum, and cement as synthetic binders
- Cellulose and starch derivatives

The content level of volatile organic compounds (VOCs) in dry earth masonry mortar is limited to a maximum of 100 ppm. Limits for absorbable organic halogen compounds (AOXs), pH values, metals/metalloids as well as radioactivity are defined in the corresponding testing procedures. If recycled earthen materials are used (Sects. 2.2.1.3 and 6.2.2.1), the products need to be tested for possible absorbable content which could pose a health hazard, particularly asbestos fibers, heavy metals, and aromatic hydrocarbons (PAHs).

Earth plaster mortars must not have increased levels of radioactivity and need to stay within the limits shown in Table 6.1.

According to [35] and [34], the product group "Clay Paints and Clay Thin-Layer Finishes" must consist of a minimum of 99 % mineral and renewable resources by weight (including chemically modified natural materials) plus water. Clay minerals must be the main binding agent. The products are specifically not allowed to contain the following materials:

- Organohalogen materials
- Preservatives which are not approved as food additives or for cosmetics
- Biocides, provided they do not require in-can preservation due to their product characteristics (e.g., very high alkalinity)

If primers contain solvents which emit harmful substances, plasticizers, and/or preservatives, they must adhere to the limits listed in Table 6.1. If methyl cellulose is used as an aggregate/additive, it must be verified that waste water is not polluted during its production.

Application

In recent years, the field of earth plaster mortars has diversified into areas of various specialized applications. This has resulted in the development of plasters with special properties and product names.

Base Coat Plaster Mortars

Base coat plaster mortars are used for the bottom layer(s) of a multilayer plaster system, for filling in uneven substrates, and for attaching insulation panels. Because this layer is intended for applications up to several centimeters thick (typically 10–20 mm) the plaster mix is often rather rich. This can lead to cracks which are covered during subsequent steps, for example, with the application of a fine-finish plaster. Base coat plasters contain relatively coarse aggregates, such as fibers up to a length of 30 mm and fine-grained gravel up to a grain diameter of 4 mm.

Top Coat Plaster Mortars

Top coat plaster mortars are used for the top layers of a plaster system. They can also be applied as a single coat with a maximum thickness of about 12 mm. In contrast to base coat earth plaster mortars, they form the final plaster surface, and therefore finer fiber aggregates and grain sizes are used ($d \leq 2$ mm). Industrially produced dry mineral mortars for use as top coat plasters are also referred to as "fine-finish plaster mortars" or use the abbreviation "CR" (colored rendering mortar) according to DIN EN 998-1.

Earth top coat and base coat mortars have similar compositions and create relatively rough surface textures. Therefore, many manufacturers do not differentiate between base coat and top coat mortars but offer the same product for both types of application.

Thin-Layer Earth Plaster Mortars are earth plaster mortars with fine-grained mineral and/or fiber aggregates used as the finishing coat of a multilayer plaster system. They are applied in thin layers of 3–5 mm and create a very fine and dense surface texture.

Clay Adhesives are used to connect large-format (earth) blocks and to attach (clay) panels in dry construction. They are applied at a maximum thickness of 5 mm with the help of a putty knife. In addition to sand and fine organic fibers they contain clay powder and other binders such as cellulose and talcum resulting in relatively high cohesion.

Clay adhesives can also be used as thin-layer earth plaster mortars.

Clay Thin-Layer Finishes

The product group of clay thin-layer finishes constitutes the transition from thin-layer plasters to paints. Clay thin-layer finishes consist of construction soil, possible mineral aggregates, fibers as well as possible stabilizing additives. According to their use, they are applied at a thickness of <3 mm.

This group includes the following products [34]:

Colored finish earth plaster mortars are applied at paste-like to plastic consistency as a final surface. They contain color pigments which are added to neutral-colored soil mixes. Coloring can also be achieved by adding special colored clays as binding agents. Both methods can be used to produce a colored surface finish for creative effects in interior spaces.

Earth putty coats are sandable clay thin-layer finishes with a particularly fine grain size. They are applied as level substrates for finish earth plasters in a very thin layer at paste-like to plastic consistency using a putty knife.

Earth paint coats comprise brushable earth plasters and clay paints which are applied as finish coats at paste-like to plastic consistency.

Brushable earth plasters are "granulated paints" which create a textured surface through specific grain sizes contained in the plaster. They are applied like paints and can be compared to fine-finish earth plasters in terms of their finished surface structure. In addition to clay powder, cellulose or starch can also be used as binders.

Clay paints are paints composed of a combination of clay powder and cellulose/starch as the binding agent but without a visibly grainy texture.

Dry Clay Plaster Boards

Dry clay plaster boards are unique among earth plasters as an alternative to the wet application of plaster mortars. They are thin clay panels reinforced with reed, currently available at a thickness of 16 mm and dimensions of 62.5×62.5 cm^2. The board's surface is covered with coarsely woven jute fabric. Dry clay plaster boards are glued like tiles to a level and dry substrate and covered with a thin fine-finish earth plaster.

Characteristic Properties/Requirements

Earth plaster mortars must adhere sufficiently to the substrate. Top coat plasters should be free of cracking and meet the visual requirements placed on them (Sect. 4.3.6.3).

Concerning the characteristic properties which need to be declared by the manufacturer, earth plaster mortars must meet the following requirements according to DIN 18947:

Maximum Grain Size/Oversized Grain

The oversized grain of earth plaster mortars must be smaller than the minimum plaster coat thickness given by the manufacturer.

Dry Bulk Density

The bulk density of hardened earth plaster mortar is divided into classes according to Table 3.7. Earth plaster mortars of the bulk density class ≤1.2 can be referred to as light-clay plaster mortars.

Strength and Abrasion

The strength properties of earth plaster mortar are classified according to DIN 18947 into different strength classes. Based on the stresses they are exposed to, they consist of compressive strength, flexural strength, and adhesive strength. Earth plaster mortars must meet the minimum requirements for the individual strength classes listed in Table 3.10. Concerning cohesive strength and abrasion, clay thin-layer finishes must meet the requirements of strength class S II.

Abrasion must not exceed the specific values listed for the respective strength classes (Table 3.10). The abrasion of colored finish earth plaster mortars and earth putty coats should not exceed 0.70 g, of brushable earth plasters and clay paints 0.20 and 0.03 g, respectively.

Degree of Linear Shrinkage

The degree of linear shrinkage should not exceed 2.0 %, or 3.0 % for fiber-reinforced earth plaster mortars. Fiber-reinforced and mineral mortars for thin-layer plaster can exhibit a degree of linear shrinkage up to 4 %. Here, the material-specific workability must be ensured for the specific layer thickness defined by the manufacturer.

Table 3.10 Strength classes of earth plaster mortar

No.	Strength class	Compressive strength [N/mm^2]	Flexural strength [N/mm^2]	Adhesive strength [N/mm^2]	Abrasion [g]
1	S I	≥1.0	≥0.3	≥0.05	≤1.5
2	S II	≥1.5	≥0.7	≥0.10	≤0.7

Resistance to Water Vapor Diffusion

It can be assumed as $\mu = 5/10$ without testing or can be determined using the test method according to Sect. 3.6.3.4.

Water Vapor Adsorption

At a minimum, earth plaster mortars should meet the requirements of water adsorption class WS I according to Table 3.32. Clay thin-layer finishes should not significantly reduce the water vapor adsorption of the underlying earth plaster mortar. This requirement is met as long as the clay thin-layer finishes reduce the water vapor adsorption of earth plaster mortars of water adsorption class WS III

- More than 3 g/m^2 after 1 h.
- More than 7 g/m^2 after 6 h [34].

Thermal Conductivity

It must be determined according to Sect. 3.6.3.4 and after rounding up the dry bulk density values ρ_d to the next 0.1 kg/dm^3 according to DIN 4108-4.

Designation

According to DIN 18947:

Earth plaster mortar—main DIN number—letter symbol with maximum grain size and organic fiber/mineral aggregate—strength class—bulk density class.

Example Designation of an earth plaster mortar according to DIN 18947 with a maximum grain size of 2 mm and an organic fiber aggregate, strength class SII, bulk density class 1.6 kg/dm^3:

Earth plaster mortar—DIN 18947—LPM 02 f—SII—1.6.

Monitoring of the Constancy of Performance and Certification

The constancy of performance during the production of earth plaster mortar according to DIN 18947 is monitored according to the systems defined by the DIBt which are shown in Fig. 3.32.

System S4: Initial type testing and in-factory production control by the manufacturer; after successful testing, Declaration of Conformity by the manufacturer. Earth plaster mortars do not receive a conformity mark as is the case with all mineral plaster mortars.

Before in-factory production control takes place, the manufacturer must conduct an initial type testing which needs to correspond to the listed requirements specific to earth plaster mortar. Testing must include the characteristic properties listed in Table 3.11. In-factory production control by the manufacturer cannot commence until initial type testing has been successfully completed.

Table 3.11 Test system for monitoring the constancy of performance of earth plaster mortar according to DIN 18947

No.	Characteristic property of earth plaster mortar	Initial type testing	In-factory production control	Scope of testing of in-factory production control[b]	Declaration: Building material name/product data sheet[c]	Chapter
1	Oversized grain	○	○	400 tons	Product data sheet	2.2.3.1
2	Bulk density class of hardened mortar	○	○	400 tons	Building material name Product data sheet	3.6.1.3
3	Linear degree of shrinkage	○	○	400 tons	Product data sheet	3.6.2.1
4	Strength class according to Table 3.10				Building material name Product data sheet	
5	Dry compressive strength	○	○	400 tons	Product data sheet	3.6.2.2
6	Flexural strength	○			Product data sheet	3.6.2.2
7	Cohesive strength	○			Product data sheet	3.6.2.2
8	Fire performance, building material classes A1, A2, B1	○	○	Once per year	Product data sheet	3.4.8
9	Harmful salts (suspicion)	○				2.2.3.4
10	Grain size fraction (upper sieve size D)				Building material name Product data sheet	2.2.3.1
11	Mineral/organic aggregates				Building material name Product data sheet	3.4.2
12	Type of earth mortar in terms of form of supply				Product data sheet	3.5.6
13	Thickness of layer min./max.				Product data sheet	
14	Thermal conductivity				Product data sheet	3.6.3.2
15	Water vapor diffusion resistance factor				Product data sheet	5.1.2.2
16	Abrasion (suspicion)[a]	○			Product data sheet	3.6.2.2
17	Water vapor adsorption class[a] according to Table 3.32			Once every 2 years	Product data sheet	3.6.3.1 5.1.2.5
18	CO_2 equivalent[a]				Product data sheet	1.4.3.1
19	Activity concentration index[a]				Product data sheet	5.1.6.1

Inspection initiated by manufacturer ○: initial type testing; in-factory production control
[a]Voluntary
[b]At a yearly production volume ≤ 1600 tons or 4× per year at a production volume > 1600 tons; for earth mortar used in thin-layer plasters, the following applies: 200 tons or 8× per year in-factory production control at yearly production volumes of ≤ or > 1600 tons
[c]Integrated into: building material name, product data sheet

The packing slip must include the following information:

- Manufacturer and plant with manufacturer's logo.
- Designation, quantity, and form of supply of the earth plaster mortar in delivery.
- Recipient and date of delivery.

3.5.6.3 Sprayed Earth Mortar

Terminology

Sprayed earth mortar is earth mortar used for filling in skeletal framework using a spraying technique. Sprayed earth mortars are not considered earth plaster mortars according to Sect. 3.5.6.2. Sprayed earth mortars with a bulk density class ≤ 1.2 according to Table 3.7 can be referred to as sprayed light-clay mortars.

Construction Soil

Suitable construction soils are semi-rich to rich and low to semi-cohesive.

Processing, Aggregates and Additives

It is especially important for sprayed earth mortar mixes, and in particular for their aggregates, to be suitable for use in machines. Sand is a suitable mineral aggregate and suitable organic aggregates include sawdust as well as finely chopped straw fibers [23].

Application

Sprayed earth mortar is used as half-timber infill, for the construction of linings and steel-reinforced walls (Fig. 4.32), and as ceiling infill.

Characteristic Properties/Requirements

Sprayed earth mortars which are applied in single or multiple layers need to adhere firmly to the substrate or the formwork.

- Dry bulk density according to Sect. 3.6.1.3
- Linear shrinkage of the final mix (or a building element specimen if necessary) according to Sect. 3.6.2.1

Designation

Based on DIN 1894-47 the following designation is proposed for sprayed earth mortar:

Sprayed earth mortar—Lehmbau Regeln—letter symbol with mineral/organic aggregate—bulk density class.

Example Designation of a sprayed earth mortar mix according to the Lehmbau Regeln with an organic fiber aggregate, bulk density class 1.4 kg/dm^3:

Sprayed earth mortar—LR—LSM f—1, 4.

3.5.7 Earth Blocks

Terminology

Earth blocks are typically rectangular-shaped earth building materials which are made from unshaped earthen materials according to Sects. 3.5.1, 3.5.3, and 3.5.4. Based on DIN EN 771-1, the following terms are defined for earth blocks and are used in the corresponding DIN 18945.

Solid earth blocks are blocks with an overall perforation surface of <15 % running perpendicular to the horizontal bedding side.

Perforated earth blocks have an overall perforation surface of ≥15 % running perpendicular to the horizontal bedding side. Earth blocks with a bulk density class of <1.2 according to Table 3.7 are referred to as *light-clay blocks* based on DIN 18945.

A further distinction is made between the terms *block bulk density* (= gross dry bulk density, the perforation surface is ignored) and *material bulk density* (= net dry bulk density, equivalent to the bulk body density of fired bricks).

Stabilized earth blocks contain chemically modifying additives which alter the blocks' water solubility and strength.

In historical earth building, earth blocks were divided into the categories of earth blocks and green unfired bricks: *Green unfired bricks* were bricks intended for firing by industrial brick producers but used in their unfired state.

Construction Soil

The soil used for making earth blocks must meet the quality requirements of unshaped earth building materials, particularly in regard to cohesive strength and plasticity as well as grain size distribution. For unstabilized blocks, Houben/Guillaud [14] recommend the following range of plasticity for construction soil: PI=0.17–0.33 and LL=0.32–0.50. According to the authors, it is difficult to narrow down this range.

Table 3.12 Usage classes of earth blocks according to DIN 18945

No.	Usage class	Application
1	Ia	Plastered exterior masonry infill of timber-frame construction, exposed to the weather
2	Ib	Fully plastered, exterior masonry walls, exposed to the weather
3	II	Cladded, weather-protected exterior masonry walls, interior masonry walls
4	III	Dry earth block construction (e.g., ceiling infill or stacked wall lining)

Processing, Shaping, and Aggregates/Additives

The processing and shaping of earth blocks is carried out following the methods described in Sects. 3.1 and 3.2.2. Similarities can be found in the ceramics industry, with the exception of the final firing process.

There are three different *shaping methods* for making earth blocks according to DIN 18945:

f: Thrown—shaped by forceful (manual or) mechanized throwing or pouring into a formwork followed by a striking pass *without* further compaction
p: Compression molded—made by pressing or ramming into a form
s: Extruded—cut from an extruded block which has been forced through a nozzle

The shaping method largely influences the mechanical properties of earth blocks. It is declared as a characteristic property in the designation system of DIN 18945 using the letter symbols "f," "p," and "s."

Earth blocks of the usage classes I (and II) according to Table 3.12 must form a (largely) homogeneous structure without a shell-like texture (often found in historical "green" bricks) and without highly compressed areas close to the surface.

Earth blocks contain the aggregates and additives of the unshaped earth building materials they were made from in the shaping process. The following aggregates are permissible for earth blocks according to DIN 18945. The corresponding letter symbols for their designation are "m" and "f":

Mineral (*m*): Natural aggregates according to DIN EN 12620, crushed brick made from mortar-free bricks, expanded perlite/expanded clay/expanded glass (see Sect. 3.4.2)/expanded slate/natural pumice according to DIN EN 13055.
Organic (*f*): Plant parts and fibers, animal hair, chemically untreated chopped wood (no wooden composites).

According to DIN EN 12878, the addition of inorganic pigments is also permissible. All aggregates which have been added to the soil mix must be fully declared.

In many developing countries, but also in the USA and Australia, the use of synthetic binders (particularly cement, but also asphalt emulsion) in the production of earth blocks is common practice. The use of these binders is regulated by local codes.

Application

Earth blocks according to DIN 18945 are classified into usage classes based on their application and the type of load they are exposed to (Table 3.12). Earth blocks of the usage classes Ib and II can be used in load-bearing walls if they meet the strength requirements.

Characteristic Properties/Requirements

According to DIN 18945 earth blocks must meet the following requirements:

Formats, Perforation, and Dimensions

Earth blocks are rectangular-shaped building materials. Their ends can be smooth or with tongue and groove joints. Their sides can be smooth or with shaped profiles. Deviations from the rectangular shape of earth blocks are possible but need to be clearly declared by the manufacturer.

The dimensions of earth blocks are stated according to DIN 105-100 in the following order: length (l) × width × (w) × height (h) in mm or as format symbols (Table 3.13). Common earth block formats are the standard format NF 240 × 115 × 71 mm and the thin format DF 240 × 115 × 52 mm, or multiples derived from the thin format up to a maximum of 12 DF (Fig. 3.35). Deviating dimensions are also permissible.

For earth blocks produced according to DIN 18945, the following nominal sizes, minimum and maximum sizes as well as dimensional tolerances for length l, width w, and height h apply (Table 3.14). The nominal size is the standard size, the minimum and maximum sizes are the permissible dimensional deviations per block. The dimensional tolerance defines the permissible variation in block dimensions within a batch delivered to the building site.

Table 3.13 Nominal sizes and symbols for earth blocks according to DIN 18945

		Nominal sizes [mm]		
No.	Format symbol	Length l	Width w	Height h
1	1DF (thin format)	240	115	52
2	NF (standard format)	240	115	71
3	2DF	240	115	113
4	3DF	240	175	113
5	4DF	240	240	113
6	5DF	240	300	113
7	6DF	240	365	113
8	8DF	240	240	238
9	10DF	240	300	238
10	12DF	240	365	238

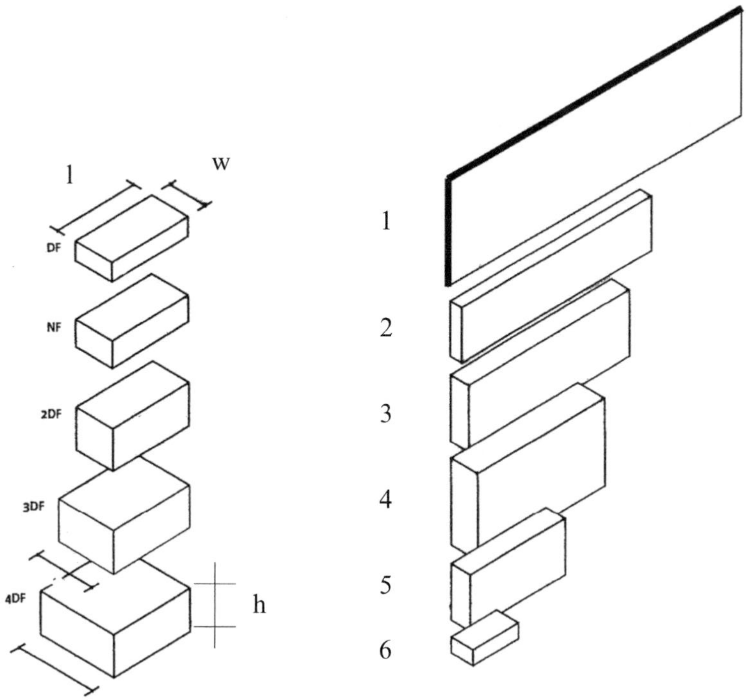

earth block formats according to DIN 105-100

l x	w x	h	[cm]
24 x	11.5 x	5.2	DF
24 x	11.5 x	7.1	NF
24 x	11.5 x	11.3	2DF
24 x	17.5 x	11.3	3DF
24 x	24.0 x	11.3	4DF

clay panel formats non-standardized

1	150 x	62.5 x	2.5
2	100 x	24.5 x	6.0
3	85 x	29.5 x	9.5
4	67 x	67.0 x	14.0
5	50 x	25.0 x	12.0 (10.0)
6	NF for comparison		

Fig. 3.35 Common formats of earth blocks and clay panels, according to [9]

Earth blocks for load-bearing masonry construction need to meet the requirements listed in Table 3.15.

Perforations in earth blocks of the usage classes I and II according to Table 3.12 should run perpendicular to the bedding side and should not fall below the minimum values for perforations, face shell thickness, and web thickness given in Table 3.15. Perforations should be distributed evenly across the bedding side. Their cross sections can take any shape but should not exceed 6 cm^2 with the exception of handling recesses and grip holes. Individual handling or grip holes should not exceed 25 cm^2 and are only intended for formats ≥3DF and special formats.

Recesses and grip holes for handling the block should be positioned in the center and only where necessary. The distance between two handling recesses or grip holes

Table 3.14 Dimensions of earth blocks according to DIN 18945

No.	Dimension	Nominal size [mm]	Minimum size [mm]	Maximum size [mm]	Dimensional tolerance [mm]
1	*l* and *w*	90	84	95	6
2	*l* and *w*	115	108	120	7
3	*l* and *w*	145	137	148	8
4	*l* and *w*	175	166	178	9
5	*l* and *w*	240	230	245	12
6	*l* and *w*	300	290	308	14
7	*l* and *w*	365	355	373	14
8	*h*	52	48	54	4
9	*h*	71	66	74	5
10	*h*	113	107	118	6
11	*h*	155	149	160	6
12	*h*	175	169	180	6
13	*h*	238	231	243	7

Table 3.15 Permissible perforation, minimum face shell, and web thickness of earth blocks of different usage classes according to DIN 18945

No.	Usage class	Permissible perforation of bedding side [%]	Face shell thickness [mm]	Web thickness [mm]	Direction of perforation in relation to bedding side
1	Ia	Unperforated[a]	≥50	≥70	Perpendicular
2	Ib	≤15	≥30	≥20	Perpendicular
3	II	≤15 (≤30)[b]	≥20	≥20 (≥4)[b]	Perpendicular
4	III	No requirements			Any direction

[a]With the exception of two centrally arranged grip holes for formats ≥3DF and special formats with an overall surface of ≤15 %

[b]Values for non-load-bearing earth masonry

should be at least 70 mm and there should not be any additional perforations between them. Grip holes are counted as part of the overall perforation surface.

The perforation volume used for determining the net volume for material bulk density can be calculated according to DIN EN 772-9.

Perforations of earth blocks are listed as a characteristic property in the DIN 18945 designation using the letter symbol "g" for "perforated."

Concentration of Harmful Salts

According to Sect. 3.5.6.

Table 3.16 Compressive strength classes of earth blocks according to DIN 18945

		Compressive strength [N/mm²]	
No.	Compressive strength class	Average value	Lowest single value
1	2	2.5	2.0
2	3	3.8	3.0
3	4	5.0	4.0
4	5	6.3	5.0
5	6	7.5	6.0

Dry Bulk Density

Earth blocks are divided into classes of block bulk density according to Table 3.7, starting with bulk density class 0.5. Earth blocks for load-bearing masonry construction must conform to a minimum bulk density class of 1.4. For bulk density classes ≤1.0 individual values should not be 0.05 kg/dm³ above or below the limit, for bulk density classes >1.0 not more than 0.1 kg/dm³ above or below.

Strength

The strength properties of earth blocks are defined by compressive strength classes according to Table 3.16. The compressive strength class corresponds to the smallest permissible single value. The average value of a series (which consists of at least three test specimens according to the Lehmbau Regeln) must be at least 25 % above the value of the compressive strength class. Test results should not be lower than the defined average values and minimum single values according to Sect. 3.6.2.2.

Earth blocks for load-bearing masonry construction must conform to a minimum compressive strength class of 2. Earth blocks for non-load-bearing masonry construction need to be sufficiently strong for processing and their intended use. This is generally the case if they have a compressive strength of at least 1 N/mm².

Deformation Behavior

Under Load. Earth blocks for load-bearing applications must have a minimum modulus of elasticity of 750 N/mm². Earth blocks of compressive strength class ≥2 generally meet this requirement. When in doubt, a test initiated by the manufacturer must be conducted according to Sect. 3.6.2.2.

Exposure to Moisture and Frost. Earth blocks must have sufficient moisture and frost resistance for their respective area of application and must meet the requirements of Table 3.17. Any swelling in earth blocks of usage classes I and II must not affect their use and surface finishing. The respective tests are conducted according to Sect. 3.6.3.1.

Table 3.17 Moisture and frost behavior requirements for earth blocks according to DIN 18945

No.	Usage class	Dip test loss of mass [%]	Contact test	Suction test [h]	Frost test [cycles]
1	Ia	≤5	No cracking or swelling	≥24	≥15
2	Ib			≥3	≥5
3	II	≤15		≥0.5	No requirements
4	III	No requirements	No requirements	No requirements	

Resistance to Water Vapor Diffusion

It can be assumed as $\mu = 5/10$ without testing or can be determined using the test method according to Sect. 3.6.3.4.

Thermal Conductivity

It must be determined according to Sect. 3.6.3.4 and after rounding up the dry bulk density values ρ_d to the nearest 0.1 kg/dm^3 according to DIN 4108-4.

Fire Performance

Earth blocks according to DIN 18945 can continue to be classified as building material class A1 without additional testing according to DIN 4102-4 as long as the organic aggregates are homogeneously distributed and do not exceed 1 % of the block's mass or volume, whichever is higher (Sect. 3.4.8). Building material classes A1, A2, and B1 must be certified once a year as part of compliance monitoring by an authorized body. Building material class B2 must be verified as part of in-factory production control by the manufacturer (Table 3.18).

Designation

According to DIN 18945 the designation of earth blocks is implemented as follows:

Earth block load-bearing/non-load-bearing—DIN number—letter symbol of earth block and production method, perforation (if applicable) and compressive strength class—usage class—bulk density class—letter symbol of format.

Example Designation of a load-bearing compressed earth block without perforations of the compressive strength class 3, usage class Ib, bulk density class 1.6, with a length of 240 mm, a width of 115 mm, and a height of 71 mm ("standard format NF"):

Earth Block—load-bearing—DIN 18945—LS p 3—Ib—1.6—NF.

Table 3.18 Test system for monitoring the constancy of performance of earth blocks according to DIN 18945

No.	Characteristic property of earth block	Initial type testing	In-factory production control/ compliance monitoring	Scope of testing of in-factory production control[d]	Declaration: Building material name/product data sheet[e]	Chapter
1	Usage class (UC I[a]) and type of use "load-bearing[a]/ non-load-bearing"		●		Building material name Product data sheet	3.5.7
2	Production methods				Building material name Product data sheet	3.2.2
3	Format, perforation and dimensions	○	○/●	250 m³ LS	Building material name Product data sheet	3.5.7
4	Bulk density class according to Table 3.7	○	○/●	250 m³ LS	Building material name Product data sheet	3.6.1.2
5	Compressive strength class according to Table 3.16	●	○/● for strength class ≥2	250 m³ LS	Building material name Product data sheet	3.6.2.2
6	Deformation behavior in the presence of moisture/frost Usage class I Usage class II	○	●			3.6.2.1
7	Mineral/organic aggregates	○			Product data sheet	3.4.2
8	Harmful salts (suspicion)					3.5.6
9	Thermal conductivity				Product data sheet	3.6.3.2
10	Water vapor diffusion resistance factor				Product data sheet	5.1.2.2
11	Fire performance, building material class B2[b] A1, A2, B1[a]	○	○ ○/●	Once per year	Product data sheet	3.4.8

(continued)

Table 3.18 (continued)

No.	Characteristic property of earth block	Initial type testing	In-factory production control/ compliance monitoring	Scope of testing of in-factory production control[d]	Declaration Building material name/product data sheet[e]	Chapter
12	CO_2 equivalent[c]				Product Data Sheet	1.4.3.1
13	Water vapor adsorption class[c]			Once every 2 years	Product Data Sheet	3.6.3.1 5.1.2.5
14	Activity concentration index[c]				Product Data Sheet	5.1.6.1

Inspection initiated by:
Manufacturer ○: initial type testing; in-factory production control
Authorized body ●: initial monitoring; compliance monitoring
[a]Certificate of conformity by authorized certifying body
[b]Declaration of conformity by manufacturer
[c]Voluntary
[d]Characteristic properties must be inspected for usage class Ia every 500 m^3, for usage classes Ib and II every 1000 m^3
[e]Integrated into: building material name, product data sheet

This information must be clearly legible and attached to the packaging and/or included in the enclosed product information sheet along with the manufacturer's logo.

Monitoring of the Constancy of Performance and Certification

The constancy of performance during the production of earth blocks according to DIN 18945 is monitored according to the systems defined by the DIBt which are shown in Fig. 3.32.

Load-Bearing Earth Blocks and Usage Class I

System 2+: Initial monitoring of factory and in-factory production control, continuous monitoring of in-factory production control (compliance monitoring) by the authorized certifying body; after successful testing, Certificate of Conformity issued by authorized certifying body; initial type testing and in-factory production control carried out by the manufacturer; after successful testing, Declaration of Conformity by manufacturer.

Non-load-Bearing Earth Blocks

System S4: Initial type testing and in-factory production control by the manufacturer; Declaration of Conformity by the manufacturer.

Before the in-factory production control takes place, the manufacturer must conduct an initial type testing which consists of the testing of the characteristic properties listed in Table 3.18. In-factory production control cannot commence before initial type testing has been successfully completed.

Compliance monitoring of in-factory production control and of personnel and technical requirements to ensure adequate production and correct labeling of the earth blocks must be carried out at least twice a year by an authorized body.

The packing slip must include the following information:

- Manufacturer and plant with manufacturer's logo
- Designation and number of earth blocks in delivery
- Recipient and date of delivery

3.5.8 Clay Panels

Terminology

Clay panels are basically flat, panel-shaped building materials which are loaded *perpendicular* to the plane. Compared to their surface dimensions, their thickness d is small. There are no defined dimensional limits, the line between "earth blocks" and "earth panels" is blurred.

Wall panels made of earth building materials are flat, panel-shaped building elements which are loaded *parallel* to the plane. They need to be sufficiently rigid to prevent buckling.

Clay panels with a dry bulk density of $\rho_d < 1.2$ kg/dm^3 can be referred to as light-clay panels.

Format

Currently there are no standardized formats for clay panels. Their dimensions can vary greatly and are not subject to regulations (Fig. 3.35). Panel thicknesses depend on the intended use:

- Thin clay panels (16–50 mm thick)
- Thick clay panels (>50–100 mm thick)
- Heavy clay panels (>100 mm thick)

Thin clay panels are similar to drywall panels in terms of their size and usually require a substructure. *Thick* clay panels are block-like and their formats are similar to those of earth blocks. Like heavy clay panels, they are self-supporting and might have perforations parallel to the panel plane.

Heavy clay panels are designed as hollow-core sheets or solid panels (wall panels) depending on their application.

Construction Soil

Clay panels are produced from prepared, unshaped earth building materials according to Sects. 3.5.1, 3.5.3, and 3.5.4. The construction soil used in the production of clay panels must meet the quality requirements of unshaped earth building materials, particularly in terms of cohesive strength and plasticity as well as grain size distribution. Cohesive clay powders or dry soils amended with sand are used for "thin" clay panels while rammed earth is used in the production of solid panels (wall panels).

Processing and Shaping

Clay panels are produced using special shaping technologies (Sect. 3.2.2). For example, belt presses are used in the production of "thin" clay panels (Fig. 3.27c). "Thick" clay panels are produced using a standard extrusion pressing method as well as other methods used in the production of building elements. On-site manual shaping is also possible. Prefabricated, large-format rammed earth wall panels for load-bearing wall construction require an adequate formwork system.

The panel edges are either smooth or have a tongue and groove design (around all the edges). A mortise and tenon joint design is also possible.

Aggregates and Additives

Dry soil has a limited capacity to absorb tensile forces and flexural tensile forces. The unshaped earth building materials which are used in the production of clay panels should therefore be reinforced with suitable fiber. Reinforcement can also be achieved using integrated mats or webbing made of plant fibers.

The organization natureplus e.V. has developed a "Clay Panels" guideline for the certification process of its quality seal [35]. This guideline applies to industrially produced clay panels and the section on "Composition, Banned and Restricted Materials" states the following:

The "clay panel" product must contain 99 % mineral and renewable resources using clay minerals as the main binder. Natural materials which have been synthetically modified (such as waxes and derivatives of cellulose and starch) must not exceed 10 % of the mass.

Clay panels are specifically not allowed to contain the following materials:

- Biocides
- Organohalogen materials
- Synthetic materials and fibers (e.g., acrylates, polyvinyl acetates), with the exception of waxes and chemically modified natural materials such as methyl cellulose

Application

Clay panels can be installed in a number of ways: using masonry methods, butt-jointed, or installed as dry construction. They can be attached using standard fasteners or glued. Clay panels have many applications which determine their dimensions and composition, for example:

- *Thin panels* for lining and facing interior building elements and non-load-bearing partition walls with a substructure (Sect. 4.3.6.2), also as permanent formwork or interior lining in multilayer wall construction or as a "substitute" for plaster through the use of dry clay plasterboards (Sect. 3.5.6.2); they can also be used as dry flooring panels and for lining the interior of pitched roofs.
- *Thick panels* for non-load-bearing partition walls without a substructure or as lining panels in existing structures, as infill in pitched roofs or as ceiling linings and infill. Clay panels with integrated heating coils or hypocaust elements for wall heating as hollow-core sheets (Sect. 4.3.7.3) present a special type of application.
- *Heavy panels* as ceiling infill panels (Sect. 4.3.4.4). Prefabricated large-format rammed earth wall panels for load-bearing wall construction require adequate assembly techniques.

Characteristic Properties/Requirements

- Formats, dimensions, and perforations/cavities if applicable
- Dry bulk density according to Sect. 3.6.1.3
- Flexural strength
- Shrinkage and swelling behavior (joints and connections)

For the production of clay panels, the organization natureplus e.V. requires adherence to the following ecological reference values per m^3 of finished product in connection with their certification process [35] (Table 3.19). If individual reference values are exceeded, it must be determined (on a case-by-case basis) if these are permissible for achieving an overall optimization of production. The product needs to be tested for pesticides (Table 6.1), if there is any suspicion of their presence, and for increased radioactivity levels (Sect. 5.1.6.1).

3.5.9 Other Earth Building Materials

In addition to the earth building materials mentioned in Sects. 3.5.1–3.5.8 other earth building materials for special building projects, e.g., for renovation and restoration, can be produced and sold by companies according to respective specifications. One example is the production of straw-clay reels for use as infill in timber-frame walls or ceilings (Sect. 4.3.4.1).

Table 3.19 Ecological parameters for the production of clay panels

Indicator	Reference value	Test method
Primary energy, nonrenewable [MJ/m^3]	4000	
Primary energy, total, incl. renewable [MJ/m^3]	9000	Life cycle inventory analysis according to DIN EN ISO 14040
Global warming potential [kg CO_2 equiv./m^3]	450	Impact categories according to CML[a] 2001
Photochemical smog [kg ethylene equiv./m^3]	0.1	Primary energy demand according to Frischknecht 1996
Acidification potential [kg SO_2 equiv./m^3]	1.0	Global warming potential 1994/100 years
Ozone depletion potential [kg CFC-11 equiv./m^3]	5 E-05	Up to the ready-to-ship product
Over-fertilization potential [kg PO_4 equiv./m^3]	0.2	

[a]Inst. of Environmental Sciences Leiden

3.6 Parameters and Tests for Earth Building Materials

As a result of the lack of technical developments in the field of earth building from the 1950s to the 1980s (Sect. 1.3), scientific research is in its early stages compared to the mass-produced mineral building materials of concrete and fired brick. There is particular demand for the development of earth-specific test methods and the systematic identification of parameters of building materials and building elements. Issues concerning the sustainability of earth building also pose complex problems and testing criteria still need to be developed.

New approaches to test methods have been developed in recent years in connection with the creation of new DIN standards for earth building materials (Sects. 3.5.6.1, 3.5.6.2, and 3.5.7). These approaches, however, still need to prove themselves in the practical earth building field.

The most important parameters and tests for earth building materials are classified according to the following main groups (Table 1.1):

- Mass and structural parameters
- Deformation parameters
- Strength parameters
- Building physics parameters

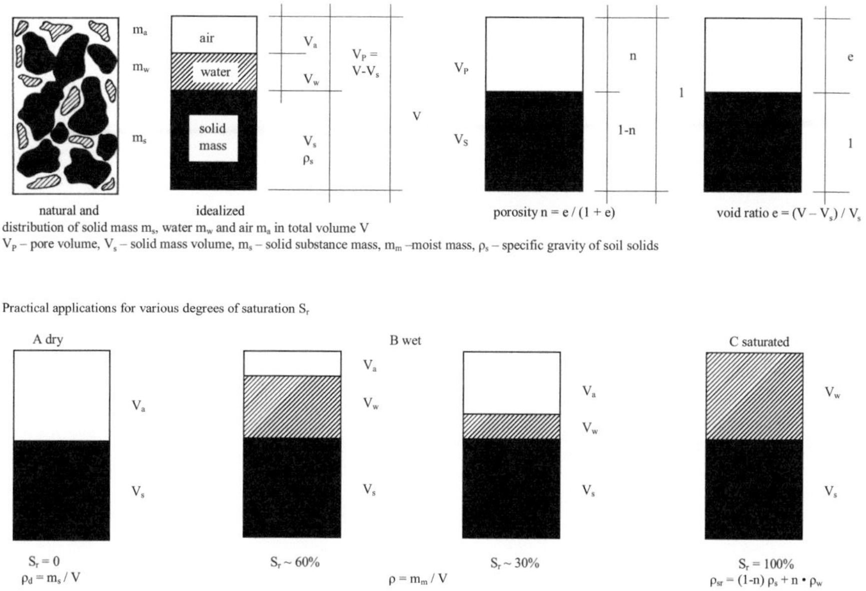

Fig. 3.36 Distribution of solid mass and pores in earth building material according to [81]

3.6.1 *Mass and Structural Parameters*

Every earth building material consists of a solid mineral mass and voids or pores which are formed by the mineral particles and aggregates (Fig. 3.36). These pores are either filled with *air* or partially or completely filled with *water* (Sect. 2.1.1.2). The distribution and spatial arrangement of these three components—the structural elements or phases—have a significant effect on the processing and deformation properties of wet earth building materials as well as on the strength and building physics properties when the finished structure is in use.

3.6.1.1 Porosity and Void Ratio

In order to describe the void content of a soil sample, two parameters are used in a model of idealized distribution of solid mass, water and air in the total volume V. Here, the void content V_p is based on either the volume of the total sample V (porosity n) or the volume of the solid mass V_s (void ratio e) as a proportion by volume. Both parameters are non-dimensional. The porosity n is often given as a percentage. Both parameters can be expressed through each other:

Table 3.20 Filling conditions of the pores of earth building materials for practical applications

Case	Filling of voids	Processing method of the earth building material (example)	Consistency of the earth building material	Bulk density of the earth building material
B	With air or water vapor, and water	"Wet" processing or shaping of earth building materials (rammed earth), drying starts after formwork is removed	Semisolid, stiff, soft and paste-like	$\rho = m_m/V$
C	With water, completely filled	Earth mortar (earth plaster)	Paste-like to liquid	$\rho_{sr} = (1-n) \bullet \rho_s + n \bullet \rho_w$ $S_r = 1$
A	With air or water vapor, completely filled	"Dry" processing (earth blocks)	Solid	$\rho_d = m_s/V$ $S_r = 0$

$$\textit{void ratio } e = V_p / V_s = n / (1-n)$$
$$\textit{porosity } n = V_p / V = e / (1+e).$$

The model of the idealized distribution of the three phases of solid–liquid–gas can be used to show practical cases which are significant for the processing of earth building materials into earth building elements and structures (Fig. 3.36) (Table 3.20). Based on these cases of phase distribution in the total volume of the earth building material, different types of density can be differentiated between.

3.6.1.2 Bulk Density ρ/Saturated Bulk Density ρ_{sr}

Terminology

The *bulk density* ρ of a wet sample of earth building material (also called wet bulk density) is typically expressed as the ratio of its moist mass m_m to its volume V (Case B):

$$\rho = m_m / V \left[g / cm^3 \right]$$

The ratio of the void volume filled with water to the total void volume of the specimen is also called the *saturation level* S_r (also known as degree of saturation).

The *saturated bulk density* ρ_{sr} refers to the density at which all voids are filled with water (Case C):

$$\rho_{sr} = \rho_d + n \times \rho_w \left[g / cm^3 \right]$$

with $\rho_w = 1.0$ g/cm³ (= density of the water).

In this case, the degree of saturation S_r of a sample is 1.

$$S_r = V_w / V_p \left[-\right].$$

The *unit weight* γ of a wet earth building sample is generally expressed as the ratio of its dead load G, allowing for the gravitational acceleration g as a permanent load, to its volume V:

$$\gamma = G / V = \rho \times g \left[kN / m^3 \right].$$

For this purpose, the gravitational acceleration is estimated at $g = 10$ m/s^2. In this manner, the corresponding weights (strength parameters and assumed loads for structural calculations, 1 kg–10 N) can be derived from the densities (mass parameters).

Test Methods

To test the bulk density ρ, the moist mass m_m of an earth building material specimen is determined through weighing and its volume is determined through measuring (dip test) or with the help of a known volume, such as by using a tube soil sampler. These tests are based on DIN 18125-1.

Lab and Calculation Values

The specific density ρ_s of the earth building material specimen is dependent on the true density ρ_s of the solid mineral material (of the construction soil), the amount of mixing water as well as the drying progression. DIN 1055-2 specifies unit weights γ of naturally moist construction soils based on empirical values (Table 3.21).

These numbers apply to the characteristic weights of naturally formed cohesive soils. They can also be applied to loose and compacted cohesive soils as long as the compaction degree is $D_{pr} \geq 0.97$. For soils with particularly high coefficients of uniformity C_u (glacial marl, clay-rich soils, mixed-particle soils of the soil groups GU, GT, SU, ST and GU*, GT*, SU*, ST*) the weights must be increased by 1.0 kN/m^3.

3.6.1.3 Dry Bulk Density ρ_d

Terminology

According to DIN 18125-1 the dry bulk density ρ_d is determined by the ratio of the dry mass m_d of a sample of earth building material to its volume V (Case A: drying has been completed).

Table 3.21 Unit weights γ of construction soils for assumed loads

No.	Soil type with letter symbol according to DIN 18196	Consistency	γ [kN/m³]
1	Slightly plastic silts UL (LL<0.35)	Soft Stiff Semisolid	17.5 18.5 19.5
2	Semi-plastic silts UM (LL=0.35–0.5)	Soft Stiff Semisolid	16.5 18.0 19.5
3	Slightly plastic clays TL (LL<0.35)	Soft Stiff Semisolid	19.0 20.0 21.0
4	Semi-plastic clays TM (LL=0.35–0.5)	Soft Stiff Semisolid	18.5 19.5 20.5
5	Highly plastic clays TA (LL>0.5)	Soft Stiff Semisolid	17.5 18.5 19.5

Test Methods

For the experimental identification of the dry bulk density ρ_d of a soil sample, the mass of a moist specimen is determined through weighing while the corresponding volume is established through immersion weighing (e.g., a specimen enclosed in paraffin). The water contained in the void volume of the wet sample is then extracted by drying the specimen at +105 °C. What remains in the specimen is the water bound in the capillary water films surrounding the clay minerals.

The dry bulk density ρ_d can also be determined mathematically using the known value of the wet bulk density ρ and the corresponding moisture content w. Three subsamples are obtained from the wet specimen (2 from the edge w_{e1} and w_{e2}, and one from the center w_c). These subsamples are weighed and the corresponding water content is determined by drying them at +105 °C.

$$w=\left(w_{e1}+2w_c+w_{e2}\right)/4.$$

$$\rho_d=m_s/V=m_m/V\left(1+w\right)=\rho/\left(1+w\right).$$

The Lehmbau Regeln [22] define that the specimen used for determining the dry bulk density ρ_d of *unshaped earth building materials* should be shaped into cubes with an edge length of 200 mm and prepared in the same manner as "on the building site."

For testing *earth mortars* according to DIN 18946-47, mortar prisms with dimensions of 160×40×40 mm are prepared according to DIN EN 1015-11 (Table 3.25). Depending on when final shrinkage is reached (Sect. 3.6.2.1), the specimens are removed from their molds after 2–7 days and stored on paper on top of a grate.

For testing, a series of three mortar prisms is conditioned under standard atmospheric conditions (23 °C/50 % RH) until constant weight is reached. Constant weight has been reached when the results of two consecutive weighings at 24 h intervals differ by no more than 0.2 mass in % based on the smaller measured value. For in-factory production controls, the following fluctuations are tolerated: air temperature ±5 °C, RH ± 15 %. The bulk density is calculated from the mass in relation to the external volume of the mortar prism.

For *rammed earth* it is nearly impossible to prepare specimens in the same manner as "on the building site." In order to determine the achieved level of compaction, obtaining specimens with the aid of a tube sampler during actual construction is recommended as an alternative. The sample should be taken from the upper layer of soil which has just been compacted and is still wet. It is also important to ensure that the specimen is "undisturbed" in terms of grain composition, compaction density, and moisture content. The moisture content of the extracted specimen must remain constant until testing begins (quality class 2, Sect. 2.2.2.2).

For testing *shaped earth building materials*, the materials themselves can be used along with specimens cut to a suitable size. These specimens should be in a dry state (= dry mass m_d).

To determine the block bulk density of earth blocks according to DIN 18945 (Sect. 3.5.7), a series of three earth blocks is conditioned under normal atmospheric conditions. The block bulk density is calculated from the determined mass of the earth blocks in relation to their respective volume (including any perforations) and then an average of the series is obtained.

For testing perforated earth blocks, suitable unperforated rectangles are cut out of the block and conditioned as described above. They are weighed, coated with heated paraffin (applied with a brush) in order to make them watertight and then weighed again. Next, the specimens are weighed under water (using a suitable scale) in a container filled with distilled water. Alternatively, the so-called sand filling method according to DIN EN 772-9 can be used.

Lab and Calculation Values

The Lehmbau Regeln [22] give dry bulk densities ρ_d of earth building materials as calculation values for the time when the material is in use (Table 3.22):

Dry earth building materials are hygroscopic which means that the surfaces of the clay minerals can bind water molecules contained in the air. This results in a moisture content which is referred to as the "equilibrium moisture content" (Sect. 5.1.2.4) for the time when the material is in use (40–70 % RH, +20 °C). The corresponding "dry" bulk density differs numerically from the dry bulk density determined under laboratory conditions at +105 °C. This is taken into account with the help of conversion factors used for determining the thermal conductivity λ in the laboratory.

Table 3.22 Dry bulk densities of earth building materials

No.	Earth building material	Dry bulk density ρ_d [g/cm³]
1	Rammed earth	1.700–2.400
2	Cob	1.400–1.700
3	Straw clay, clay with added fibers	1.200–1.700
4	Light clay	300–1.200
5	Earthen loose fill	300–2.200
6	Earth mortar	600–1.800
7	Earth blocks	600–2.200
8	Clay panels	300–1.800

The dry bulk density ρ_d is a characteristic property in the designation of earth blocks and earth mortars according to DIN 18945-47. It is listed in the form of classes according to Table 3.7.

3.6.1.4 Proctor Density ρ_{pr}

Terminology

For every earth building material, the dry bulk density which can be attained through compaction depends on the compaction work and the material's moisture content. With an *optimal moisture content* w_{pr} and predefined compaction work, the highest dry bulk density—the *standard or Proctor density* ρ_{pr},—can be attained for every soil.

Test Methods

The experimental determination of the Proctor density is defined in DIN 18127. It requires a minimum of four individual tests with different moisture contents. The prepared soil specimen is placed into a steel cylinder in three equally thick layers and compacted using evenly distributed blows from a free-falling rammer (Fig. 3.37). The ramming mass, drop height, dimensions, and rotation of the specimen are predefined using mechanical compaction. These specifications result in a compaction level of 60 Ncm/cm³. Compaction can also be carried out using manual rammers in which case the predefined value cannot be exactly set.

Lab and Calculation Values

Placed in an orthogonal coordinate system (w; ρ_d), the single values form a characteristic parabolic compaction curve with the highest attainable dry bulk density ρ_{pr} at w_{pr} as the vertex. Table 3.23 [11] shows empirical values of ρ_{pr} and w_{pr} for different construction soils.

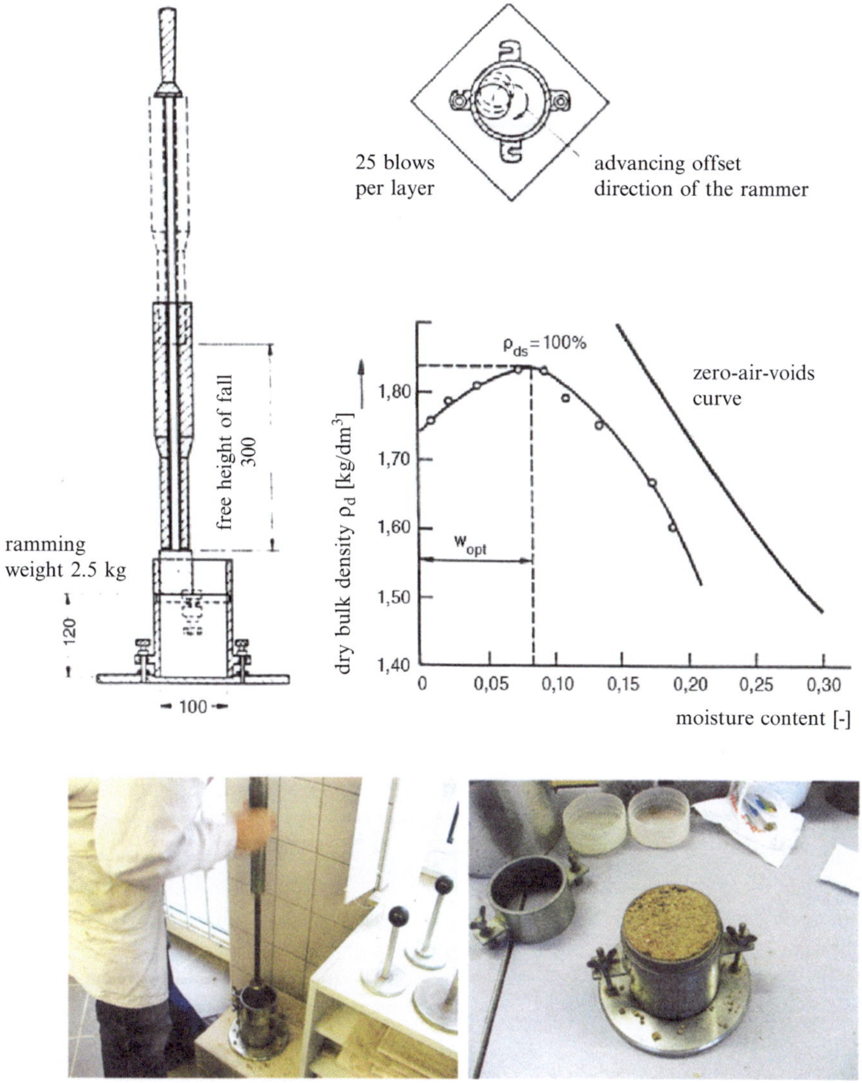

Fig. 3.37 Determining the Proctor density [11, 43]

Table 3.23 Proctor densities of different construction soils

No.	Soil	Proctor density ρ_{pr} [g/cm³]	w_{pr} [–]
1	Loess, loess soil	1.70–1.85	0.18–0.13
2	Glacial till, glacial marl	1.80–2.00	0.14–0.11
3	Eluvial soil	1.75–1.95	0.17–0.10
4	Slope-wash soil	1.65–1.85	0.21–0.14
5	Fluvial soil	1.50–1.75	0.27–0.16
6	Pure clay	<1.50	>0.25

For practical compaction work, the attained bulk density D_{pr} is given as the dry bulk density ρ_d in percentages of the Proctor density ρ_{pr} (=100 %) as a quality requirement:

$$D_{pr} = \rho_d / \rho_{pr} [-]$$

As part of the construction monitoring process of the "Himmelsleiter" project in Nordhausen, Germany [36], the following values were determined for the rammed earth used in the project:

$$\rho_{pr} = 1.81 g / cm^3 \quad at \quad w_{pr} = 12.66\%.$$

The attained bulk density of the compacted rammed earth was in the range of D_{pr}=0.95–0.98. This corresponded to values most commonly required for earth works.

Influencing Variables

The Proctor test was developed to verify the attained quality of compression work in road construction. In principle, it can also be employed for comparable applications using impulse and ramming compaction (Sect. 3.2.2.1) in earth building, e.g., for rammed earth building elements.

Particularly for load-bearing earth building elements, it is important to achieve a high density of the mineral grains through appropriate compaction, thereby minimizing the void volume. While this is relatively easy for non-cohesive gravels and sands due to their comparatively large pores, the air contained in clay soils can only be pushed out of the larger pores. As a result of low porosity, air contained in the finer pores, which are partly enclosed by water, is very difficult to remove or cannot be removed at all.

The *moisture content at the time of installation* has the following effect on compaction: If the vertex is approached from the "dry" side of the compaction curve, the capillary strength of the soil sample obstructs compaction: the applied compaction work is not high enough to completely break up the crumb structure of the soil. On the "wet" side, the pore water or capillary tension limits compaction: the compaction tool "bounces." The compaction curve runs roughly parallel to the zero-air-voids curve which specifies the saturated bulk density of the soil ρ_{sr} (Fig. 3.37).

A higher *liquid limit LL* or *plasticity PI* results in an increase in the water binding quality of the soil, which is based on mineral-chemical conditions, and the optimal moisture content w_{pr}. At the same time, the maximum dry bulk densities ρ_{pr} decrease (Fig. 3.38). It can also be observed that a decreasing maximum dry bulk density ρ_{pr} and an increasing optimal moisture content result in a flatter compaction curve [11].

Figure 3.39 shows the relationship between compaction work, achievable dry bulk density, and moisture content at the time of installation based on the Proctor test [11]. This relationship has been established for road and embankment construction

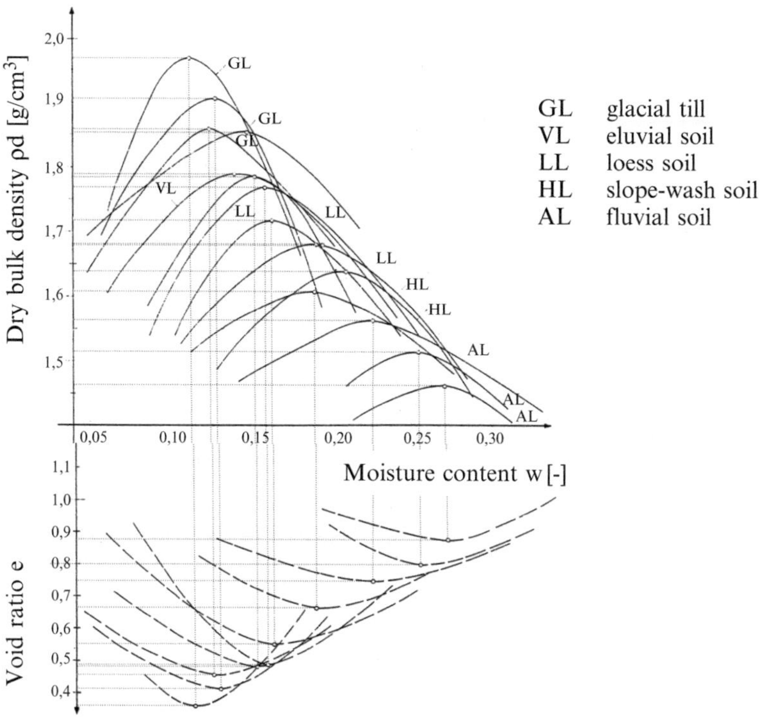

Fig. 3.38 Proctor curves for different construction soils [11]

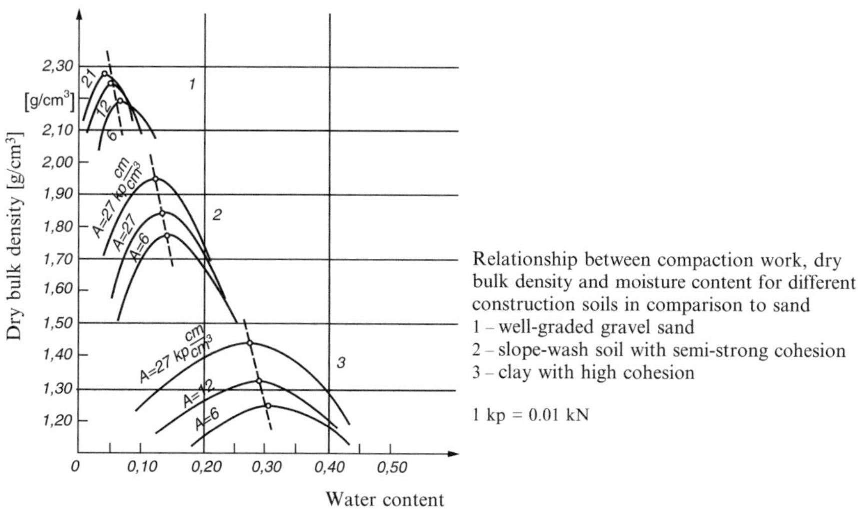

Fig. 3.39 Influence of compaction work A on the Proctor density [11]

and, in principle, can also be applied to earth building: If the *compaction work* is increased while the moisture content at the time of installation remains the same, the maximum dry bulk densities ρ_{pr} increase in a more or less logarithmic manner. An increase in compaction work thus makes it possible to use (nearly) dry, unshaped earth building materials. This virtually solves the problem of shrinkage deformations. Applications in the earth building field include the "dry compression" of earth blocks and clay panels [12] (Sect. 3.2.2.1) and high-pressure spraying of nearly dry light-clay mixes as infill for timber-frame construction [24].

On the other hand, a direct comparison between standard compaction work applied in road construction and the specific compaction work applied in rammed earth construction is not possible. Compared to rammed earth construction, the level of compaction in road construction seems to be higher because the maximum dry bulk densities are achieved at a relatively low moisture content at the time of installation. This might be an explanation for the fact that the maximum dry bulk density of rammed earth samples is achieved at moisture contents $w > w_{pr}$ [26].

It also becomes obvious that pure clays clearly exhibit lower dry bulk densities compared to construction soils with much wider grading (here: slope-wash soil) when compacted at the same level. This can be explained by the micropores in the clay which are enclosed by water. When compared to construction soils, considerably higher pressure needs to be applied to pure clays to extract water in order to achieve higher dry bulk densities.

3.6.1.5 Specific Density ρ_s

Terminology

The term *specific density* ρ_s (also: specific gravity G_s) refers to the average density of the minerals of the grain mixture contained within the nonporous solid substance m_s (Fig. 3.36)

$$\rho_s = m_s / V_s \left[g / cm^3 \right].$$

Test Methods

DIN 18124 describes the pycnometer method used to determine the specific density ρ_s. For this test, a pycnometer (a glass flask with a stopper) is filled about two thirds full with distilled water. Next, a prepared, dry soil sample is added and weighed at +20 °C. The air in the water was removed earlier through boiling. The mixture is vaporized and the mass of the soil sample is determined.

Table 3.24 Specific densities of different construction soils

No.	Soil	Specific density ρ_s [g/cm³]
1	Loess, loess soil	2.65–2.70
2	Glacial till	2.68–2.72
3	Eluvial soil	2.68–2.74
4	Fluvial soil (clay)	2.69–2.75
5	Pure clay	2.70–2.78

Lab and Calculation Values

Specific densities ρ_s can be found in the range of approximately $\rho_s = 2.65–2.80\ g/cm^3$ and can be used for calculations as tabular values or according to Table 3.24 [11].

Influencing Variables

Soils with high proportions of Al or Fe in their clay mineral substance (e.g., laterite soils, Sect. 2.1.2.6) exhibit much higher single values whereas soils with organic or lime portions (Sect. 2.2.3.4) are below the values listed above.

3.6.2 *Structural Parameters*

Structural parameters describe the behavior of an earth building material or earth building element under load as a result of external stresses. It is essential to know these structural parameters in order to ensure the structural integrity of load-bearing earth building elements.

A distinction is made between

- Deformation parameters
- Strength parameters

In general, *strength parameters* (β) are used to describe the resistance of a building material to deformation caused by external forces (stress). It is important to know the stress limits of the material in order to prevent its failure. The stress which a building material can be permanently exposed to is referred to as creep resistance.

Deformation parameters, on the other hand, describe the path leading to the point where failure occurs. This path is shown as the relationship between stress (σ, τ) and deformation (ε: compression/expansion or s: displacements). In soil mechanics, the relationship is described with the help of Hooke's law or the Mohr—Coulomb failure criterion. These principles generally deal with subgrade or hydraulic engineering in connection with soil material with a moist consistency in its finished state.

(Load-bearing) building elements and structures made of earth building materials are in a "dry" state once they are in use. For calculation purposes, this state needs

Table 3.25 Deformations of building materials, overview

	Load dependent		
Deformations	Immediate	Time dependent	Load independent
Reversible	Elastic ε_{el}	Delayed elastic $\varepsilon_{v,el}$	Thermal expansion ε_T Moisture expansion ε_f
Non reversible (permanent)	ε_{bl} settling	Flow viscous, plastic ε_{fl}	Chemically induced expansion ε_c Cracking

to be described using corresponding parameters (e.g., dry bulk density ρ_d, Sect. 3.6.1.3). Once the building elements are in use, the remaining moisture is defined as the equilibrium moisture content (Sect. 5.1.2.4). Load transfer within the building element occurs via the "grain-to-grain pressure" mechanism (Fig. 2.34).

For earth building, it is also important to consider the state of the building during construction (before drying has been completed), as well as possible water damage during the building's lifetime. In such situations, the load-bearing earth building elements are (still) wet and load transfers with significant deformations are only possible to a limited extent.

3.6.2.1 Deformation Parameters

Deformations ε of materials are generally defined as the ratio of the volume change ΔV to the initial volume V as a result of exposure to various external stresses

$$\varepsilon = \Delta V / V.$$

The accompanying sign specifies the type of deformation: expansion (+), compression (–). Vertical compression is also called settling.

Deformation parameters can be classified according to Table 3.25.

Load-Independent Deformations

Terminology

Load-independent deformations of earth building materials are a result of volume changes caused by different exposures or mechanisms:

- *Thermal strains* ε_T are caused by a change in temperature ΔT of the solid mineral substance.
- *Moisture strains* ε_f are caused by the release or absorption of physically bound pore water and are referred to as *shrinking* (–) and *swelling* (+) (Sect. 2.2.3.3). They are *reversible*. Expansions caused by freezing pore water (+) constitute a special type.

– *Chemically induced strains* ε_c. For chemically stabilized earth building materials (lime, cement), *shrinkage* (–) can also be of importance. "Chemical shrinkage" results in a permanent decrease in volume caused by the chemical binding of water. Here, the volume of the new formation is always smaller than the sum of the volumes of the binding agent and water. However, when gypsum hardens the volume of the new formation is larger than the sum of the volumes of the initial material and water. These deformations are referred to as *chemical expansion* (+).

It is important to point out that the term "shrinkage" describes two different types of volume decrease which have to be clearly distinguished between: it can refer either to a loss in volume caused by the physical act of drying (shrinkage limit SL, Sect. 2.2.3.2) or to a loss in volume caused by a chemical reaction.

The ceramics industry distinguishes between the terms "shrinkage during drying" and "firing shrinkage." *Shrinkage during drying* describes the decrease in volume of the unfired pieces caused by the evaporation of the physically bound water before firing (Sect. 3.5.7 "unfired bricks"). *Firing shrinkage* refers to an additional and irreversible decrease in volume which is caused by a loss of the chemically bound water within the clay during the sintering process at the time of firing ([6], *Bd. 3: Thermische Prozesse*). In earth building, firing shrinkage does not occur. Therefore, it is sufficient to use the general terms "shrinkage" and "degree of shrinkage" in the following sections.

If a building element is unable to move freely during load-independent deformations, stresses occur which can result in *cracks* once the strength of the building material is exceeded. Although these cracks generally do not affect the stability of the building element, they restrict its usability. There are external and internal obstructions of movement. External obstructions can occur, for example, in building elements which are fixed in place while internal obstructions are caused by fluctuations in temperature and moisture across the building element (e.g., different degrees of drying in a rammed earth wall).

Test Methods

Linear Degree of Shrinkage: The linear degree of shrinkage (or linear shrinkage) of earth building materials is determined using a similar method as for testing construction soils. However, the moisture content at the time of installation and the dimensions of the prism-shaped specimens are different for the individual earth building materials (Table 3.26). Mortar prisms for earth mortar are prepared according to DIN EN 1015-2.

The demolded specimens are stored on a piece of plastic wrap and dried until the final degree of shrinkage under standard atmospheric conditions (for earth mortar 23 °C/50 % RH) is reached. The constant weight of earth mortar according to DIN 18946-47 has been reached when the results of two consecutive weightings taken at 24 h intervals differ by no more than 0.2 mass in % based on the smaller value. The linear shrinkage is the change in the specimen's length compared to its initial length in %. A test consists of a series of three specimens from which the average change in length is calculated and given as the result.

Table 3.26 Specimen dimensions for determining the linear shrinkage, overview

Earth building material	Dimensions (l × w × h) [mm]	Distance between measuring marks [mm]	Moisture content/ consistency at time of installation	Source	Chapter
Construction soil	220 × 40 × 25	200	Standard consistency	DIN 18952-2	2.2.3.3
Rammed earth	600 × 100 × 50	500	From ready-to-use mix	Lehmbau Regeln [22]	3.5.1
Earth mortar	160 × 40 × 40	–	Slump 175 ± 5 mm	DIN EN-1015-3 and 11 or DIN 18946-47	3.5.6

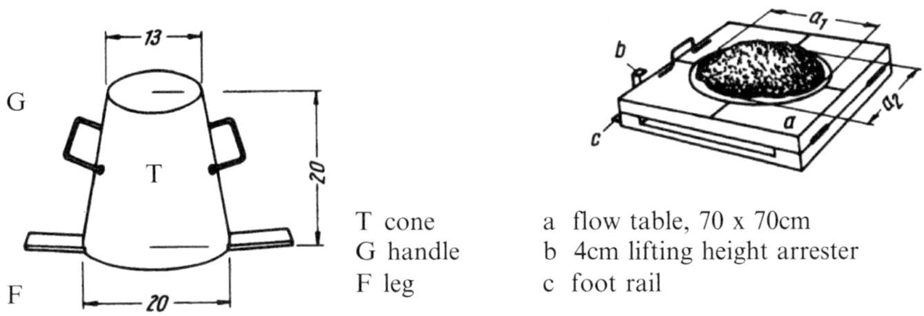

Fig. 3.40 Determining the slump of fresh earth mortar [37]

Volumetric shrinkage deformations of earth building elements can be tested for different earth building materials with the help of a building element sample.

Slump of Fresh Earth Mortar: This test requires a flow table and a truncated cone with predefined dimensions. The cone is placed exactly in the center of the flow table and filled with 1.5 L of the test earth mortar. The cone is then removed by slowly pulling it upward. Next, the flow table is lifted all the way to the arrester (4 cm) and dropped 15 times at 1-min intervals. During this process, the mortar should not separate or crumble (Fig. 3.40, [37]).

Afterwards, the diameter of the mortar is measured in two perpendicular axes a_1 and a_2 with the help of a sliding caliber and the arithmetic mean is calculated. The consistency of the mortar is time dependent, the elapsed time after mortar preparation (when water is added) is therefore given in the form of an index, e.g., a_{15} = slump after 15 min.

The abovementioned mortars must conform to the declared properties. Sampling needs to be conducted according to DIN EN 1015-2.

Lab and Calculation Values

For linear shrinkage of construction soils, see Sect. 2.2.3.3.

Influencing variables

The shrinkage deformations is influenced by a number of factors: the respective consistency of the earth building material which is attained by adding water, the structure and proportion of clay minerals in the construction soil (cohesive strength), the pore structure of the earth building material and aggregates as well as the drying conditions. Shrinkage deformations typically increase with a higher moisture content at the time of installation, an increase in cohesive strength, a denser pore structure, and faster and uneven drying.

Load-Dependent Deformations

Terminology

Load-dependent deformations are caused by dead loads, other permanent loads, and live loads. Depending on the duration of the load stress, these deformations are divided into *immediate* or *long-term* deformations with elastic and plastic, or plastic and delayed-elastic portions (Table 3.25).

The term elastic means that deformations caused by exterior strains develop shortly after the loads are applied and reverse immediately after load removal (swelling).

Plastic deformations remain permanent after load removal. After a specific elastic limit has been reached the so-called "plastic flow" occurs. This means that the deformation continues to increase as time progresses without a further increase in stress. This condition is referred to as viscous behavior. Liquids as well as solid materials generally resist deformation. Their molecules are bound by the Van der Waals forces. When the plastic flow sets in, these forces are continuously overcome and bonds are newly formed. In earth building, this characteristic becomes relevant in practical situations, for example, when preparing earth mortar to the required consistency (Fig. 3.40).

In the first case of linear-elastic material behavior, the relationship between normal stresses σ and compression ε forms a straight line. The slope of this straight line between two normal stresses σ_1 and σ_2 is the uniaxial *modulus of elasticity in compression E* (also called: elastic modulus E, E modulus, Young's modulus) based on Hooke's law (Fig. 3.41):

$$E = \Delta\sigma_z / \Delta\varepsilon_{z,el} \left[N / mm^2\right]$$

$$actual\,\sigma = E \cdot \varepsilon_{el}.$$

For the spacial state of stress for an elastic body, this results in

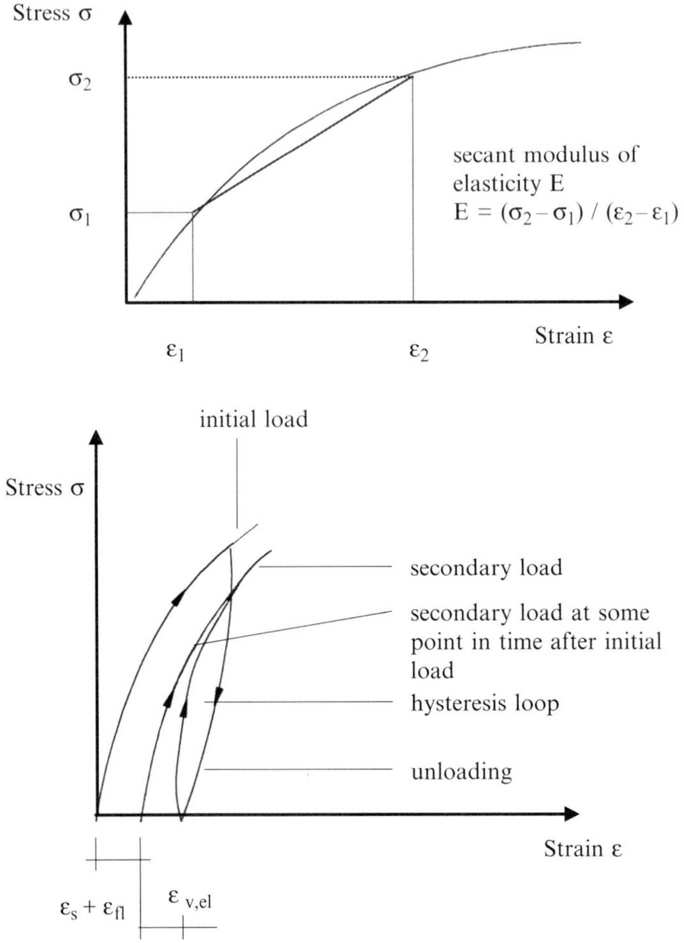

Fig. 3.41 Deformations of elasto-plastic materials: stress-strain diagram and determination of the secant modulus of elasticity, according to [42]

$$\varepsilon_x = 1/E\left[\sigma_x - v\left(\sigma_y + \sigma_z\right)\right],\ for\,\varepsilon_y, \varepsilon_z accordingly.$$

Here, ν (also μ) is the Poisson's ratio. In the elastic range, it expresses the ratio between the lateral strain ε_x and the longitudinal compression ε_z

$$v = \varepsilon_x / \varepsilon_z\,[-].$$

As a dimensionless material constant, its values range from 0 to 0.5 and are often between 0.1 and 0.4. For clay-rich soil, the Poisson's ratio is $\nu = 0.30–0.45$ [38].

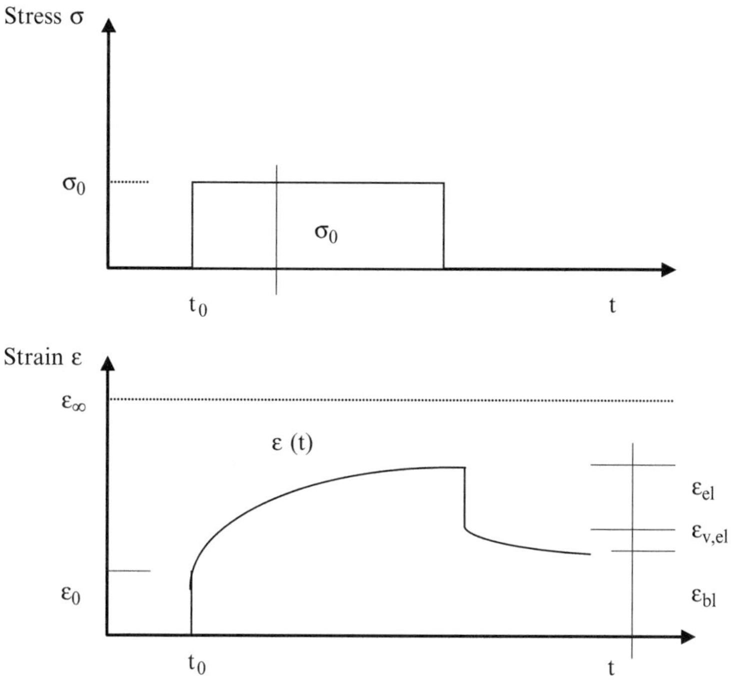

Fig. 3.42 Deformations of visco-elastic materials: creep, according to [42]

The linear-elastic behavior of a specimen caused by the shear force Q or the shear stress τ is described by *shear modulus G*. For isotropic material it is connected to the modulus of elasticity E as follows:

$$G = E/2(1+v)\left[N/mm^2\right], \left[MN/m^2\right].$$

"Linear elastic" behavior is the ideal situation. In reality, transitional forms are more likely with an overlapping of elastic deformations and plastic or viscous behavior after loading and unloading.

Materials exhibiting *elasto-plastic deformation behavior* (for example, concrete but also clay-rich soils) have a curved σ–ε–line. In order to be able to apply Hooke's law to these materials as well, the E modulus is defined as a secant or tangent modulus (Fig. 3.41). Corresponding to the variable rise of the σ–ε–line, the E modulus is also variable and must be specified for the respective stress interval σ_2–σ_1.

After unloading has occurred, the deformation reverses immediately by the amount of the elastic portion ε_{el}. If the unloading is permanent, the delayed elastic expansion $\varepsilon_{v,el}$ is also reversed. The permanent expansion ε_{bl} now consists of *settling* ε_s and *flow* ε_{fl}. The intersection of the unloading curve with the x-axis corresponds to the permanent expansion. With delayed reloading, the expansion decreases further by the delayed elastic portion $\varepsilon_{v,el}$. The loading and unloading curves form the hysteresis loop.

An elastic expansion ε_{el} occurs immediately upon loading, and a continued permanent load σ_0 results in an increase in deformation called *creep* (Fig. 3.42). The total strain increases over time aiming for a final state $\varepsilon_{k\infty}$ which consists of delayed elastic and delayed permanent portions. For concrete and masonry, creep is almost entirely finished after 3 to 5 years, provided that stress, temperature, and humidity remain constant.

If the expansion ε_0 is constant, the deformation-causing stress σ_0 decreases over time. Under constant load, the initial stress σ_0 is decreased by viscous structural changes until the residual stress becomes too small to allow further flow. The stress drop $(\sigma_0 - \sigma_t)$ relates to the initial stress and is referred to as *relaxation* ψ: $\sigma_t = \sigma_0(1 - \psi_t)$.

Due to the proportionality between creep stresses σ_k and creep expansions ε_k a stress-independent creep coefficient Φ or final creep coefficient Φ_∞ is introduced as a parameter to describe this property.

$$\Phi = \varepsilon_k / \varepsilon_{el} \quad or \quad \Phi_\infty = \varepsilon_{k\infty} / \varepsilon_{el} = \varepsilon_{k\infty} \cdot E / \sigma_k .$$

Test Methods

E Modulus. It is possible to determine the deformation behavior of earth building materials using a test which is based on the standard test method of the elasticity modulus E in soil mechanics. In this test, cylindrical specimens with a specimen to height ratio of 1:1.5 are used [39]. The three test methods in Table 3.27 (according to [38]) show different possibilities of lateral expansion of the loaded specimens based on their applications in soil mechanics.

Applied to the deformations of a vertically loaded, load-bearing wall made of earth building materials, different deformations perpendicular to the plane of the load application within the axes can occur: the lateral expansions in the wall's longitudinal axis are more confined than in the lateral axis. The constrained modulus E_s based on the oedometer consolidation test would correspond to the "prevented" expansions in the longitudinal axis, whereas the uniaxial or E modulus based on the unconfined compression test would more likely describe the "unconfined" lateral expansion.

It would be possible to model such a situation in the triaxial test according to DIN 18137-2. The condition of use which is described in soil mechanics as "moist" corresponds, in the field of earth building, to a situation of water damage or to building materials during construction. In order to examine the behavior of earth building materials in a "dry" condition of use, specimens with a solid consistency would have to be tested. No known tests exist.

Dierks and Ziegert [27] determined the E modulus based on DIN 1048-5 (concrete) using dry, prism-shaped rammed earth specimens with the dimensions $150 \times 150 \times 300$ mm and unconfined lateral expansion.

The E modulus for *earth blocks* according to DIN 18945 is determined using a compression test machine of at least class 2 according to DIN EN ISO 7500-1.

Table 3.27 Load-dependent deformation parameters of earth building materials, experimental determination, overview

Test	Unconfined compression test	Triaxial test	Oedometer consolidation test
DIN	18136	18137-2	18135
Lateral expansion	Unconfined	Confined	Prevented
Strains/ deformations	$\sigma_x = \sigma_y = 0$; $\sigma_z \neq 0$ $\sigma_z = F/A$ A—specimen cross section F—applied compression force d—specimen diameter h—specimen height	1. Hydrostatic phase $\sigma_x = \sigma_y = \sigma_z = \sigma$; $\varepsilon_x = \varepsilon_y = \varepsilon_z = \varepsilon$; $\varepsilon_z = \Delta h/h$ 2. Shear phase $\Delta\sigma_z = \Delta\sigma_1 > 0$; $\varepsilon_z = \Delta h/h$; $\varepsilon_x = \varepsilon_y = \Delta d/d$	$\varepsilon_x = \varepsilon_y = 0$; $\varepsilon_z \neq 0$; $\sigma_z \neq 0$ $\sigma_z = F/A$; $\varepsilon_z = \Delta h/h$ $\sigma_x = \sigma_y$ radial stresses
State	Uniaxial state of stress (sign) • Longitudinal compression $\varepsilon_z = \Delta h/h$ (−) • lateral extension $\varepsilon_x = \varepsilon_y = \Delta d/d$ (+)	Rotationally symmetric stress and strain state	Uniaxial deformation state
E-Modulus	$E = \Delta\sigma_z/\Delta\varepsilon_z$ (based on Hooke, also Young's modulus)	$E = \Delta\sigma_z/\Delta\varepsilon_z$ (based on Hooke, also Young's modulus)	Constrained modulus $E_s = \Delta\sigma_z/\Delta\varepsilon_z$
Diagram	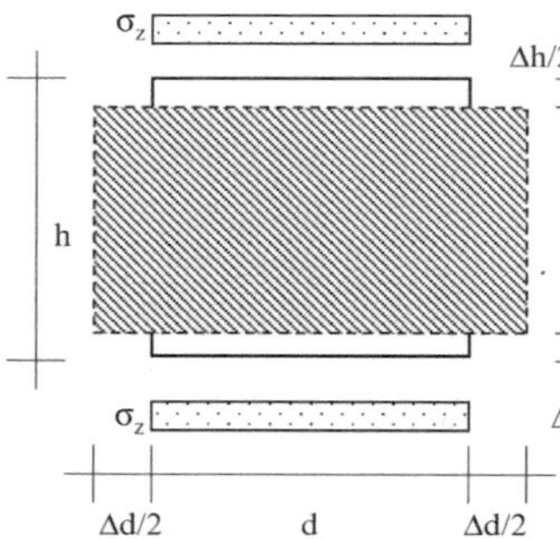	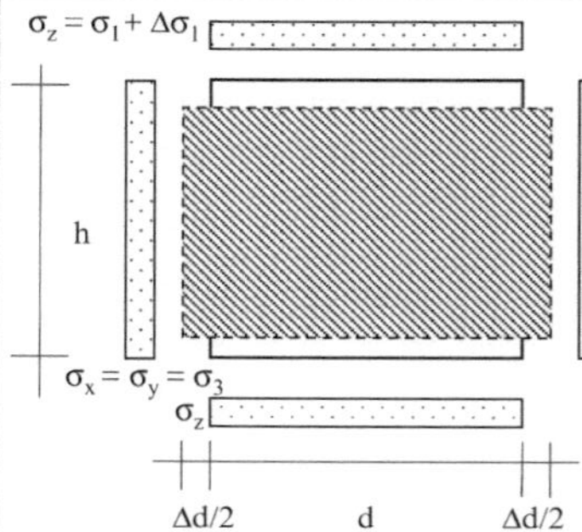	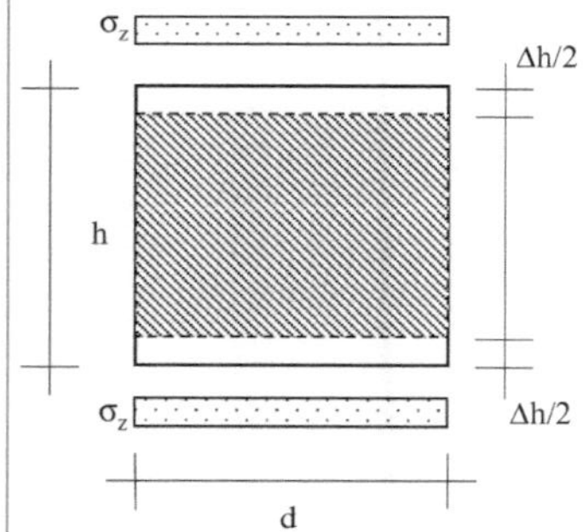

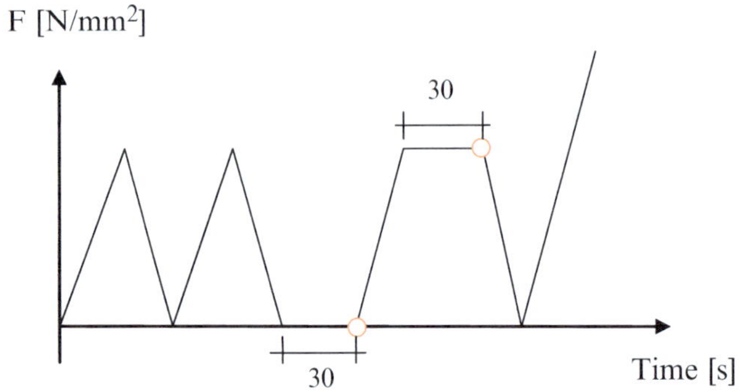

Fig. 3.43 Time sequence for the measurement of the *E* modulus for earth blocks according to DIN 18945

To produce the test specimens, earth blocks with a nominal height of ≤71 mm are sawed in half perpendicular to their longitudinal axis. The halves are laid on top of each other and joined with mortar. The cut surfaces should face away from each other in opposing directions. Cement mortar (mix: 1 part cement of strength class 42.5 to 1 part washed natural sand 0/0.1) is used for joining the blocks or leveling the pressure-loaded areas. The slenderness ratio of the height/width of the specimen must be ≥1.

The blocks do not have to be evened out with mortar if the specimens can be sanded to make them plane parallel. In this case, the pressure-loaded area must form a continuous surface. It is important to ensure that the sanding does not pull aggregates >1 mm out of the surface.

The prepared specimens are conditioned under normal atmospheric conditions (23 °C/50 % RH) until a constant weight is reached. This can be confirmed by weighing the specimens two times at 24 h intervals. Their difference should not exceed 0.2 mass in % based on the smaller value.

For earth blocks with nominal heights >71 mm, the *E* modulus must be tested using a full block.

The *E* modulus is determined in the third load cycle at a load of 1/3 of the ultimate load. First, the specimen is loaded twice and then once again after a 30 s interval using 1/3 of its ultimate load each time. Each load is maintained for 30 s (Fig. 3.43). If needed, the dry compressive strength can also be tested after a further unloading phase. The same test setup can be used for determining the *E* modulus and the dry compressive strength. The test is shown in Fig. 3.44 with earth blocks which have been cut using a chisel according to ARSO [40].

A third of the ultimate load means a load selected for all specimens which corresponds to a minimum of 1/3 and a maximum of 0.4 times the ultimate load.

A special problem is posed by the deformation behavior of earth building materials under dynamic loads. This is particularly important for earth building structures in seismic areas (Sect. 5.2.4.2). Olivier and Velkov [41] used three different soil samples, one of them cement-stabilized, to determine the *dynamic E modulus*. Here,

Fig. 3.44 Determining the *E* modulus and the dry compressive strength of earth blocks [40]

a dynamic load is applied to the specimens in the form of dynamic vibration (0.5, 1.0, and 2.0 Hz) after the individual levels of the designed load path have been reached. The load levels were in the elastic, non-linear plastic range and near the ultimate load.

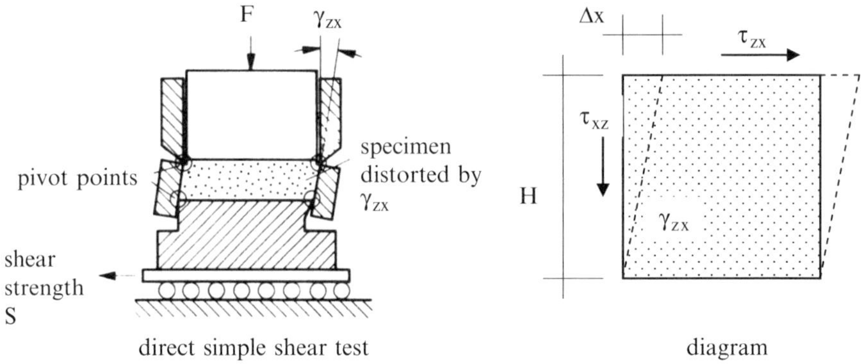

Fig. 3.45 Deformation behavior of earth building materials: lateral expansion and shear distortion, according to [38, 78]

A comparison of the static E moduli, which were determined at the same time as the dynamic E moduli, led to the following result: It appears that a considerate solidification took place during the dynamic load application in all specimens. The dynamic E modulus at the highest load level (near the ultimate stress) was higher than the corresponding static E moduli by a factor of 1.75–1.80 in all specimens.

Shear Modulus: The so-called direct simple shear test is used to describe the shear modulus G as the quotient of the shear stress τ_{zx} and the shear distortion γ_{zx} (Fig. 3.45)

$$G = \tau_{zx} / \gamma_{zx}.$$

The shear displacement diagram is used to compare the applied displacement distances s to the measured shear stresses τ. For clay-rich soils and pure clays of at least stiff consistency and under normal stresses σ, the ultimate limit state τ_{Br} (= shear strength β_s) is attained after only short displacement distances and, after loosening the specimen, falls to a low degree of residual shear strength or residual strength. In soft pure clays, the ultimate limit state is attained after greater displacement distances.

Lab and Calculation Values

Elastic Material Behavior

Table 3.28 [42] lists E moduli for the elastic material behavior of some of the main building materials.

For rammed earth, Dierks and Ziegert [27] specify a time-dependent function of creep to describe the *elasto-plastic material behavior* as follows:

$$\varepsilon_k(t) = 0.0654 \ln(t) + 0.62 \quad for \quad t > 0.25d$$

d = number of days after load application.

Table 3.28 *E* moduli of selected building materials

No.	Building material	*E* modulus [N/mm^2]	Comment
1	Glass	50,000–85,000	
2	Masonry	500–15,000	
3	Standard concrete	15,000–60,000	
4	Steel	200,000–210,000	
5	Wood, along the grain Wood, across the grain	7000–18,000 300–1500	
6	Rammed earth	550–960 700–7000 $300 \times \beta_D$ 500	[27, 60]; after 10 load cycles of up to 1/3 of the failure load [14]; declaration without load range NZS 4297 [50] [80]
7	Earth blocks according to DIN 18945	≥750	Stress diagram Fig. 3.43 and Sect. 3.5.7

According to this function, a permanent load of 0.4 N/mm^2 after 200 days of use would result in a creep deformation of $\varepsilon_{k\infty} < 0.1$ %. This means that due to its optimized grain composition and low shrinkage deformation (<0.1 %) the rammed earth used in this test is comparable to concrete.

Viscous Material Behavior

According to the Lehmbau Regeln [22], *slump* $a = 140$ mm for mortar prisms is used to determine the linear degree of shrinkage (Sect. 3.5.6, Fig. 3.40).

3.6.2.2 Strength Parameters

Depending on the type of stress a building element is subjected to, different strength types can be divided into compressive strength, tensile strength, flexural strength, buckling strength, shear strength, and torsional strength. In an actual stress situation, different stress types are likely to overlap in the building element.

In terms of strength parameters in earth building, testing has so far been largely limited to compressive strength. In most cases, it is sufficient to know the values for compressive strength because the test methods "automatically" measure the strength properties across the axis of the load application.

To test the compressive strength of earth building materials, methods common to concrete and masonry construction have been adopted or modified. As the applications of earth building materials become more diversified, the need for testing and regulation of other stress types in earth building increases. New standardized testing procedures need to be developed or the methods used in concrete or masonry construction must be appropriately modified.

Structural strength parameters are typically determined using short-term tests which means that it takes approx. 1 min to achieve the maximum load.

For dynamic stresses (wind, earthquakes), a dynamic portion, in addition to structural strength, needs to be taken into account.

Dry Compressive Strength

The compressive strength β_D of an earth building material is generally expressed as the stress which leads to the failure of the building material caused by a force F acting vertically on the loaded cross section A

$$\beta_D = \max . F / A \left[N / mm^2 \right].$$

According to the Lehmbau Regeln [22], the compressive strength needs to be verified for dimensioning load-bearing earth building elements. For the earth block and earth mortar product groups according to DIN 18945–47, the dry compressive strength is a characteristic property and must be declared accordingly.

Rammed Earth and Cob

Safety Concept. When producing rammed earth specimens in a laboratory setting, the compaction conditions are different from those of the actual building element on site. This results in higher dry bulk densities as well as higher compressive strengths. Furthermore, the tested strength reflects the short-term strength which is higher in terms of numbers compared to the creep strength of the actual building element. This is caused by the comparatively fast load application until failure load is reached during the test.

For the mathematical verification of the compressive strengths within the load-bearing wall construction, only a fraction of the compressive strength determined in the laboratory is used (Sect. 4.2.3.1).

Distinctions are generally made between:

- The *arithmetic* mean of a minimum of three individual tests
- The *characteristic value* taken under consideration of a specific fractile of the statistical distribution of the individual test results
- The *design or calculation value* including a safety factor

In the Lehmbau Regeln [22], this "global" margin of safety between the *dry compressive strength* β_D, which is material specific and is determined in the laboratory, and the *permissible compressive stress* σ_p, which is proven mathematically for the earth building element, is defined as approx. seven times the permissible compressive stress. The characteristic values determined for a compressive strength test are assigned to a strength class which is used to derive the permissible compressive stress in walls as a calculation value according to Table 3.29.

Table 3.29 Compressive strength classes and permissible compressive stresses for rammed earth and cob

Compressive strength class β_D [N/mm²]	1[a]	2	3	4
Permissible compressive stresses in walls σ_p [N/mm²]	0.2	0.3	0.4	0.5

[a]For cob only

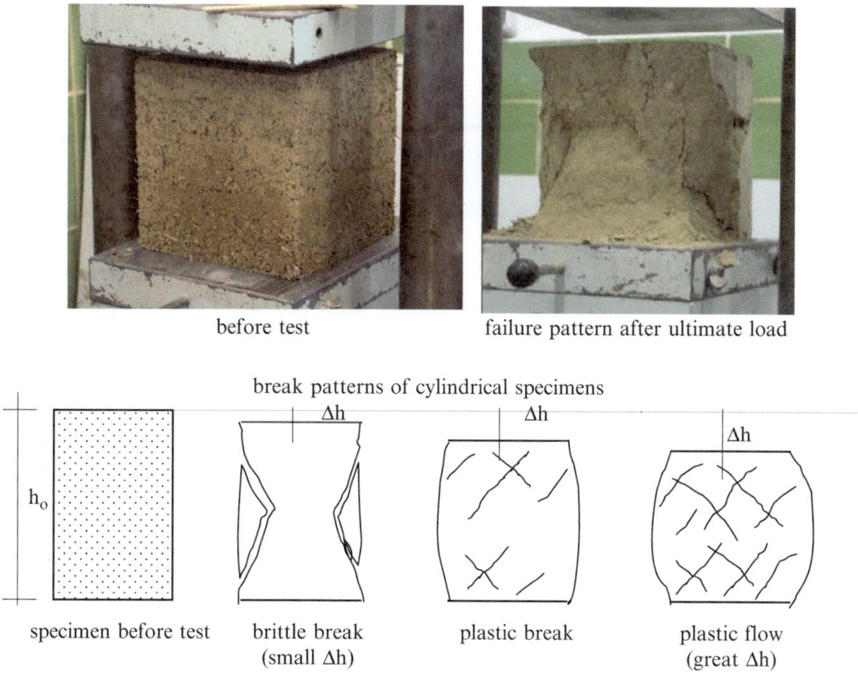

Fig. 3.46 Testing the dry compressive strength of rammed earth [43]

The smallest value of the three single values which need to be determined per test should not be lower than the value $\beta_D = 2.0$ N/mm². Accordingly, the minimum value for the calculated compressive stress in walls is set at $\sigma_p = 0.3$ N/mm². On the other hand, the maximum value for strength class 4 is limited to a compressive stress in walls of $\sigma_p = 0.5$ N/mm².

For pillar-like walls, the permissive stresses up to 1.5 times the minimum cross section of the wall must be reduced by a factor of 0.8.

Test Methods. According to the Lehmbau Regeln [22], the dry compressive strength β_D for rammed earth and cob is determined using a minimum of three specimens per test. The specimens are made in the laboratory with the aid of steel cube molds with an edge length of 20 cm. The direction of the applied load during testing must be identical to the direction of the applied compaction work during the production of the specimens. When the specimens are installed in the test apparatus, their bedding sides are leveled using a cement mortar layer no thicker than 5 mm. Load is applied to the dry specimen in a press until failure occurs (Fig. 3.46). After the ultimate load has been reached, the specimen exhibits a failure pattern which is typical for unconfined lateral expansion: instead of compressing, the soil is displaced at a 45° angle to the longitudinal axis which is a shear stress break as shown by the remainder of the specimen in the illustration. This failure pattern is typical for brittle materials which have much lower shear strength than compressive strength.

In order to be able to determine the dry compressive strength in a laboratory setting, the specimen needs to achieve the prevailing equilibrium moisture content of the building element in its finished state. The specimens are dried under standard atmospheric conditions (+20 °C, 65 % RH) until a constant weight has been reached. This process should not be artificially accelerated which means that drying times of at least 6 weeks need to be scheduled.

With an increase in water content in wet specimens, a brittle break turns into a plastic break and finally into plastic flow. The "wet" state corresponds to situations when earth building elements are exposed to water damage.

There are no uniform regulations on an international scale concerning the test methods for determining the dry compressive strength of rammed earth.

Lab and Calculation Values. At the Bauhaus University in Weimar, Germany, research was conducted on the strength development of *rammed earth* while drying [26, 43]. A total of 105 cube specimens with an edge length of 20 cm each were made using manual compaction. In the subsequent tests, three different mixes (rammed earth, rammed earth with straw fibers, loess soil as construction soil for rammed earth and rammed earth with straw fiber mixes) were used to compare the attained cube compressive strengths. Different moisture contents at the time of installation w ($w < w_{pr}$, w_{pr}, $w > w_{pr}$) and different drying times t (t=7, 14, 28, 45, 90 d) were included (Fig. 3.47). Overall, values in the range of β_D=0.90–3.89 N/mm^2 were attained. The dry compressive strength β_D was determined as a continuous test run along with the dry bulk density ρ_d (Sect. 3.6.1.3).

Dierks and Ziegert [27] established cube compressive strengths of β_D=2.4–3.5 N/mm^2 for cubes with an edge length of 20 cm and ρ_d=2.24 g/cm^3. Samples with flax fiber aggregates attained the highest value.

The cube compressive strengths of rammed earth samples with edge lengths of 10 cm (5.6 N/mm^2) and 15 cm (2.9 N/mm^2) determined by Fischer et al. [44] are not directly comparable. In these tests, the specimens with a moisture content of w–w_{pr} (Sect. 3.6.1.4) had the highest strengths.

The strengths β_D=2.6–4.2 N/mm^2 specified by Minke [45] refer to cylindrical specimens with a diameter of 7.6 cm and a height of 10 cm and can therefore not be directly compared. Maniatidis and Walker [46] achieved similar results with an average value of β_D=2.46 N/mm^2 for cylindrical specimens with a diameter of 10 cm and a height of 20 cm. The specimens were made with $w = w_{pr}$ and dried for about 4 weeks under standard atmospheric conditions until a constant weight was achieved.

Ziegert [47] conducted oedometer consolidation tests on 15, 20, and 30 cm test cubes made of *cob*. The specimens were cut out of three different existing wall structures and planed. The scale of the average compressive strengths of the cubes was found in the range of β_D=0.63–1.12 N/mm^2. The tendency of rammed earth samples to experience a decrease in uniaxial compressive strength with an increase in specimen size could not be confirmed for the tested cob samples. It appears that local discontinuity within the soil structure or mechanical influences during the production of the specimens played a more important role.

The Lehmbau Regeln [22] continue to include "empirical values" for the dry compressive strength of earth building materials as shown in Table 3.30.

Table 3.30 Dry compressive strength of earth building materials, empirical values

Building material	Dry bulk density ρ_d [kg/dm³]	Dry compressive strength β_D [N/mm²]
Rammed earth with mineral aggregate	2.0–2.2	3–5
Rammed earth with plant fiber aggregate	1.7–2.0	2–3
Cob	1.4–1.7	1
Earth blocks	1.6–2.2	2–4

Table 3.31 Permissible compressive stresses in walls and specimen dimensions for determining the dry compressive strength of rammed earth, specifications in different national regulations

Country	Source	σ_p [N/mm²]	Specimen dimensions [mm]	Comment
Australia	[70]	0.7	Cylinder/prism $d=150$; $h=110$ $h=150$; $l=150$; $w=1.3$ h	Unstabilized (cement-stabilized 5.2), margin of safety using coefficient of variation
	[49]	1.0	Shape factors	Curing time 28 days, unstabilized (cement-stabilized 2.5)
New Zealand	[50]	0.5	$h/d=0.4$–5 with shape factors	Curing time 28 days air-dried, unstabilized, smallest single value of a series of a minimum of 5 individual tests must be $\beta_D>1.3$ N/mm² at h/d = 1
USA	[71]	2.07[a]	Cube $h=l=w=102$	Five samples per test, one of the samples is allowed to be below the value $\beta_D=2.07$ N/mm², but must reach at least 1.725 N/mm²
Switzerland	[72]	0.3–0.5	Cube $h=l=w=200$	Unstabilized; $\beta_D=2$–4 N/mm² (0.5 N/mm² for light clay)
India	[73]	1.4	Cylinder $d=100$; $h=200$	Cement-stabilized (moist 0.7)

[a]Unclear if σ_p or β_D

Very little information can be found in the different national earth building regulations concerning permissible compressive stresses in walls σ_p (Table 3.31).

Influencing Variables. Influencing variables for the dry compressive strength of earth building materials are: grain distribution and quality; quantity and quality of the clay minerals (binding agents) and the resulting cohesive strength; drying conditions; quality of soil preparation; amount of mixing water; compaction work; aggregates and additives.

In experiments conducted at the Bauhaus University in Weimar, Germany [26, 43] (Fig. 3.47), the influence of the moisture content at time of installation w on the attained compressive strength of the cubes manifested itself in the following manner: after 90 days of drying under standard atmospheric conditions all mixes with $w>w_{pr}$ attained higher strengths than those with $w \leq w_{pr}$. The maximum value achieved was 3.89 N/mm² at $\rho_d=1.92$ g/cm³ in the "loess soil" series, $w>w_{pr}$. The

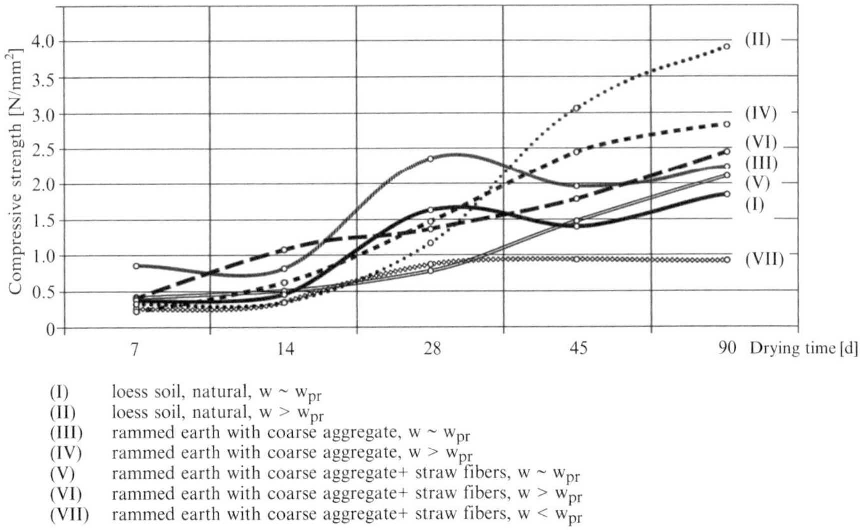

Fig. 3.47 Testing the dry compressive strength of rammed earth [26, 43]

lowest value achieved was 0.90 N/mm² at $\rho_d = 1.53$ g/cm³ in the "rammed earth with straw fibers" series, $w < w_{pr}$. There was no clear evidence that the straw fibers increased the strength.

This suggests that the Proctor test (Sect. 3.6.1.4) is only of limited use as a criterion for producing rammed earth structures. Based on the tests, working on the "wet" side of the Proctor curve would result in higher strengths but also lead to more shrinkage deformations. It would also considerably prolong the drying times until the equilibrium moisture content is reached. Therefore, the mixes used in the field are more likely to be "dry" mixes with a moisture content of $w < w_{pr}$. However, in order to achieve sufficient strength within the building element, the static *compaction work* applied in the Proctor test must be modified when installing rammed earth (Sect. 3.2.2.1).

Rischanek [48] has proven that a prolonged aging time of the construction soil (Sect. 3.1.1.3) results in a considerably higher dry compressive strength, an effect which was already known in traditional earth building practice in China and Central Asia.

Specimen dimensions used for testing the dry compressive strength vary considerably in literature sources and different national standards (Table 3.31). Specimen shapes range from cylinders to prisms and cubes. There are no conversion factors for different specimen sizes used for determining the uniaxial compressive strength of rammed earth or other earth building materials (with the exception of [49] and [50]). A decrease in specimen dimensions leads to an increase in dry compressive strength under the condition that the material properties, the compaction work, and the specimen shapes remain unchanged. (Example: In [46], cylindrical specimens

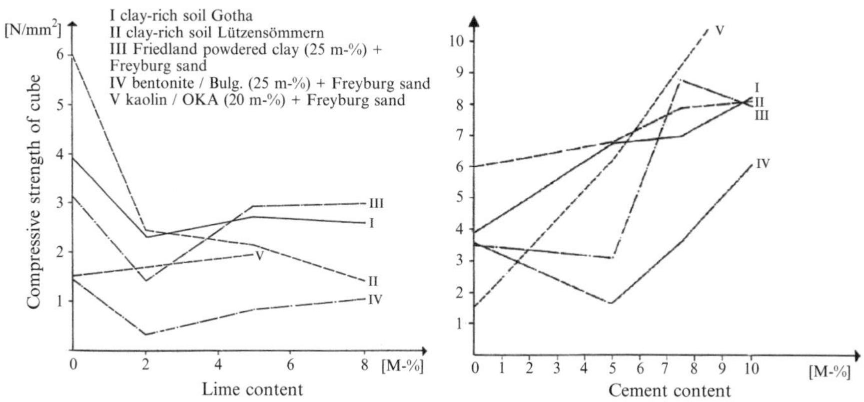

Fig. 3.48 Dry compressive strength of stabilized rammed earth specimens [51]

with a height of 60 cm and a diameter of 30 cm achieved a dry compressive strength value ($\beta_D = 1.9$ N/mm^2) which was 23 % lower than for specimens with a diameter of 10 cm. For prism-shaped specimens of the same height and with cross-sectional areas $A = 30$ and 10 cm^2, the smaller specimen achieved a β_D value which was approximately 50 % higher.)

The same can be observed in compact (e.g., cubical) specimens compared to prisms and cylinders.

Research activities are, to a large extent, dedicated to the role additives and aggregates play in increasing the dry compressive strength and weather resistance of rammed earth. In this context, work conducted at the HAB Weimar, Germany [51], tested the effects of the binders lime and cement. The compressive strength was determined with the help of four 200 mm sample cubes per binder. The results essentially verified the links which have already been established in the literature (Fig. 3.48).

The addition of *lime* is primarily suitable for very rich or highly cohesive soils. The cation exchange which occurs on the surface of the clay colloids causes an extensive change in the affinity for water as an immediate reaction. The structure of the soil becomes crumbly and loose, and its natural moisture content decreases by 1–2 % in relation to how much lime is added. This change results in an effective compactability of the rich soil. An additional long-term effect, due to the very slow hydraulic reaction between the lime and the clay minerals, is an increase in strength as long as optimal compaction is achieved. Hydrated lime and quick lime are suitable lime types.

The addition of *cement* is suitable for lean soils with a small clay mineral portion or less expansive clay minerals. The cement forms a water-insoluble, rigid cement gel in the soil which envelopes the mineral particles, binds them together, and solidifies into a rigid and continuous matrix. A second, water-soluble strength matrix is formed by the clay minerals within the soil. The strength of the soil–cement-mix is determined by the question if both strength matrices can develop freely or if mutual interferences occur. Such interferences, for example, can be caused by a high clay mineral content. The shrinkage reaction of the stabilized soil caused by the clay fraction is reduced by the rigid cement matrix. With an increase in cement, the cement matrix dominates resulting in "soil cement."

Heavy earth building materials (rammed earth with *gravel as a coarse aggregate*, $\rho_d = 2.0–2.4$ g/cm^3) possess an increased dry compressive strength. With values of $\beta_D = 3.0–5.0$ N/mm^2, it can be in the range of fired bricks of the lower compressive strength classes which is sufficient for one- and two-story residential construction.

The addition of ground *pottery shards* (in traditional African earth building) or crushed fired bricks results in an increase in strength of rammed earth building elements through the development of pozzolanic effects.

Small additions of *fiber material* increase the transverse tensile strength and, with it, the dry compressive strength of rammed earth. If the fiber content is increased further, determining the point of failure becomes less clear: the voids between the fibers form "crush spaces" and the fibers themselves act as tensile reinforcement.

Long-term *changes in the relative humidity* over the lifetime of the structure also affect the dry compressive strength of rammed earth. Using core samples of archaeological findings of rammed loess soil, Utz/Micoulitsch [52] have shown that with an increase in the relative humidity from 30 to 98 % the equilibrium moisture content (Sect. 5.1.2.4) rises from 2 to 6 % while the dry compressive strength decreases by approx. 30 %. Under the same test conditions the flexural strength even drops by approx. 70 %. Dierks and Ziegert [27] have observed similar reductions in dry compressive strength under comparable test conditions: an increase in the permanent relative humidity from 65 to 88 % results in a rise of the equilibrium moisture content from 0.7 to 1.3 % and a corresponding decrease in dry compressive strength by 35 %.

It should be noted that the abovementioned results were obtained under artificial test conditions making it almost impossible to directly apply them to real life situations. However, the tendencies described above highlight the importance of carefully considering all possible stresses during the design phase if a building, particularly for load-bearing earth building structures.

Earth Mortars

The testing of earth mortars according to DIN 18946-47 does not only consist of dry compressive strength tests but also the testing of additional strength types which are determined by the specific demands placed on the individual products. According to the specific requirements placed on earth mortars, this "combined" strength testing series is grouped into strength classes according to Tables 3.8 and 3.10.

Test Methods. The determination of the *dry compressive strength* β_D of earth mortars according to the Lehmbau Regeln [22] and DIN 18946-47 is based on DIN EN 1015-11 and DIN EN 998-1,2. These standards specify that dry compressive strength and flexural strength can be determined with the help of a continuous test series.

For the tests, 160 × 40 × 40 mm specimen prisms are prepared at the required working consistency of the earth mortar. The specimens are dried until a constant weight is achieved and conditioned under standard atmospheric conditions. The consistency is checked using the flow table test according to DIN EN 1015-3 by measuring the diameter of a fresh mortar sample (Sect. 3.6.2.1).

The dry compressive strength can also be tested with the help of the mortar prisms which were used in the flexural strength test and were broken in half (Sect. 3.6.2.2). The 16 mm × 16 mm load application plate is placed on the end face of the demolded mortar prism at a distance of 16 mm from the edge. A load is applied at a predefined speed until failure occurs. The final result is the smallest value of a minimum of three tests.

Lab and Calculation Values: Minke [53] specifies dry compressive strengths in the range of $\beta_D = 1.00–3.04$ N/mm^2 tested in 14 commercially available *earth plaster mortars*. The flexural strengths, which were determined at the same time, are in the range of $\beta_f = 0.18–0.69$ N/mm^2 and amount to approximately 1/10 of the dry compressive strength.

Dettmering and Kollmann [54] provide an overview of the scale of compressive strengths of plaster mortars which are used in restoration and preservation work. The compressive strength of lime plasters is classified as "low" at $\beta_D = 1–1.5$ N/mm^2. "Rigid" cement plasters are around $\beta_D = 10–30$ N/mm^2. There is no information about earth plasters.

For this reason, the German Association for Building with Earth (DVL) initiated dry compressive strength and adhesive strength tests (Sect. 3.6.2.2) for five different commercially available earth plaster mortars. The results range from $\beta_D = 0.7–1.8$ N/mm^2 [55].

A plaster mortar's tendency to develop cracks is typically also assessed by the quotient of compressive and flexural strengths [54]: in general, a flexural strength of 1/3 of the compressive strength is desirable. There is not enough reliable data to apply this observation to earth plaster mortar.

The dry compressive strength of *earth masonry mortar* is based on the strength of the earth blocks used in the specific construction project according to Sect. 3.5.7.

Earth Blocks

Earth blocks according to DIN 18945 are grouped into compressive strength classes (Table 3.16) based on the specific individual requirements they must meet.

Test Methods: The dry compressive strength β_D test for earth blocks should be conducted according to DIN 18945. The preparation of the specimens is the same as for the *E* modulus test. The test needs to be carried out within 1 h of the specimens being removed from the climate cabinet.

The test load is applied at a constant speed perpendicular to the bedding joint of the earth block until failure occurs after 30–90 s. A test series consists of a minimum of six specimens.

This test can be combined with the *E* modulus test. Here, the stress diagram shown in Fig. 3.43 is followed by a subsequent unloading. The test is then carried out until the failure load is attained.

There are no uniform regulations on an international scale for test methods of the dry compressive strength of earth blocks. For cement-stabilized earth blocks the "wet" compressive strength of hydrated specimens is determined as well, for example, using a test method according to ARSO [40].

Tensile Strength

The tensile strength β_T of a building material can generally be expressed as the tension determined in a tensile test as the quotient of the maximum amount of tensile force *F* and the original cross section *A*

$$\beta_T = \max . F / A \left[N / mm^2 \right].$$

Based on their cohesive strength, earth building materials also possess tensile strength. However, compared to their compressive strength, it is low and therefore not included in calculations for load-bearing building elements.

Based on the different uses of earth building materials, tensile strengths for typical load conditions include:

- Axial tensile strength in the form of cohesive strength according to Niemeyer (Sect. 2.2.3.2)
- Splitting tensile strength
- Tensile adhesion strength
- Flexural strength

Axial Tensile Strength

Terminology: The cohesive strength according to Niemeyer (Sect. 2.2.3.2) is comparable to the axial tensile strength β_z of a *moist* construction soil (earth building material) as a processing parameter for a predefined "standard" testing consistency (Fig. 2.24).

The "*dry*" tensile strength was determined at the Bauhaus University in Weimar, Germany, with the help of the same test apparatus which is used to determine the cohesive strength [56].

Test Method: An apparatus for the determination of the cohesive strength according to Niemeyer was used to load dry "figure-eight-shaped" specimens of 13 different construction soils. The load was applied until brittle fractures occurred.

Lab and Calculation Values: *"Wet" Tensile Strength* (*Niemeyer*). β_{TW} = 50–360 g/cm^2 or 0.005–0.036 N/mm^2, higher for pure clays (Table 2.5). The corresponding moisture content at the time of installation w_N (test consistency) is near the plastic limit PL according to Sect. 2.2.3.2 (w_N = 1.19 PL—3.37 with r_{xy} = +0.79; Table 2.7 [57]).

"Dry" Tensile Strength [56]. The tensile strength measured at the moment the brittle fracture occurred increased to 21–67 times the specimens' values at standard consistency. The absolute values were the highest for rich and very rich soils:

Lean soils	<0.4 N/mm^2
Semi-rich soils	0.4–0.6 N/mm^2
Rich soils	0.6–0.9 N/mm^2
Very rich soils	>0.9 N/mm^2

The relative values of dry and wet tensile strengths were considerably higher for lean and semi-rich soils than for rich and very rich soils.

Influencing Variables. The practical significance of "wet" tensile strength lies in the preparation of earth building materials: it is needed to turn the mix of construction soil, water, and aggregates and additives which improve the soil's properties, into a workable, homogeneous mass suitable for shaping. Tensile strength values provide information about the processing qualities of construction soils. Here, processing refers to changing the shape of a mix exhibiting viscous behavior (Sect. 3.6.2.1).

Strong cohesion, well-graded grain sizes, angular grain shapes, and a rough grain surface of the mineral components have proven to increase strength. The tensile strength increases as the earth building material dries.

Splitting Tensile Strength

Terminology: For the indirect determination of the tensile strength of materials with brittle fracture tendencies, such as solid rock, firm pure clays, marl as well as some clay-rich soils, the *splitting tensile* strength test (also known as the Brazilian Test) can be used with the help of cylindrical specimens. The splitting tensile strength of concrete is tested according to DIN EN 12390-6.

A practical application of the splitting tensile strength, in terms of a field test, is the *shock resistance* of earth blocks (earth block drop test). According to earth building standards and recommendations in New Zealand and Australia [49, 50], this test should be carried out as a quality control measure during construction.

Test Method. For the *splitting tensile strength test*, a cylindrical specimen with $h/d \sim 1$ is mounted in a rigid test frame between fixed plates. Two opposing parallel lines of the specimen's surface (fiber board strips) are then loaded with a constant load increase of 0.05 N/mm^2 per second until failure occurs (Fig. 3.49).

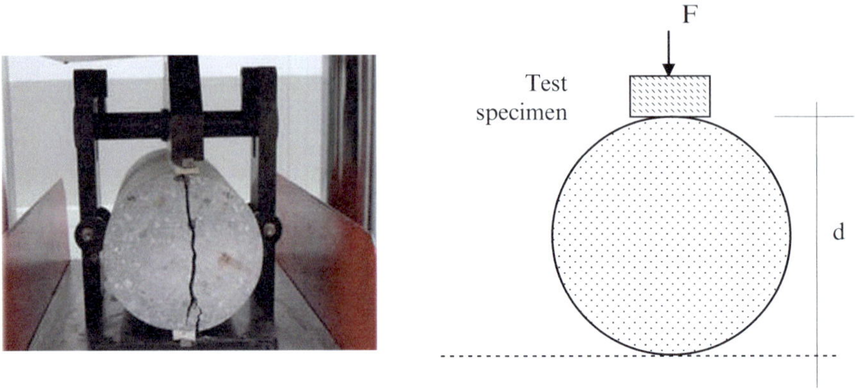

Fig. 3.49 Splitting tensile strength: test setup, according to [40, 78]

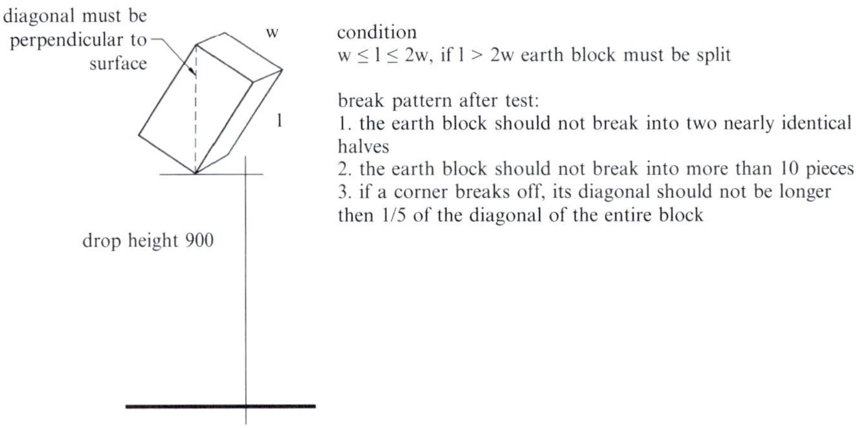

Fig. 3.50 Splitting tensile strength: test setup, according to [49, 50]

In order to determine the *shock resistance of earth blocks*, a dry earth block is dropped from a height of 900 mm onto a solid, flat surface and the resulting failure pattern is assessed. Before the earth block is dropped, it is turned in such a manner that the diagonal of the block's bedding side is perpendicular to the impact surface and the lowest edge of the block is at a height of 900 mm [49]. For testing the shock resistance of earth blocks, the permissible and impermissible failure patterns after impact are described in Fig. 3.50. The test should be carried out on 5 in every 2500 earth blocks.

Figure 3.51 [40] shows a test method for earth blocks with a rectangular cross section which was developed for determining the splitting tensile strength according to ARSO. Two plastic or hardwood strips with a 1 cm^2 cross section are mounted in a press together with the test earth block. Both strips are positioned vertically

Fig. 3.51 Splitting tensile strength test of earth blocks according to ARSO [40]

above one another and loaded with a constant speed of 0.02 mm/s until failure occurs. The test is repeated on each half block obtained from the initial break so that every full block provides three test results.

Lab and Calculation Values. The splitting tensile strength for cylindrical specimens is determined by using the equation

$\beta_{ST} = 2 \cdot \max F / \pi \cdot d \cdot h$ (or $w \cdot h$ for rectangular cross sections)

d—cylinder diameter

h—height

max$F$—stress along the surface line, ultimate load.

According to DIN EN 12390-6, the following applies to concrete and can serve as a guide for comparing the splitting tensile strength β_{ST} with the axial tensile strength β_{AT}

$$\beta_{ST} = 1.2\beta_{AT}.$$

No corresponding values for earth building materials have been found to date.

possible failure patterns according to DIN 18555-6 (diagram)

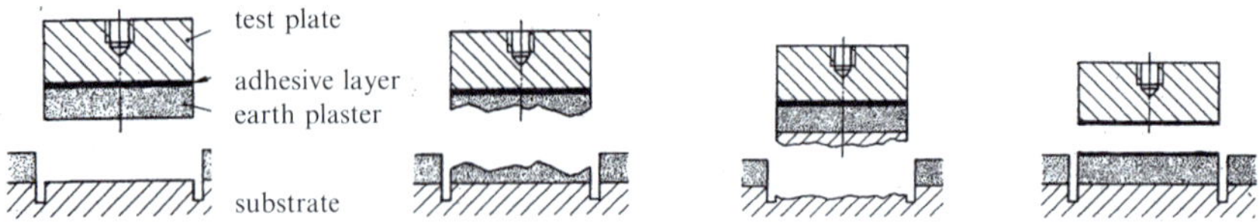

a) adhesive surface substrate /mortar failure (adhesion failure)

b) mortar failure (cohesion failure)

c) substrate failure (cohesion failure)

d) adhesive layer failure

adhesion failure (a) after testing [55]

Fig. 3.52 Adhesive strength test of earth plasters

Tensile Adhesion Strength

Terminology: The tensile adhesion strength β_{TA} indicates the amount of tensile stress acting perpendicular to the adhesive surface required to break the bond between mortar and plaster substrate. The test shows if the adhesion of the top coat to the base coat or the entire plaster system to the substrate is sufficient.

Furthermore, failure can occur in masonry systems if the adhesive strength between the earth block and the masonry mortar in the bedding joint is exceeded. Failure can also be caused by the block tensile strength if the adhesive strength of the bedding joint mortar is high and the tensile strength of the earth blocks in the direction of the block height is low.

Test Methods: The tensile adhesion strength β_{TA} of earth plaster mortars is tested according to DIN 18947 on the basis of DIN EN 1015-12 (Fig. 3.52 [58]). The earth plaster is prepared to test consistency, applied to the substrate (a horizontal concrete slab) and the plaster surface is worked. The test surfaces are then stored for a minimum of 14 days making sure the final 7 days are under standard atmospheric conditions ((23 ± 2) °C/(50 ± 5) % RH).

Next, a minimum of five specimens with a diameter of $d = 50$ mm (also 70 mm) are cut out of the test surface by drilling core holes through the earth plaster to a depth of approx. 3 mm into the substrate.

After the contact surfaces have been cleaned with a brush and loose particles and dust have been removed with compressed air, test plates are attached to the dry earth plaster using a suitable glue. After the glue has hardened, the tensile strength of the earth plaster samples is determined using suitable devices (such as from Dynatest or the HP 850). The duration of the test should not exceed 60 s.

The test can result in four different failure patterns (Fig. 3.52) [DIN 18555-6]):

- Adhesive failure in the contact surface mortar/substrate
- Cohesion failure in the mortar
- Cohesion failure in the substrate
- Failure in the adhesive layer

Lab and Calculation Values: Minke (Boenkendorf) [45] specifies the criterion of $\beta_{TA} \geq 0.05$ N/mm^2 as a general requirement placed on "earth plaster on an earthen substrate." Riechers and Hildebrand [59] consider a general value of $\beta_{TA} = 0.08$ N/mm^2 for plasters as "sufficient for general applications."

Dettmering and Kollmann [54] provide numerical values for the tensile adhesive strength β_{TA} of standard plasters used in restoration and conservation projects in the range of 0.1–0.5 N/mm^2, for gypsum and cement plasters 0.4–0.9 and 1.0–2.0 N/mm^2. There are no corresponding values for earth plasters.

For this reason, the DVL initiated adhesive strength tests for five different commercially available earth plasters [55]. The numerical values of the tests were in the range of $\beta_{TA} = 0.03$–0.12 N/mm^2. Smooth concrete was used as the substrate (which is unfavorable in terms of the expected result). The specimens were obtained from a single-layer earth plaster through "core drilling" or by using a "cutter" according to DIN EN 1015-12. The earth plaster was applied at standard working consistency. The resulting failure patterns were "cohesion failure" and "adhesion failure" as well as transitional types (Fig. 3.52).

These systematically conducted tests are the first to prove that the test method described in DIN EN 1015-12 can also be applied to earth plasters. Based on these tests, numerical values have been specified for the tensile adhesive strength classes of earth plaster mortar in DIN 18947 (Table 3.10).

Influencing Variables: The cohesive strength and composition of the plaster mortar, the consistency as well as the properties of the plaster surface and the substrate exert a particular influence on the tensile adhesive strength of earth plaster mortar.

Flexural Strength

Terminology: Flexural strength β_F is activated in earth building materials when loads are applied perpendicular to the plane causing the building element to act as a panel. It indicates how much load must be applied in order to break a building material when bending. Typical stress situations resulting in flexural tension are found in

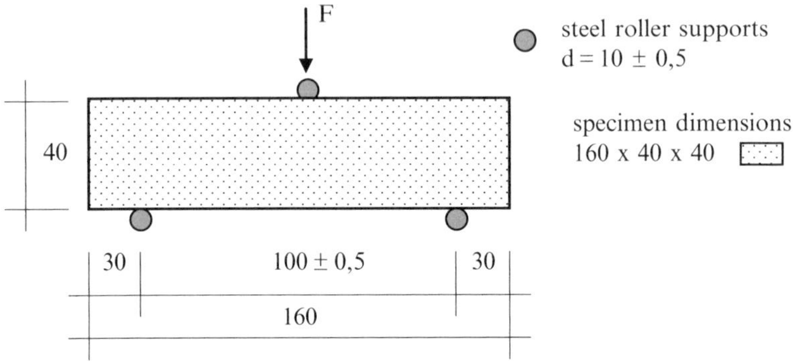

Fig. 3.53 Flexural strength test of earth plaster according to DIN EN 1015-11

earth plasters but also through a dynamic stress placed on earth building materials, for example, during earthquakes. Other examples include non-load-bearing building elements as well as those which do not experience a significant degree of superimposed load (infill, wall linings, and self-supporting walls).

Test Methods: The specimens used for determining the flexural strength of *earth mortars* are produced in the same manner as for the dry compressive strength test according to DIN 18947. Both tests can also be combined in a test series.

A specimen is placed on two steel rollers ($d = 10$ mm) which are 100 mm apart. A third roller is used to apply a load in the center until breakage occurs (Fig. 3.53, DIN EN 1015-11).

Dierks and Ziegert [60] used 600 × 150 × 150 rectangular-shaped *rammed earth* specimens to determine the flexural strength based on DIN 1048-5. In order to achieve a "diaphragm action" the specimens were "placed on their sides" (turned 90° around their longitudinal axis) so the ramming joints became vertical and thus parallel to the direction of loading.

According to NZS 4298 [50], the flexural strength of *earth blocks* is determined using a field test following the procedure described in Fig. 3.54. The test block is supported as a "beam" along the edges of its wide faces in a linear fashion. The load is applied by stacking "earth" blocks on top until the failure load has been reached. A linear stacking load is applied to the test earth block in the center and to one wide face. The test must be carried out on 5 out of every 5000 earth blocks as a quality control measure during construction if the blocks did not pass the "drop test" (Fig. 3.50).

Jagadish et al. [61] determine the flexural strength of *stabilized earth block masonry* with the help of the test shown in Fig. 3.55. In the test, lime-cement mortar or stabilized earth masonry mortar was used. The illustration shows two possible ways of applying the horizontal force: via pull rope (a) or via press (b) using a specimen which has already been broken.

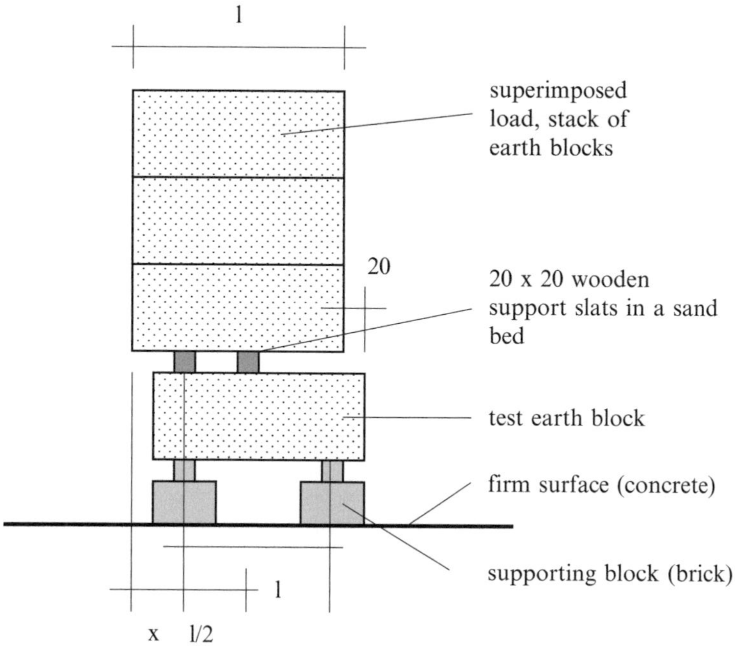

Fig. 3.54 Flexural strength test of earth blocks according to NZS 4298 [50]

Lab and Calculation Values: The average values of the flexural strength of *rammed earth* samples according to [60] were in the range of $\beta_F = 0.36–0.63$ N/mm^2. As expected, the values for the samples containing fiber aggregates were considerably higher.

NZS 4297,-8 [50] requires the following values for the flexural strength of *earth blocks*:

$\beta_F > 0.25$ N/mm^2 as the minimum value of five individual tests according to Fig. 3.54
$\beta_{F,C} = 0.1$ N/mm^2 as the calculation value for planning

The values of the flexural strength for stabilized *earth block masonry* (Fig. 3.55 [61]) were in the range of $\beta_F = 0.031–0.414$ N/mm^2.

For standard *plasters* used in renovation and restoration work, Dettmering and Kollmann [54] specify numerical values for the flexural strength β_F in the range of 0.2–1.0 N/mm^2, and for gypsum and cement plasters 1.0–2.0 and 2.0–7.0 N/mm^2. There are no corresponding values for earth plasters. Hardened earth mortar [35] requires a flexural strength of $\beta_F \geq 0.4$ N/mm^2 after a setting time of 28 days.

Numerical values for the flexural strength classes for *earth plaster mortar* are specified in DIN 18947 (Table 3.10).

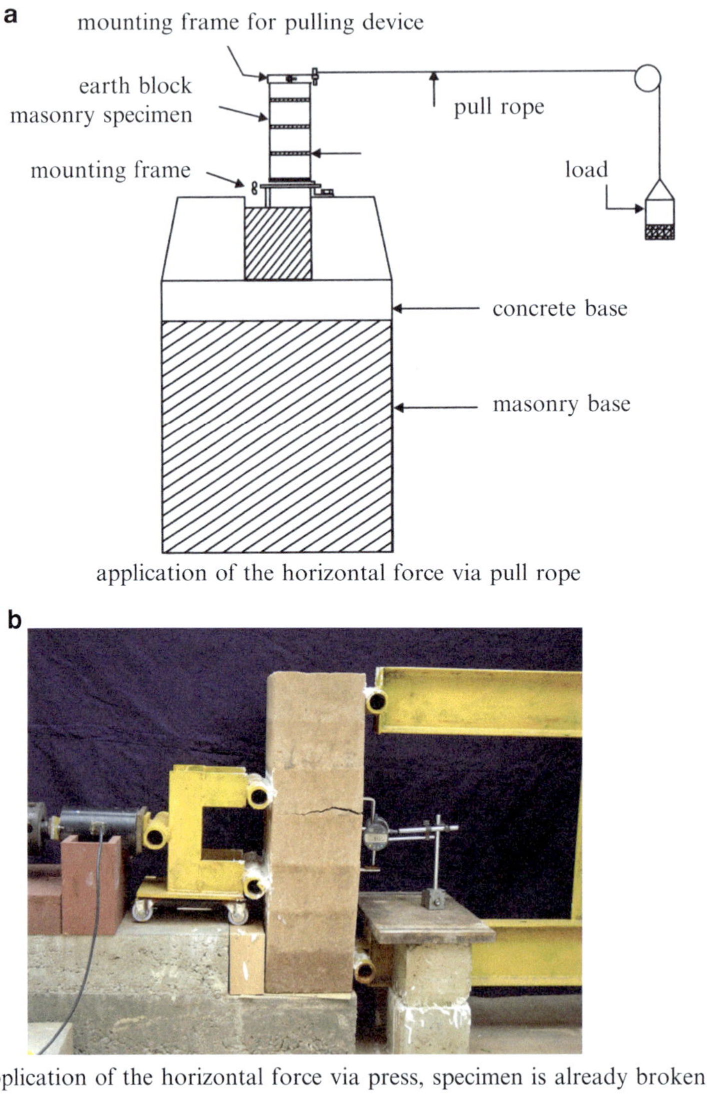

Fig. 3.55 Flexural strength test of stabilized earth block masonry, according to [61]

Shear Strength and Friction Coefficient

Terminology

The activation of the shear strength β_S in an earth building element correlates to the stresses within the element during the transfer of horizontal loads. Failure occurs along horizontal surfaces which are typically predefined by processing.

The shear strength β_S of an earth building material is generally expressed as the stress which leads to failure of the building material caused by a shear stress F acting on the loaded cross section A

$$\beta_s = \max . F / A\left[N / mm^2\right].$$

In soil mechanics, the stress application is formulated on the basis of the Mohr/Coulomb general failure criterion:

$$\tau = \beta_{AS} + \mu \cdot \sigma_D.$$

Influencing parameters are the degree of the perpendicular acting compressive stress σ_D, the roughness of the surfaces in the sliding plane, which is expressed with the help of a material-independent friction coefficient μ, as well as the adhesive shear strength β_{AS}, which is a result of the bond (= cohesion c) with its determining factors (surface roughness, pore structure, moisture content) and the strength of the earth building material.

Test Methods

In soil mechanics, the shear strength of clay soils and pure clays is determined by the direct simple shear test shown in Fig. 3.45 as well as the triaxial test (Table 3.27).

In rammed earth or cob structures, individual blocks are clearly visible. Their borders are defined by the horizontal ramming joints between the stacked rammed earth or cob layers (caused by the specific technology used) and by their vertical edges. The vertical "butt joints" (which in the case of cob are slightly slanted in the opposing direction of the work) are staggered like those found in block masonry. The edges of the individual blocks butt against each other (without "joint mortar") and are potential weak spots in actual structures.

Dierks and Ziegert [60] determined the shear strength of *rammed earth* with the help of a 150 × 150 × 300 mm specimen which was loaded parallel to the ramming joints. At first, the failure patterns showed slanted cracks perpendicular to the main stresses at approx. 60 % of the applied failure load. This was followed by a sudden vertical shear failure. This failure pattern seemed to indicate that the compaction created an "interlocking" of the individual rammed earth layers resulting in a largely isotropic behavior of the material in terms of shear strength.

The horizontal joints in earth block masonry form potential sliding planes. Therefore, the adhesive shear strength of *earth masonry mortar* according to DIN 18946 needs to be determined on the basis of DIN EN 1052-3. In the test, sand-lime bricks are used as masonry. They are first stored under standard atmospheric conditions ((23 ± 2) °C, (50 ± 5) % RH) until a constant weight is reached. The constant weight has been reached when the results of two consecutive weightings at 24 h intervals differ by no more than 0.2 mass in % based on the smaller value. The sand-lime bricks should not be prewetted before they are laid.

To prepare the specimens, three blocks (which have been prepared under standard atmospheric conditions) are laid and joined on top of each other using the earth masonry mortar to be tested. They are then stored for 2 weeks, ensuring the second week is under standard atmospheric conditions. The test is conducted using the

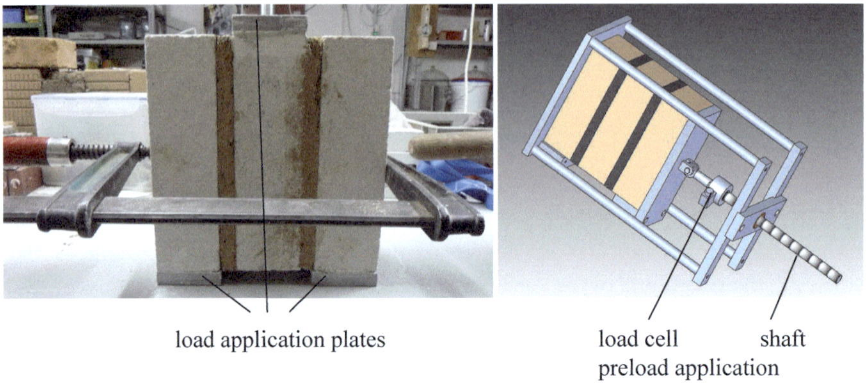

Fig. 3.56 Adhesive shear strength test of earth masonry mortar according to DIN 18946 [62]

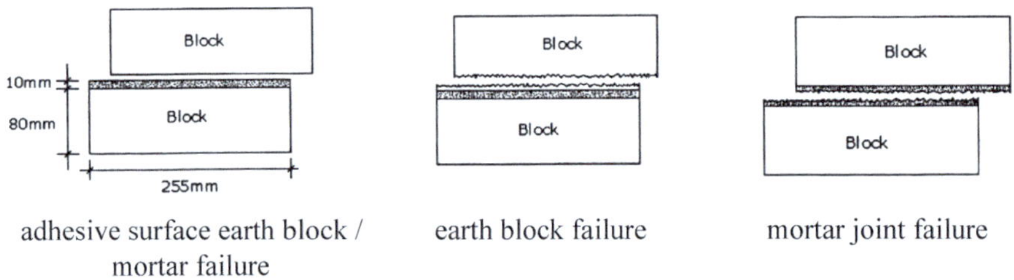

Fig. 3.57 Earth bock masonry under shear stress: possible failure patterns [63]

setup shown in Fig. 3.56 (Fontana [62]): First, the shaft is used to load and unload the specimens perpendicular to the bedding joint with load increments of 0.05, 0.10, and 0.20 N/mm^2. Next, the test load for determining the adhesive shear strength is applied to the specimens via the central load application plate. Failure should occur at 20–60 s after load application.

For cement-stabilized earth block masonry, Venkatarama Reddy and Uday Vyas [63] examined the impact of the adhesive shear strength on the degree of compressive stress acting perpendicular to the surface. Based on the principle of the direct simple shear test shown in Fig. 3.45, two earth blocks (with 5 and 14 % cement added) were laid on top of each and joined other using a lime-cement mortar. The lower block was fixed in place in a steel box to restrict movement. Shear force was applied to the upper block via a steel frame which encased the block. The bedding sides of the connected earth blocks formed the potential sliding planes with varying degrees of roughness.

Figure 3.57 shows the three resulting failure patterns. A fourth pattern was a partial break along the mortar joint and in the earth block. These failure patterns are similar to those found in the adhesive strength test of earth plasters in Fig. 3.52.

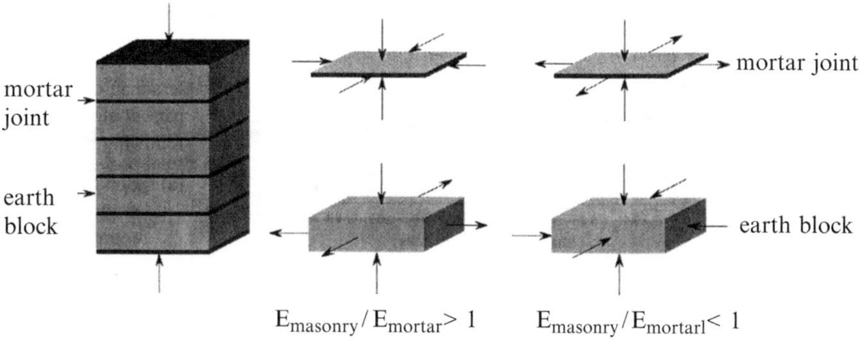

Fig. 3.58 Earth block masonry under compressive stress: stress state in earth block/mortar joint [63]

Lab and Calculation Values

Using the test setup according to DIN 18946 shown in Fig. 3.56, Fontana [62] determined the adhesive shear strength of earth masonry mortars for the first time. Depending on the earth masonry mortar used and the pretreatment of the blocks (wet or dry), the values were in the range of 0.042–0.135 N/mm^2 which is above the minimum values required by DIN 18946 (Table 3.10).

The test setup described in [60] resulted in shear strength values in the range of $\beta_S = 0.55$–0.89 N/mm^2. It was also proven that concrete and rammed earth are similar in terms of material behavior: for the concrete and rammed earth samples tested in [60], the ratio between shear strength and compressive strength β_D and between shear strength and flexural strength β_F is:

Concrete: $\beta_S \sim 0.23\,\beta_D$; $\beta_S \sim 1.6\,\beta_F$
Rammed earth: $\beta_S \sim 0.27$–$0.33\,\beta_D$; $\beta_S \sim 1.41$–$1.52\,\beta_F$.

In NZS 4297,-8 [50], the following calculations are applied for shear strength: $\beta_{S,C} = 0.09$ N/mm^2.

Influencing Variables

The strength properties of earth block masonry are affected by the ratio between the *E* moduli of the earth blocks E_B and the masonry mortar E_M (Fig. 3.58 [63]).

At a ratio of $E_B/E_M > 1$, the mortar in the horizontal joint, which is less stiff, shows a stronger tendency to "deform transversely" than the block. However, the bond with the earth block confines this deformation causing further tensile stresses within the earth block due to the confined transverse deformations in the mortar. The higher the difference in stress between block and mortar, the lower the compressive strength of the masonry structure. For sandy (earth) mortars with a dense structure, the mortar's influence on the compressive strength of the masonry structure can be expressed by the compressive strength of the mortar with sufficient precision. Earth masonry mortars with lightweight aggregates, on the other hand, can exhibit a higher potential for transverse deformation. This would further reduce the compressive strength within the masonry structure.

According to [63], the adhesive strength between block and mortar in the horizontal joint only substantially affects the masonry's compressive strength if the masonry mortar is stiffer than the earth block ($E_B/E_M < 1$). For this situation, the following relationship has been determined for the ratio between shear strength (β_S) and compressive strength (β_D):

$$\beta_D = 1.457 + 5.01\ \beta_S \left(r_{xy} = 0.89\right).$$

The fracture mechanism of stiff mortar to soft block depends on the adhesive strength in the joint between mortar and block. A high degree of strength means that the horizontal compressive forces within the masonry structure increase as long as the shear strength in the horizontal joint shows resistance. Once the tensile strength along the surface of the mortar and block joint fails, the horizontal compressive forces applied by the shear forces disappear and vertical cleavage cracking develops as the typical failure pattern.

Friction coefficients μ are also important in load-bearing earth construction where horizontal loads in earth building elements are transferred vertically and have to overcome joints made of different materials, such as the transfer of wind loads via roof structures in copings. Friction coefficients for individual material combinations are specified in [60]:

Wood (rough)/earth mortar: $\mu_G = 0.30–0.54$
Wood (planed)/earth mortar: $\mu_G = 0.26–0.53$
Fired bricks/earth mortar: $\mu_G = 0.37–0.56$
Concrete/rammed earth: $\mu_G = 0.41–0.64$

Wear Resistance

Surfaces made of earth building materials are subject to various types of mechanical wear during their lifetime:

- Surface abrasion (plasters, wall surfaces, floors)
- Scratches/cracks (plasters, wall surfaces, floors)
- Impacts (corners of wall openings, wall surfaces)
- Grooves/nicks (floors)

These different types of wear typically appear together making them rather complex. It is therefore important to simulate the actual wear conditions as accurately as possible during testing. The development of reproducible and standardized test

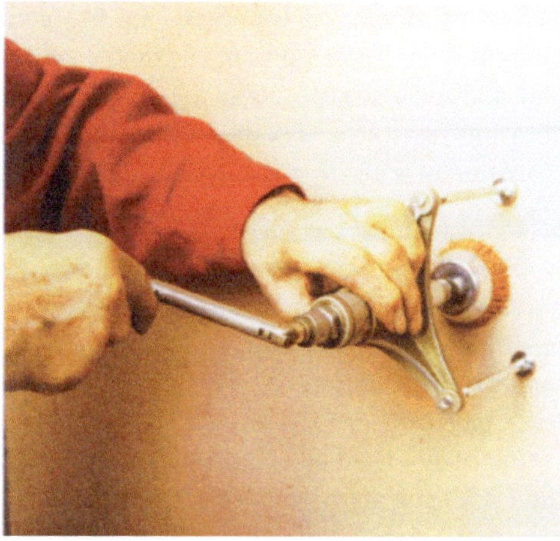

Fig. 3.59 Determining the abrasion of earth plaster, manually operated brush

methods for determining wear resistance has only recently begun. Accordingly, few tests with conclusive results have been conducted.

Nevertheless, the application of earth building materials, particularly earth mortars, should be carefully considered in the planning stages of areas where a high degree of mechanical wear is to be expected. This includes parts of buildings which are highly frequented by the public.

Abrasion Resistance

Terminology: A predefined testing method is used to determine the abrasion dust quantity (in g) of building element surfaces made of earth building materials. This quantity serves as a measure of the mechanical strength against surface abrasion.

Test Methods: For determining abrasion resistance, Minke [53] developed a test which is based on the Böhme grinding wheel standardized in DIN 52108. In this test, the earth plaster surface receives final compression using a wooden or plastic float. Next, a hard, rotating brush with a diameter of 7 cm is pressed against the earthen surface using a weight with an applied pressure of 2 kg. After 20 brush rotations, the abrasion dust quantity is weighed. This test can also be carried out using a manually operated testing device (Fig. 3.59 [64]).

The abrasion test of earth plaster mortars has been included in DIN 18947 as a mandatory test. Proof of abrasion strength in earth plaster mortars is also a requirement for obtaining the quality seal from the organization natureplus e.V. [35].

Lab and Calculation Values: Using the method described above, the abrasion dust quantities of 15 commercially available earth plasters containing different aggregates and additives were established in [53] to provide a standard of comparison for abrasion resistance. For the tests, the mortars were prepared at a 140 mm slump consistency according to DIN EN 1015-3 (Fig. 3.40) and applied to a substrate. The abrasion dust amounts ranged from 0.1 to 7.0 g.

The DVL commissioned tests based on the same procedure to determine the abrasion resistance of five additional commercially available earth plasters which resulted in abrasion dust quantities of 0.3–6.7 g [55]. The tested earth plasters exhibited fluctuations of more than one decimal power in terms of their abrasion resistance.

DIN 18947 and the technical information sheets 01 and 06 [34, 65] (published by the DVL) specify permissible abrasion dust quantities for earth plaster mortars and clay thin-layer finishes (Table 3.10).

Influencing Variables: The resistance of an earth building element to surface abrasion is influenced by the degree of the grinding force, the strength and smoothness of the surface and the properties of the construction soil (cohesive strength, grain distribution, grain shape, and angularity) and its aggregates.

Earth plasters are often prepared as lean mixes in order to minimize shrinkage cracks. This also decreases the cohesive strength which is needed to bind sand grains at the surface of the plaster. This results in an undesirable "dusting" of the plaster even when only lightly touched.

Edge Strength

Terminology: The edge strength is a measurement of the stability of protruding edges of door and wall openings made of earth building materials which are exposed to mechanical stresses in the finished building.

Test Methods: Minke [53] developed a test method for determining the edge strength of earth plaster mortar, earth panels, and earth blocks. In this test, a weight is dropped from a predetermined height (125 mm for earth mortar) onto the edge of a specimen which has been mounted at a 60° angle. The lower part of the drop weight consists of a steel ball which hits the specimen 10 mm from the edge. The test determines the weight at which a chipping failure of the specimen occurs.

In connection with this test, it is important to point out that it is common practice to use plaster profiles and corner guards to protect edges and corners prone to shocks.

Impact Resistance

Terminology: To test the impact resistance of a building material or building element its resistance to impact and shock stresses is measured. The more energy the building material absorbs, the stronger it is.

Test Methods: Based on the testing of hardened concrete according to DIN 1048-2, Dierks and Ziegert [60] conducted tests using rebound hammers to determine the compressive strength of rammed earth walls.

This test only measures the resistance on the surface of the building element. However, for assessing the strength properties of a building element, it is important to determine the condition of the entire cross section.

Scratch Resistance

Terminology: A specific test method is used to determine the scratching effect of pointy or sharp-edged objects on the surface of earth building elements as a measure of their mechanical stability when facing such stresses.

Test Methods: Depending on the function of a building element and the demands placed on it in terms of use, testing the mechanical strength of surfaces made of earth building materials might require a finer differentiation of the strength criteria than can be determined using the test method described here. One example is testing the remains of earth building structures as part of archaeological digs. Here, it is of particular importance to stabilize the very frail surfaces made of earth building materials, typically with the aid of chemical stabilization.

In this context, UTZ [52] has developed a qualitative test for determining the scratch resistance of archaeological earth building elements. First, cylindrical core samples with a length of 5 cm were taken from the building element which needed to be stabilized. Then, a nail loaded with a predetermined weight and mounted in a pulling device was pulled across the specimen at a speed of 0.025 m/s. The results compared the qualitative differences of the scratch patterns in untreated and stabilized samples at different degrees of relative humidity.

3.6.3 Building Physics Parameters

3.6.3.1 Hygric Parameters

Earth building materials and elements are not waterproof. Until recently, there were no parameters or suitable test methods for determining the water resistance of earth building materials. In connection with the development of DIN 18945, related research was conducted from 2009 to 2011 at Germany's Federal Institute for Material Testing (BAM). This research focused on the experimental simulation of possible moisture stresses on earth building materials and elements, in particular earth blocks and earth mortar as well as entire building elements made from them.

Capillary Water Absorption

Terminology

In general, the experimental determination of the capillary water absorption (absorptive capacity) of building materials and building elements is conducted according to DIN EN ISO 15148:

$$m_w = A \cdot \sqrt{t}$$

Here, A is the water absorption coefficient, a material parameter subject to the building material structure, porosity, bulk density, temperature, and initial moisture content. It is expressed as a function of the absorbed mass of water m_w [kg/m^2] per unit of area and of time (t).

Test Methods

Dip Test: A qualitative assessment of the resistance of earth building materials to water ("suspension capability"), instead of water absorption, can be conducted using the following test according to DIN 18952-2 (withdrawn) (Fig. 3.60): The bottom of a specimen with the dimensions 220×40×25 mm (prism, degree of shrinkage test for construction soil, Table 3.26) is lowered 50 mm into a container of water and its appearance is visually evaluated after 45 and 60 min. A complete separation of the bottom part of the specimen after 45 min is a sign of a soil with low water resistance. A specimen which holds its shape for more than 60 min indicates that the soil is difficult to suspend, has good water resistance and is therefore suitable as an earth building material.

The dip test for earth blocks according to DIN 18945 is based on the principle of the "suspension test" according to DIN 18952-2: with the help of a mounting device, the end of a test earth block is lowered 10 cm deep into a water bath where it remains for 10 min. The material loss is determined by filtering out the residue which is left in the water (Fig. 3.61) [66]. It is then dried at 40 °C, conditioned under normal atmospheric conditions and weighed. A test consists of a series of three earth blocks.

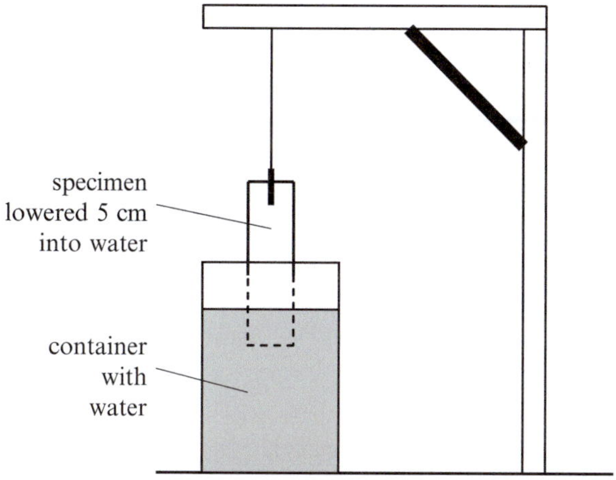

Fig. 3.60 Suspension test of earth building materials according to DIN 18952-2

Fig. 3.61 Dip test of earth blocks of different usage classes [66]

The material loss is determined by the ratio of the mass of the filtered material from the three earth blocks to their initial mass. After the dip test, the earth blocks are evaluated based on DIN 18945 in terms of their respective usage classes according to Table 3.17. According to the table, the mass loss of earth blocks of usage class I should not exceed 5 %. Earth blocks of usage class II should not exceed 15 %.

Fig. 3.62 Contact test of earth blocks, earth blocks before and after the test [66]

The dip test for usage classes I and II should not cause cracks or permanent swelling deformations.

Contact Test: The contact test according to DIN 18945 determines the behavior of earth blocks subjected to short contact with water during the application of plaster or masonry mortar.

An absorbent (cellulose) cloth, the size of the long narrow side of an earth block, is placed on the block's surface and moistened with an amount of water which corresponds to the average amount of water contained in a 15 mm thick mortar layer (0.5 g/cm^2 of block surface). The specimens are then stored for 24 h in a closed container placed on a rack above water. Finally, they are exposed to normal atmospheric conditions for 2 days. Afterwards the surface of the earth blocks is assessed. Earth blocks of usage classes I and II should not exhibit cracks or permanent swelling deformations (Table 3.17).

Figure 3.62 [66] shows an earth block before and after the contact test. The earth block does not conform to the requirements for usage classes I and II.

Suction Test: The suction test according to DIN 18945 determines the behavior of earth blocks when exposed to a temporary excess supply of water. Such stresses include those on exterior timber-frame walls during driving rains (Sect. 5.2.1.2) with water entering and collecting between the frame construction and the earthen infill.

For this test, three earth block halves are first conditioned under standard atmospheric conditions ((23 ± 2) °C, (50 ± 5) % RH) until a constant weight is reached. Next, fired bricks which support capillary action (or other porous blocks) are placed closely together in a shallow pan forming a continuous plane. The pan is then filled with water, stopping 1–5 mm below the upper edge of the fired brick layer. The water which is later absorbed by the halved earth blocks during the test needs to be continuously refilled up to this original level. Sponge cloths are laid on top of the fired brick layer upon which the earth blocks are placed bedside down. The condition of the specimens is visually assessed after 30 min, 3 h, and 24 h. Swelling should not cause any visual cracking on the top and sides of the specimens within the time periods listed in Table 3.17. Swelling itself is not a criterion for failure. Figure 3.63 [66] shows spalling in earth blocks after the suction test.

Fig. 3.63 Earth blocks after suction test, spalling [66]

Lab and Calculation Values

Lustig-Rössler [67] examined the "suspension capability" according to DIN 18952-2 using three earth samples (lean soil, pure clay, earth mortar). Complete separation of the immersed sections occurred in all specimens after 2 h (lean soil, pure clay) and 2.5 h (pure clay). This indicates their suitability as earth building materials.

For light clays with different bulk densities and different aggregates and additives as well as for soils and clay-rich soils, Minke specifies values of capillary water absorption as a function of the suction time which has been determined using the modified test setup described in [45].

Influencing Variables

The ability to absorb water within a specific time differs greatly from soil to soil: lean soils can only absorb a relatively small amount of water but in a short period of time. Rich soils and pure clays, on the other hand, have a high water absorption capacity due to their comparatively high portion of clay minerals. However, the absorption takes much longer because of their increased swelling potential (DIN 18132, Sect. 2.2.3.3, Fig. 2.19). The intensity of the swelling process is influenced by the quality of the clay minerals (Sect. 2.2.3.4) which can additionally interfere with water absorption.

Applied to the mechanism of moisture transport in earth building materials and building elements (Sect. 5.1.2.1), this indicates that, due to the relatively small pores in rich soils and pure clays, moisture transport caused by capillary action reaches further than in leaner soils. However, as a result of the increased swelling deformations found in rich soils and pure clays, the moisture transport is effectively shorter within the reference time period because the swelling clay minerals inhibit a further advancement of the moisture ([68], Sect. 5.2.1.2).

Other influences on this mechanism are exerted by the bulk density, with its corresponding pore structure, as well as possible aggregates and additives contained within the earth building materials used.

Frost Test

Terminology

Moisture-penetrated earth blocks are not frost resistant. Freezing water inside the pores expands by approx. 10 % which can result in deformations destroying the structure of the earth blocks. Because of their very limited resistance to moisture, the German Federal Institute for Material Testing (BAM) developed its own method for testing the behavior of earth blocks during freeze–thaw cycles.

Test Methods

For this test, three full earth blocks are prepared under standard atmospheric conditions and set up on their long and narrow sides. An absorbent (cellulose) cloth, the size of the long and narrow side of an earth block, is placed on the block's surface and evenly moistened with 0.5 g of water/cm^2 of block surface. The specimens are stored in a closed container at 23 °C for 24 h and then placed in a freeze–thaw cabinet at a temperature of at least −15 °C. A sensor is used to test the temperature in the center of the block. The timing sequence for a freeze–thaw cycle in the freeze–thaw cabinet is as follows:

- The conditioned specimen is placed in the climate cabinet
- Cooling phase of 6 h
- Freezing phase (temperature −15 °C or lower) of 34 h
- Thawing phase of 8 h
- The specimen is stored in a closed container under standard atmospheric conditions and moistened via cloth for 24 h every two cycles

In addition to the requirements of the dip, contact, and suction tests, earth blocks of the usage class I also need to be exposed to the number of freeze–thaw cycles given in Table 3.17. The blocks may be used in their respective usage class if they do not show any cracking or swelling deformations within the minimum number of required cycles.

Water Vapor Sorption

Terminology

Water in a gaseous state is called water vapor and is part of the surrounding air. In adjacent building elements, water vapor is subject to the various mechanisms of moisture transfer (Sect. 5.1.2.1). The balancing process between the humidity in the air and the humidity in the material of adjacent building elements is referred to as sorption. Water sorption can be plotted as a sorption isotherm (Sect. 5.1.2.5).

Test Methods

Water vapor sorption of earth plaster mortars is determined by installing 15 mm thick specimens into a steel-plated mold which is sealed on five sides. This ensures that the sorption of water vapor only takes place on the sixth, unsealed side. The test surface is 1000 cm^2. Thin-layer earth plasters are applied on top of an earth base coat in order to attain the required test thickness of 15 mm. The specimens are conditioned under standard atmospheric conditions ((23 ± 2) °C/(50 ± 5)% RH) until a constant weight is reached.

For the test, the air humidity is raised to (80 ± 5) % RH while maintaining a constant temperature. The increase in weight of the specimens is measured after 0.5 h, 1 h, 3 h, 6 h, and 12 h. The measured amounts of adsorbed humidity are expressed as g/m^2. The mean value of at least three single tests is determined and no value must differ more than 20 % from the mean value. Earth plaster mortars adsorbing more than 60 g of humidity after 12 h are classified as highly active in relation to their adsorption potential.

This test was included in DIN 18947 as voluntary test.

Lab and Calculation Values

According to DIN 18947, A.2 (for information purposes), water vapor sorption for earth plaster mortar is divided and designated into classes WS I–III (Table 3.32).

Table 3.32 Water vapor adsorption classes for earth plaster mortar

No.	Water vapor adsorption class	Water vapor adsorption [g/m^2] after x [h]				
		0.5	1	3	6	12
1	WS I	≥3.5	≥7.0	≥13.5	≥20.0	≥35.0
2	WS II	≥5.0	≥10.0	≥20.0	≥30.0	≥47.5
3	WS III	≥6.0	≥13.0	≥26.5	≥40.0	≥60.0

3.6.3.2 Thermal Parameters

Thermal Conductivity

Terminology

Thermal conductivity λ, also known as the coefficient of thermal conductivity is defined as the quantity of heat conducted through 1 m^2 of a 1 m thick layer of a material in 1 s at a constant temperature difference of 1 K between both surfaces. The unit is expressed as W/mK.

Test Methods

Thermal conductivity λ is tested according to DIN 52612-3 using the guarded hot plate method. In this test, a flat, homogeneous specimen of the earth building material is placed in a testing apparatus between a heating plate and a cooling plate. The specimen is thermally insulated along its edges. The thermal conductivity of the specimen is determined by measuring the heat flow from the heating to the cooling plate and the temperature differences between the plates in a steady state.

Lab and Calculation Values

The Lehmbau Regeln [22] list calculation values for thermal conductivity λ of earth building materials (Table 3.33). They are based on a compilation by Volhard [69] listing the most unfavorable values taken from earlier standards, bibliographical references, and individual test results. They have now been included in DIN 4108-4.

For comparison purposes: The thermal conductivity of conventional building materials is between 0.02 (polyurethane) and 200 W/mK (aluminum).

Influencing Variables

Thermal conductivity λ is one of the most important starting values for thermal calculations (Sect. 5.1.1.3). In the practical building process, the dry bulk density ρ_d, the moisture content w, and the temperature largely determine the thermal conductivity of the earth building material. Thermal conductivity λ decreases with an increase in porosity, in other words, a decrease in dry bulk density. Thermal conductivity λ increases with a rise in moisture content of the building material. Water conducts heat much better than air. Metals are also good thermal conductors while many mineral building materials, including earth, conduct heat poorly. Materials with $\lambda < 0.15$ W/mK are considered to be insulation materials.

Table 3.33 Coefficients of thermal conductivity λ of earth building materials

No.	Dry bulk density ρ_d [kg/dm^3]	Coefficient of thermal conductivity λ [W/mK]	Earth building materials (according to Tables 3.3 and 3.22)
1	2.2	1.40	Rammed earth, earthen loose fill
2	2.0	1.10	Rammed earth, earthen loose fill
3	1.8	0.91	Rammed earth, earthen loose fill, earth mortar, clay panels
4	1.6	0.73	Cob, straw clay, clay mixed with fiber, earthen loose fill, earth mortar, earth blocks, clay panels
5	1.4	0.59	Cob, straw clay, clay mixed with fiber, earthen loose fill, earth mortar, earth blocks, clay panels
6	1.2	0.47	Light clay, earthen loose fill, earth mortar, earth blocks, clay panels
7	1.0	0.35	Light clay, earthen loose fill, earth mortar, earth blocks, clay panels
8	0.9	0.30	Light clay, earthen loose fill, earth mortar, earth blocks, clay panels
9	0.8	0.25	Light clay, earthen loose fill, earth mortar, earth blocks, clay panels
10	0.7	0.21	Light clay, earthen loose fill, earth mortar, earth blocks, clay panels
11	0.6	0.17	Light clay, earthen loose fill, earth mortar, earth blocks, clay panels
12	0.5	0.14	Light clay, earthen loose fill, clay panels
13	0.4	0.12	Light clay, earthen loose fill, clay panels
14	0.3	0.10	Light clay, earthen loose fill, clay panels

Specific Heat Capacity

Terminology

The specific heat capacity c_p is defined as the quantity of heat Ws which is required to change the temperature of 1 kg of material by 1 K. The unit is expressed as Ws/kgK or kJ/kgK.

In addition, the volumetric heat capacity $S = c_p \bullet \rho$ [Ws/m^{3}K] is given as a volume-specific value.

Lab and Calculation Values

The Lehmbau Regeln [22] list calculation values for the specific heat capacity c_p of earth building materials (Table 3.34).

For comparison purposes: The specific heat capacity c_p of inorganic building materials and air is approx. 1.0 kJ/kgK, of wood 2.1 kJ/kgK, of water 4.2 kJ/kgK, of aluminum 0.8, and other metals 0.4 kJ/kgK (DIN 4108-4, Table 7).

Table 3.34 Specific heat capacity c_p of earth building materials

No.	Dry bulk density ρ_d [kg/dm^3]	Aggregates, mineral [kJ/kgK] Sand, gravel, lightweight aggregates	Aggregates, organic [kJ/kgK] Straw	Fine fibers	Wood chips
1	≥1.6	1.0	1.0	1.0	1.0
2	1.4	1.0	1.0	1.1	1.1
3	1.2	1.0	1.0	1.1	1.2
4	1.0	1.0	1.1	1.1	1.3
5	0.8	1.0	1.1	1.2	1.4
6	0.6	1.0	1.1	1.3	1.5
7	0.4	–	1.2	1.4	-

Influencing variables

In light clays, the value c_p rises with an increase in the percentage of organic fiber aggregates.

References

1. Abduraschidov, K.S.; Tulaganov, A.A.; Pirmanov, K. et al: Baukonstruktionen aus Massivlehm in seismischen Gebieten Mittelasiens—Bewertung und Vorschläge zur Verbesserung der Erdbebensicherheit. In: LEHM 2004 Leipzig, Beiträge zur 4. Int. Fachtagung für Lehmbau, pp. 248–257, Weimar: Dachverband Lehm e.V. (2004)
2. International Labour Organisation ILO (Ed.): Techn. Series TM No. 6. Small scale brickmaking; Techn. Series TM No. 12. Small-scale manufacture of stabilized soil blocks. ILO, Geneva (1984–1987)
3. v. Bodelschwingh, G.: Ein alter Baumeister und was wir von ihm gelernt haben—Der Dünner Lehmbrotebau, 3. Aufl. Bünde: Heimstätte Dünne(1990)
4. Mukerji K, Wörner H (1991) (CRATerre): Soil preparation equipment—Product information. German Appropriate Technology Exchange Gate/Basin, Eschborn
5. Schroeder, H. et al: Bauingenieurpraktikum Havanna 1990. Weimar: Hochschule f. Architektur u. Bauwesen, Fak. Bauingenieurwesen, Arbeitsmaterialien WB Tropenbau, Heft 1 (1991)
6. Krause, E.; Berger, I.; Nehlert, J.; Wiegmann, J.: Technologie der Keramik, Bd. 1: Verfahren—Rohstoffe—Erzeugnisse.; Krause, E.; Berger, I.; Plaul, T.; Schulle, W.: Technologie der Keramik, Bd. 2: Mechanische Prozesse.; Krause, E.; Berger, I.; Kröckel, O.; Maier, P.: Technologie der Keramik, Bd. 3: Thermische Prozesse. Berlin: VEB Verlag f. Bauwesen (1981–1983)
7. Kärlicher Ton- und Schamottewerke Mannheim & Co. KG, KTS—einen Ton besser. Mülheim-Kärlich 2003—company brochure
8. Dachverband Lehm E.V/Abu Dhabi Authority for Culture & Heritage (Ed.): Planning and Building with Earth—Course handbook for architects, engineers, conservators & archaeologists, Weimar: unpublished (2009)
9. Dachverband Lehm e.V. (Hrsg.): Kurslehrbuch Fachkraft Lehmbau. Weimar: Dachverband Lehm e.V. (2005)
10. Miller, T.; Grigutsch, A; Schultze, K.W.: Lehmbaufibel—Darstellung der reinen Lehmbauweise. Weimar, Schriftenreihe d. Forschungsgemeinschaften Hochschule, Heft 3, Reprint der Originalausgabe von 1947, Verlag Bauhaus-Universität (1999)

11. Striegler W, Werner D (1973) Erdstoffverdichtung. VEB Verlag f. Bauwesen, Berlin
12. Schüler G, Wagner B (1990) Hochfeste Lehmsteine. In: Baustoffindustrie 5:160–61
13. Kapfinger, O.; Rauch, M.: Rammed earth—Lehm und Architektur—Terra cruda. Basel: Birkhäuser (2001)
14. Houben, H.; Guillaud, H.: Earth construction—A comprehensive guide. London: Intermediate Technology Publications (1994)
15. Зубкин, В. Е.; Коновалов, В. М.; Королёв, Н. Е.: Зонное нагнетание сыпучных сред (Zonal compaction of fill material). Москва: изд. РУСАКИ (2002)
16. Tulaganov, B.A.; Schroeder, H.; Schwarz, J.: Lehmbau in Usbekistan. In: Zukunft Lehmbau 2002–10 Jahre Dachverband Lehm e.V., pp. 91-101; Weimar: Dachverband Lehm e.V./ Bauhaus-Universität (2002)
17. Suske, P.: Hlinené domy novej generácie. Bratislava: vyd. Alfa (1991)
18. Pollack, E.; Richter, E.: Technik des Lehmbaus. Berlin: Verlag Technik (1952)
19. Schrader M (1997) Mauerziegel als historisches Baumaterial—Ein Materialleitfaden und Ratgeber. Edition Anderweit, Suderburg
20. Tate, D.; Park, K.; Lawson, W. D.; Ertas, A.; Maxwell, T. T.: Developing Large-Scale Compressed Earth Block (CEB) Building Systems: Strategies for Entrepreneurial Innovation. In: EARTH USA 2011 Albuquerque, Proc. 6th Int. Earth Building Conference, pp. 270–277, Albuquerque, NM: Adobe in Action (2011)
21. Easton, D.: Industrielle Formgebung von Stampflehm. In: LEHM 2008 Koblenz, Beiträge zur 5. Int. Fachtagung für Lehmbau, pp. 90–97. Weimar: Dachverband Lehm e.V. (2008)
22. Dachverband Lehm e.V. (Hrsg.): Lehmbau Regeln—Begriffe, Baustoffe, Bauteile. Wiesbaden: Vieweg+Teubner I GWV Fachverlage, 3., überarbeitete Aufl. (2009)
23. Kortlepel, U. u. a.: Lehmspritzverfahren. Aachen: Landesinstitut f. Bauwesen u. angewandte Bauschadensforschung NRW 2.23-1994 (1994)
24. Dingeldein, T.: Konzeption einer neuen Lehmbauweise mit maschineller Einblastechnik. In: "Moderner Lehmbau 2002", pp. 9-13, KirchBauhof Berlin (Hrsg.), Stuttgart: Fraunhofer IRB (2002)
25. Wilson, Q.: Die Produktion von in der Sonne getrockneten Adobe Ziegeln in den USA. In: LEHM 2008 Koblenz, Beiträge zur 5. Int. Fachtagung für Lehmbau, pp. 98–103. Weimar: Dachverband Lehm e.V. (2008)
26. Schroeder, H.; Schnellert, T.; Sowoidnich, T; Heller, T.: Moisture transfer and change in strength during the construction of rammed earth walls. In: Living in Earthen Cities—Kerpic `05; Proc. 1st Int. Conference, pp. 117–125, Istanbul Technical University ITU (20005
27. Dierks, K.; Ziegert, C.: Tragender Stampflehm—ein betonverwandtes Konglomerat mit Vergangenheit und Zukunft. In: Avak, R.; Goris, A., pp. G.39–G.65. Berlin: Stahlbeton Aktuell (2001)
28. Thurm C (2001) Analyse der Möglichkeiten der ökologisch verträglichen Modifizierung von Baulehmen mit Zuschlägen und Zusatzstoffen. Bauhaus-Universität, Fak. Bauingenieurwesen, unpublished diploma thesis, Weimar
29. Maniatidis, V.; Walker, P.: A review of rammed earth construction. Bath UK, University Report (2003)
30. Niemeyer, R.: Der Lehmbau und seine praktische Anwendung. Grebenstein/Staufen: Ökobuch Verlag 1982/1990 (Reprint der Originalausg. von 1946)
31. Scholz, J.; Zier, H.-W.: Schäden an den Fresken Moritz von Schwinds im Palas der Wartburg und Möglichkeiten der Restaurierung. In: Europäischer Sanierungskalender 2009 (4. Jahrgang), pp. 343-359, Berlin-Wien-Zürich: Beuth Verlag (2009)
32. Dachverband Lehm e.V. (Hrsg.), Lehm-Mauermörtel—Begriffe, Baustoffe, Anforderungen, Prüfverfahren. Technische Merkblätter Lehmbau, TM 03, Weimar: Dachverband Lehm e.V. (2011)
33. Dachverband Lehm e.V. (Hrsg.), Lehm-Putzmörtel—Begriffe, Baustoffe, Anforderungen, Prüfverfahren. Technische Merkblätter Lehmbau, TM 04, Weimar: Dachverband Lehm e.V. (2011)

34. Dachverband Lehm e.V. (Hrsg.), Lehmdünnlagenbeschichtungen—Begriffe, Anforderungen, Prüfverfahren, Deklaration. Technische Merkblätter Lehmbau, TM 06, Weimar: Dachverband Lehm e.V. (2015-03)
35. Natureplus e.V. (Hrsg.) Richtlinien zur Vergabe des Qualitätszeichens "natureplus". RL 0607 Lehmanstriche und Lehmdünnlagenbeschichtungen (September 2010); RL 0803 Lehmputzmörtel (September 2010); RL 0804 Stabilisierte Lehmputzmörtel (proposed); RL 1006 Lehmbauplatten (September 2010); RL 1101 Lehmsteine (proposed); RL 0000 Basiskriterien (May 2011); Neckargemünd: 2010
36. Schroeder, H.; Bieber, A.: Neue Stampflehmprojekte in Thüringen. In: LEHM 2004 Leipzig, Beiträge zur 4. Int. Fachtagung für Lehmbau, pp. 190-201, Weimar: Dachverband Lehm e.V. (2004)
37. Backe, H.; Hiese, W.: Baustoffkunde für Berufs- und Technikerschulen und zum Selbstunterricht, 8. Aufl. Düsseldorf: Werner Verlag (1997)
38. Rütz, D., Witt, K. J. u. a.: Wissensspeicher Geotechnik, 18. Aufl. Weimar: Bauhaus-Universität, Fak. Bauingenieurwesen Studienunterlagen Geotechnik (2011)
39. Fuchs, E.; Klengel, K.J.: Baugrund und Bodenmechanik. Berlin: VEB Verlag f. Bauwesen (1977)
40. Center for the Development of Industry ACP-EU/CRATerre-Basin: Compressed earth blocks, No. 16 Testing procedures (1998) Brussels (1996–1998)
41. Olivier, M.; Velkov, P.: Dynamic properties of compacted earth as a building material. 3rd Int. Conference STREMA 93, Southampton, Wessex Inst. of Technology (1993)
42. Stark; J.; Krug, H.: Baustoffkenngrößen. Weimar, Schriften der Bauhaus-Universität No. 102, F.A.-Finger—Institut f. Baustoffkunde (1996)
43. Schnellert, T.: Untersuchung von Transportprozessen der Einbaufeuchte in Baukonstruktionen aus Stampflehm während der Austrocknung, unpublished diploma thesis. Weimar: Bauhaus-Universität, Fak. Bauingenieurwesen(2004)
44. Fischer, F.; Mönnig, H.-U.; Mücke, F.; Schroeder, H.; Wagner, B: Ein Jugendklub aus Lehm—Baureport Herbsleben. Weimar, Wiss. Z. Hochsch. Archit. Bauwes.—A.—35 (1989) 3/4, pp. 162-168
45. Minke, G.: Handbuch Lehmbau. Baustoffkunde, Techniken, Lehmarchitektur, 7. erw. Aufl. Staufen: Ökobuch Verlag (2009)
46. Maniatidis, V.; Walker, P.: Structural capacity of rammed earth in compression. J. Mater Civil Eng.; ASCE; March 2008, pp. 230-238
47. Ziegert C (2003) Lehmwellerbau—Konstruktion, Schäden und Sanierung. Berichte aus dem konstruktiven Ingenieurbau, Heft 37. Fraunhofer IRB, Stuttgart
48. Rischanek, A.: Sicherheitskonzept für den Lehmsteinbau. Wien: Technische Universität, Fak. Bauingenieurwesen (2009)
49. Earth Building Association of Australia EBAA (Ed.): Building with earth bricks & rammed earth in Australia. Wangaratta: EBAA (2004)
50. Standards New Zealand: NZS 4297: 1998 Engineering Design of Earth Buildings; NZS 4298: 1998 Materials and Workmanship For Earth Buildings; NZS 4299: 1998 Earth Building Not Requiring Specific Design; Wellington: Standards New Zealand (1998)
51. Mönnig, H.U.; Fischer, F.; Schroeder, H.; Wagner, B.; Mücke, F.: Spezifische Eigenschaftsuntersuchungen für Erdbaustoffe einschließlich der Eigenschaftsverbesserung durch Modifikation und Möglichkeiten des Feuchtigkeitsschutzes für in Erdstoff errichtete Gebäude. Weimar, Hochschule f. Arch. u. Bauwesen, WBI, WB Tropen- u. Auslandsbau, unpublished research report (1988)
52. Utz R (2004) Stabilisierung von Lösslehmoberflächen in archäologischen Grabungen am Beispiel der Terrakotta-Armee des Qin Shihuangdi. Universität, Fak. Geowissenschaften, Diss, München
53. Minke, G.: Lehmputze—ihre Eigenschaften und deren Veränderung durch Zusätze, Verarbeitung und Oberflächenbehandlung. In: Moderner Lehmbau 2005, Beiträge, pp. 67-81, Berlin: Umbra GmbH (2005)
54. Dettmering T, Kollmann H (2001) Putze in Bausanierung und Denkmalpflege. Verlag für Bauwesen, Berlin

55. Schroeder, H.; Ziegert, C.: Haftfestigkeitsprüfungen an Lehmputzen. In: LEHM 2008 Koblenz, Beiträge zur 5. Int. Fachtagung für Lehmbau, pp. 202–205, Weimar: Dachverband Lehm e.V. (2008)
56. Steinmann M (1995) Studie zu korrelativen Bezügen zwischen geotechnischen, mineralogischen und lehmbautechnischen Parametern von Lehmbaustoffen, unpublished diploma thesis. Bauhaus-Universität, Fak. Bauingenieurwesen, Weimar
57. Schroeder, H.: Klassifikation von Baulehmen. In: LEHM 2000 Berlin; Beiträge zur 3. Int. Fachtagung für Lehmbau des Dachverbandes Lehm e.V., pp. 57-63, Berlin: Overall-Verlag (2000)
58. Schroeder, H; Volhard, F.; Röhlen, U.; Ziegert, C.: Die Lehmbau Regeln 2008–10 Jahre Erfahrungen mit der praktischen Anwendung. In: LEHM 2008 Koblenz, Beiträge zur 5. Int. Fachtagung für Lehmbau, pp. 12–21, Weimar: Dachverband Lehm e.V. (2008)
59. Riechers, H.J.; Hildebrand, M.: Putz—Planung, Gestaltung, Ausführung. In: Mauerwerk-Kalender 2006, pp. 267–300, Berlin: W. Ernst & Sohn (2006)
60. Dierks, K.; Ziegert, C.: Materialprüfung und Begleitforschung im tragenden Lehmbau. In: LEHM 2000 Berlin, Beiträge zur 3. Int. Fachtagung für Lehmbau des Dachverbandes Lehm e.V., pp. 46-56. Berlin: Overall Verlag (2000)
61. Jagadish KS, Venkatarama Reddy BV, Nanjunda Rao KS (2007) Alternative building materials and technologies. New Age International Publishers, New Delhi
62. Fontana, P.; Grünberg, U.: Untersuchungen zur Haftscherfestigkeit von Lehmmauermörtel. In: LEHM 2012 Weimar, Beiträge zur 6. Int. Fachtagung für Lehmbau, pp. 104–113 (Beilage), Weimar: Dachverband Lehm e.V. (2012)
63. Venkatarama Reddy BV, Uday Vyas CV (2008) Influence of shear bond strength on compressive strength and stress-strain characteristics on masonry. Mater Struct 41:1697–1712
64. Dachverband Lehm E.V/Abu Dhabi Authority for Culture & Heritage (Ed.): Building with Earth—Course handbook for building trades: masons, carpenters, plasterers, Weimar: unpublished (2009)
65. Dachverband Lehm e.V. (Hrsg.): Anforderungen an Lehmputz als Bauteil. Technische Merkblätter Lehmbau, TM 01, Weimar: Dachverband Lehm e.V. (2014-06)
66. Röhlen U, Ziegert C (2010) Lehmbau-Praxis, Planung und Ausführung. Bauwerk Verlag, Berlin
67. Lustig-Rössler U (1992) Untersuchungen zum Feuchteverhalten von Lehm als Baustoff. GH/ Universität, Diss, Kassel
68. Künzel, H.: Regenbeanspruchung und Regenschutz von Holzfachwerk-Außenwänden. In: Statusseminar "Erhaltung von Fachwerkbauten", Sonderheft Verbundforschungsprojekt "Fachwerkbautenverfall und Fachwerkbautenerhaltung, pp. 32–39, Fulda: ZHD (1991)
69. Volhard, F.: Leichtlehmbau, alter Baustoff—neue Technik, 5. Aufl. Heidelberg: C. F. Müller (1995)
70. CSIRO Australia, Division of Building, Construction and Engineering, Bull. 5: Earth-Wall Construction. North Ryde, NSW, Australia: 4th edn (1995)
71. Regulation & Licensing Dept., Construction Industries Div., General Constr. Bureau: 2006 New Mexico Earthen Building Materials Code. CID-GCB-NMBC-14.7.4. Santa Fe, NM (2006)
72. Schweizer Ingenieur- und Architekten-Verein SIA (Hrsg.): SIA Dokumentationen. D 077 (1991), Bauen mit Lehm; Schaerer, A.; Huber, A.; Kleespies, T.; D 0111 (1994), Regeln zum Bauen mit Lehm; Hugi, H.; Huber, A.; Kleespies, T.; D 0112 (1994), Lehmbauatlas; Hugi, H.; Huber, A.; Kleespies, T. Zürich: SIA 1991 u. (1994)
73. Bureau of Indian Standards: Indian Standard 2110-1980: Code of practice for in situ construction of walls in buildings with soil-cement. New Delhi: 1981 (1991 1st rev., 1998 reaffirmed)
74. Rigassi V (1995) Compressed earth blocks, Vol. 1: Manual of production.; Guillaud, H. u.a.: Compressed earth blocks, Vol. 2: Manual of design and construction. Gate/CRATerre-EAG. Vieweg-Verlag, Braunschweig
75. Dachverband Lehm e.V. (Hrsg.): Lehmbau Verbraucherinformation. Weimar: Dachverband Lehm e.V. (2004)

76. Schreckenbach H, Abankwa JGK (1982) Construction technology for a tropical developing country. Gesellschaft für Technische Zusammenarbeit GTZ, Eschborn
77. Röppischer K, Grund K, Hiersemann T (2000) Lehmbaupraktikum 2000, unpublished internship report, Weimar. Bauhaus-Universität, Fak. Architektur, Weimar
78. Smoltczyk H. (Hrsg.): Grundbau-Taschenbuch, Bd. 1 Grundlagen, 3. Aufl. Berlin: Ernst & Sohn (1980)
79. Walker, P.; Standards Australia: The Australian Earth Building Handbook. Sydney: Standards Australia International (2002)
80. Bobe, R.; Hubaček, H.: Bodenmechanik, 2. Aufl. Berlin: VEB Verlag f. Bauwesen (1986)

4

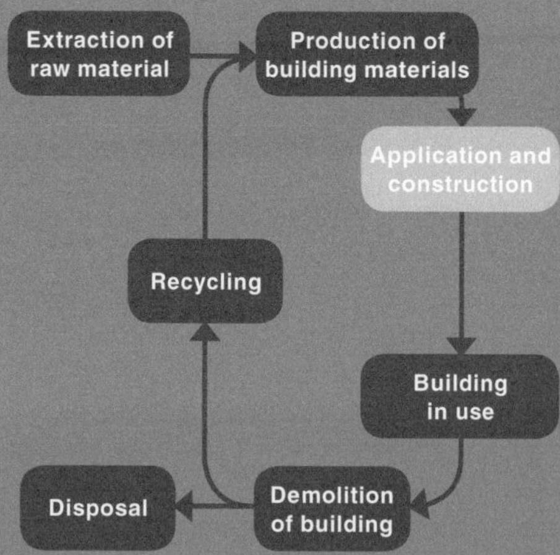

Earthen Structures – Planning, Building and Construction Supervision

After the usability of earth building materials has been verified, they can be processed into building elements and structures.

Like all structures, earth buildings must fully meet general requirements in terms of structure stability, function, design and the users' needs. The building envelope has to ensure that an optimal connection is made between the environmental conditions on the exterior and the living and working conditions inside the building.

According to the terminology of DIN EN 1990, introduced in 2012, the *construction type* describes the primary load-bearing materials used in a structure, such as timber construction, masonry construction and earth building. For practical construction work, however, it has proven useful to apply the term "earth building" to non-structural elements, non load-bearing infill, partition walls and earth plasters as well. The term *construction method* (or building technique) describes the building method, e.g. cast-in-place concrete construction or rammed earth construction.

© Springer International Publishing Switzerland 2016
H. Schroeder, *Sustainable Building with Earth*,
DOI 10.1007/978-3-319-19491-2_4

4.1 Function and Design

It is the creative work of architects and planners to give an outward appearance to a building with defined functions. In this context, the building needs to meet certain aesthetic and design requirements while integrating visually into its surroundings. The exterior appearance of traditional buildings follows regional principles which have been passed down from generation to generation. These principles are based on experiences with regional, climate-appropriate building styles which make use of locally available materials. They should, however, not be seen as inflexible as they also always reflect changes in terms of new functions and new building materials or in the owner's economic situation.

The layout of the functions in a building as well as the structure's exterior appearance are the result of an optimization process which considers aspects of the proposed building site and site-specific natural (physical) parameters as well as anthropogenic (sociocultural) factors. These considerations include the prevailing climate (connected to the seasonal change of room functions within the building), natural disasters, available building materials, family structures, businesses, religious beliefs and local traditions, safety aspects, and, finally, the owner's financial resources. With regard to existing earth buildings, this optimization process has created an extraordinarily varied range of structural and design solutions, both in the earth building traditions of different cultures and climates as well as in new construction.

The *exterior appearance* of earth buildings is primarily influenced by the prevailing climate and the chosen building method. Figures 1.1 and 1.16 show how these influences were already reflected in optimized structural designs thousands of years ago: In hot and dry climates, compact structures using high-mass construction methods prevailed. The buildings were arranged in clusters, standing close to each other to provide better mutual shading. Due to its scarcity, wood was used sparingly and reused whenever possible. In moderate and wet–dry climates with extensive forests, a combination of earth and wood as building materials was common. This combination allowed for thinner wall systems. Steeply pitched roofs, which helped rainwater drain quickly, were a prominent feature of these buildings.

The exterior appearance of a building is also determined by its placement in relation to adjacent structures and the shapes of its main building elements. When it comes to buildings made of earth building materials, various aspects of exterior building design can be distinguished between:

- Spatial urban layout: stand-alone, row house, individual groups, building clusters
- Number of stories/access: one story, multistory/from the inside or outside
- Wall axis: linear, curved, round
- Roof: flat roof, pitched roof, curved roof (Sect. 4.3.5)

In Germany, buildings are divided into building classes based on their exterior design (height, floor space) according to the Model Building Code (MBO) (Table 4.1):

Table 4.1 Building classes according to the Model Building Code (MBO)

Building class	Description
1a	Stand-alone buildings with a height[a] up to 7 m containing not more than two building units with a maximum of 400 m^2 total
1b	Stand-alone agricultural or forestry buildings
2	Buildings with a height[a] up to 7 m containing not more than two building units with a maximum of 400 m^2 total
3	Other buildings with a height[a] up to 7 m
4	Buildings with a height[a] up to 13 m containing building units with a maximum of 400 m^2 each
5	Other buildings including underground buildings

[a]Height of the floor level of the highest habitable space based on the average ground level

The requirements placed on building elements and building materials differ depending on the building class. For example, the higher the building class, the higher the fire resistance rating requirements of the building elements.

The construction type or the primary building materials used in the structure must be taken into consideration for a possible risk of damage evaluated by building insurance companies. In this context, the fire performance of the building materials has priority (Sect. 3.4.8). The moisture resistance of the materials must be considered as well.

The MBO provides the framework for the classification of buildings. The framework is then regulated by the individual German states in their respective state building codes (Landesbauordnungen—LBO).

Other important aspects of the design of earth buildings are surface textures and colors.

Today, general building functions in terms of traditional and modern earth building can be grouped into the following categories:

- Residential buildings/dormitories/hotels/residences
- Education (daycare facilities for children, schools)
- Public buildings (public agencies, administration, museums)
- Cultural and recreational facilities (playgrounds, zoos, art spaces, open space structures)
- Religious/sacred buildings
- Commercial buildings/warehouses
- Buildings for the spa and health sector
- Buildings for agriculture/horticulture
- Facilities/buildings providing protection

According to DIN EN 1990, the function or intended use of a building is referred to as the *type of building*, such as residential buildings or daycare facilities for children. The type of building is included in the form of applicable safety factors in the semi-probabilistic dimensioning method for load-bearing structures (Sect. 4.3.3.1).

4.1.1 Residential Buildings

Residential buildings make up the building category which probably reflects all of the above building design influences in their most diverse forms. In Germany, with its moderate climate and former extensive forests, half-timber houses with earth infills have shaped the appearance of rural areas as well as urban settlements for centuries. Most of the buildings consist of a lower and an upper floor. The walls continue upward as slanted walls which start at a "knee wall" and are determined by the pitch of the roof.

Figures 4.1 and 4.2 exemplify that, even under similar climatic conditions and with the same local building materials available, regional traditions shaped the design of a building considerably. The "Gotische Haus" in Limburg/Lahn, Germany (started in 1289), is a magnificent five-story half-timber building with stake work and earth infill (see also Fig. 1.19). It sits in the middle of a densely developed urban landscape of half-timber houses. The traditional Korean "Hanok" house was constructed using the same building technique but looks completely different as it is a stand-alone, one-story building which appears to blend into the countryside [1].

Figures 1.15a and 4.3 show two examples of residential buildings in hot and arid climates: Traditional courtyard houses of the Arabic-Islamic building tradition form a compact spatial structure with mutually shading facades. While wood and other vegetal building materials are scarce, soil suitable for rammed earth building is readily available. The traditional design of the rammed earth houses in Ait Benhaddou, Morocco, shown in Fig. 1.15a, exemplifies the defensive character of

Fig. 4.1 "Gotisches Haus" in Limburg/Lahn, Germany, half-timber construction consisting of vertical stakes and an infill of straw-clay clumps

Fig. 4.2 Traditional Korean "Hanok" house, half-timber construction with bamboo slats and daub [1]

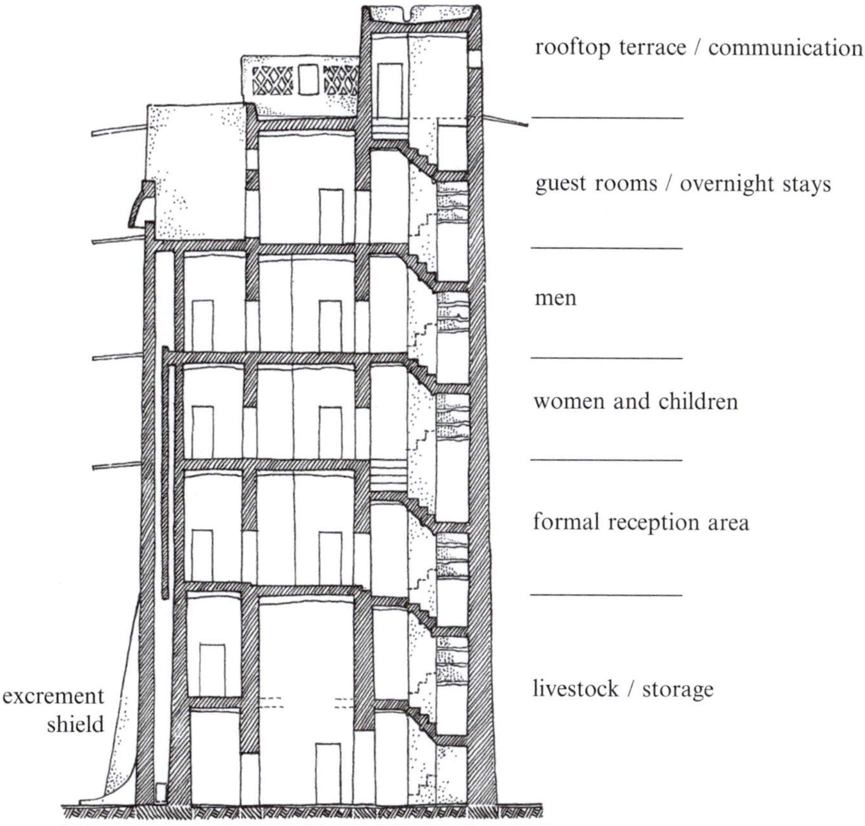

Fig. 4.3 Sectional view of a Southern Arabian earth block (tower) house with a distribution of functions, Shibam, South Yemen [2]

the buildings in addition to their housing function (Tighremt = individual, fortified residential building; Ksar = fortified settlement of various residential buildings). Flat roofs prevail as a result of low rainfall levels (Sect. 4.3.5.2).

A distribution of functions different from contemporary European concepts can be seen in Fig. 4.3 [2]. It shows the cross section of a six-story rammed earth "tower" in Shibam, South Yemen. A strict gender segregation within the house as well as ground floor storage and livestock functions is clearly visible.

Humans and livestock living under one roof was also a common concept in specific examples of traditional European architecture which combined living quarters with stables for animals. Even after World War II, the homes of new settlers in Germany were designed in this manner (Fig. 1.24). During the winter months, the animals' body heat helped save heating costs.

The tower-like design of the building shown in Fig. 4.3 is a reflection of the local topographical situation: the narrow Wadi (dry valley) Hadramaut limits building activity and forces people to construct higher structures. The densely clustered buildings (up to 30 m high) have been depicted in numerous photographs. The city of Shibam is a UNESCO world heritage site (Sect. 1.2). Because of its dense layout with earth block buildings up to ten stories high, it is also called "the Manhattan of the desert."

A contrast to traditional rammed earth construction in North African and Arabic cultures is the freestanding "modern" house made of cement-stabilized rammed earth nearby Sydney, Australia. It was built in 2005 under similar climatic conditions. For shading purposes, the roofs have large overhangs, and the exterior walls have narrow window slits to protect the house from the intense sunlight (Fig. 4.4). In both examples, the enormous thermal storage capacity of the exterior walls buffers the extreme daily air temperature amplitude and creates a comfortable indoor climate.

Fig. 4.4 Load-bearing rammed earth walls stabilized with cement

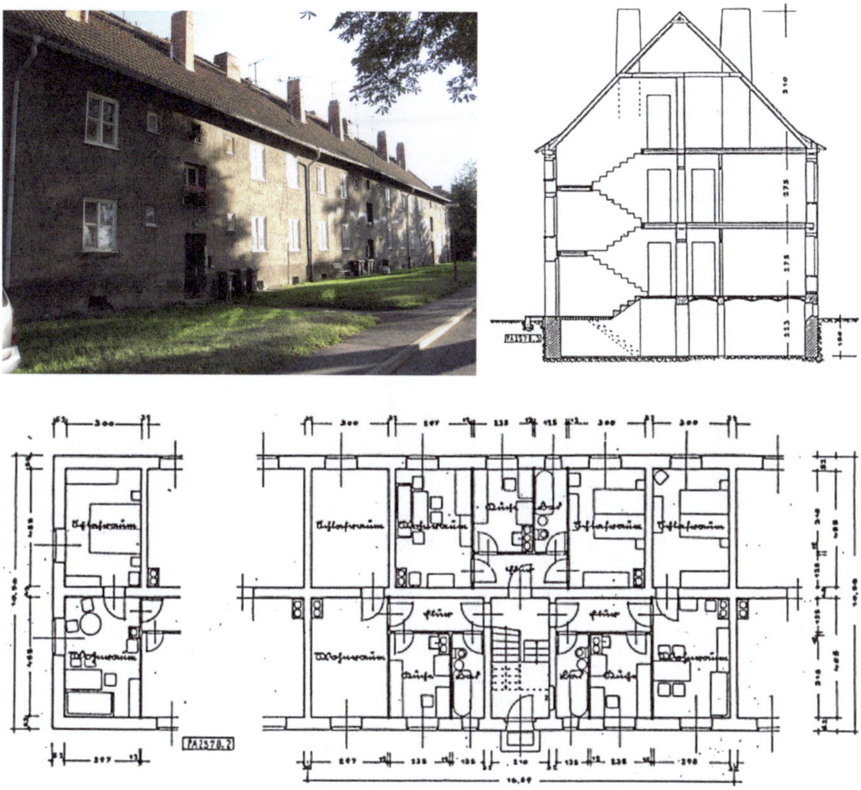

Fig. 4.5 Two-story row house for 18 families in Gotha, Thuringia, Germany [1, 3]

Figure 4.5 shows a two-story, rammed earth row house for 18 families built in 1951 in Gotha/Thuringia, Germany. The building was constructed as part of the German Democratic Republic's reconstruction program [3]. Similar residential buildings using the same technique were built in the German states of Saxony and Saxony-Anhalt up to around 1960. Despite scarce resources in postwar Germany, the designers added some sparse decorative elements to the exterior such as in Mücheln near Merseburg (Fig. 1.26).

The three-story duplex shown in Fig. 4.6 is a modern residential rammed earth building constructed in a moderate climate. At the beginning of the 1980s, approx. 60 single- and multifamily homes were built in Ile d'Abeau, Villefontaine, near Lyon, France, using various earth building techniques. Today, this housing development is considered the starting point of modern earth building in Europe which is characterized by a new architectural language: on focus on the aesthetics and originality of the structure, a clear and functional division between the building elements, visibility of the (traditional) building techniques, harmony, and sensuality created by the interplay of color, texture, and the feel of the surfaces.

Thousands of residential buildings constructed using a load-bearing cob technique are among Central Germany's historical buildings (Sect. 4.3.3.1). They are

Fig. 4.6 Residential rammed earth building, three-story duplex, Ile d'Abeau, Villefontaine/Lyon, France

Fig. 4.7 Residential building (Schillerhaus) in Leipzig, Germany, cob construction [1]

Fig. 4.8 Residential buildings, Tulou type in Tianloukeng, Fujian Province, China [1]

oftentimes covered with a lime (cement) plaster. One way to recognize the original building technique used is by their overall compact design, their seemingly small wall openings in relation to their wall thickness, as well as the tapering of their exterior walls toward the top of the structure. An example can be seen in Fig. 4.7 [1]. Built in 1717 as a home for the petty bourgeoisie, this modest-looking house witnessed a great moment in history: Friedrich Schiller wrote his world-famous "Ode to Joy" here in 1785, later adapted by Beethoven for the final chorus of his ninth symphony.

Figure 4.8 shows a building design and spatial arrangement which is very unusual for a residential building. Around 400 years ago, the Hakka people in the southwestern Chinese province of Fujian developed these fortress-like buildings called "Tulou." The buildings were originally designed to house an extended family (clan structure) consisting of 100 people or more. The structure of the building consists of a one- to four-story high ring (or square) made of rammed earth with a diameter of up to 80 m. The inner courtyard is entered through a single gate which can be locked. From here, the rooms and apartments located in the rammed earth part of the structure can be accessed via separate balconies. The outbuildings are located in the inner courtyard.

Santa Fe is the capital of New Mexico located in the Southwest of the USA. The city's architecture is characterized by traditional earth block (adobe) building which can be traced back to the indigenous Native American population (Pueblos) (Fig. 1.14b). The exterior design of the primarily one- to two-story buildings is characterized by flat roofs and protruding roof beams (vigas).

Fig. 4.9 Applegate Estate, residential adobe building in Santa Fe, NM, USA

Figure 4.9 shows a residential building constructed in the typical "Pueblo-style" architecture. It was built as a farmhouse around 1700 and purchased in 1845 by a Spanish officer. In 1920, artist Frank Applegate became its owner and had the house renovated according to his vision. The current owners have adapted the house to modern standards and listed it for sale with "Santa Fe Properties/Luxury Portfolio International" in 2013.

4.1.2 *Education Buildings*

Recent construction projects of schools and daycare facilities which combine contemporary architectural design with different earth building techniques have brought earth as a building material back into focus with architects and planners. Many of these often unique projects have won national and international architecture awards. Above all, the occupants value the buildings for their balanced indoor climate and the health benefits for children, adolescents, and teachers.

The load-bearing earth block domes by architect and university professor Gernot Minke (from Kassel, Germany) are a rather uncommon sight in the Central European culture. From the exterior, they form a green, hilly landscape, while the interior appears solemn and almost sacred. These structures convey the unexpected design possibilities of the earth block building technique. An example in Fig. 4.10 [1] shows the Waldorf Kindergarten in Sorsum/Hannover, Germany, built in 1997. The central multipurpose room is covered by a self-supporting dome with a height of 7 m and a diameter of 11 m.

Fig. 4.10 Daycare facility Sorsum, Germany, earth block dome [1]

The Waldorf School in Weimar, Germany, follows an entirely different design concept. The school was designed as a new construction project within an existing row of attached buildings from a former monastery. The structures date back to various time periods and were built in different styles (Fig. 4.11 [1]). The load-bearing structure of the school is rather unusual: it consists of a three-story reinforced concrete frame with walls created by a rammed earth infill.

In 1997/1998, the daycare facility “Perlboot” in Gera/Thuringia, Germany, was built using the traditional half-timber technique with earth infill in the middle of a neighborhood of prefabricated concrete slab buildings. The special feature of this facility is its shape: architects Maria Hoffmann and Franz Wilkowski used the structure of the nautilus shell (which in German can also be referred to as a “Perlboot”—pearl boat) as inspiration (Fig. 4.12). The floor plan was modeled after the spiral shape of the shell. The load-bearing exterior walls and partition walls were designed as half-timber walls with an earth infill. Some of the interior infill is exposed and consists of manually shaped earth clumps.

The self-help school project METI (Modern Education and Training Institute) is a new construction project built in Rudrapur/Dinapur, Bangladesh, in 2005/2006. It was constructed using a cob-like technique [4]. The ground floor consists of load-bearing cob walls, while the upper floor is designed as a light, airy bamboo structure with a roof using extended overhangs. The three classrooms on the ground floor are connected to an organically shaped “cave room” through holes in the walls. Instead

reinforced concrete support structure with rammed earth infill

Fig. 4.11 Waldorf School in Weimar, Germany, new construction within an existing row of attached buildings [1]

of using "modern," energy-intensive building materials such as concrete, steel, and fired brick, the project gave preference to local materials and building techniques which required a high amount of manual labor. In 2007, the architects Anna Heringer and Eike Roswag received the prestigious Aga Khan Award for Architecture for their school design shown in Fig. 4.13.

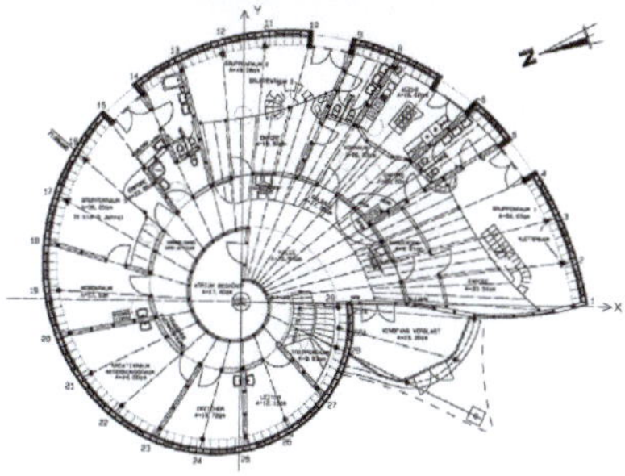

floor plan ground floor and section

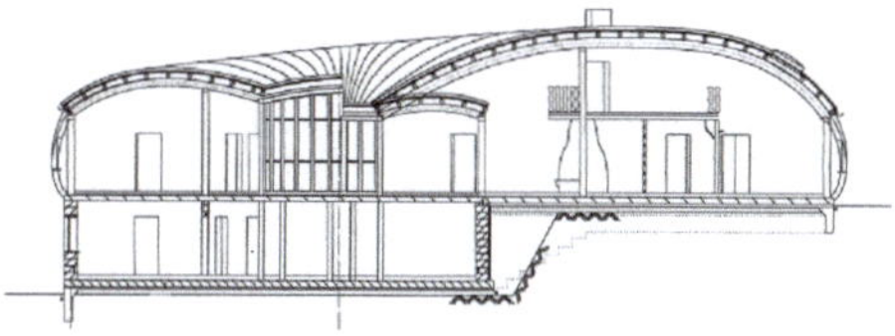

Fig. 4.12 Daycare facility "Perlboot" in Gera, Germany [1]

Fig. 4.13 METI School, cob/bamboo, Rudrapur, Bangladesh [4]

4.1.3 Sacred Buildings

Through the ages, religious or sacred places have played an important role in people's lives. In the past, designing structures for worship and, later, religious purposes always posed particular challenges for builders: this included a highly visible building site, unusual dimensions of the floor plans and walls, orientation of the building axis, illumination of the interior space, and roof shapes which dominated the landscape. When it came to choosing materials, the European Christian culture relied on durable, water-resistant building materials such as natural stone or fired brick early on. Earth as a building material played a minor role in contrast to other regions, such as Africa and pre-Colombian or present-day Latin America.

Sacred places were called "huaca" by the people of pre-Colombian America. These places were carefully chosen by the priests. Terraced earth block pyramids with monumental dimensions are one example of these ceremonial structures. Numerous earth block pyramids have been preserved as ruins in Latin America. The Pachacamac pyramid complex on the Pacific coast near Lima, Peru, consists of more than 10 "huacas," constructed of earth blocks by successive civilizations over a large area. Figure 4.14 shows a ceremonial road pointing toward the tip of a pyramid. Priests wanted to convey to pilgrims from nearby mountains that this place was

processional road to the tip of the earth block pyramid, the ocean is directly beyond

Fig. 4.14 Huaca de Pachacamac, earth block pyramid near Lima, Peru

Fig. 4.15 San Esteban Church, Acoma Pueblo, NM, USA

the gateway to the underworld. The afterlife would begin beyond the infinite expanse of water. For most of the pilgrims, it was the first time in their lives that they had seen the ocean.

The earth block pyramids (Ziggurats) in the Middle East represent a very similar type of building. Examples can still be found in present-day Iraq and Iran (Fig. 1.11).

The San Esteban church at Acoma Pueblo, approx. 100 km west of Albuquerque, New Mexico, USA, is located on a mesa (table mountain) rising approx. 110 m above the desert plain. In her book "Holy Adobe" [5], Leonore Harris called this mystical place "Island in the Sky." Construction of the adobe church began in 1629. The local Native American residents of Acoma Pueblo were forced by the military and the church to transport the required earth from the canyon floor to the mesa. Fortress-like on the exterior, the church has a very plain interior, and its earth floor appears to connect it to the underlying mountain (Fig. 4.15).

In Berlin, the rammed earth Chapel of Reconciliation (Kapelle der Versöhnung), built in 1999/2000 on Bernauer Straße along the former Berlin Wall, has become a modern "ceremonial site." The location is just as significant as the examples mentioned above: in 1891, the Reconciliation Church (Versöhnungskirche) was built on the same site. It later obstructed the line of fire of GDR border guards and was therefore demolished in 1985. After the fall of the wall only 4 years later, the congregation decided to build a new church in the same location. They voted against an early design which suggested the use of concrete because memories connected to this location seemed too burdened by this building material. The solution which was presented by the architects Peter Sassenroth and Rudolf Reitermann envisioned

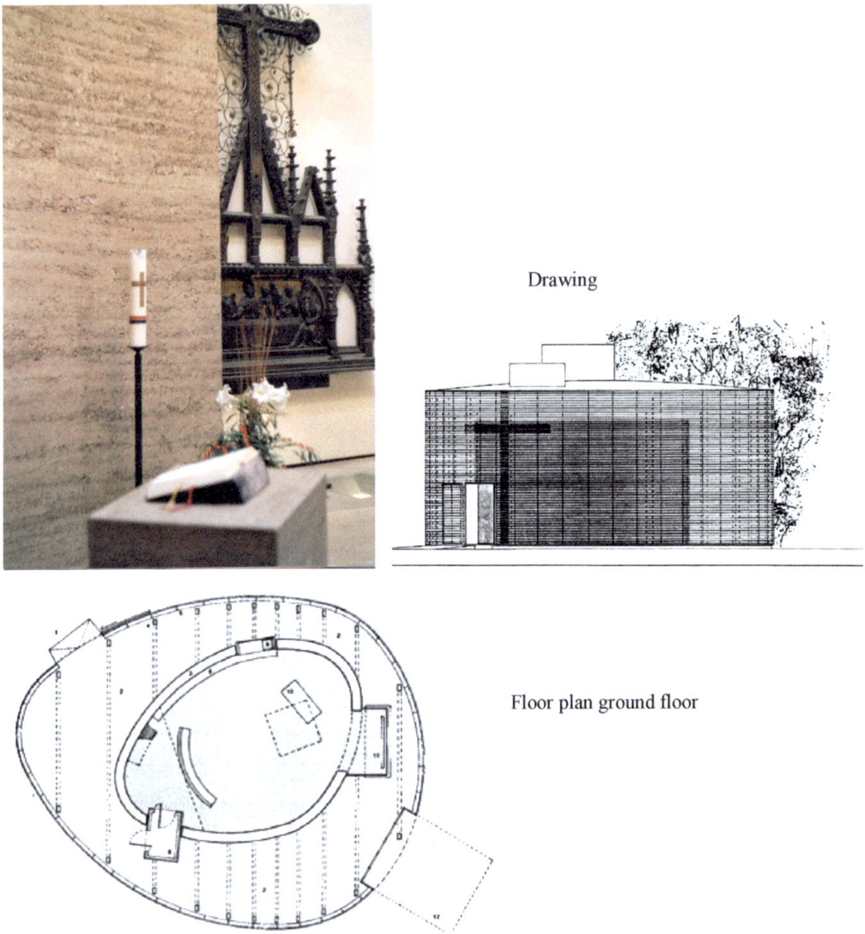

Fig. 4.16 Chapel of Reconciliation Berlin, load-bearing rammed earth wall construction [107]

the use of a rammed earth mix with the addition of crushed brick salvaged from the destroyed church. This allowed the old church to be "resurrected." Together with the adjacent Berlin Wall Memorial, it has become a place of remembrance as well as an annual meeting place for Germany's political elite.

There are other remarkable aspects of this church. One example is the "organic" geometry of the floor plan for the load-bearing rammed earth wall (7.2 m high and 0.6 m thick) which encloses the church interior. It was partially built on top of the foundation of the previous church, and its unusual geometry presented a particular challenge for the design of the formwork. The rammed earth core of the church is encased by a light, transparent timber structure constructed of a horizontal frame. The resulting space between the outer shell and the solid interior earthen wall is used as a passageway. This forms a stark contrast which is further enhanced by the interplay of natural light and shadows on the surface of the earthen wall (Fig. 4.16).

Fig. 4.17 Prince Ahmad Bin Salman—Mosque in Riad, Saudi Arabia [6]

Roofs in traditional Arabic-Islamic sacred buildings are often associated with the image of a dome (Sect. 4.3.5) which, according to religious beliefs, represents the vault of heaven. Domed structures are also common in Christian religious architecture, such as in St. Peter's Basilica in Rome. The Prince-Ahmad-Bin-Salman-Mosque in Riyadh, Saudi Arabia, depicted in Fig. 4.17, is somewhat unusual for an Islamic religious building: its design is reduced and very modern, with a bitumen-sealed flat earth roof and supported by interior limestone columns. The load-bearing exterior walls consist of a 7-m-high masonry structure made of cement-stabilized earth blocks. The minaret is 16 m high and also made of earth block masonry [6]. The building is of particular importance for earth building in the Arab world because it was constructed from 2008 to 2010 based on a special decree issued by Prince Faisal bin Salman. The use of earth building materials for the construction of the new mosque was meant to draw attention to the roots of Arabic architecture which modern Islamic architecture is hoping to build on.

4.1.4 Agricultural Buildings

Agricultural buildings made of earth building materials are among the "everyday" buildings which are rarely mentioned in research literature, apart from accounts in local or family histories. In Central Germany, there are still a large number of rammed earth or cob barns (likely a few thousand) in a sometimes good but often bad state of preservation. The original function of these barns has been lost: in summer, they were used for storing the harvested sheaves of grain which were threshed later in the winter months when there was no work in the fields. This activity required a large, open room which offered a maximum amount of floor space. If at all, the space was braced by an open support structure integrating a level of ceiling joists (with only a few vertical posts) which could be covered with planks as required. The resulting structure often had a slender design (ratio of wall thickness to wall height) which was not suitable for earth building materials and typically led to cracking (Sect. 5.2.5). The structural condition of these barns, which was already unstable, was further weakened by altering the buildings' functions. Additional interior retrofitting was often installed improperly and damaged the earth walls even more.

4.2 Fundamental Principles of the Building Trades

In Germany, the planning and building of earthen structures is carried out in accordance with standard regulations and fundamental principles of the building trades.

4.2.1 Standards and Regulations

4.2.1.1 VOB and BGB

The *German Construction Contract Procedures* (VOB, Vergabe-und Vertragsordnung für Bauleistungen) regulate the legal relationship between the client (owner) and the contractor (construction company). If the contractual parties do not explicitly agree to use the VOB, the provisions of the *German Civil Code* (BGB, Bürgerliche Gesetzbuch) automatically apply to the contractual relationship.

The VOB is divided into three parts:

- Part A: General provisions relating to the awarding of construction contracts.
- Part B: General contract conditions relating to the execution of construction work.
- Part C: General technical specifications in construction contracts (ATV).

Part C includes the "General technical specifications in construction contracts (ATV)" as a list of applicable DIN regulations for the most important trades.

The BGB states that, where applicable, a building must be confirmed as free of faults on the basis of generally accepted rules of technology. According to the VOB, contractors must perform their services on their own responsibility and as stipulated in the contract, following mandatory *generally accepted rules of technology*. These rules are evaluated based on three facts. They must be:

- Scientifically proven to be correct
- Technologically recognized by qualified persons
- Confirmed in practice through sufficient experience

These conditions are met through the development of standards which are agreed upon by consensus and then applied by all parties concerned. In comparison, *accepted engineering practice standards* are, for example, technical documents concerning special topics developed by professional associations which serve as recommendations.

4.2.1.2 DIN Standard

The DIN standards are used as a basis for creating technical specifications for the individual trades, for defining fields of application, for determining construction work regulations, and for describing the difference between associated services and additional services (which is important for pricing). Associated services are calculated based on standard prices, while additional services are calculated using "special" prices which depend on the specific situation.

Historical Development

In 1944, the "Lehmbauordnung" (Earth Building Code) was issued as the first technical regulation for building with earth in Germany. Because of World War II, it was not until 1951 that the code was introduced by the building authorities as DIN 18951. It consisted of Part 1 "Regulations for construction work" and Part 2 "Notes." Previously, the Earth Building Code was introduced into the state building codes of different German federal states with minor amendments. An example of one such state was Schleswig-Holstein.

The following earth building DIN standards did not make it beyond the pre-standard stage:

DIN 18952: Construction soil

- Part 1: Terminology, types (1956–2005)
- Part 2: Testing construction soil (1956–2010)

DIN 18953: Construction soil and earth building elements

- Part 1: The use of construction soil (1956–2005)
- Part 2: Earth masonry walls (1956–2005)
- Part 3: Rammed earth walls (1956–2005)
- Part 4: Cob walls (1956–2005)
- Part 5: Light-clay walls in frame construction (1956–2005)
- Part 6: Earth floors (1956–2005)

DIN 18954: Construction of earth buildings, guidelines (1956–2005)
DIN 18955: Construction soil, earth building elements, moisture protection (1956–2008)
DIN 18956: Plastering earth building elements (1956–2008)
DIN 18957: Earth shingle roofing (1956–2005)

In addition, a former DIN standard regulated the use of earth mortars for masonry and plaster but only in connection with building elements made of building materials other than earth:

DIN 1169: Earth mortar for masonry and plaster (1947–06)

In 1971, these DIN standards were withdrawn due to being "outdated and economically irrelevant" and were not replaced by new standards. However, they continued to be regarded as "generally accepted engineering standards" by the building authorities. This meant that, if required, the building methods defined in these standards did not need to be verified on an individual basis.

At the same time, separate earth building regulations were developed in the German Democratic Republic because "the regulations for earth buildings from 4 October 1944 no longer met the development requirements of earth building technology in the GDR" (Earth Building Code of the GDR, "Lehmbauordnung—LBO").

These regulations included:

- Regulations for building with earth (February 23, 1953)
- Terms, application, and processing of earth as a building material (Earth Building Code of the GDR—Lehmbauordnung der DDR) (December 23, 1953)
- Regulations for building with earth and for training earth building personnel (November 24, 1955)

The term "earth building personnel" in connection with a professional qualification is of particular interest. Corresponding to the fields of construction, design, and building material testing, the three different levels of qualification available were "earth building professional," "earth building designer," and "earth building expert."

When compared to the DIN standards mentioned above, the "Earth Building Code of the GDR" included more precise regulations on the dimensioning of load-bearing earth building elements (walls, pillars).

Furthermore, there was a draft of "Building regulations for the construction of earth buildings" (April 1962), published by the Special Inspection Group for Agricultural Construction of the State Building Inspection in Potsdam. The regulations are based on the abovementioned "Earth Building Code" and define further specifications for the dimensioning of load-bearing earth building elements. It is not clear if these regulations ever made it past the draft stage.

Careful research in archives would be required to find out how long these regulations were in force. A project in Herbsleben/Thuringia received a building permit in 1987/1988 [7] issued by the former State Building Inspection of the District of Erfurt based on the "Earth Building Code of the GDR" from 1953.

Earth Building Regulations: Lehmbau Regeln

In Germany, the development of regulations in the building trade sector is the responsibility of the "Construction Standardization" expert commission. This commission is part of the conference of the 16 State Building Ministers (ARGEBAU) who are in charge of construction, housing, and settlements. In 1995, the group decided to examine the earth building standards which were withdrawn by the German Institute for Standardization e. V. (DIN) and use them as the basis for developing current technical regulations for earth building work. This decision was made due to the considerable rise in the number of earth building activities, both for restoration as well as new construction work. The German Association for Building with Earth e. V. (DVL) was invited to serve as a professional organization in the project group formed by representatives of ARGEBAU and the German Institute of Construction Technology (DIBt—Deutsches Institut für Bautechnik).

Procedure

Over time, the DVL formed its own project group consisting of experienced specialists. The work of the DVL project group was funded by the German Federal Environmental Foundation (Deutsche Bundesstiftung Umwelt—DBU). The ARGEBAU "Construction Standardization" specialist commission decided to include the *Lehmbau Regeln* [8] in the Model List of Technical Building Regulations and to recommend their introduction as technical building regulations in Germany's individual state building codes.

In order to attain the quality of a technical building regulation approved of by the building authorities, the draft presented by the DVL needed to pass through the following procedural steps:

1. Identification of the need for current earth building regulations through the ARGEBAU "Construction Standardization" specialist commission as a result of an increased demand coming from the practitioners working in the field. The DVL appointed an expert project group which developed a work program and determined an organizational structure.
2. Development of a draft of regulations based on existing national and foreign norms and standards and first-hand experiences in accordance with the defined structure on three levels:

   Level 1: Authors (Volhard/Röhlen + assistance by Ziegert for the third edition 2009).

   Level 2: Authors + further experts selected by the DVL.

   Level 3: Editorial committee: Level 2 + expanded group of earth building specialists including some from outside the DVL. These groups needed to reach a consensus within a predetermined time frame.

3. Presentation of the draft to the broader professional public, inclusion of ideas, suggestions, final discussion, etc.
4. Submission of the revised draft for ratification/recommendation by ARGEBAU as national technical regulations.
5. Certification by the responsible EU department.
6. Publication in government gazettes or similar official bulletins leading to implementation.

This procedure was also applied during the revision of the Lehmbau Regeln in 2006/2007. The project was once again funded by the DBU.

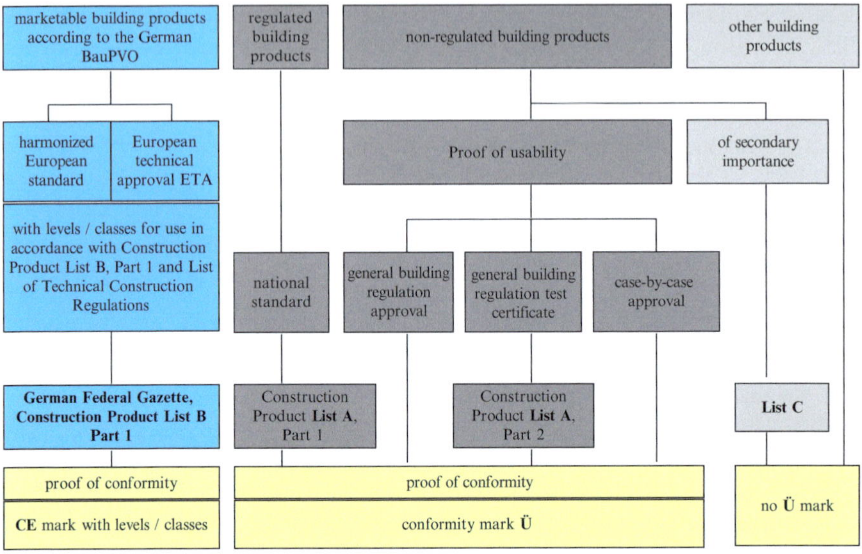

Fig. 4.18 Structure of the building regulation classification of building products according to LBO and BauPG [www.dibt.de]

Classification as a Building Regulation

Building products which need to conform to the structural requirements of MBO §20 and European Regulation No. 305/2011 [9] (BauPVO) (Sect. 1.4.1) must be tested and monitored for their usability (Sect. 3.4.9). Based on their application and the required verification, building products are classified into construction product lists: Construction Product List A, Parts 1 and 2, and List C. These lists are administered by the DIBt.

The state building codes in Germany govern the use of building products on a national level. They differentiate between regulated, nonregulated, and other building products (Fig. 4.18).

There are technical regulations for r*egulated building products*, mainly in the form of DIN standards, DIN EN standards, or DIN-ISO standards, which are published in Construction Product List A Part 1. *Nonregulated building products* are listed in Construction Product List A, Part 2. These products typically only have building test certificates.

Regulated and nonregulated building products display a conformity mark ("Ü-mark" on the national German level, CE mark on a European level) which can be found either on the product itself or on the packaging or packing list.

Other building products are products which play a minor role. Although these products are used in the building industry and generally accepted technical regulations which apply to them exist, they are not listed in Construction Product List A, Parts 1 and 2.

For classifying the Lehmbau Regeln [8] as building regulations, the ARGEBAU "Construction Standardization" specialist commission needed to consider a special characteristic of earth building materials: they can be produced on site as well as in an industrial setting. The suitability of earth building materials produced on site can only be confirmed using manual testing methods, while industrially produced earth building materials need to undergo testing based on respective product standards. The Lehmbau Regeln do not represent product standards. Therefore, a procedure applying to all earth building materials was selected as a compromise: earth building materials were included in the DIBt Model List of Technical Building Regulations and classified as "other building materials," exempt from the obligation to provide verification.

Furthermore, the application of the Lehmbau Regeln was limited to the construction of residential buildings of Building Classes 1 and 2 (Table 4.1) up to two full stories and not more than two units. For further-reaching uses, the verification of suitability required by building inspection authorities remains a requirement. For fire protection as well as sound and thermal insulation verification, the respective standards in their most current versions must be adhered to. With regard to thermal insulation, the coefficients of thermal conductivity for earth building materials specified in DIN 4108-4 have now been updated based on the Lehmbau Regeln. This classification has also been maintained for the third, revised edition of the Lehmbau Regeln 2009 [10].

Except for the states of Hamburg and Lower Saxony, the Lehmbau Regeln have been adopted into all German state building codes (as of June 2012). In Hamburg and Lower Saxony, earth building is classified as a "nonregulated construction method." This means that a permit has to be obtained on a case-by-case basis. The DIBt Model List and the introduction of the Lehmbau Regeln in other states can be used as a reference for these cases. In all other German states, the Lehmbau Regeln are a "generally accepted engineering standard," and their application in the planning and construction of buildings is mandatory.

The publication of the Lehmbau Regeln in 1999 closed a 30-year gap in the assessment of earth building construction by the building inspection authorities. This also resulted in a significant improvement in legal certainty in the field of earth building. It is now possible to go through the standard application process when applying for a building permit for an earth building structure within the territory covered by the Lehmbau Regeln. Before the introduction of the Lehmbau Regeln, the complex process of "case-by-case approval" which involved the next highest approval agency had to be selected. This change has been crucial for the development of earth building into a small, independent sector within Germany's building industry since the mid-1990s.

Content

Building on the DIN standards which were withdrawn in 1971, the Lehmbau Regeln are based on the current state of technology and numerous new developments in the field of earth building materials. They define construction soils, earth building

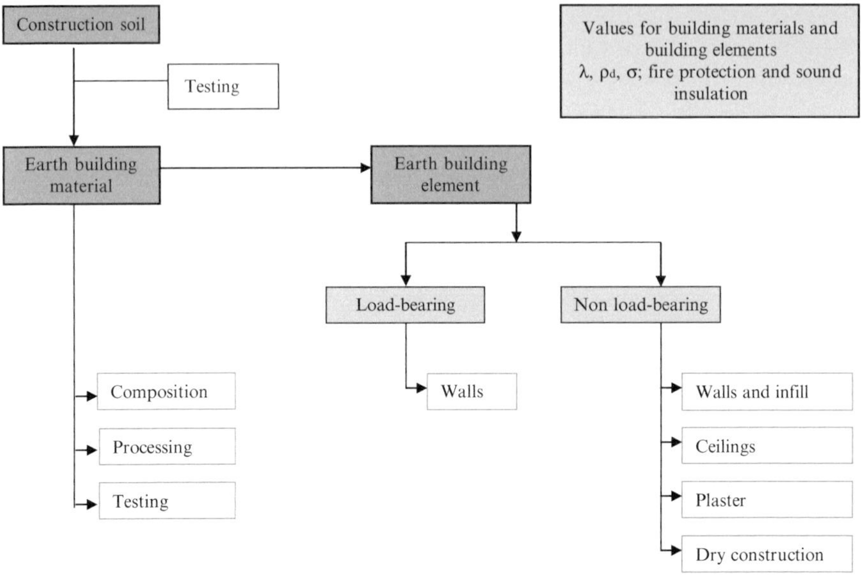

Fig. 4.19 Content structure of the Lehmbau Regeln

materials, as well as resulting building structures and apply only to the use of clay mineral-bound earth building materials. Although the Lehmbau Regeln do not exclude the use of chemically "stabilized" earth building materials, such materials are not part of their scope of validity.

The Lehmbau Regeln consist of three parts: construction soil (Chap. 2), earth building materials (Chap. 3), and earth building elements (Chap. 4) (Fig. 4.19). *Construction soil* defines soils which are suitable for the production of earth building materials with the help of testing criteria. The individual *earth building materials* are described in terms of their composition, processing, properties, and tests. For *earth building elements*, there is a differentiation between load-bearing walls and non-load-bearing building elements (walls and infill, ceilings, plaster, dry construction). Chapter 5 summarizes building material and building component values. Chapter 6 assigns the main earth building services to the respective VOB trades and standards (Table 4.2). This classification makes it possible to define the boundaries between associated services and additional services for calculation purposes.

The New DIN Standards

The Lehmbau Regeln permit the industrial production of earth building materials as well as individual on-site production. Industrially produced earth building materials—which represent the large majority of products in Germany—are building product in terms of the MBO or Regulation EU No. 305/2011 [9]. They are thus

Table 4.2 Earth building services according to the trades defined in the VOB and DIN standards

No.	Section in LR [8]	Earth building service	VOB trade	Corresponds to DIN
1	4.1.3	Earth block walls	Masonry	18330
2	4.1.4	Rammed earth walls	Concrete and reinforced concrete work	18331
3	4.1.5	Cob walls	Masonry	18330
4	4.2	Vaulted ceilings	Masonry	18330
5	4.3.1	Earthen infill of half-timber walls	Masonry	18330
6	4.3.2	Non-load-bearing rammed earth walls	Concrete and reinforced concrete work	18331
7	4.3.3	Non-load-bearing earth block masonry	Masonry	18330
8	4.3.4	Light-clay wall construction, wet	Concrete and reinforced concrete work	18331
9	4.3.5	Clay panel walls	Masonry	18330
10	4.3.6	Dry-stacked earth walls	Masonry	18330
11	4.3.7	Sprayed earth walls	Concrete and reinforced concrete work	18331
12	4.4	Beam ceilings	Masonry	18330
13	4.5	Earth plaster	Plaster and stucco work	18350
14	4.6	Rammed earth floors	Screed work	18353
15	4.7	Dry construction	Dry construction work	18340

subject to the relevant requirements (performance tests) of usability and of the proof of suitability (Ü-mark, Sect. 3.4.9). The Ü-mark is issued by the DIBt for factory and industrially produced building products based on a production control procedure defined in DIN 18200. This procedure, however, is not covered by the Lehmbau Regeln.

Procedure

Immediately after the publication of the third, revised edition of the Lehmbau Regeln, the DVL appointed an "Earth Building Standardization" advisory committee consisting of qualified DVL members, external experts, as well as representatives of the DIBt and the German Institute of Materials Research and Testing (BAM). The goal of the committee was to draw up texts for draft standards which could later be developed into DIN standards. Of high priority was the development of product standards for earth blocks and earth mortar.

As part of a 3-year research project called "StandardLehm" (Earth Standard) which was funded by Germany's Federal Ministry for Economics and Technology (BMWi), BAM carried out numerous tests on building materials and building elements. The test results were included in the draft standards. These drafts were put before the "Earth Building Standardization" advisory committee and an "Extended Advisory Committee for Standardization" for voting. The extended advisory committee was also appointed by the DVL and included additional outside experts from various organizations, with a focus on representatives from product manufacturers.

On May 31, 2011, the DVL applied for the formation of an "Earth Building" working committee with the German Institute for Standardization (DIN). As part of the application, the drafts (which had been developed and agreed upon by the "Earth Building Standardization" advisory committee) were given to DIN. They were intended to serve as the basis for discussion for the development of the draft standards. These drafts were published by the DVL as "Technical Information Sheets for Earth Building" in June 2011:

Sheet 02	Earth blocks—terms and definitions, building materials, requirements, test methods [11]
Sheet 03	Earth masonry mortar—terms and definitions, building materials, requirements, test methods [12]
Sheet 04	Earth plaster mortar—terms and definitions, building materials, requirements, test methods [13]
Sheet 05	Quality monitoring of construction soil as the raw material for industrially produced earth building materials—Guidelines [14]

The Technical Information Sheet DVL 01 (2008) "Requirements of Earth Plasters" was revised and published in 2014 with the title "Requirements of Earth Plaster as a Building Element" [15].

The scope of application of the Technical Information Sheets covers industrially produced earth building materials. They are regarded as the "state of technology" and are a recommendation.

On September 23, 2011, the constituent meeting of the DIN working committee "Earth Building" took place (NA 005-06-08 AA "Lehmbau"). The DIN drafts under development were given the following titles:

E DIN 18945	Earth blocks—terms and definitions, building materials, requirements, test methods
E DIN 18946	Earth masonry mortar—terms and definitions, requirements, test methods
E DIN 18947	Earth plasters—terms and definitions, building materials, requirements, test methods

Construction Soil (Sheet 05) was not subject of the DIN standardization procedure.

The abovementioned DIN drafts were finished in July 2012 and made available to the interested public for commenting up to 2012. At the same time, the Technical Information Sheets DVL 02–04 were withdrawn. The standards DIN 18945–47 were published in April 2013. They are considered "generally accepted engineering standards" and their application is mandatory. As with all plaster mortars, earth plaster mortars according to DIN 18947 are classified as "other building products" and are not included in Construction Product List A, Part 1. They therefore do not receive an Ü-mark.

Content

DIN 18945–47 adhere to the general requirements of form and content for the development of DIN standards for building products. They consist of the following sections:

- Application area/normative references/terms
- Application classes/requirements
- Designation/labeling
- Testing
- Proof of conformity and inspection of the declared information
- Packing slip/product information sheet
- Appendix (for information)

The scope of application of DIN 18945–47 covers industrially produced earth building products without chemical stabilization. The Lehmbau Regeln continue to apply for the use and processing of earth building products.

Perspectives

With regard to the development of new DIN standards for earth building, the following trends can be observed. Their development, however, is always subject to the availability of the necessary resources:

1. Building material standards need to be developed for more building products. Clay panels (see [16]) and ready-to-use mixes, for example, for rammed earth, are currently being considered.
2. The building material standards need to be followed by processing standards. Here, an important aspect will be to define partial safety factors for load-bearing structures made of earth building materials (Sect. 4.3.3.1). With the official introduction of the Eurocodes EC 1–9 as DIN EN 1990–1999 on July 1, 2012, this approach has become mandatory for the dimensioning of load-bearing structures using earth building materials.
3. The DIN standards for earth building must be integrated into corresponding European or national application standards (which might still exist concurrently) as separate chapters on earth building. These are:

   - Earth blocks and earth masonry mortar in DIN EN 1996-1-1
   - Earth plasters in DIN 18550-2 or DIN EN 13942-2

   This is also true for the addition of corresponding sections on earth building for the Standard Services Book for the Building Trades (STLB-Bau, Standardleistungsbuches Bau) (Sect. 4.2.2.1).
4. The Lehmbau Regeln need to be adjusted to reflect current changes. They will continue to cover "earth building activities on site" in terms of building inspection, whereas the DIN standards apply to the production and processing of industrially manufactured earth building materials.

4.2.1.3 Foreign Earth Building Regulations

Types of Documents

The International Organization for Standardization (ISO) differentiates between standards and normative documents: a *standard* is a document which has been reached by consensus among experts, published by a national organization for standardization, and confirmed by a national authorized body. It establishes rules and guidelines for activities and their results for general and repeated use, with the intention of achieving an optimal degree of regulation in the appropriate context. Standards always need to be based on the field's current state of technology and contribute to the benefit of the general public. As a rule, their application is obliged.

A *normative document* is different from a standard in that it does not meet the requirements placed on form and content and is not issued by an (national) organization of standardization. However, a normative document can attain the quality of a "standard" and be accepted by a national authorized body by meeting the necessary requirements.

In addition, specialist associations and organizations publish *Technical Recommendations*, leaflets, and other documents which are generally also based on the field's state of the art but can only be considered recommendations.

Overview

For an analysis of existing earth building standards, 39 different regulations from 19 countries and regions have been identified [17, 18]. Compared to other conventional building materials, the number of earth building regulations is low. Table 4.3 gives an overview of the countries which have standards or normative documents for earth building. In some countries, earth building regulations are part of the general building regulations.

Content

The identified standards and normative documents generally cover the following points:

Building materials/building techniques: earth blocks made using different shaping processes, rammed earth building, cob building, earth block masonry, non-load-bearing walls and infills

Material characteristics: texture, plasticity/cohesion, chemical stabilization, natural additives, classification, shrinkage and compactibility, test methods

Local conditions: impacts of earthquakes

Table 4.3 Overview of countries with standards and normative documents for earth building

No.	Country	Document: Title	Document: Type	Contents: Construction soil	Contents: Earth building materials	Contents: Earth building method	Contents: Geographical level	Source
1	Africa	ARS 671-683 (1996)	S		EB	EBM	R	[30]
2	Australia	CSIRO Bulletin 5, 4th ed. (1995)[a]	ND	CS	EB, CSEB, EMM	RE, EBM	N	[46]
3	Australia	EBAA (2004)	ND	CS	EB, EMM	EBM, RE	N	[47]
4	Brazil	NBR 8491-2, 10832-6, 12023-5, 13554-5 (1984–1996)	S		CSEB		N	[43]
5	Brazil	NBR 13553 (1996)	S			CSRE	N	[43]
6	Columbia	NTC 5324 (2004)	S		CSEB		N	[44]
7	France[b]	AFNOR XP.P13-901 (2001)	S		EB		N	[22]
8	Germany	Lehmbau Regeln (2009)	S	CS	C, LC, EB, EM, CP	RE, C, EBM, EP, EI, WL	N	[8]
9	Germany	RL 0607 (2010)	ND		EPM		N	[16]
10	Germany	RL 0803 (2010)	ND		EPM		N	[16]
11	Germany	RL 1006 (2010)	ND		CP		N	[16]
12	Germany	TM 01 (2014)	ND		EPM		N	[15]
13	Germany	TM 02 (2011)[c]	ND		EB		N	[11]
14	Germany	TM 03 (2011)[c]	ND		EMM		N	[12]
15	Germany	TM 04 (2011)[c]	ND		EPM		N	[13]
16	Germany	TM 05 (2011)	ND	CS			N	[14]
17	Germany	TM 06 (2015)	ND		EPM		N	[109]
18	Germany	DIN 18945 (2013)	S		EB		N	Appendix

(continued)

Table 4.3 (continued)

No.	Country	Document: Title	Document: Type	Contents: Construction soil	Contents: Earth building materials	Contents: Earth building method	Contents: Geographical level	Source
19	Germany	DIN 18946 (2013)	S		EMM		N	Appendix
20	Germany	DIN 18947 (2013)	S		EPM		N	Appendix
21	India	IS: 2110 (1998)	S	CS, CSS		RE	N	[37]
22	India	IS: 13827 (1998)	S		EB	EBM, RE[d]	N	[36]
23	India	IS 1725 (2013)	S		CSEB		N	[38]
24	Kenya	KS02-1070 (1999)	S		CSEB		N	[35]
25	Kyrgyzstan	PCH-2-87 (1988)	S	CS, CSS		RE[d]	N	[29]
26	New Zealand	NZS 4297-9 (1998)[e]	S		E, EB	RE, EBM, EP[d]	N	[45]
27	Nigeria	NIS 369 (1997)	S		CSEB		N	[110]
28	Nigeria	NBC 10.23 (2006)	BC	CS		EBM, RE	N	[34]
29	Peru	NTE E.080 (2000)	S		EB	EBM[d]	N	[42]
30	Spain	MOPT Tapial (1992)	ND	CS		RE	N	[23]
31	Spain	UNE 41410 (2008)	S		CEB		N	[25]
32	Sri Lanka	SLS 1382 part 1-3 (2009)	S		CSEB	EBM	N	[39]
33	Switzerland	Regeln zum Bauen mit Lehm (1994)	ND	CS	EB, C, EM	EBM, RE, EI, WL	N	[27]
34	Tunisia	NT 21.33, 21.35 (1998)	S		CEB		N	[31]
35	Turkey	TS 537, 2514, 2515 (1985–1997)	S		CSEB		N	[28]
36	USA	UBC, Sec. 2405 (1982)	BC			EBM[d]	L	[111]
37	USA	14.7.4 NMAC (2009)[f]	BC		EB, EMM	EBM, RE[d]	L	[40]
38	USA	ASTM E2392/E2392M (2010)	S	CS	EB, EM,	C, EBM, RE, EM, WL[d]	N	[41]

39	Zimbabwe	SAZS 724 (2001)	S	CS		RE	N	[32]
		SADCSTAN/TCI SC5-001 (2012)	S					

[a]Withdrawn in 2008

[b]In 2010 the French national earth building association As Terre began developing a French earth building standard which is planned to include the main earth building techniques used in France: earth block masonry LSM, rammed earth STL, cob WL, light-clay LL, and earth plaster LP

[c]Withdrawn in 2012 and replaced by the corresponding DIN

[d]Taking into account the effects of earthquakes on earthen structures

[e]Currently being revised

[f]There are a number of local standards (L) for earth blocks similar to the NMAC (San Diego/CA, Tucson/AZ, Marana/Pima/AZ, Boulder/CO)

S Standards, published by national organizations for standardization or by specialized organizations through coordination of content (using the principle: "consensus among experts") and acceptance by an authorized organization for standardization; *BC* partial chapter on earth building which is part of a building code published by a national organization for standardization; *ND* normative document, published by a specialized organization through coordination of content, but *without* acceptance by an authorized organization for standardization; *L* local; *N* national; *R* regional; *C* cob; *CP* clay panels; *CS, CSS* construction soil, construction soil stabilized with cement; *EB* earth blocks, compressed earth blocks CEB, compressed and stabilized earth blocks CSEB, poured earth blocks PEB; *EBM* earth block masonry; *EM* earth mortar, earth plaster mortar EPM, earth masonry mortar EMM, sprayed earth mortar SEM; *EI* earth infill, wattle and daub WD, poured earth infill PEI; *LC* light clay; *RE* rammed earth, rammed earth stabilized with cement CSRE; *WL* wall lining

In Germany, the predominant earth building techniques employ combinations of earth and wood in half-timber or timber-frame construction (in addition to some load-bearing earthen structures). Foreign earth building regulations, on the other hand, primarily apply earth as a building material for load-bearing structures.

Numerous countries only have regulations for individual earth building techniques which are typical for the respective region, for example, earth block construction. Additional, specific natural impacts, for example, earthquakes, are taken into consideration. The use of synthetic stabilizers (lime, cement) and waste products is also included in certain regulations. Some countries such as Italy [19], Chile [20], and Morocco [21] have standards for the restoration of historical earthen buildings which were published by local organizations and national standardization agencies.

European Union/Switzerland

European standardization is becoming increasingly important as the European Single Market develops. The European Committee for Standardization (CEN) is the common European standards institution. Its members consist of the national standardization institutions of all EU countries plus Switzerland, Iceland, and Norway. Germany's member organization is the DIN. These standardization institutions are also members of the International Organization for Standardization (ISO). European standards (EN) need to correspond with the respective international standards (ISO). In Germany, European standards are published as DIN EN standards and international standards as DIN-ISO standards. Corresponding national standards will eventually have to be withdrawn. This process is taking place at the moment.

In the building industry, additional Eurocodes (EC) exist which have been published alongside ENs and EN-Vs (preliminary standards). The Eurocodes define standardized rules for the design, dimensioning, and construction of buildings and engineered structures in Europe. First, the ECs were published as European preliminary standards EN-V. They were introduced by way of so-called National Application Documents for application on a trial basis by the building supervisory authorities. Starting in 1997, the EN-V standards have continuously been converted into European Standards EN. The expert commission for building technology of Germany's Conference of the State Building Ministers introduced the (former) ECs through the building supervisory authorities as DIN EN 1990–1995 and, when completed, DIN EN 1996, 1997, and 1999 by July 1, 2012. The standards have the following titles:

EC 0—Basis of structural design—DIN EN 1990
EC 1—Actions on structures—DIN EN 1991
EC 2—Design of concrete structures—DIN EN 1992
EC 3—Design of steel structures—DIN EN 1993
EC 4—Design of composite steel and concrete structures—DIN EN 1994
EC 5—Design of timber structures—DIN EN 1995
EC 6—Design of masonry structures—DIN EN 1996

EC 7—Geotechnical design—DIN EN 1997
EC 8—Design of structures for earthquake resistance—DIN EN 1998
EC 9—Design of aluminum structures—DIN EN 1999

The partial safety concept is included in these standards (Sect. 4.3.3.1).

The Regulation EU 305/2011 [9] was introduced by the building supervisory authorities as the new European Construction Products Regulation (CPR) by July 1, 2012. It replaces the previously applicable Construction Products Directive (CPD) for trade within Europe which was implemented in Germany as the Construction Products Law (BauPG—Bauproduktengesetz). The CPR applies to all construction products which are permanently installed in structures above and below ground or which are relevant for meeting the requirements for usability listed in Sect. 4.2.1.3.

The use of construction products with harmonized European specifications is still possible on the basis of the CPR. They are included in Construction Product List B which was also published by the German Institute of Construction Technology (DIBt—Deutsches Institut für Bautechnik).

Harmonized European specifications are (Fig. 4.18):

- Harmonized European standards
- European technical approvals
- National specifications approved in Europe

They are developed on behalf of the European Committee for Standardization (CEN) using a specific process which involves the standardization institutes of the European member states.

In *France*, regulations for building with earth blocks have been introduced by the building supervisory authorities [22], and a standard for rammed earth is being developed.

Spain has a normative rule for building rammed earth structures [23, 24]. The Spanish standard UNE 41410 [25] is the first product standard published by a national standardization authority in Europe which adheres to the principle of the general regulations for harmonized marketing conditions in the EU according to Regulation EU 305/2011 [9] as described in Sect. 3.4.9.

In *Italy*, national regulations for building with earth are currently being developed [26]. In other countries, the German Lehmbau Regeln have been translated into respective national languages and published (Hungary 2005, Romania 2010).

In 1994, the *Swiss* Society of Engineers and Architects SIA published the "Regeln zum Bauen mit Lehm" ("Regulations for Building with Earth" (D 0111)) [27]. These regulations have been supplemented by an "Earth Building Atlas" showing completed examples and technical details (D 0112). This was preceded by a comprehensive description of earth building as it relates to Switzerland (D 077). The Swiss "Regulations for Building with Earth" have the status of a recommendation.

The rules were developed by a team at the ETH Zurich as part of a 2-year research mandate by the Swiss Federal Office of Energy. As in Germany, the material combination of earth and wood is typical for Switzerland. Therefore, the building regulations not only deal with earth building techniques for load-bearing construction but also include infill with earth building materials.

The Swiss earth building regulations inspired the development of corresponding regulations in Germany 3 years later, and there was an extensive exchange of information between both teams.

Turkey

Between 1995 and 1997, the national Turkish Standards Institution TSE published standards for the production of cement-stabilized earth blocks [28].

CIS

As the only republic out of 15 former Soviet republics, Kirghizstan published a "Republic Standard" for the construction of low-rise buildings made of cement-stabilized straw clay in 1988 [29].

Africa

Today, earth building materials are still being used on a daily basis on the African continent. In various countries, technical regulations for the quality of earth building materials and their use date back to colonial times, for example, in the form of "Technical Notes." It has therefore become necessary to define modern quality standards and create legal provisions for these materials.

As the African continent merges into the "African Union," it has become increasingly important to create individual national and harmonized African building standards.

This was the goal of a project which helped develop a number of standards for the production, use, and testing of compressed earth blocks (CEB). The project was initiated by the organization CRATerre in cooperation with the African Regional Organization for Standardization ARSO and supported by additional European funders as part of the ACP development cooperation. In 1996, these standards were confirmed and approved by ARSO as ARS African Regional Standards [30]. In preparation, drafts had been developed by an international team of experts from eight countries which were discussed and agreed upon at an ACP-EU seminar. The standards are in the process of being implemented into national building laws.

In *Tunisia*, there are standards for the production of CEB, NT 21.33, and 21.35. They were published in 1998 by the national Tunisian standardization organization INNOPRI [31].

The *Zimbabwe* Standard Code of Practice for Rammed Earth Structures (SAZS 724:2001) should also be mentioned [32]. In 2012, this standard was introduced as a regional standard SADCSTAN/TCI SC5-001 in the countries of the Southern African Development Community (SADC).

In 2006, the new National Building Code (NBC) went into effect in *Nigeria*. The NBC consists of four parts, and part 2 includes paragraph 10.23 with earth building regulations. These regulations apply to structures made of sun-dried earth blocks (adobes), rammed earth, and cement-stabilized earth blocks [33, 34].

In 1999, the national standardization institution of *Kenya*, the Kenya Bureau of Standards KEBS, published a standard for the production of cement-stabilized earth blocks [35].

In *Morocco*, a technical regulation for earthquake-resistant building with earth was published in 2012. Three ministries were involved in the development of the text. In addition to binding guidelines for building material properties and design, the document also contains recommendations and comments [21].

In *Egypt*, a national standard for CEB is currently under development.

India/Sri Lanka

In 1993, the Bureau of *Indian* Standards published a national regulation which deals with the seismic retrofitting of structures made of earth building materials [36]. The building methods mentioned in the regulation are the earth clump, earth block, and rammed earth techniques. The regulation applies to earth building materials without synthetic additives (lime, cement, etc.).

The incentive for developing this regulation was based on the observation that approx. 50 % of all residential structures in India have earthen walls which exhibit inadequate earthquakes resistance.

The regulation was drawn up by a group of experts from the fields of construction and architecture, industry, geophysics, and earthquake engineering and has been confirmed as a national building standard by the responsible standardization organization.

India also has a national regulation for building rammed earth structures [37].

Venkatarama Reddy submitted a draft for an Indian building regulation for the production, use, and testing of stabilized CEB which has been implemented by the Bureau of Indian Standards BIS as IS 1725 [38].

After the 2006 tsunami disaster, a building standard draft for construction with stabilized earth blocks was developed and officially introduced in *Sri Lanka* in 2009 [39].

USA

The first earth building standards were published in the 1940s by the National Bureau of Standards. In the 1970s, these regulations were modified for the states of Texas, New Mexico, Utah, Arizona, California, and Colorado and published as the Uniform Building Code (UBC). The dominant building method in all regulations is the earth block (adobe) technique. The regulations take the impacts of earthquakes into account.

There have been efforts to update the existing regulations. New Mexico [40] and California have already completed this process.

The "Standard Guide for Design of Earthen Wall Building Systems ASTM E2392" [41], published by the American Society for Testing and Materials (ASTM) in 2010, touches on aspects of sustainable building—even if only in a nonbinding manner. The mere fact that the world's leading industrial nation has developed such a standard should be an incentive for other industrial nations to take action in this field.

South America

A national earth building standard was published in *Peru* in 2000. An English translation was also produced [42].

The standard describes the design and construction of earth block (adobe) structures, taking the seismic conditions in Peru into account. It was developed by a team of representatives from architecture and engineering organizations as well as universities and the building industry and has been confirmed by the responsible standardization organization as a national building standard.

Between 1984 and 1996, the *Brazilian* National Standards Organization ABTN published a group of standards for the production of cement-stabilized earth blocks and rammed earth [43].

In 2004, the *Colombian* Institute of Technical Standards and Certification ICONTEC published a standard for the production of cement-stabilized earth blocks [44].

New Zealand/Australia

In 1998, the *New Zealand* Standard Council, which is responsible for building standardization, published three standards which regulate earth building on a national level [45]:

NZS 4297: 1998 Engineering Design of Earth Buildings
NZS 4298: 1998 Materials and Workmanship for Earth Buildings
NZS 4299: 1998 Earth Buildings Not Requiring Specific Design

NZS 4297 defines the basic principles for the design and dimensioning of earth building structures. NZS 4298 regulates the requirements placed on building materials and their application in rammed earth, poured earth and earth block construction. Earth block construction is divided into hand-molded blocks (adobe) and CEB with or without the addition of extra binders. NZS 4299 defines structures made of earth building materials which do not require specific design. They are restricted with regard to building height and floor plan as well as in terms of live loads and additional design parameters.

For the development of the standard, the Standard Council appointed a technical group consisting of representatives from universities in New Zealand, architect and engineering organizations, as well as the Earth Building Association of New Zealand (EBANZ).

The group was formed in 1994, originally as a project between Australia and New Zealand, with the objective of publishing a common earth building standard. However, a consensus could not be reached, and both countries went their separate ways as far as the standardization of earth building is concerned.

Currently, a revision of the 20-year-old earth building standards is being prepared.

In *Australia*, the first national building regulation for earth building was published in 1952. It regulated construction with rammed earth, CEB, and hand-molded earth blocks (adobe). The fourth revised edition was published in 1987 by CSIRO Australia, Division of Building, Construction and Engineering, which was the responsible standardization organization at that time [46]. In 2008, this regulation was withdrawn.

In 2004, the Earth Building Association of Australia (EBAA) published a normative regulation for building with earth in Australia taking the current state of building practices into account [47]. The regulation is the result of work conducted by a group of Australian earth building practitioners under the leadership of the EBAA.

The regulation has the status of a recommendation because it has apparently not (yet) been approved as a national standard by the standardization organization currently responsible, the Building Code of Australia BCA. In addition, the organization called Standards Australia published the manual *The Australian Earth Building Handbook* [48] in 2002 which also serves as a summary of the current state of earth building in Australia. This manual has not undergone the process described in Sect. 4.2.1.2 which involves a period of public comment. This makes its status as a normative document questionable.

Trends

A development toward an international harmonization of national earth building regulations is barely visible due to the fact that only a few attempts have been made so far. However, it is exactly this type of harmonization of regulations, also on an international level, which is a prerequisite for liberating earth building (with its diverse possibilities in terms of material, technology, construction, and design) from its prevailing classification as a traditional, owner-builder material. A future-oriented development of earth building can only gain traction on the basis of an "engineered construction technique" comparable to masonry or concrete.

The worldwide exchange of information via electronic media has enabled the earth building community to discuss national regulations for building with earth. International conferences present opportunities to share achievements and talk about existing problems. In this context, different assessments of the role of earth building have become visible between industrial nations and developing countries:

In *industrial* nations, earth as a building material is increasingly being used in everyday construction work because of its ecological and design qualities. Existing earth building traditions, which were often buried for decades in the wake of the industrialization of the building trade, are now being revived and developed further in accordance with current technical standards.

By contrast, in many *developing* countries, an interruption in the use of earth as a building material has never occurred. Here, earth building is still part of the everyday building practice which, however, is largely characterized by traditional building methods, do-it-yourself construction, and neighborly help. This is the reason why earth building is frequently equated with backwardness, a stigma which has to be overcome. Earth is affordable and readily available in most regions. Concrete and reinforced concrete, on the other hand, are regarded as "modern" building materials and seen as a measure of development. However, for the majority of the population, these building materials are unaffordable. In many developing countries, particularly in areas with a high risk of natural disasters (such as earthquakes), the necessity to establish earth building regulations is slowly being recognized. This is based on the realization that earth will remain a building material for the foreseeable future. In some developing countries, earth building has recently also been linked to energy conservation and sustainable development.

Compared to other conventional building materials, earth building lacks an internationally binding use of terminology. Standardized test methods for determining values of earth building materials and building elements are also largely missing. Various test methods have been adopted from "related" fields (e.g., concrete, soil mechanics, ceramics) and modified for earth building purposes. The specific material properties of soil have only been partially determined and documented. However, the existence of standards for earth building materials and elements which have been published by national standardization organizations forms the prerequisite for mutual understanding between all parties involved in the building process. This also includes the drafting of contracts.

4.2.2 Calls for Bids and Commissioning of Construction Work

4.2.2.1 Call for Bids

The VOB differentiates between open, restricted, and negotiated calls for bids. Open calls for bids follow a prescribed procedure to publicly invite bids from an unlimited number of companies. The same procedure applies to the restricted call for bids, but here only a limited number of companies are invited to submit bids. Negotiated calls for bids do not follow a formal procedure and are commonly used by private clients to commission building work.

Call for bid documents and work specifications are typically drawn up by architects or civil engineers. Bids for itemized building services can generally also be called for by the clients themselves. Trying to save money at this stage, however, often leads to misunderstandings and, potentially, to higher costs. An experienced earth building company will point out mistakes or misunderstandings in the call for bids and, if necessary, submit an alternative bid.

The call for bid documents can also be drawn up by the building company planning to carry out the work. This represents a considerable time investment which the company can invoice for retroactively once the contract has been awarded.

The call for bid documents for building work include a list of specifications for all building work which the client has asked to be carried out by a competent, reliable, and efficient construction company at a precisely calculated market price. This is only possible if the expected building work and its level of quality are described as detailed as possible. When formulating the individual items on the list, the entire building project needs to be considered in its technological sequence, and all partial steps need to be recorded.

Comprehensive and up-to-date call for bid documentation for earth building services is still rare. Therefore, the use of already-established specification texts is recommended. These can be found in various standard references, such as [49, 50]. In [49], call for bid specifications are assigned to individual "earth building trades" based on the trade structure of the Standard Services Book for the Building Trades (STLB-Bau):

912	Earth masonry work
913	Earth building work, wet
923	Plaster work
925	Screed work
934	Paint work
939	Dry construction work

The second and third digits refer to the commonly known trade numbers, e.g., 013 concrete work. The listed prices are average net prices before sales tax.

The STLB-Bau contains product-independent VOB- and DIN-compliant call for bid specifications for the European market which can be found in an online database. Since 2011 and starting with earth plaster, relevant texts for earth building materials and earth building elements have gradually been integrated into the STLB-Bau and made available online.

In addition, producers of earth building materials also provide call for bid specifications which are, however, usually tailored to their own products.

The call for bids should also clearly specify the expected quality of the earth building work. Frequent disputes include cracks in earth plaster, deviations in the color and texture of earthen finishes, or the weathering of rammed earth surfaces.

4.2.2.2 Price Calculation

In the specifications, the respective quantities of the individual items are determined, and a unit price is given by the building contractor. The unit price includes: the labor costs, material costs, and expenses. Material costs include the actual costs of the building materials (with and without transport) as well as expenses for special tools and equipment needed, services and special insurance for the project, water and electricity. In addition to unit prices or hourly wages, the calculation needs to allocate general and overhead costs such as expenses for machines, tools and small parts, equipment, vehicles, rental fees, bookkeeping and tax consulting fees, insurances as well as business losses and profits.

Labor costs include not only direct hourly wages but also costs for housing, per diem rates and other expenses. Labor costs per unit of work to be carried out can be calculated with the help of typical work times for earth building work. Contractors typically develop their own individual targets for their work. The German Association for Building with Earth (Dachverband Lehm e. V.) has drawn up work time guidelines for the building elements of walls, ceilings and plasters to assist with price calculations for earth building work [51] (Table 4.4):

Table 4.4 Calculation of typical work times for earth building services

No.	Building element	Work time [min/m, m^2, m^3]
1	*Walls*	
1.1	Rammed earth	8–12 h/m^3
1.2	Masonry, unfired "green" brick 2DF/11.5	48–92
1.3	Masonry unfired "green" brick 3DF/17.5	53–110
1.4	Masonry, 10 cm, large-format elements	35–55
1.5	Extra charge: masonry infill between timber construction	10
1.6	Extra charge: exposed masonry work, single face	30
1.7	Extra charge: masonry work above openings, approx. 1 m	20–25
1.8	New earth infill	55–90
1.9	Extra charge: exposed timber framing	18
1.10	Interior earth block wall lining, light-clay blocks, $d = 11.5$ cm	55–70
1.11	Interior earth block wall lining, light-clay blocks, $d = 10$ cm; (large-format elements + cavity filling of approx. 3 cm)	40–55(+16)
1.12	Interior light-clay wall construction, 30 cm	100–160
1.13	Interior wall lining, wood-chip light clay, 15 cm	65–85
1.14	Interior wall lining reed panel, 5 cm	30–38
1.15	Interior wall lining wood fiber insulation panel, 6 cm	27–35
1.16	Interior wall lining wood wool panel, 5 cm	27–35
1.17	New earth infill using historical techniques	120–135
1.18	Repair of infill panel, straw clay, larger areas	35–80
1.19	Repair of infill panel, minor repairs	15–22
1.20	Dry-stacked wall lining, "green" unfired bricks, DF	25–30
1.21	Dry-lining sub-construction, wall (wooden slats)	35
1.22	Dry-lining with drywall clay panels	35
1.23	Earth plaster boards, attached with mortar, wall	28–36
2	*Plasters*	
2.1	Exterior lime plaster, 2 layers	40
2.2	Extra charge: exterior exposed timber framing	25
2.3	Substrate priming	6–8
2.4	Clay slip adhesion coat	6–12
2.5	Earth base coat, wall	13–17
2.6	Earth finish coat, wall	14–19

(continued)

Table 4.4 (continued)

No.	Building element	Work time [min/m, m^2, m^3]
2.7	Earth finish coat, wall, single layer	20–25
2.8	Earthen fine-finish coat, wall	12–17
2.9	Colored earth plaster	19–25
2.10	Clay paints	6–10
2.11	Leveling coat, wall	12–16
2.12	Reed mats applied to wooden wall elements	6
2.13	Reed mats covering full wall surface	11–16
2.14	Slip coat priming	6–12
2.15	Pre-wetting	6
2.16	Plaster reinforcement larger area, burlap	8–10
2.17	Plaster reinforcement larger area, fiber glass mesh	3–7
2.18	Corner bead	7
2.19	Plaster edge finishing	15 (–30)
2.20	Plaster surfaces in exposed timber-frame construction	5–7
2.21	Special finishes	5–7
3	*Ceilings*	
3.1	Slatted timber ceiling + 6-cm straw clay	95
3.2	Earth reels + 6-cm straw clay	145

4.2.2.3 Awarding of Contracts

In order to promote fair business practices, obtaining bids from a number of construction companies is recommended. The construction work should be contracted with competent, efficient and reliable companies at market prices. Here, the principle applies that the most convincing technical offer should win the contract as long as it lies within the mid-range of the offers received from competing companies. "Dumping" offers appear attractive at first glance but carry the risk of uncertainties and must therefore be reviewed carefully (e.g., through additional meetings with the bidder). In order to prevent price projections by bidders or price fixing, the names of competing contractors should never be disclosed to a bidder.

Contracting authorities often raise questions regarding the professional competence of earth building companies. Contractors competing for earth building contracts must be confident that they are able to carry out the work at the required level of quality and should be able to provide adequate references. In this context, the specialized qualification of employees is crucial and gives a company a competitive advantage.

With the professional continuing education course "Specialist for Building with Earth," the DVL teaches the required technical knowledge based on the Lehmbau Regeln (which were published by the association) [51]. Graduates of the course receive a certificate from a Chamber of Trade which entitles them to register their business with the respective Chamber of Trades and Crafts based on

Article 8 of Germany's Trade and Crafts Code. The graduate's construction company is also granted the right to display the DVL's "round seal" designating them as a "company specialized in building with earth." A list of these specialized companies is provided at www.dachverband-lehm.de and can provide guidance for contracting authorities.

After a contract has been awarded, the contracting parties can agree on a *contract for work based on the German Civil Code* (*BGB*) or a *construction contract based on the VOB*. The VOB contract has a warranty period of 4 years and is better suited to the specific requirements of construction work. The BGB contract includes a warranty period of 5 years and applies automatically if the VOB has not been specifically agreed on. It is also possible to agree on a construction contract based on the VOB with an additional clause covering a 5-year warranty period according to the BGB. The main legal principles are defined in Part B of the VOB, Sections 1–18.

4.2.3 Execution of Construction Work

4.2.3.1 Construction Management

In Germany, the role of a *construction site manager* is clearly defined in the Federal State Building Codes. According to these building codes, the construction site manager's responsibilities toward the building inspection authority include the following:

- Ensuring that all companies involved in the building project adhere to relevant regulations, particularly in terms of occupational health and safety as well as fire protection
- Obtaining all required permits for a smooth construction process
- Ensuring that the construction work is carried out in accordance with the building plans and permits

Construction management by an *architect* or *structural engineer* is often equated with Phase 8 "site supervision" according to the German Fee Schedule for Architects and Engineers (HOAI). However, site supervision does not necessarily entail the same responsibility toward the building inspection authority. The task of a construction site manager, as defined by the state building codes, can or must be contractually agreed upon as an "additional service."

The Lehmbau Regeln [8] stipulate that the preparation of building materials and the execution of building work must be carried out by a person experienced in earth building. This is especially true for load-bearing earth building elements, particularly vault construction, and if owner-builder contributions are planned. This responsibility also includes initiating the production of test specimens. "Experienced," in this context, means that the person has sufficient theoretical knowledge of earth building as well as relevant practical experience in the execution of earth building projects.

Architects or engineers, who are contracted as construction managers or site supervisors of an earth building site but have no earth building experience, first need to familiarize themselves with the specific material properties of earth as a building material and the particular requirements placed on earth building structures. This particularly applies to questions of drying times of the building or the protection of other building elements from moisture penetration. The permanent protection of the building from the elements during construction and drying as well as after its completion is of primary concern.

Contractors act as construction managers within their fields of responsibility. This means that they are only responsible for the partial services carried out by their company. This particularly includes ensuring the safety of everyone involved in the building process as well as preventing damage to the environment. If necessary, they need to coordinate their activities with all other companies involved in the project. If contractors are required by the client to carry out tasks which go well beyond their responsibilities, separate contractual agreements and fees need to be negotiated.

Building management tasks can also be carried out by the clients themselves for their own construction projects. In this case, the guidelines set forth by the Lehmbau Regeln apply, particularly for clients who have no experience in executing earth building projects. All agreements between contractors and clients should be written and countersigned by the client.

4.2.3.2 Construction Work

For the construction process, all work should adhere to the guidelines of the Lehmbau Regeln [8] and the instructions given by the manufacturers of the building materials. In the event of conflicts with other regulations, the technological "state of the art" applies. If necessary, the manufacturers of the respective building materials can also be consulted. If serious doubts persist, it may be necessary to give "Notification of Reservations" according to the VOB, Part B Section 4.

Clients are often interested in carrying out some of the work themselves in connection with the preparation and use of earth building materials. This wish is often further encouraged by privately offered workshops in which the lay person is promised that "everything" about earth building can be learned in one weekend.

The concept of "DIY construction" has a long tradition in earth building. In developing countries, earth building is classified as a "nonengineering building method" and is executed almost entirely by the owners themselves (or with the help of neighbors) without contracting a construction company.

In Germany, as well, it was common for "owner-builders" to construct their own homes out of earth building materials, particularly in rural areas. Farmers were not only experts in agriculture and animal husbandry, they were also familiar with the local techniques of house construction and earth building. In this manner "downtimes" in farming could be filled with construction work.

Urban homes built of earth were constructed by "earth builders" ("Kleiber") in past centuries (Fig. 1.20). The workers had a low status within the building guilds, and the status was then transferred to the building material itself. Most literature on the history of construction focuses on stone and bricks, wood, and half-timber structures. Descriptions of earth buildings are rare as they appeared too mundane and not worth mentioning.

Urban homes built of fired bricks were never a topic of DIY construction. They were planned by master builders and executed under their supervision. This is the reason why no one in Germany would consider offering weekend workshops for lay people on the topic of masonry construction with fired bricks today.

Owner-builder construction with earth was also very important after World War I and World War II. This resulted in the "postwar" image attached to earth building which persisted for decades.

These topics were also the focus of German earth building enthusiasts at the beginning of the 1990s. At that time, earth as a building material was "rediscovered" based on its ecological aspects, and its future prospects were the subject of heated discussions. Many earth builders wanted to preserve the freedom they possessed in their construction work and feared that rules such as those applied to masonry or concrete construction might regulate earth building "to death."

Subsequent developments have shown, however, that earth as a building material can only be accepted by society if it is seen as a "normal" building material. This requires the existence and application of current building regulations.

Clients should consider these points if they decide to work with earth building materials as owner-builders. They should always request detailed instructions from the manufacturer regarding the use of the materials and only execute the construction work under professional guidance. The work carried out by the owner-builder must be clearly distinguished from the work executed by a contractor, in order to settle issues of warranty in the case of construction defects.

4.2.3.3 Completion of Construction Work

Final acceptance of the construction work should be documented in an inspection record, such as according to the VOB, Part B Section 4. Final acceptance can also take place in the form of a "tacit acceptance," for example, when a client moves into a finished house. The warranty period begins with the official hand over and acceptance of the work. Within this period, the contractor guarantees that all executed work will be free from defects. If defects arise and the contractor is notified by the client, the defects must be remedied at the contractor's expense. The warranty period depends on the type of construction contract (Sect. 4.2.2.3).

Many earth building companies have been faced with the situation of being required to offer particularly long warranty periods [51]. In such cases, clients often perceive earth building materials as still being in their "experimental stages" which imposes an unacceptably high risk on them. Experienced specialized earth building companies can counter such arguments by stressing the high quality of their work. Modern earth building has come of age, not least of all, due to the quality standards set forth in building regulations which contractors can refer to and which position earth building in the same category as other types of construction work.

Other experiences made by earth building contractors have included cases where the earth builder's warranty period extended beyond the warranty period given by other contractors involved in the building project. Such situations should be avoided because, in the event of damages, clients might try to hold those contractors accountable who are still legally "accessible," independent of who is responsible for the defects.

Should defects arise during the warranty period, they must be remedied by the contractor immediately. Defects should not be remedied after expiration of the warranty period. Here, a gesture of good will could be seen as an "admission of defective work."

The final inspection and acceptance of remedied defects should again be put in writing.

4.3 Planning and Execution of Earth Buildings

When planning earth building structures, general requirements of usability according to Section 3, Paragraph 2, of the MBO as well as special requirements according to the Lehmbau Regeln [8] and DIN 18945–47 must be met (Sect. 4.2.1.2). Furthermore, a number of general material-specific principles need to be observed:

1. If earth building materials are applied in a wet state, especially those with a high organic fiber content, it must be ensured that they can *dry as quickly as possible* in order to prevent the development of mold or rotting. This can be achieved through cross ventilation or artificial drying (Sect. 3.3.3).
2. *Loads* should only be applied to earth building elements designed for load-bearing use after they have dried sufficiently and deformations from settling and shrinkage have been largely completed.
3. During construction, suitable *weather protection* must be ensured. All earth building materials which are stored on site as well as all earth building elements which are currently under construction or already completed (especially if made of hollow blocks) need to be protected from precipitation using suitable covers. Standing water on impermeable ceilings and floors must be prevented.
4. In general, standard *moisture protection construction practices* must be adhered to. For earth building elements, it is particularly necessary to initiate:

   - The appropriate design of moisture barriers in order to prevent contact with rising ground moisture or moisture coming from the sides, as well as splash back
   - The prevention of contact with standing water during the entire construction period and the useful life of the building, e.g., in cases of accidental water damage

Table 4.5 Use of earth building materials in different earthen building elements, overview

Building material								
	Unshaped						Shaped	
Building element	Rammed earth	Cob	Straw clay	Light clay	Earthen loose fill	Earth mortar	Earth blocks	Clay panels
Floor	■				■			
Wall, load bearing	■	■				■	■	■
Wall, non-load bearing	■	■	■	■		■	■	■
Ceiling and roof			■	■	■		■	■
Drywall construction							■	■
Plaster			■	■		■		

5. In order to ensure the high-quality processing of earth building materials into earth building elements, the *tests* defined in the Lehmbau Regeln [8] must be conducted to an appropriate extent. If necessary, monitoring during construction must verify adherence to the criteria. These requirements must be met to guarantee the unrestricted use of the building during its intended useful life.

Due to the special properties of the material, architects and planners should have a general knowledge of the processing of earth building materials into building elements. Product-specific instructions are usually provided by the individual manufactures.

Table 4.5 shows a simplified comparison of earth building materials and their processing into building elements [52].

4.3.1 Foundations, Cellar Walls, and Stem Walls

Building elements which come into direct contact with the ground, such as foundations and cellar walls, should never be made of earth building materials. Instead, water-resistant materials (concrete, fired bricks, natural stone) should be used. A stem wall constructed out of water-resistant materials should be installed on top of the foundation or cellar walls and should extend to a minimum height of 50 cm above ground level. In addition, a water-repellent plaster or coating should be applied to the outside of the stem wall if necessary.

The joint between the stem wall and the beginning of the earth wall must be sealed against rising damp using a horizontal moisture barrier. Exposed-lip stem walls should be avoided as they cause rainwater runoff to collect at the foot of the earth wall leading to moisture penetration in this area.

Earth walls erected on water impermeable ceiling panels need to be constructed on top of a horizontal moisture barrier made from water-resistant materials with a minimum thickness of 5 cm.

4.3.2 Floors

Rammed earth floors were very common in traditional architecture in Central Germany. They formed the lower enclosure of the "parlor" but were also used in domestic areas of the household, for example, as barn floors. Rammed earth floors in cellars proved to be suitable for storing fruits and vegetables.

For a traditional rammed earth floor, a barrier layer of clay-rich soil of approx. 10 cm was first applied to a level plane and compacted. This was followed by a layer of coarse to medium gravel with a thickness of 20–25 cm which acted as a capillary break. The final installation of rammed earth was carried out in layers which were approx. 6–7 cm thick forming a total thickness of approx. 20 cm. The consistency of the rammed earth material at the time of installation was semisolid to rigid. Each layer was thoroughly compacted and left to dry before the next layer was added. Any cracks which formed were covered up by subsequent layers.

After leveling and compaction, the last layer (earth screed) was additionally tamped with a flat board to tighten the "grain on grain" structure. Pore water was released, giving the surface a shiny appearance and increasing the mechanical strength of the floor. To this end, special patterns made of bricks or gravel were laid into the upper layer to add more strength to the floor. In order to increase the wear resistance of the screed, various materials were applied and incorporated into the upper layer during the tamping process. Such materials included cow's blood, animal bile and urine, anvil residue (metal oxide), tar, and bitumen [53].

Many projects in modern rammed earth construction once again include the installation of rammed earth floors, such as in the "Chapel of Reconciliation" in Berlin [54] and the "Chapel" of the Central Hospital in Suhl, Germany [55] (Fig. 4.20). In both projects, the surface of the floor was stabilized with a penetrating wax in order to fulfill various requirements such as the use of power sweepers or to allow people to walk on the floor with wet shoes.

A very special category of buildings with rammed earth floors is made up of historical orangery buildings which were an integral part of castle grounds, particularly in the Baroque period up to around 1870. They served, and still serve, as the winter home for tropical plants which are displayed in the historical gardens in the summer months.

The original rammed earth floors were eventually replaced by concrete floors as a result of the development of modern transportation and irrigation techniques but also due to a growing number of visitors. In the course of extensive restoration measures over the past 20 years, existing concrete floors have been removed in some of the orangery buildings and replaced with "new" rammed earth floors. In Germany, this includes orangeries of the castles and parks in Schwetzingen, Großsedlitz, Potsdam-Babelsberg, and Weimar-Belvedere [56]. Reasons for a "return" to rammed earth floors can primarily be found in the desire to restore the original appearance of the spaces as well as to create a special indoor climate for wintering the plants.

When installing rammed earth floors, an expansion joint between the foundation (or cellar wall) and the earth floor needs to be included.

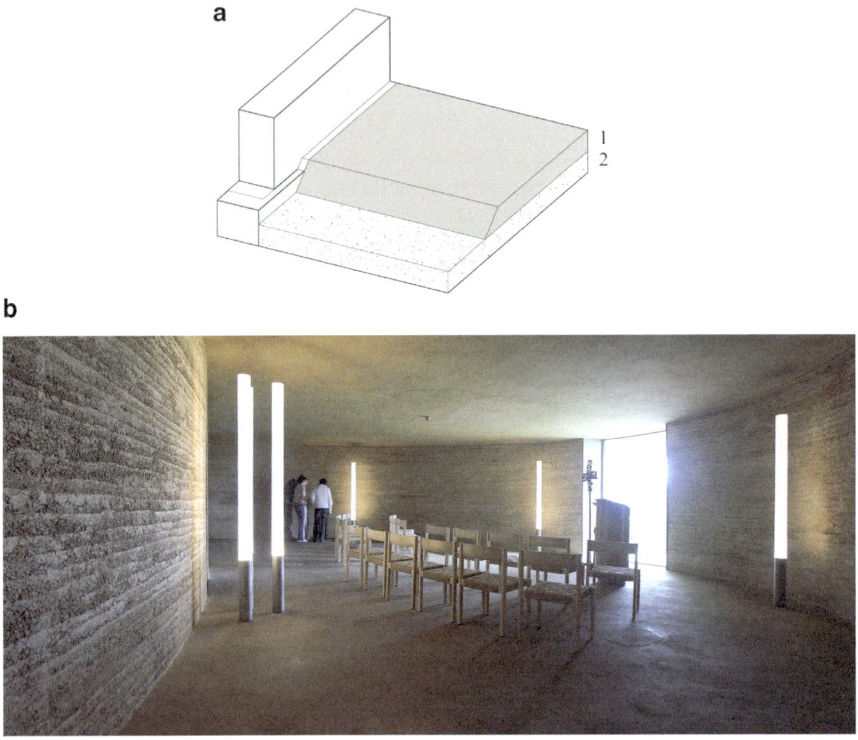

Fig. 4.20 Rammed earth floor (**a**) basic structure [52]. (*1*) Rammed earth applied in layers (6–7 cm) ca. 15–20 cm. (*2*) Gravel–sand capillary break layer ca. 15–20 cm. (**b**) Chapel in the Central Hospital in Suhl/Germany [55]

As a rule, rammed earth floors are not suitable in regions with a high water table. In addition, all general requirements must be met in terms of stem walls (≥50 cm above ground level), moisture barriers in the masonry walls, and building drainage.

The Lehmbau Regeln [8] do not specify requirements of rammed earth floors in terms of strength properties. Although they are classified as "non-load-bearing" building elements, rammed earth floors are exposed to a high degree of wear, particularly in public buildings. Therefore, the same minimum requirements concerning compressive strength and degree of shrinkage as those used for load-bearing rammed earth walls should be applied (Sect. 3.6.2.2).

4.3.3 Wall Construction

Terminology. The German word "Wand" (wall), in reference to a building element, originates in the Old High German/Indo-Germanic language and roughly describes a vertical "woven structure covered with mud" (Wikipedia). In contrast, the German

Fig. 4.21 Boundary wall made of rammed earth, around Grenoble, France

word "Mauer" (masonry wall) generally refers to a solid, self-supporting structure used to enclose land. Earthen boundary walls as well as town fortifications can still be found in rural regions and historic town centers (Fig. 4.21). The Great Wall of China also belongs to this building category (Fig. 1.6).

Walls can be classified according to a variety of criteria, such as material, building technique, or function.

In the context of a building, there are interior and exterior walls. When designing structures, it is important to consider the structural function of a wall, which means differentiating between load-bearing and non-load-bearing functions.

Applications. In traditional earth building in Central Europe, the building materials of rammed earth, cob, and earth blocks (earth clumps) were primarily applied in the construction of load-bearing walls using specific building techniques. Another widespread technique was the combination of a load-bearing wooden frame with earth as a non-load-bearing infill material for half-timber construction.

In developing countries in arid and semiarid regions but also in Australia and the Southwestern USA, earth building is closely connected to the concept of load-bearing walls making this its primary field of application [40, 46].

The main focus of modern earth building in Germany lies in the possibilities of combining earth building materials with other building materials (load-bearing frames, insulation materials) as well as their specific applications in non-load-bearing areas based on the individual building element. So far, load-bearing applications have only slowly started to be reintroduced.

Table 4.6 Use of earth building materials in wall construction—overview

Building material							
Type of wall	Rammed earth	Cob	Earth block	Straw clay	Light clay	Clay panels	Earth mortar
Wall, load bearing	■	■	■			■	■
Wall, non-load bearing							
• Partition walls, interior	■			■			
• Historical wall panels		■	■	■	■	■	■
• Thermal insulation layers			■	■	■	■	
• New timber frame construction, infill			■	■	■	■	

Table 4.6 gives an overview of the application areas of earth building materials for wall construction based on their functions.

In addition to their structural role, walls determine the shape and room layout of a building and perform functions of thermal comfort and indoor climate. Nowadays, wall surfaces also need to meet specific requirements in terms of aesthetics and design.

4.3.3.1 Load-Bearing Walls

Walls and wall sections are classified as load-bearing if they support vertical and/or horizontal loads and/or serve as stiffening elements for load-bearing walls.

Due to the lack of systematic research, our knowledge of the strength and load-bearing characteristics of earth building materials is limited. This has resulted in uncertainties which have led to higher safety margins in earth building than in concrete or brick masonry construction. If the dimensioning of earth building elements could become more economical, the range of applications in load-bearing areas would most likely expand. A conversion of the dimensioning procedure toward partial safety factors according to DIN 1055-100 was completed for all load-bearing construction with the introduction of DIN EN 1990 (Eurocode) on July 1, 2012. This resulted in the need to develop appropriate dimensioning procedures for load-bearing earth building elements.

Dimensioning

Overview of Dimensioning Concepts

A dimensioning concept generally consists of a calculation method and a safety concept.

Calculation methods. There are two calculation methods: the simplified and the detailed analysis method. In the *simplified method*, the calculation assumptions (e.g., linear stress curve) and the dimensioning process itself are easier. The safety margin is not explicitly expressed in the equations but has instead been integrated into the permissible stresses. In addition, the stresses on the walls (which would require more complicated verification) can be neglected. They too have already been covered by the safety margins, a reduction of permissible stresses or by constructive requirements and regulations. The application of the simplified method, however, requires adherence to certain limits, such as the number of stories, distances between transverse walls, and wall heights. This guarantees that the dimensioning result is always on the safe side and, at the same time, not too uneconomical or too far removed from the results of a more detailed calculation.

The *detailed analysis method* is used wherever the application limits of the simplified method cannot be met or where verification is required concerning the structural stability of the entire building or of individual stories or building elements. The detailed analysis method, for example, covers the bracing effect between wall and ceiling and buckling behavior in a more realistic way. This generally results in more complex calculations for individual verification. The detailed analysis method can be used to derive the rules of the simplified method. This guarantees that the safety of the building elements used based on the simplified method is not below the values which would have been attained through a more detailed calculation.

Safety concept. In terms of the safety concept, there are three different methods of demonstrating structural stability:

- Verification of permissible stresses using the simplified method
- Verification of the ultimate load using the detailed analysis calculation method
- Application of the partial safety factor method contained in the Eurocodes

The structural stability test by means of the *permissible stresses* is carried out using the simplified method and is subject to the condition

$$\text{actual}\ \sigma \leq \text{permissible}\ \sigma.$$

The actual stresses must be determined for the finished state as verification levels which need to be compared to the permissible stresses. The permissible stresses defined in standards and the constructive limits already contain the required safety margin for load-bearing capacity.

The permissible stresses are based on a predetermined and proven global safety margin between the calculation value and the mean value of the compressive strength β_k which was determined using an accelerated lab test based on a slenderness ratio of the wall construction at $h/d = 10$.

In the *limit analysis*, the more detailed dimensioning method is used to prove that under the condition

$$\gamma \cdot S \leq R_k(\beta_k),$$

the γ-fold working loads S in the state of failure can be supported by the calculation values for strength R as a verification level.

In DIN EN 1990–1999 (the former Eurocodes), which were introduced on July 1, 2012, verification is established using the *partial safety method*. It specifies partial safety factors for load-bearing capacity and structural resistance which are used to record actual conditions more precisely and to make dimensioning more economical. Verification of the condition

$$S_{\mathrm{d}}\left(\gamma_{\mathrm{f}} \cdot S\right) \leq R_{\mathrm{d}}\left(\beta_{\mathrm{k}} / \gamma_{\mathrm{M}}\right)$$

is used to increase the imposed working loads S by the partial safety factors to an impact dimensioning value S_{d} and to reduce the load-bearing capacity by the partial safety factor γ_{M} for the building material property to get the dimensioning value R_{d} of the material resistance. This means that the verification level lies between the load-bearing capacity side and the structural resistance side.

Limit states are defined for the impacts S (stresses) which indicate the level (capacity to withstand stresses) above which the support structure of the building can no longer meet the design requirements:

- Limit state of usability: occurrence of deformations, vibrations, dislocations, cracking.
- Limit state of structural safety: loss in strength, loss in stability, irreversible material creep resulting in structural collapse or other forms of structural failure.

The impacts are included in the characteristic properties of the building materials (in mathematical terms: basis variable, e.g., Table 3.9 for earth masonry mortar, Table 3.11 for earth plaster mortar, Table 3.18 for earth blocks) and, with the help of statistical methods, are described with their characteristic values using distribution functions.

The safety factor on the structural resistance R side (capacity to withstand stresses) describes the reliability of a structure in the form of consequence and reliability classes in the event of structural failure according to DIN EN 1990, Appendix B. First, buildings are classified based on their intended use/function (Sect. 4.1) in terms of consequences for human life in the event of damage. In the event of damage, a residential building has greater damage potential for human life and therefore a higher protection value than an agricultural building, for example, a barn without regular human activity. The consequence classes are linked to the intended useful life of the building in order to define the reliability class or the reliability index. It is provided as a numerical value in the respective table and can be included in the safety verification process.

By describing the characteristic values on the impact side and the resistance side with the help of statistical methods, it is possible to estimate the probability of the values being exceeded or undercut. These methods are also called probabilistic or semi-probabilistic methods.

Considerations dealing with partial safety factors for earth building materials are fairly new. The problem was first approached by Walker et al. [57] with the introduction of a partial safety factor γ_{M} for the production of a rammed earth mix. This factor ranges from the values 3 to 6. It is divided into classes which evaluate the experience of the manufacturer, the quality of production monitoring, and the consistency of the test results. The most unfavorable value of $\gamma_M = 6$ roughly corresponds to the global safety factor for dry compressive strength tests of rammed earth according to the Lehmbau Regeln [8].

Based on the partial safety method specified in DIN EN 1990, Rischanek [58] has developed a safety concept for earth block construction. Here, the values of the characteristic properties of earth building materials are first determined in a series of tests with the aim of improving strength properties. These values are then included in the dimensioning process of a "prototype construction" made of earth block masonry.

Dimensioning Concept of the Lehmbau Regeln

In the Lehmbau Regeln [8], the simplified method is used for the dimensioning of load-bearing walls made of earth building elements. Existing stresses in the condition of use need to be determined as the verification level and compared with the permissible stresses. The permissible compressive stresses (Sect. 3.6.2.2), which are defined in the Lehmbau Regeln, contain a "global" safety margin in connection with the structural resistance: they are reduced to approx. 1/7 of the *compressive strength* β_D, determined through an accelerated test in the lab.

The compressive strength β_D of rammed earth and cob as well as earth blocks is determined based on the conditions outlined in Sect. 3.6.2.2.

The Lehmbau Regeln assume that loads are centric. However, this principle is indirectly broken by structural specifications in terms of story heights, distances between transverse walls, lengths of bearing surfaces, etc., because such specifications include off-center loads. In addition, wind loads, which are applied vertically to the wall surface, are present. Eccentricity is therefore unavoidable and limited to $e \leq b/6$ of the core cross section. This means that for the application of eccentric loads, the permissible stress distribution of the loaded surface is trapezoidal or at most triangular at the pulled edge when the failure load is reached.

After the introduction of DIN EN 1990 and the semi-probabilistic dimensioning concept associated with it, the Lehmbau Regeln now need to be adapted for dimensioning load-bearing walls made of earth building materials. Due to the significant "global" safety margin, there is no risk in continuing to use the present method until the development and introduction of the respective partial safety factors have been completed.

Models of Load-Bearing Behavior

Load-bearing walls made of earth building materials are exposed to lateral *loads* as *planes* (e.g., wind loads) and vertical loads as *plates* (e.g., dead loads). They therefore need to be able to absorb compressive, shear, tensile, and flexural stresses or any combination thereof. The tensile and flexural strength of earth building materials is low. In their load-bearing capacity, they are therefore primarily used in building elements which are exposed to compressive stresses.

Deformations in earth building materials are not linear (elastoplastic); the σ–ε-line is curved. The deformation moduli are therefore determined as tangent or secant moduli and, strictly speaking, are not building material constants. They can only be determined and specified for defined areas of stress $\Delta\sigma$. For these areas, a linear-elastic material behavior and the validity of HOOKE's law are assumed, which is based on a proportionality between stresses and expansions (Sect. 3.6.2.1).

The state of stress in load-bearing earth walls when *reaching the limit state* can be described in its simplest form (linear-elastic) using the MOHR/COULOMB failure criterion (Sect. 3.6.2.2)

$$\tau = \mu\sigma + c.$$

Corresponding to the geotechnical condition of use, the stresses are typically determined in connection with moist soils. However, structures made of earth building materials are dry in their finished state. Accordingly, the adhesive shear strength (c) would have to be determined for the dry building material.

The failure criterion describes the failure of the material. It can be applied as appropriate to earth building materials which are used in the construction of load-bearing walls (Sect. 3.6.2.2).

Rammed Earth. Dierks/Stein [59] have introduced a model of load-bearing behavior of rammed earth which is based on an analogy to cast-in-place concrete. This is supported by the following arguments:

- Rammed earth is a mix which consists of mineral grains of different sizes with varying portions of clay minerals as the binder = "clay-bound conglomerate."
- The break patterns of concrete and rammed earth specimens are similar. Exposed to uniaxial pressure in an accelerated test, both materials fail in the same manner by exceeding the transverse tensile strength (Fig. 3.46).

Arguments against an analogical model to concrete are:

- The different characters of cement and clay mineral binders: Cement forms a *rigid*, water-insoluble gel in the concrete which causes the conglomerate to harden irreversibly. In contrast, the binding forces in rammed earth are based on the electrochemical attraction between the clay minerals (cohesion) and the friction between the coarser mineral grains. They form a *plastic*, water-soluble bond within the conglomerate which can also lead to differences in long-term behavior: due to the preserved activity of the clay minerals, the potential of interactions between mineral grains and water films and of the redistribution of water in the construction soil is much higher than in concrete or fired bricks. This property can also be used to explain the long-term stability of the "earthen tower houses" in Yemen or Southern Morocco which have survived for centuries even though the material strength at the base of the wall has been exceeded (based on calculations).
- Differences in processing: Today, ceiling-high formwork is used in the construction of walls made of cast-in-place concrete, and the fresh concrete is continuously

poured in. Rammed earth construction is also a monolithic process. However, as a result of the rammed earth technique, construction joints are formed along the formwork sections or appear based on daily progress. This can lead to cracks and a redistribution of stresses.

Earth blocks. Earth blocks are laid like fired bricks to form earth block masonry following standard masonry bonding rules. The bond increases the load-bearing capacity under compressive and shear stresses. This also enables the transfer of horizontal loads through adhesion and/or friction between the earth block and the joint mortar, under the condition that the earth blocks have been laid in a full bed of mortar.

The tensile and flexural strength of earth block masonry is only about 10–20 % of its compressive strength. In this context, load transmission from block to block is primarily carried out by the masonry mortar in the bedding joint. Partially filled bedding joints cause peak stresses in the block.

Transverse tensile stresses develop in the earth blocks if the masonry is exposed to compressive stresses perpendicular to the bedding joints. Block failure occurs when the transverse tensile strength of the block $\beta_{T,B}$ is exceeded. Due to its generally higher level of lateral strain, the mortar in the bedding joint increases the transverse tensile stresses in the block as the block constrains the expansion of the mortar (Sect. 3.6.2.2).

Construction

Based on the Lehmbau Regeln [8], the following structural requirements must be met in cases which use the simplified calculation method to prove the structural stability of load-bearing walls made of earth building materials with the help of a comparison between existing and permissible stresses. If the requirements are met, a verification of spatial stiffness is not needed. For greater story heights and distances between transverse walls, a verification of spatial stiffness must be supplied as used in DIN 1053-1, taking the slenderness of the wall or the influence of lateral support into account.

Wall Height and Minimum Wall Thickness

The Lehmbau Regeln specify the following minimum thicknesses for load-bearing walls based on the specific earth building material according to Table 4.7:

Table 4.7 Load-bearing walls made of earth building materials, minimum wall thicknesses

No.	Earth building material	Wall thickness, exterior [cm]	Wall thickness, interior [cm]	Minimum cross section for pillar-like walls [cm^2]
1	Earth blocks	36.5	24.0	1300
2	Rammed earth	32.5	24.0	1600
3	Cob	40.0	40.0	3200

Table 4.8 Load-bearing walls made of earth building materials, distances between transverse walls

No.	Thickness of the wall to be stiffened [cm]	Story height [m]	Minimum thickness of stiffening transverse walls [cm]	Maximum distance between centers [m]
1	24.0–36.5	≤3.25	11.5	4.5
2	>36.5–49.0	≤3.25	17.5	6.0
3	>49.0–61.5	≤3.50	24.0	7.0

The values apply to story heights ≤3.25 m. The minimum thickness of the exterior walls can be decreased to 24 cm for one-story buildings with a story height ≤2.5 m which are not permanently occupied by people. In this case, verification of the permissible compressive strength and the spatial stability is not required.

Interior walls must meet the following requirements:

- Story height ≤2.75 m.
- Live loads including any supplements for partition walls ≤0.275 N/mm^2.
- Only permissible as intermediate bearing supports for continuous ceilings with a bearing distance of ≤4.50 m or up to 6.0 m if a centering bar has been used on the bond beam.

If these conditions cannot be met, the interior walls must have the same thicknesses as the exterior walls.

Stiffening Building Elements

As with construction using other building materials, load-bearing earth walls require stiffening building elements (rigid diaphragms: transverse walls and ceilings) to be able to absorb and transfer horizontal loads (wind, earthquakes). The following wall thicknesses and distances apply (Table 4.8).

Stiffening transverse walls need to be constructed at the same time as the load-bearing exterior walls. They should be raised from the stem wall or the foundation at the same time as the exterior walls without any structural weakening or significant shifts in alignment. If stiffening transverse walls are to be constructed using a different technique or at a later point in time, a suitable structural connection between the transverse walls and the load-bearing exterior walls must be guaranteed. In earth block masonry, a toothing technique can be used to connect transverse walls as long as the same earth building materials are used. Rammed earth walls are connected to each other or to other masonry walls using a 5-cm-deep keyway which is provided in the wall to be stiffened (Fig. 4.22 [60]).

Bearing Surfaces of Ceilings and Walls

Lintels above doors and windows need to be supported by bearing surfaces of a minimum length of 24 cm. Where calculations require longer bearing surfaces, lintel deflection should be limited to 1/500.

Fig. 4.22 Connecting a transverse earth block wall to an exterior load-bearing rammed earth wall [60]

Bearing surfaces of ceiling joists should be arranged in a manner which facilitates a symmetrical and even distribution of the ceiling load across the entire cross section of the wall. If the required dry compressive strength of the earth building material in the area of the bearing surface is not sufficient, *bond beams* made of materials with a higher compressive strength can be installed. The following materials have proven to be suitable: reinforced concrete (prefabricated or cast-in-place concrete), steel (T-beams), wood (boards), or bonded masonry made of blocks with a higher compressive strength (Fig. 4.23 [60]). This also applies to support elements which are exposed to tensile stresses. When using concrete and steel, thermal bridging through exterior insulation layers should be prevented.

Tie Anchors

Ceilings and transverse walls need to be sufficiently anchored to load-bearing, enclosing walls to be able to resist tensile stresses. In earth walls which are only stiffened on one side, tie rods need to be installed at ceiling height and at each third of the wall height. These rods must tie into the transverse walls at a minimum depth of 1.5 m.

Figure 4.22 shows a wire tie which connects a transverse earth block wall to an exterior rammed earth wall in the upper third of the wall height. The wire is attached to a vertical rod within the rammed earth wall. Today, mesh-like webbing of

Fig. 4.23 Firebrick bond beams used as a bearing surface for a reinforced concrete beam ceiling, installed on top of an exterior rammed earth wall and on top of a transverse earth block wall, shown without and with a reinforced concrete beam ceiling [60]

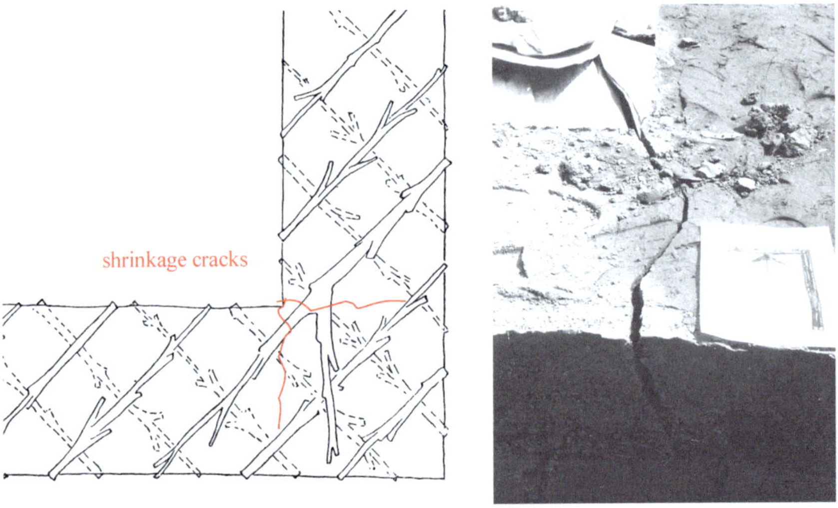

reinforcement with the help of branches [61]

shrinkage crack running through the entire thickness of a rammed earth wall [60]

Fig. 4.24 Cracking in the corners of rammed earth and cob walls [61]

reinforced plastic can be used as tensile-resistant tie material. Such meshes are commonly used in foundation engineering, for example, to stabilize embankments (so-called geogrids, see also Fig. 5.66).

The same technique is carried out in wall corners. Figure 4.24 shows branches which are laid into the corners of rammed earth and cob walls to prevent shrinkage

Fig. 4.25 Wooden anchors in an earth block wall for attaching door frames [60]

cracks caused by tensile stresses in traditional earth building [61]. The second picture shows a rammed earth wall with a crack running through the entire thickness of the wall caused by building material which was too clay rich and too wet at the time of installation [60].

The tensile-resistant anchoring of ceilings and transverse walls to the load-bearing enclosing walls is particularly important for earthen structures in seismic areas (Sect. 5.2.4.2). It prevents a "falling over" of the walls caused by horizontal stresses. Often, this requirement is not met for financial reasons or out of ignorance and can have fatal consequences in the event of an earthquake.

Tie rods can also be used for repairing cracks in load-bearing earthen walls (Sects. 5.2.5 and 5.3.3.2).

The frames around window and door openings can be attached to the walls using standard anchors (Fig. 4.25 [60]).

Mixed Building Techniques

Earth building materials which are used in a wet state should not be mixed with other building materials (such as fired bricks, concrete elements, natural stone) within the individual layers. This is especially true for door and window reveals which often receive a cladding of different materials for design purposes.

Different settling behaviors during the drying period of the earth building materials can lead to cracking. However, continuous horizontal layers of different building materials are permitted.

Channels and Recesses

Channels and recesses for technical installations (Sect. 4.3.7.1) are permitted in load-bearing walls made of earth building materials without further verification as long as their layout and dimensions conform to the limits specified in DIN 1053-1, Table 10. If these limits are exceeded, channels and recesses need to be included in the structural stability verification.

Building Procedures

The Lehmbau Regeln [8] stipulate that the construction of load-bearing walls made of earth building materials should only be executed by professionals or under professional supervision.

Rammed Earth

Application. A number of factors have contributed to rammed earth construction being seldom used in Germany until recently: long drying times, a comparatively high amount of manual labor, the risk of cracking in improperly installed rammed earth, and the required weather protection during construction. In the last few years, however, a number of new projects have been completed which show that rammed earth can offer very interesting design possibilities and, by using modern construction techniques, can also be used as an alternative to concrete in load-bearing applications [54, 62, 63] (Fig. 4.16).

A special feature of modern rammed earth construction is the prefabrication of wall elements which can be assembled into wall structures on the building site with the help of cranes (Fig. 4.26 [54]) (see also Sect. 3.5.8 and Fig. 3.18). This shortens construction and drying times.

In developing countries in hot and dry climates but also in the Southwestern USA and in Australia, rammed earth construction continues to be used for load-bearing walls, often with the addition of cement as a synthetic binder (Fig. 4.4).

Processing. Rammed earth is a monolithic building process. Layers of the prepared rammed earth mixture are poured into formwork and compacted. This process shapes the building element in place (Sect. 3.2.2.1, Fig. 3.17). The individual layers of loose material should not exceed a height of 15 cm. The material is sufficiently compacted when the height of the poured layer has been reduced by about 1/3 in its compacted state. This means that 1 m^3 of rammed earth mix produces approx. 0.67 m^3 of compacted rammed earth wall. Markings on the inside of the formwork can be used as a guideline. Compaction should begin at the outer edges of the wall, working parallel to the wall axes.

Fig. 4.26 Load-bearing wall construction using prefabricated rammed earth elements [54]

A rammed earth mixture which is ready to be installed has a fine and crumbly, pourable consistency with evenly distributed moisture. It can therefore also be lifted in concrete buckets and poured into the formwork where it is distributed.

The optimal moisture content of rammed earth material at the time of installation depends on the respective clay mineral portion and its composition as well as the grain size distribution. When testing the materials by hand, the following should be observed: after pressing the material together and opening the hand, the soil sample should barely hold its shape without falling apart. If the soil is too dry, the crumb structures might not be broken down by the applied compaction resulting in insufficiently compacted sections in the lower part of the layer. After drying, these sections could separate from the wall (Fig. 4.27). If the rammed earth mix is too wet at the time of installation, the pore water acts as a cushion during compaction: the compaction tool "bounces" inside the formwork. During the hand test, the soil "smears" on the surface of the hand [7].

For larger construction projects, it is recommended to build a test wall section on the building site using the rammed earth mixture intended for the project. The test wall section should be built at a scale of 1:1 (Fig. 4.28). This section can be used to test the attained compaction quality based on the number of compacting passes and the effects of colored soils. The final rammed earth mix can then be prepared in a mixer according to tested recipes. A ready-to-use factory mix stored on the building site should be protected from the weather until it is used to avoid potentially affecting the moisture content at the time of installation.

Fig. 4.27 Rammed earth, installation with insufficient compaction [60]

Fig. 4.28 Rammed earth, test wall for determining installation criteria

Construction. The most important structural features for the construction of load-bearing walls made of earth building materials are described in Sect. 4.3.3.1. For rammed earth construction, these particularly include:

- Construction of a foundation or base (see Sect. 4.3.1)
- Construction of ceiling and wall bearing surfaces or bond beams
- Anchoring of ceilings and transverse walls to load-bearing enclosing walls to ensure tensile strength (this includes the reinforcement of building corners, particularly in seismic areas)
- Anchors, channels for technical installations (Sect. 4.3.7.1)
- Mixed building techniques

Horizontal reinforcement with tensile-resistant materials (e.g., geogrid made of synthetic materials but also tensile-resistant local building materials used in traditional earth building, Fig. 4.24) can increase dry compressive strength and reduce the formation of shrinkage cracks.

The rammed earth wall must be sufficiently dry, and settling must be completed, before loads (ceilings, roof system) can be applied to the wall. In this context, it is important to facilitate the natural drying process, e.g., through cross ventilation (Sect. 3.3.3). Today, drying is primarily done artificially. This, however, has a negative impact on the energy balance for the construction of the building (Sect. 1.4.3.2).

After removing the formwork, fresh wall surfaces should be protected from wind-driven rain, splash back, and direct sunlight.

Rammed earth wall surfaces do not make good plaster substrates. Measures to improve plaster adhesion include the integration of horizontal bands of brick or concrete into the individual layers (Fig. 4.28). When used on exterior wall surfaces, these bands also inhibit erosion caused by draining rainwater. During the construction of recent rammed earth projects in Europe, surfaces were generally left unplastered in order to highlight the coarse-grained structure of the soil as a design element.

Verification and construction monitoring. The Lehmbau Regeln [8] stipulate the following tests and respective scopes for required construction monitoring of load-bearing rammed earth construction:

- Dry bulk density according to Sect. 3.6.1.3:
  The dry bulk density is commonly determined in combination with the dry compressive strength in the form of a continuous test series.

- Dry compressive strength according to Sect. 3.6.2.2:
  The tests must be carried out in a timely manner before construction commences. Subsequent testing during construction for on-site mixes for every 10 m^3 of earth building material started or for factory mixes for every 50 m^3 of material started.

- Linear degree of shrinkage according to Sect. 3.6.2.1:
  One test for on-site mixes for every 10 m^3 of earth building material started or for factory mixes for every 50 m^3 of material started.

Due to the use of different soils as well as colored soils during the construction of the rammed earth projects in Suhl and Nordhausen [55], the following additional tests were carried out:

- Grain size distribution according to Sect. 2.2.3.1:
  One test for every 50 m^3 of earth building material started.
- Dry bulk density with moisture content at time of installation according to Sect. 3.6.1.3 (sample taken from the upper compacted layer with the help of a metal nozzle):
  One test for every 10 m^3 of earth building material.
- Lime content and organic additives according to Sect. 2.2.3.4:
  One test for every rammed earth material used for the entire building project.

Additional tests are necessary if the quality and/or the mix ratio of the ingredients (construction soil, aggregates) are changed.

Currently, there are no mandatory test methods for testing the strength of existing load-bearing rammed earth walls.

Cob

Application. Due to the high degree of manual labor and long drying times, load-bearing cob construction is currently not used for new construction in Germany. However, the existence of a large number of historical cob structures, particularly in the new German states, has created a need for information about suitable restoration techniques.

Building techniques which require a high degree of manual labor are suitable construction methods in developing countries as they provide new jobs and livelihoods to many people. One example is the new construction of the Meti School in Bangladesh which was executed using a cob technique adapted to local conditions (Fig. 4.12 [4]).

Processing. As a result of the construction technique, cob walls are approx. 60 cm thick at the base and grow thinner toward the top. Using a pitchfork, the prepared cob mix is stacked into layers approx. 80 cm high *without* formwork and compacted with a board. The side surfaces are shaped by vertically trimming off the partially dry cob with the help of a pointed spade (Fig. 4.29 [61]). After the first layer has dried for about one week, the next and all subsequent layers are added in the same manner until the intended wall height is reached. Loads can only be applied after the final layer has dried.

There are a number of local building techniques which do not require formwork and use earth building materials similar to cob:

In Austria, cob construction is called "*g'satzter Bau*" (stacked construction) [64].

Historical cob building was also particularly common in southwest England where, due to the specific geological situation, rocky soils were used as well (Fig. 4.30 [65, 66]).

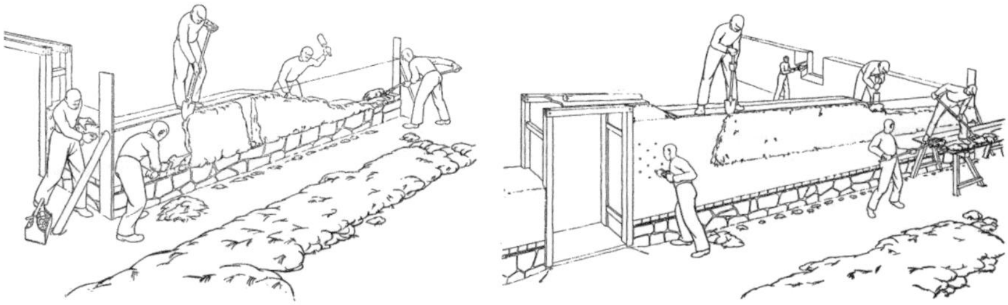

Fig. 4.29 Cob, wall construction [61]

Fig. 4.30 Local cob technique in England [66]

To a certain extent, the *sod houses* in Great Britain, Ireland [67], and Scandinavia can also be seen as local versions of cob. Using a spade, patches of sod were cut from the ground and, while still moist, stacked rootside-up into load-bearing wall structures.

In contrast to the cob technique, the *massoni* in Central Italy [68] were not stacked as loose material using a pitchfork. Instead, straw-clay reels (massoni) were made and manually stacked to form a wall (Fig. 4.31). As with cob construction, the wall surfaces were then trimmed with a pointed spade.

Construction. Structural features according to Sect. 4.3.3.1.

Verification and construction monitoring. The Lehmbau Regeln [8] stipulate the following tests and respective scopes for required construction monitoring of load-bearing cob construction:

Fig. 4.31 Local "massoni" cob technique in Italy [68]

- Dry bulk density and moisture content at time of installation according to Sect. 3.6.1.3:
  The dry bulk density is commonly determined in combination with the dry compressive strength in the form of a continuous test series.
- Dry compressive strength according to Sect. 3.6.2.2:
  The first batch is tested before construction commences. Subsequently, one test for on-site mixes for every 10 m^3 of earth building material started or for factory mixes for every 50 m^3 started.

Currently, there are no mandatory test methods for testing the strength of existing load-bearing cob walls.

- Volumetric shrinkage according to Sect. 3.6.2.1:
  This test is possible but only makes sense when carried out on a sample of the actual building element.

Sprayed Earth and Poured Earth

Application. The search for more effective and economic methods for use in load-bearing, monolithic earth construction led Easton [69] to the adaptation of a sprayed concrete technology at the end of the 1980s, called PISE. The technique can be applied in the construction of load-bearing earthen walls of one- or two-story residential buildings.

In addition, there have been attempts to adapt cast-in-place concrete technologies to earth building. One technique uses the same machine systems which are used in the concrete industry for the preparation and processing of the material. The soil is prepared to a pourable/liquid consistency (*Cast Earth*) and a synthetic binder is added (gypsum, cement). In order to control the hardening process of the soil, which is different from concrete, the use of special additives is required (www.castearth.com). There are also examples of the use of cement in poured earth [70].

A special application in Turkey is the use of a pourable, gypsum-stabilized soil, called *alker* [71]. A relatively high gypsum content of approx. 10 % causes the mix to harden after only 20 min, before the shrinkage deformations of the clay portion set in.

Processing. The *PISE technology* (pneumatically impacted stabilized earth) uses a dry earth–cement mix which is transported pneumatically to the place of installation through a flexible rubber hose. The end of the hose has a mixing nozzle which adds

Fig. 4.32 Load-bearing walls made of sprayed earth with steel reinforcement [69]

the amount of water needed to attain the required consistency as the mix leaves the hose. The spray pressure produces the required compaction of the mix (Fig. 4.32).

The Cast Earth technique uses stationary and mobile mixers, formwork systems, and concrete pumps which are common in the cast-in-place concrete industry. The poured earth is compacted in the formwork with the help of vibrators (which are also used in concrete technology).

The alker technique is executed in the same manner as the Cast Earth method.

Construction. For shaping building elements when using the PISE technique, the earth building material is sprayed (at the desired thickness of the wall) onto a one-sided, ceiling-high rigid formwork panel made of wood. The material is applied to form the full thickness of the walls which, additionally, are reinforced with steel (for seismic reasons). After spraying, the "open" side of the formwork is plumbed with a board (Fig. 4.32).

In 1 day, five workers can construct approx. 25–30 m^3 of wall (with a thickness of approx. 45 cm).

In the Cast Earth technique, shaping is carried out with appropriate formwork systems, similar to cast-in-place concrete construction. Steel reinforcement is only required under certain conditions, e.g., in seismic areas.

For the construction of load-bearing walls, the gypsum-stabilized alker mix is also applied using formwork, similar to cast-in-place concrete construction. It can also be used for producing blocks which are laid as masonry units to form load-bearing walls. Reinforcement must be used in seismic areas.

Earth Clumps

Application. In Germany, using earth clumps to build walls is a separate historical building technique, known in the region of Eastern Westphalia and the Ruhr area as the *Duenner mud loaf method*. It was developed and applied in the early 1920s by

residential building in Schweicheln / Dünne, Germany, before restoration of the exterior plaster, 1996

traditional construction according to [72]

Fig. 4.33 Load-bearing wall construction using the *Duenner mud loaf* technique

Reverend Gustav von Bodelschwingh who worked as a missionary in East Africa and based his method on traditional African earth building techniques [72]. Several hundred buildings constructed using this technique still exist today.

Due to the high amount of labor and long drying times, this method is currently only of interest in Germany in the context of restoration work of existing buildings (Fig. 4.33).

Processing. Straw clay mixed to a soft consistency is shaped into loaf-shaped elements which are stacked wet (*without* mortar) into a masonry structure to form a wall. Thin twigs are laid between the layers, particularly in the corners and at wall

Fig. 4.34 Load-bearing wall construction using the Czech earth building technique *opus spicatum* [73]

manual shaping of the earth clumps, ring-shaped stacking and striking of the joints

Fig. 4.35 Load-bearing wall construction using the *pottery technique* [74]

intersections, to serve as tensile reinforcement. After the walls have reached a height of approx. 1 m, the material has to dry for about 1 week. Loads can only be applied to the walls after they have completely dried.

A Czech version of the mud loaf method is called *opus spicatum*. Here, wet "mud loafs" are also stacked without mortar but at a 45 % slant. In the next layer, the bedding joints are slanted in the opposite direction. Visually, three layers of "mud loafs" thus create one "ear of grain" (Fig. 4.34) [73].

Even today, similar building techniques are still used for construction work in rural regions of Africa and Central Asia on a daily basis:

Pottery technique: rammed earth or straw-clay material prepared to a soft consistency is shaped into clumps which are stacked on top of each other to form a wall (Africa). The joints between the earth clumps are smoothed by hand resulting in a flat wall surface (Fig. 4.35) [74].

For similar building techniques on the Arabian Peninsula and in Central Asia, clumps made of earth building material prepared to a soft consistency are thrown onto a firm surface with force. This type of impulse compaction creates good adhesion between the earth clumps:

Zabour technique: traditional earth building technique in Yemen in which earth clumps are thrown manually from a height of approx. 1.8 m onto the base of the wall (foundation) or the wall copings [75] to form layers.

Fig. 4.36 Load-bearing wall construction using the *pachsa* technique [76]

Fig. 4.37 Load-bearing wall construction using the *guvalja* technique

Pakhsa in Central Asia: a rammed earth mix which has been prepared to a stiff consistency is cut out with a spade and stacked manually into wall structures. It can also be thrown (similar to the Zabour technique) and trimmed with a spade to create a smooth surface (Figs. 3.21 and 4.36) [76]. A similar building technique in Iran is called *tschineh*.

Guvalja in Central Asia: dry earth clumps are laid as masonry units using earth mortar or in a plastic state without mortar (Fig. 4.37).

Construction and verification. Structural features can be designed similar to rammed earth and cob construction (Sect. 4.3.3.1). Special test methods are not known.

Earth Block Masonry

Terminology. The fields of masonry (DIN 1053-1) and concrete construction use the term "single- or double-leaf" wall construction. In *double-leaf* exterior wall construction, the functions of the wall are deliberately separated and assigned to the individual leaves. The cavity between the leaves can be filled with air and/or insulating material. The interior leaf is responsible for load transfer and forms the interior boundary, while the exterior leaf provides protection from the weather and mechanical impacts, and also shapes the visual appearance of the building. *Single-leaf* (earth block) masonry combines the structural function, thermal insulation, as well as all other functions in one building element. In addition, single-leaf exterior walls can be covered with insulation materials.

Application. Currently, load-bearing earth block walls are seldom used in new construction in Germany. The existence of many historical structures, particularly in the new German states, makes this technique interesting primarily in the field of restoration (Fig. 5.19, Sect. 5.3.3.2).

In the hot and dry climates of Central Asia (Fig. 4.38), Africa, Latin America, India, and also in the Southwestern USA and Australia, this building technique continues to be used in load-bearing applications. Cement is typically added as a synthetic binder.

Processing. Dry, crack-free earth blocks are laid to form wall structures using standard masonry techniques. Compacted or CEB should be processed in a manner which guarantees that dead and live loads are applied in the direction of compaction or compression.

shell of residential building in Tashkent, Uzbekistan [52]

Fig. 4.38 Load-bearing wall construction made of earth blocks [51]

The blocks should be laid into a full mortar bed using standard masonry bonding practices (Fig. 4.38). It is important to note that earth masonry mortar needs more time to harden compare to the setting times of lime or cement mortar. To prevent the plastic mortar from squeezing out of the lower bedding joints, no more than 2 m of wall height, or a maximum of one story height, should be laid per day. The recommended maximum thickness of head and bedding joints is 1 cm. To improve plaster adhesion, cutting a 1-cm-deep channel into the wet joints is also recommended.

Construction. Structural features should be designed as described in Sect. 4.3.3.1. In earthquake-prone regions, it is of particular importance to include bond beams which serve as a shear-resistant anchoring system for connecting ceilings and transverse walls to the exteriors walls.

Verification and construction monitoring. DIN 18945 (Table 3.18) regulates the verification of the constancy of performance in terms of the characteristic properties of industrially produced earth blocks and the respective test scopes. Separate construction monitoring is therefore not required.

4.3.3.2 Non-Load-Bearing Walls and Infills

In terms of their structural function, non-load-bearing walls and infills made of earth building materials only need to carry their dead load and possible wind loads. They generally do not have any type of stiffening effect and are combined with a load-bearing frame made of materials with higher compressive and/or tensile strength. They function as room enclosing/dividing systems and meet the respective requirements in terms of building physics. The following types of non-load-bearing walls and infills can be distinguished between:

- Historical infill in half-timber construction
- Interior thermal insulation layers in older buildings
- Infill of timber-frame structures in new construction
- Non-load-bearing partition walls

Interior thermal insulation layers are discussed as a building preservation measure in Sect. 5.3.3.2.

Historical Infill

Traditional half-timber construction consists of a wooden support frame and an infill of different building materials, oftentimes earth building materials. The term *infill* refers to the complete filling of a timber-frame panel to form a space-enclosing building element. A *panel* is the space within the timber frame which is created by vertical posts and/or horizontal noggin pieces and/or slanted braces used for corner reinforcement (Fig. 4.39 [77]). Materials used as infill include straw

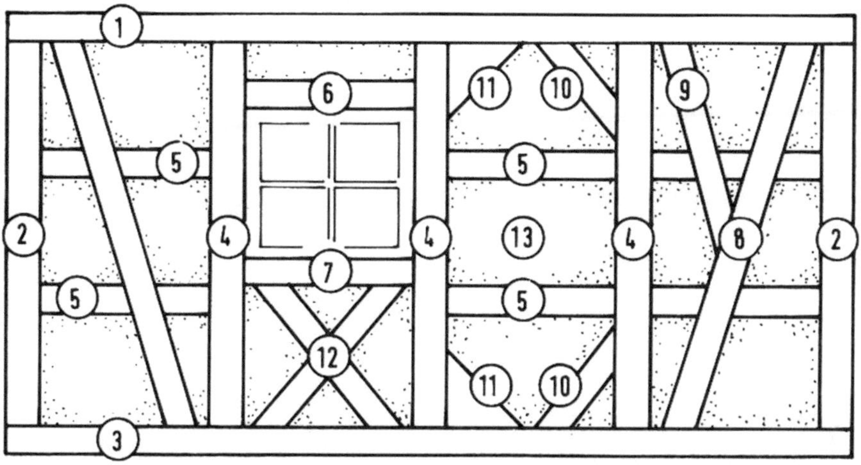

Fig. 4.39 Half-timber construction, support elements and infill [77]. (*1*) Top plate. (*2*) Corner post. (*3*) Sill plate. (*4*) Post. (*5*) Noggin piece. (*6*) Lintel. (*7*) Sill. (*8*) Brace. (*9*) Counter brace. (*10*) Knee (tension) brace. (*11*) Corner block brace. (*12*) St. Andrew's cross. (*13*) Panel

clay, light clay, earth blocks, as well as earth masonry mortar and earth plaster mortar. Today, knowledge of these historical techniques, which differ greatly from region to region, is particularly important for the restoration of half-timber structures (Sect. 5.3.3.2).

If the walls are to be plastered and finished as exposed timber framing, the panels are filled with earth building materials to within about 2 cm below the flush edge of the frame and plumbed. Using suitable tools (combs, rakes), the straw fibers contained in the straw-clay or light straw-clay mixes can be combed so that they protrude out of the wet surface to form a good substrate for the plaster. Another method of roughening the substrate is the application of puncture holes or scratch patterns.

For exposed timber framing, the members of the timber frame are not plastered. The plaster stops at the flush edge of the beam. The wood is washed with a sponge in order to clearly separate the edge between the panel and the wooden frame. After drying and shrinking of the earth plaster, this edge opens up to a gap in the millimeter range.

On the exterior surfaces of weather-exposed walls, a final coat of lime plaster (instead of earth plaster) is typically applied to the panel. Panels filled with earth building materials on weather sides can also be protected from wind-driven rain with sidings made of water-resistant building materials (such as wood shingles or slate but also with bundles of straw pushed into a bed of earth in so-called feathered walls [78]).

Two of the most common infill techniques are wattle and daub and earth block masonry.

Wattle and Daub

Stakes in combination with a woven lattice provide an even, grid-like support structure for the earth building materials when placed into the timber-frame panel (Fig. 4.40 [79]). The stakes can be spaced closely together or further apart.

The support structure is formed by stakes with pointed ends made of dry hard wood (oak) which are wedged vertically between two noggin pieces. The stakes are placed *closely together, at a distance up to approx. 6 cm.* The noggin pieces of the timber frame have grooves or notches cut into the middle of them. Straw clay or light straw clay prepared to a soft consistency is applied to both sides of the stake work and pushed through the gaps between the individual stakes (Fig. 4.40b). It is also possible to weave bundles ("braids") of straw clay and light straw clay between the stakes (Fig. 4.40d [80]) or wrap the stakes with a straw-clay or light straw-clay material (Fig. 4.40c). These wrapped stakes (while still wet) are wedged vertically into the grooves of the noggin pieces or horizontally between the frame.

The same technique was also used for the infill of ceiling panels (Sect. 4.3.4.1).

If not enough suitable stakes are available, they can also be *placed further apart, at a distance of approx. 10–15 cm.* For this technique, an uneven number of stakes is wedged between the noggin pieces. Flexible branches (hazel, wicker) which extend the width of the panel are woven horizontally between the stakes to form a dense lattice (Fig. 4.40a). Shorter, thicker branches need to span a minimum of three stakes. Before the earth building material is applied, it must be ensured that the entire lattice sits firmly in the timber-frame panel. Instead of branches, ropes or strings made of natural fibers can also be used to form the lattice.

Earth Blocks

Residential buildings were constructed using exposed half-timber or plastered half-timber techniques. In agricultural buildings, the infill usually remained unplastered.

Just like in masonry construction, earth block infill is laid into a bed of earth mortar (Fig. 4.41 [51]). The interior infill side is covered with earth plaster, while the weathered exterior surfaces are finished with lime plaster. In order to improve plaster adhesion to the substrate, 1-cm-deep channels are cut into the mortar joints of the blocks.

At times, other locally available building materials were used instead of earth blocks (e.g., tufa [81] and field stones in northern Thuringia/Germany) and laid using earth masonry mortar.

In the traditional architecture of Central Asia, earth blocks and earth clumps ("guvalja") were used as infill in historical half-timber construction. Both building materials are still used today as infill in new construction (Fig. 4.42).

Timber-Frame Infill in New Construction

Traditional half-timber construction forms the basis of modern timber-frame construction systems. They include a high degree of prefabrication resulting in significantly shorter construction times. These systems include timber stud, timber frame, and timber panel construction. Their differences lie in the structural systems of the

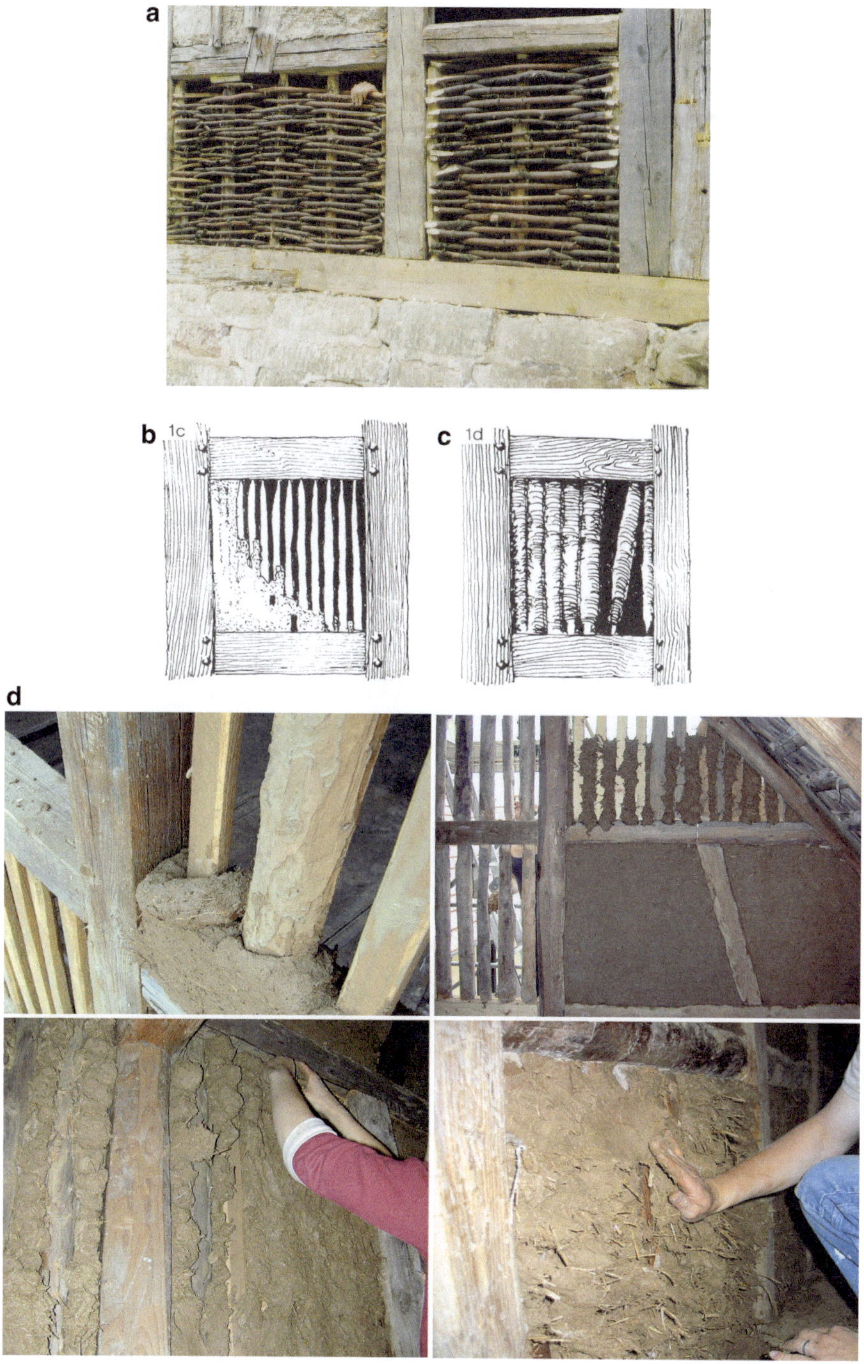

Fig. 4.40 Half-timber construction, historical infill, stake work. (**a**) Widely spaced stakes, with woven branches. (**b**) Closely spaced stakes, covered with straw clay on both sides [79]. (**c**) Stakes wrapped in straw clay ("Weller") [79]. (**d**) Closely spaced stakes, with straw-clay "braids" [80]

Fig. 4.41 Half-timber construction, infill using earth blocks [51]

Fig. 4.42 Timber-frame construction, traditional infill with earth clumps (guvalja) in Uzbekistan

load-bearing timber frame used. As in traditional half-timber construction, the openings created by the timber framework in modern timber-frame construction are also completely filled with earth building materials. The earth infill is non-load bearing and does not provide stiffening functions.

As with all new construction, these structures must meet the specifications of Germany's Energy Conservation Regulation EnEV 2009 [82] (Sect. 5.1.1.2). According to the regulation, the U-value of exterior walls must be ≤ 0.28 W/m^2 K for outside air. This requirement cannot be reliably met with currently available earth building materials using acceptable wall thicknesses up to a maximum of

40 cm. It is therefore necessary to construct walls using multilayer systems with a separate insulation layer (generally on the exterior). A revised version of the EnEV, published in 2014, defines further developments up to 2020 with the objective of stipulating that all new construction after 2021 must meet the "lowest energy consumption standard." In this context, plans to impose stricter *U* values have not yet surfaced. Inevitably, buildings will have to be equipped with high-tech solutions for artificial ventilation and climate control which, in turn, are also connected to energy consumption. It remains to be seen to what extent such buildings and technical systems can be managed by their occupants without the help of professionals.

Through the use of earth building materials, different thermal requirements can be met: in exterior walls by using light clay for thermal insulation, in interior walls and ceilings by using "green" unfired bricks for thermal mass.

Timber-Frame Construction with Light-Clay Infill

Applications. A predecessor of this building technique was called "earth frame construction." It was used especially in East Germany after World War II as an "economical building technique" for the construction of simple residential and agricultural buildings [83]. The load-bearing frame was made of peeled rounded or split timber (see Fig. 4.42, Uzbekistan).

Since the mid-1980s, this building technique has primarily been used in the new construction of residential buildings (Fig. 4.43) but has also been applied in public

new residential construction in Saarbrücken-Bous/Germany, 1990

Fig. 4.43 Timber-frame construction with light-clay infill

Table 4.9 Timber-frame construction with wood-chip light-clay infill, calculation example of *U*-value

No.	Wall-building material	ρ_d [kg/dm³]	λ [W/mK][a]	*d* [m]	d/λ [m²K/W][a]	*U* [W/m²K]
1	Wood-chip light clay	0.700	0.21	0.35	1.667	
2	Interior earth plaster	1.500	0.65	0.02	0.031	
3	Exterior lime plaster	1.500	0.65	*0.02*	0.031	
4	d_{total}			0.39		
5	R_{se+si}				*0.170*	
6	$R_T = 1/U = \Sigma d/\lambda + R_{se+si}$				1.729	
7	$U = 1/R_T$					0.578
8	*Version with additional insulation*					
9	Wood-chip light clay	0.700	0.21	0.35	1.667	
10	Reed panel	0.225	0.056	0.05	0.893	
11	Interior earth plaster	1.500	0.65	0.02	0.031	
12	Exterior lime plaster	1.500	0.65	*0.02*	0.031	
13	d_{total}			0.44		
14	R_{se+si}				*0.170*	
15	$R_T = 1/U = \Sigma d/\lambda + R_{se+si}$				2.792	
16	$U = 1/R_T$					$0.358 > U_{required} = 0.28$

[a] λ values according to the Lehmbau Regeln [8]

building projects (Fig. 4.11). The building technique, however, does not sufficiently meet current requirements in terms of thermal insulation of exterior walls according to EnEV [82] (minimum requirement EnEV 2009: $U \leq 0.28$ W/m² K; low-energy house—standard: $U \sim 0.20$ W/m² K). Even with additional insulation, this requirement cannot be met mathematically when using the numerical examples of the selected building materials listed in Table 4.9 (plane of reference: midspan) and reasonable wall thicknesses ($d \leq 40$ cm). A "calculation trade-off" with other building elements might be possible in some cases. Reed panels do not promote capillary action and therefore obstruct the drying of the wood-chip light clay which is installed wet.

Construction and processing. Light-clay building materials are used as wet, shapeless mixes or dry blocks and panels. Processing light clay in a wet state is linked to a number of disadvantages: more moisture entering the load-bearing timber structure, longer drying times, and the occurrence of settling as well as a higher mass of the material during installation, often leading to undesirable mold formation on interior walls.

Structural framework. The required cross section of the structural framework is divided into load-bearing supports made of square timber posts or boards and a non-load-bearing cavity frame made of slats, which is needed from a technological perspective (Fig. 4.44 [84]).

Fig. 4.44 Timber-frame construction with light-clay infill, new residential construction

The *load-bearing frame* must be able to support wind loads and any exterior wall cladding and insulation panels. The dead loads of the light-clay walls must be absorbed by the structural framework on each floor or after a maximum height of 4 m [8]. The distance between the support posts of the load-bearing frame is limited to 1 m which prevents the light-clay infill from falling out. Lintels and headers for wall openings are integrated into the support structure with the help of squared timbers or boards. The same is true for any horizontal timber used for attaching cabinets.

The *cavity frame* ensures dimensional accuracy of the planned wall structure during installation of the wet light-clay material. It consists of a tensile- and shear-resistant connection between the vertical slats and the load-bearing posts (at a maximum distance of 1.2 m) and to the floor and ceiling. It also serves as the sliding plane for temporary formwork or as the support structure for attaching permanent formwork (Sect. 3.2.2.1). Depending on the earth building material used and the respective formwork, the vertical slats of the non-load-bearing cavity frame are spaced at distances of approx. 35–40 cm. The corners of interior walls require stable end supports to handle the formwork which comes from two directions.

Particular care should be taken when planning the connections between foundation/floor–wall, wall–ceiling, and wall–roof. Any ceiling joists, purlins, or braces penetrating the exterior shell of the building can be potential weak spots in the form of thermal bridges and should therefore be avoided. It is difficult to close these weak spots by filling them with light-clay material or covering them at a later point in time.

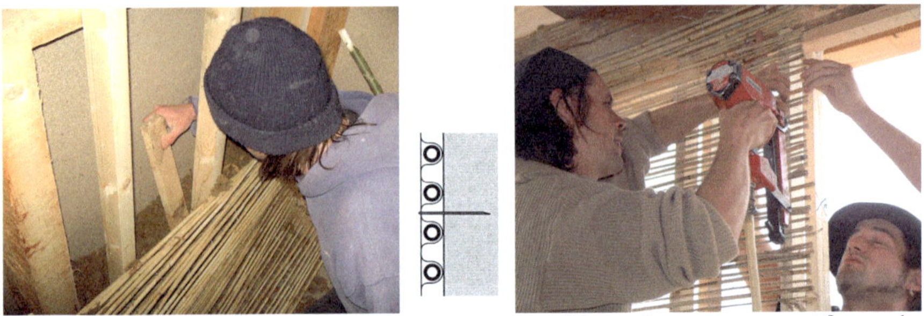

installation of light straw-clay attaching reed mats as permanent formwork

Fig. 4.45 Timber-frame construction with light straw-clay infill [51]

Additionally, the general guidelines on weather and moisture protection provided in Sect. 4.3 apply. Before commencing with the earth building work, the roof should be covered, at least temporarily, and all work on the load-bearing frame must be completed.

Installation. When installing the *wet light-clay material* manually with the help of a temporary framework, the framework elements are screwed or clamped to both sides of the load-bearing and cavity frames in the form of a sliding formwork. The earth building material is added in layers and lightly compacted (Fig. 3.19).

For the use of permanent formwork, one side (preferable the exterior side) of the framework is first shuttered from floor to ceiling. Later, the formwork can also be used as the plaster substrate. Next, the opposite side is covered with reed mats: as the work of adding the light-clay material progresses, the mats are unrolled (starting at floor level) and attached to the cavity frame. The mats should be installed with the stems running horizontally (Fig. 4.45). During compaction of the light-clay material, it must be ensured that the reeds are not pushed apart. The red mats can also be used as a plaster substrate (Sect. 4.3.7.2).

Both edges of the reed mats should be attached to the vertical slats of the cavity frame, with approx. 10 cm of overlap at the joints. The galvanized binding wire should not lay on the substructure. Instead, it should form an edge over the reeds which is used to staple the reed mat to the cavity frame at a spacing of approx. every 5 cm. The staples should also be galvanized and at least 25 mm long. Fifteen fastening points should be added per linear meter of binding wire.

The light-clay mix should be used to completely cover or fill all horizontal sections of the load-bearing frame and the cavity frame as well as corners, voids, and upper panel boundaries. In the event of settling after installation and drying, it might be necessary to insert additional light-clay material from the front or the sides. Holes or leaks in the exterior building envelope reduce thermal insulation and, as a consequence, the comfort level inside the building. They can also cause considerable structural damage if condensation occurs.

Light-clay infill for walls can also be applied using the spray method (Sect. 5.3.3.2).

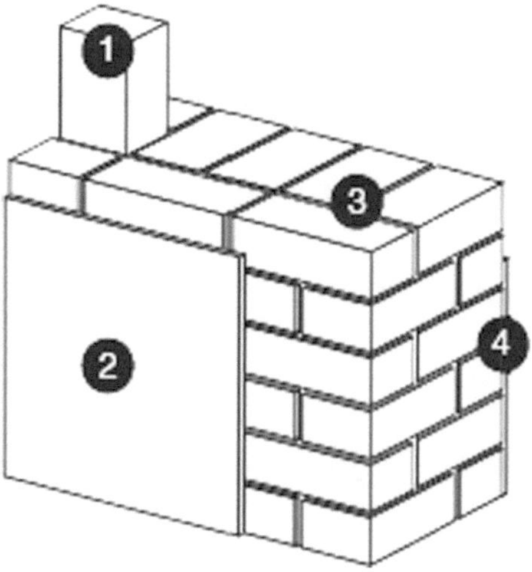

Fig. 4.46 Timber-frame construction with light straw-clay infill, single-layer, light earth block masonry. (*1*) Timber frame. (*2*) Interior earth plaster. (*3*) Light earth blocks. (*4*) Exterior lime plaster

When using *light-clay blocks* as a "dry" infill for timber-frame construction, an additional cavity frame is generally not required. The load-bearing posts are encased by the earth blocks on all sides to build up the required wall thickness. The blocks are laid following standard rules of masonry construction. On the interior wall surface, the light-clay blocks can be used as the substrate for an interior earth plaster. Lime plaster should be applied on the exterior, particularly on surfaces exposed to rain (Fig. 4.46).

Drying. Walls made of a light-clay mix with organic fibers which has been installed wet require air movement for drying. Insufficient or zero air movement can lead to rotting inside the wall's core or to the formation of mold along the interior wall surfaces. It is therefore important to ensure that drying times are short (Sect. 3.3). If this is not possible, artificial drying becomes inevitable. For this reason, the wall thickness should not exceed 30 cm for light clay which has been installed wet as long as drying on both sides is unobstructed.

If drying is only possible from one side (due to directly adjoining walls but also formwork made of wood wool panels and reed which inhibit or reduce capillary action), walls made of light clay which has been installed wet should not exceed a thickness of 15 cm. Light-clay walls which adjoin materials with good capillary action (earth, bricks) can have a maximum thickness of 20 cm [8]. Particularly for the latter, good ventilation during the drying phase must be guaranteed.

If the walls are to be plastered, the respective drying times of the specific light clay must be observed. Earth building elements with a moisture content of $w \leq 10\ \%$ are considered sufficiently dry. This can generally be verified by visual inspection [8].

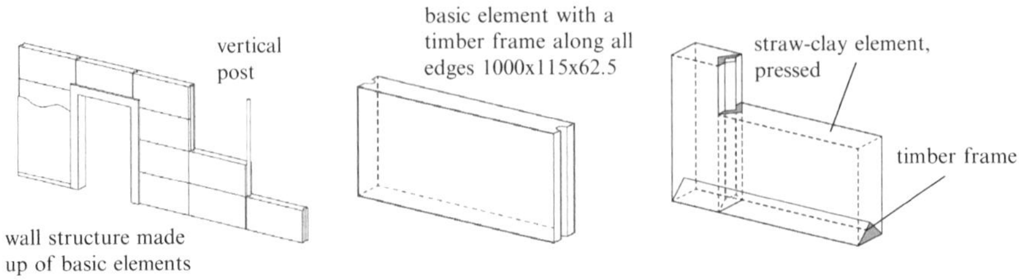

Fig. 4.47 Framed cavity technique with light straw-clay infill, single layer [85]

Timber-Frame Wall Panels with Light Clay

Application. So far, this building technique has only been used by one company in Northern Germany in the field of residential construction [85]. The attained dry bulk density of the earth building materials was specified as ≥250 kg/m³ by the manufacturer.

Construction and processing. Using special presses, clay-bound straw is compressed into panels and then pressed into a timber frame along all edges. The frames are then connected to timber posts and, in this manner, assembled into a load-bearing wall structure (Fig. 4.47). The walls are finished with a lime plaster on the exterior and with an earth plaster on the interior.

This building technique only uses small amounts of earthen material and is comparable to the straw-bale technique which is especially widespread in Scandinavia but also becoming increasingly popular in Germany [86].

Multilayer Exterior Walls Made of Timber Frames with Integrated Earth Building Materials

Reed panels with d ≤ 10 cm. For this technique, the wooden posts of the structural framework must be flush with the exterior wall surface in order to serve as the support for attaching the reed panels. The earth building materials are installed against the reed panels from the inside, completely encasing the wooden posts on three sides. Reed panels inhibit the drying process of wet earth building materials. Therefore, dry earth blocks should be used instead of wet light-clay mixes.

Cellulose fiber insulation used as infill material of a timber-frame construction with an interior layer of light-clay blocks (Fig. 4.48a [87]). On the exterior, the lateral boundaries of the panels are formed by a lining of soft wood fiber boards. On the interior, they are formed by a layer of earth blocks laid flush with the wooden posts using earth masonry mortar. This layer is anchored to the wooden post at intervals of approx. 50 cm. It serves as a good substrate for an interior earth plaster but can also be designed as exposed masonry (Fig. 4.48b [88]).

The resulting cavity is filled with cellulose fibers using pneumatic equipment. For this, the soft wood fiber panels are drilled open at the upper edge of the timber-frame panel and subsequently closed again after filling. The insulation layer on the

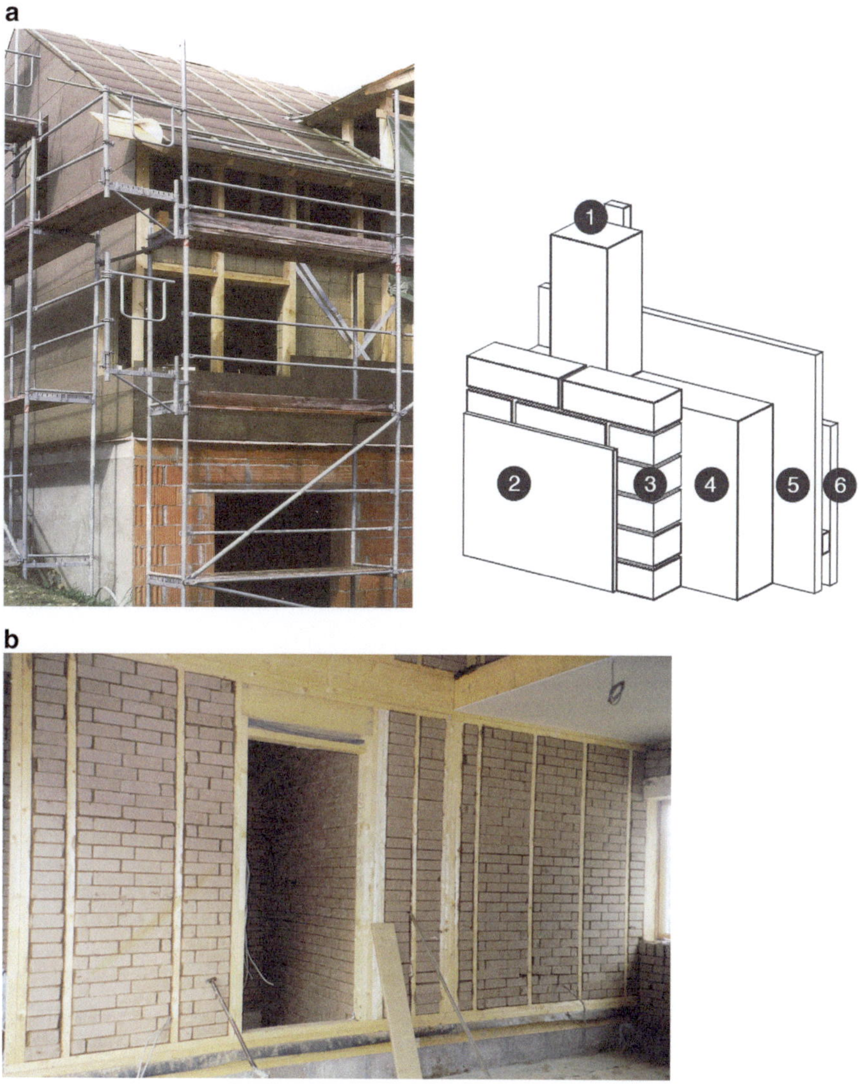

Fig. 4.48 Timber-frame construction, multilayer exterior wall. (**a**) Version A: interior earth plaster. (*1*) Timber frame. (*2*) Interior earth plaster. (*3*) Light-clay block masonry. (*4*) Cellulose fiber insulation. (*5*) Wood fiber insulation panel. (*6*) Rear-ventilated wood cladding. (**b**) Version B: exposed masonry, interior [88]

exterior is well-protected by a rear-ventilated wood cladding which is attached to the timber frame, forming the final layer of the facade.

The described wall structure meets the thermal requirements according to EnEV 2009 [82] as well as those of the low-energy house standard (plane of reference: midspan, Table 4.10).

Table 4.10 Timber-frame construction with interior earth block layer, calculation example of U-value

No.	Wall-building material	ρ_d [kg/dm^3]	λ [W/mK][a]	d [m]	d/λ [m^2 K/W][a]	U [W/m^2 K]
1	Wooden cladding	0.60	0.130	0.02	1.538	
2	Ventilation layer			0.04		
3	Wood fiber panel	0.20	0.048	0.02	0.417	
4	Cellulose fiber insulation	0.06	0.040	0.14	3.500	
5	Wood-chip light-clay masonry	1.00	0.35	0.115	0.329	
6	Interior earth plaster	1.50	0.65	*0.02*	0.031	
7	d_{total}			0.355 (0.315)		
8	R_{se+si}				*0.170*	
9	$R_T = 1/U = \Sigma d/\lambda + R_{se+si}$				5.985	
10	$U = 1/R_T$					$0.167 < 0.2$

[a]λ values according to the Lehmbau Regeln [8]

For a completely dry construction version of this technique, the interior layer of earth blocks could be replaced by earth panels. Thin clay panels ($d \sim 3$ cm) are attached to the wooden posts using standard fasteners; thick clay panels ($d \sim 8$ cm) are glued or installed dry (tongue and groove) and anchored to the wooden post at certain intervals.

Instead of cellulose insulation, other environmentally sustainable insulation materials with a similar λ value (~0.04 W/mK) can be used.

Timber panel construction with an integrated layer of earth blocks. Various manufacturers of prefabricated timber house systems have developed wall structures which integrate earth blocks into exterior walls (preferably as "green" unfired bricks) to form interior thermal-mass layers. Suitable structural systems primarily include timber panel systems which are entirely prefabricated and installed at the building site.

The wall elements can also be supplied with the interior side "open" which allows owner-builders to install the "green" bricks themselves, under professional guidance. The "green" bricks are stacked on their $l \times h$ sides without the use of mortar following standard masonry bonding practices. They are clamped in place with horizontal battens which are attached to the vertical beams at intervals of approx. 50 cm as the stacking work progresses.

If the walls are to be plastered, the earth blocks are laid with "open" head joints of approx. 5 mm. This allows the plaster to lock into the open joints creating a stable connection with the lining. The plaster is applied in two layers and reinforcement mesh should be embedded in the first plaster coat while it is still wet.

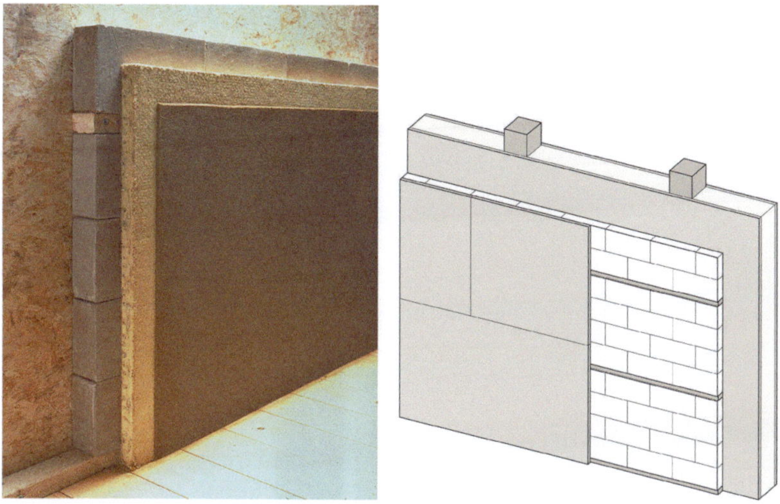

Fig. 4.49 Wood panel construction with integrated earth block layer [52]

If a drywall cladding system (e.g., using thin light-clay panels) is planned, the earth blocks should butt up against each other (Fig. 4.49 [52]). The depth of the clamping battens should be slightly less than the height of the stacked earth blocks to ensure that the cladding panels fit tightly against the blocks.

Non-Load-Bearing Partition Walls

Non-load-bearing partition walls made of earth building materials can be constructed in single or multiple layers. As shear walls, they need to be sufficiently rigid and stable and must form tensile- and shear-resistant connections with the surrounding load-bearing walls.

Earth block masonry. The earth blocks are laid in earth masonry mortar using standard masonry bonds (according to Fig. 4.38). One-centimeter-deep channels should be cut into the joints if the masonry is to be plastered. For design reasons, the walls are often left as exposed masonry.

Partition walls made of clay panels. Using standard fasteners, "thin" clay panels are attached to both sides of a support structure made of wooden slats, making sure that the panel joints are staggered. The cavity is then insulated (Fig. 4.50 [52]). Strips of reinforcement mesh are laid over the joints and brushed with fine-finish earth plaster. After drying, the entire surface is plastered (Fig. 4.70).

"Thick" earth panels are structurally self-supporting and have a tongue and groove system integrated into their edges. They are erected on top of a support board using a running bond (Fig. 4.51 [89]) and are connected without mortar. Before installation, the joints are wetted in order to activate the adhesion of the clay

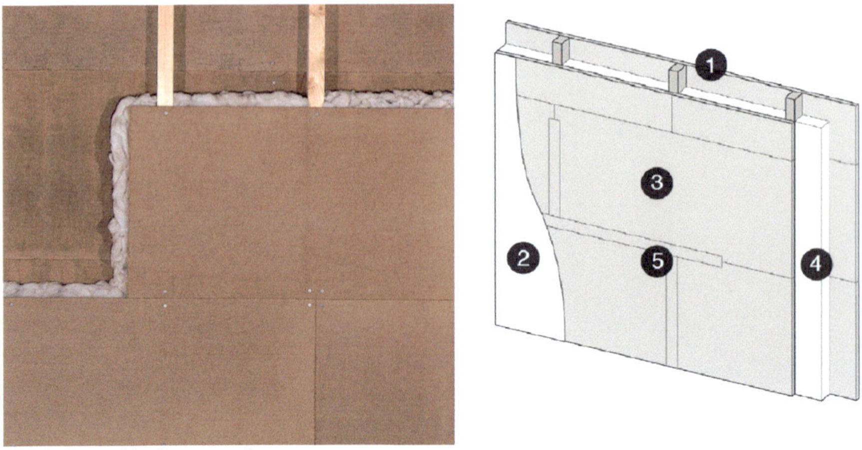

insulation with sheep wool

Fig. 4.50 Non-load-bearing partition walls, "thin" clay panels attached to a timber-frame construction, insulated [52]. (*1*) Load-bearing frame. (*2*) Earth plaster. (*3*) "Thin" clay panel. (*4*) Insulation. (*5*) Reinforcement mesh

Fig. 4.51 Non-load-bearing partition walls, "thick" self-supporting clay panels [89]

minerals. It is also possible to use glue or mortar by employing the thin-bed method. The panels can be connected to other (load-bearing) walls using standard fasteners (nails, screws).

Lining Panels

Lining panels made of rammed earth with dimensions of approx. $150 \times 80 \times \geq 8$ cm are available with an integrated mounting system. A clay adhesive is used to attach them to the existing wall structure and the joints are covered with a fine-grained rammed earth mix. This product primarily meets thermal and aesthetic requirements in connection with interior spaces.

4.3.4 Ceilings

Within a structure, ceilings generally serve as stiffening diaphragms and form the upper boundary of the room. They carry a dead load and also need to transfer live loads from the rooms above to the walls which, in turn, transfer them to the foundation. In addition to their structural functions, ceilings have to meet requirements in terms of thermal comfort.

Earth building materials were traditionally used in timber beam ceilings, either in stake panels between the joists or as continuous overlays on top of sheathing which was inserted between or laid on top of the ceiling joists. A visible lining or plaster layer forms the underside of the ceiling, while the floor (with floorboards) forms the upper side (which can also be left out in ancillary buildings).

Timber beam ceilings with earth building materials were used in connection with load-bearing wall structures made of earth, bricks or natural stones, but primarily in half-timber construction.

In connection with modern timber-frame building systems, timber beam ceilings and their traditional construction principles are becoming increasingly popular again. They are constructed using modern earth building materials according to the general principles of the planning and execution of earth building elements described in Sect. 4.3.

Timber beam ceilings are divided into:

- Slatted timber ceilings
- Insert ceilings
- Ceiling overlays

Depending on the position of the timber slat panels or the sheathing in relation to the height of the ceiling joists, "half" or "full" slatted timber ceilings and earth insert ceilings (Fig. 4.52) can be differentiated between. The term *half* ceiling refers to ceilings where the level of the structural frame lies roughly between the lower third and the middle of the height of the ceiling joist. The bottom and lower parts of the sides of the joists remain visible or can be completed covered with a cladding.

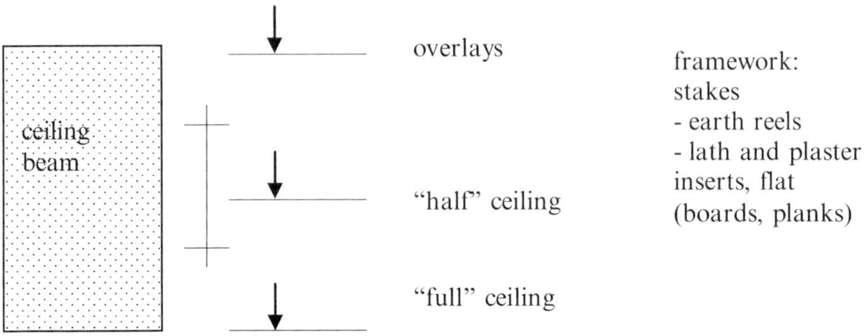

Fig. 4.52 Earth building materials in timber beam ceilings, position of the structural support frame for the earth building materials

In *full* ceilings, only the bottoms of the ceiling joists are visible. They can also be plastered or covered with cladding. In ceiling *overlays*, the timber slats or sheathing which support the earth building materials lies on top of the ceiling joists.

4.3.4.1 Slatted Timber Ceilings

Similar to panels in wall construction, the structural framework for holding the earth building materials used in slatted timber ceilings consists of stakes (rounded timbers with a diameter of 4–12 cm, made of pine or hardwood, which can also be split) and slats (3×5 to 4×6 cm). They are inserted sideways into cut grooves in the ceiling joists, laid on top of support battens which are attached to the sides of the ceiling joists, or nailed down on top of the ceiling joists.

Suitable earth building materials are straw clay and light straw clay and any pourable construction soil which can be used as loose fill material. Slatted timber ceilings should not be walked on during their drying phase.

Slatted timber ceilings can be divided into earth reel ceilings and lath and plaster ceilings.

Earth Reel Ceilings

For centuries, earth reel ceilings represented the standard ceiling construction technique in Germany. Today, knowledge and skills related to their design and installation are primarily required in the field of building restoration.

For the construction of earth reel ceilings, stakes with a thickness of 4–6 cm are wrapped in long-stemmed straw and coated with a paste-like earth mix. First, the straw is twisted into a "braid" and dipped into an earth mix prepared to a liquid to paste-like consistency (Fig. 4.53). It is also possible to lay out the straw as a mat, cover it with earth, and wrap both materials around the stakes to form a "roll." Before use, the long-stemmed straw needs to be presoaked. Soft straws including rye, oat, and barley are suitable types.

wrapping of the earth reels and insertion into the ceiling cavity

Fig. 4.53 Earth reel ceiling

The wet reels are inserted sideways into grooves which have been cut into the ceiling joists and then pushed tightly against each other. Instead of using grooves, battens can be attached to the sides of the ceiling joists and the earth reels can be laid on top. Depending on the position of the reels, the space between the reels and the upper edge of the ceiling joists can be filled with pourable construction soil (no specific requirements) or other materials.

Before wrapping the reels, the stakes are cut to the width of the ceiling cavity to fit between the ceiling joists. In order to ensure that the reels (which are installed wet and pushed against each other) fit tightly in the ceiling cavity, they can be cut slightly longer due to the shrinkage of the wood and the earth after drying. On the other hand, it is important to ensure that stakes which are too long do not change the position of the ceiling joists. For the transfer of standard loads, eight reels per linear meter are sufficient [8].

Depending on the position of the earth reels in the ceiling structure, full, half, or earth reel overlay ceilings can be differentiated between (Fig. 4.52). The distance between the ceiling joists is approx. 1 m.

In *full* earth reel ceilings, the reels are inserted or laid into the lower third of the ceiling joists and end flush with the lower edge of the joists. The final (earth) plaster or cladding covers the bottoms of the ceiling joists. The space above the reels is filled to the top edge of the joists with pourable construction soil or other materials.

In *half* earth reel ceilings, the reels are inserted or laid into the upper third of the ceiling joists and end flush with the upper edge of the joists. This position makes them lighter than full reel ceilings. The lower part of the ceiling joists remains visible. The ceiling panels are finished with cladding or *a*(*n*) (earth) plaster.

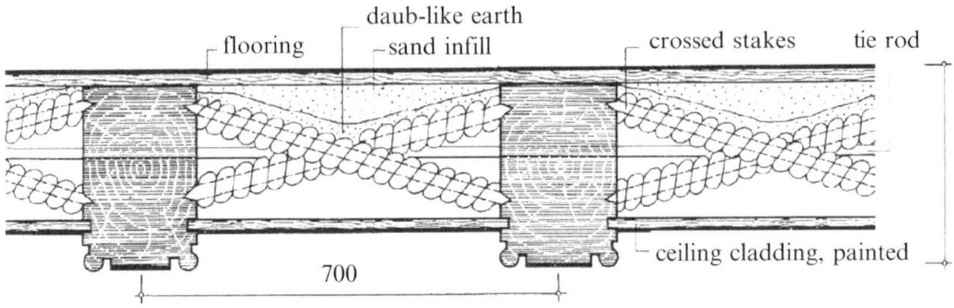

Fig. 4.54 Crossed-stake ceiling [53]

For *overlay* or *stretched* earth reel ceilings, the 8–12-cm-thick stakes are wrapped with straw-clay braids, placed on top of the ceiling joists and attached with nails. The underside of the ceiling was typically covered with a cladding.

In traditional construction, these ceilings were also used as a type of economical building technique in agricultural buildings. Distances of up to 1.7 m between the ceiling joists were used. The floor was made up of an earth screed and the ceiling joists remained visible from below. The ceiling panels were plastered with earth.

A special type of earth reel ceiling is the so-called crossed-stake ceiling which was used for ceilings exposed to high stresses and for ceilings with spans of approx. 5 m or more (Fig. 4.54 [53]). These ceilings could be found in town halls, restaurants, and representative residential buildings, often with decorative elements on the ceiling joists and panels.

The crossed stakes were wrapped with straw clay and installed at alternating angles ("crisscrossing"), spaced either at intervals of up to 2 m or more closely together. For installation, they were inserted into grooves cut into the ceiling joists or nailed down. When spaced closely together, the crossed stakes formed a depression which was brushed with daub-like material and filled up to the top edge of the ceiling joists using kiln-dried sand or pourable earthen material. The underside of the ceiling was generally finished with cladding.

The crisscross arrangement of the stakes created a better diaphragm effect in the ceiling cavities resulting in a good transverse distribution of the ceiling loads. Steel tie rods were also inserted at intervals of approx. 2 m. This further increased or helped maintain the diaphragm effect when the stakes (which had been installed wet) decreased in length after drying. The distance between the ceiling joists is <1 m.

Lath and Plaster Ceilings

A framework of thin stakes, slats, or split round timbers is attached at 2–6-cm intervals (lath) using the following methods: they are nailed to the top of the ceiling joists, wedged into grooves in the ceiling cavities, inserted on top of lateral support battens, or attached to the undersides of the ceiling joists (Fig. 4.55 [52]). As with slatted timber ceilings, lath and plaster ceilings can therefore be divided into "half," "full," and "overlay/stretched" ceilings (Fig. 4.52).

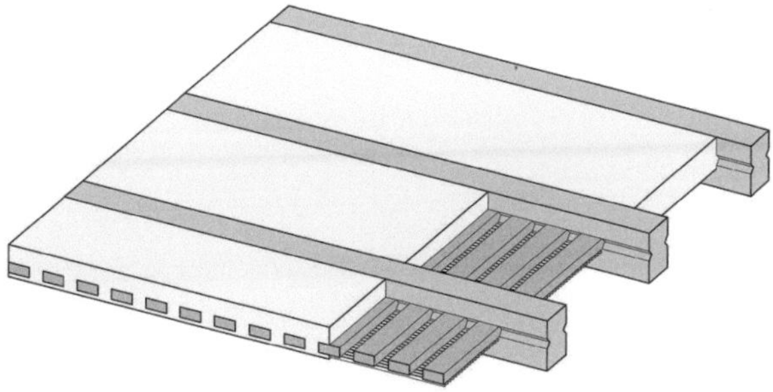

Fig. 4.55 Lath and plaster ceilings, "half" [52]

In contrast to earth reels, long-fiber straw clay or light straw clay is applied to the grid-like lath from above and pushed through the gaps between the timbers. The curls hanging through the lath are pushed up against the timbers from below and smoothed out. After drying, the underside of the ceiling or ceiling panels is finished with an earth plaster.

The underside of the lath slats can also be covered with a plaster lath material. The cavity up to the upper edge of the ceiling joists can then be filled with straw clay or light straw clay and lightly compacted ("stuffed"). The lath and plaster ceiling can also be installed using temporary formwork. This method should be used especially if no suitable fiber material is available or if the cohesion of the earthen material is not sufficient enough for smoothing out the hanging straw-clay curls. After the earthen material has dried, the underside of the ceiling is plastered.

4.3.4.2 Insert Ceilings

In the nineteenth century, insert ceilings were very common in Germany. A continuous sheathing of boards, planks or slabs was placed between the ceiling joists. The boards were inserted into cut grooves or laid on top of support battens which were attached to the sides of the joists (Fig. 4.56 [52]). As with earth reel ceilings, earth insert ceilings could also be designed as "full" or "half" ceilings depending on the position of the insert. The sheathing and joists were often decorated with ornaments. "Overlapped" sheathing was used as a design element and, at the same time, represented an economical building technique: planed boards with a width of approx. 10 cm were used as the lower layer and rough-sawn slabs as the upper layer. They were either laid next to each other with space between them or overlapped.

The insert support surface (also called "dead floor") was covered with loose fill which consisted of dry or wet earth or other materials. Before applying dry, crumbly loose fill, the sheathing was covered with wet daub-like material to stop the material from trickling down to the room below. For wet fill material, a layer of dry straw was first spread over the insert support surface.

Nowadays, these types of ceiling panels are completely covered with a strong building paper before the dry earth building materials are added.

Floorboards were typically used to finish the top of timber beam ceilings. In agricultural buildings, ceilings made of light-clay infill over inserts were finished with a continuous screed of straw clay or straw clay with added fibers [53].

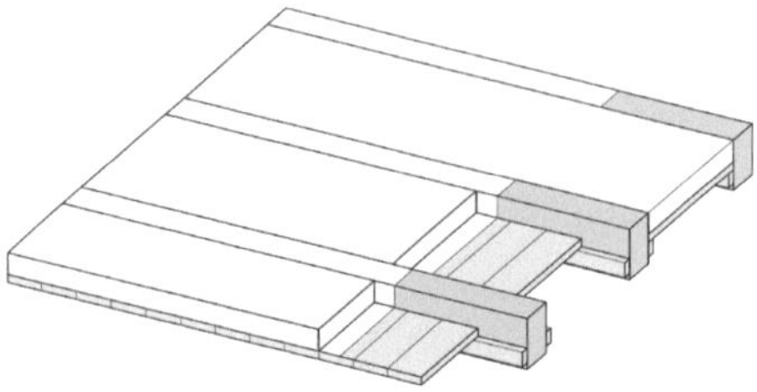

filling of the ceiling cavities with "green" unfired bricks on top of building paper

Fig. 4.56 Insert ceilings "half" [52]

4.3.4.3 Earth Overlay Ceilings

In modern earth building, the design principle of a ceiling overlaid with earth building materials has become common once again (see "stretched" earth ceiling, Sect. 4.3.4.1). Earth overlay ceilings consist of sheathing laid on top of the ceiling joists and a subsequent layer of dry or wet earth building material (Fig. 4.57a [52]). The disadvantage of using wet earth building materials is the increase in moisture entering the load-bearing timber construction, resulting in longer drying times and a greater weight on the ceiling panels during construction. The materials might also require compaction.

Today, dry earthen loose fill, "green" unfired bricks, earth blocks, or unreinforced clay panels are primarily used as overlay materials. When using shaped earth building materials, it is particularly important to completely fill the joints with mortar or sand for sound insulation purposes. Before the earth building materials are added, the sheathing should be completely covered with building paper. The use of so-called honeycomb flooring is recommended when using dry earthen loose fill (Fig. 4.57b) to ensure that the loose fill remains evenly distributed during the building's useful life.

The upper boundary of the ceiling is formed by a floor system which conforms to the strength of the earth building materials and meets all requirements in terms of sound and thermal insulation. The sheathing used as support for the earth building materials should be dimensioned according to structural requirements.

4.3.4.4 Earth Panel Ceilings

The cavities between the ceiling joists are filled with prefabricated straw-clay or light straw-clay panels. Reinforced earth panels can be used in load-bearing applications; unreinforced panels are only self-supporting.

The reinforcement rods (slats) in load-bearing earth panels rest on top of the joists or are inserted into cut notches or on top of support battens which are attached to the sides of the joists (Fig. 4.58a).

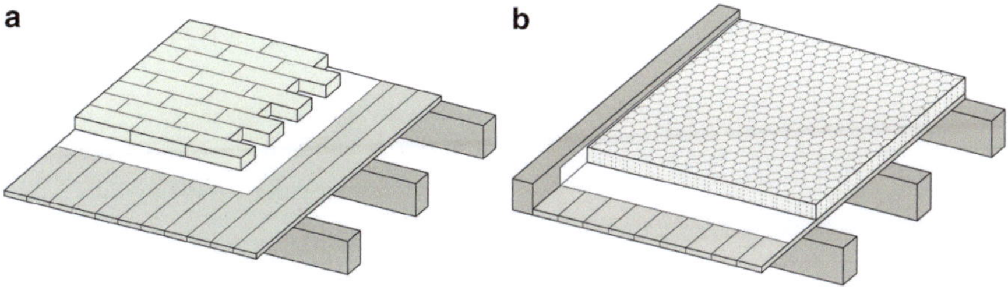

Fig. 4.57 "Overlay" ceiling using earth building material overlays [52]. (**a**) Earth blocks. (**b**) Earthen loose fill with honeycomb flooring material to prevent material shift

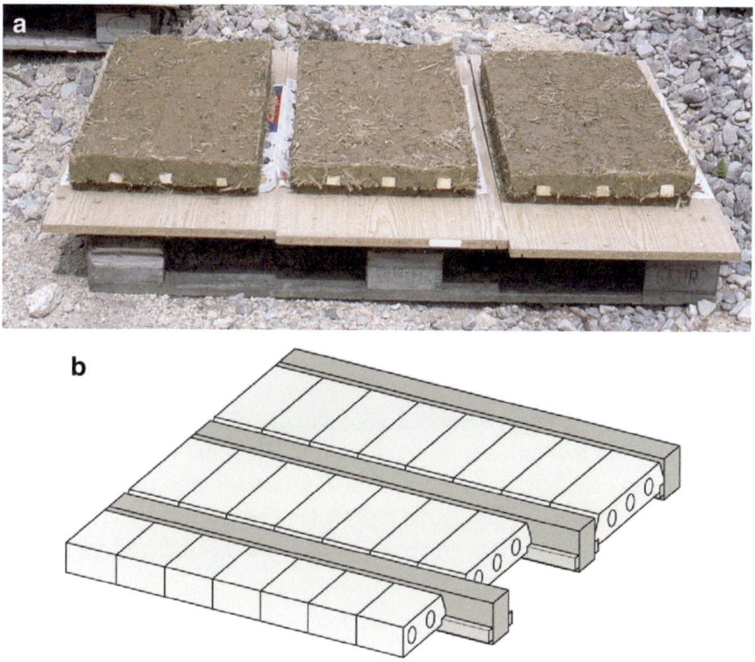

Fig. 4.58 Prefabricated earth panels for ceilings. (**a**) "Reinforced" earth panel. (**b**) Light-clay elements for inserting [52]

The clay panels should be installed dry. Holes or irregularities can be filled with earth mortar or a light-clay material. Finishing the underside of the panels with plaster is recommended.

Some manufacturers have also begun selling ceiling insert panels ($d = 125$ mm) (Fig. 4.58b).

4.3.5 *Roof Systems*

Roof systems form the upper exterior boundary of a building. They are directly exposed to the natural elements of rain, snow, wind, and sun. Therefore, the planning, design, and construction of roofs require special care, particularly in terms of thermal and sound insulation as well as fire protection.

In general, a roof system consists of two structural elements: the *supporting framework* for carrying the roof loads (outer skin, wind, rain and possible snow loads, thermal insulation) and the *outer skin* (protection from rainwater and sun insolation, wind tightness). In addition, the outer skin can be covered with a moisture protection layer. Roofs in moderate and cold climates require thermal insulation. The substructure and the outer skin can be designed as single or multiple layers.

In addition to their thermal function, roof systems also need to meet aesthetic requirements. Especially in traditional architecture, the *roof shape* and choice of material determine the visual appearance of a building and define the landscape. Different climatic conditions as well as regional building traditions have led to the development of different roof shapes which can generally be divided into:

- Flat roofs (pitch $<10°$)
- Pitched roofs (shed roof, gable roof, hip roof, pavilion roof)
- Curved roofs (vault, dome)

The material's sensitivity to water represents a special problem for roof systems made of earth building materials. It is therefore not possible to employ such roof construction systems in humid climates. In moderate climates, they can only be used to a limited extent. In both regions, roofs are typically pitched to allow large amounts of rainwater to drain quickly. In hot and dry climates, this is generally not necessary due to the low amount of precipitation. Here, flat roofs serve as suitable thermal mass and heat buffers against intense solar radiation. These roofs can be designed to incorporate a limited amount of wood. Curved roofs can be constructed entirely without the use of wood which is a great advantage for traditional construction in these climate zones where wood is scarce.

4.3.5.1 Pitched Roofs

In the traditional architecture of Central Europe, earth building materials were primarily used for the interior parts of pitched roofs which can be classified into

- Rafter insulation (Fig. 4.59b)
- Partition walls in front of the bearing surfaces for the structural roof framework (Fig. 4.59a [83])

Light-clay insulation material between the rafters can be installed using the method for timber slat ceilings and earth insert ceilings described in

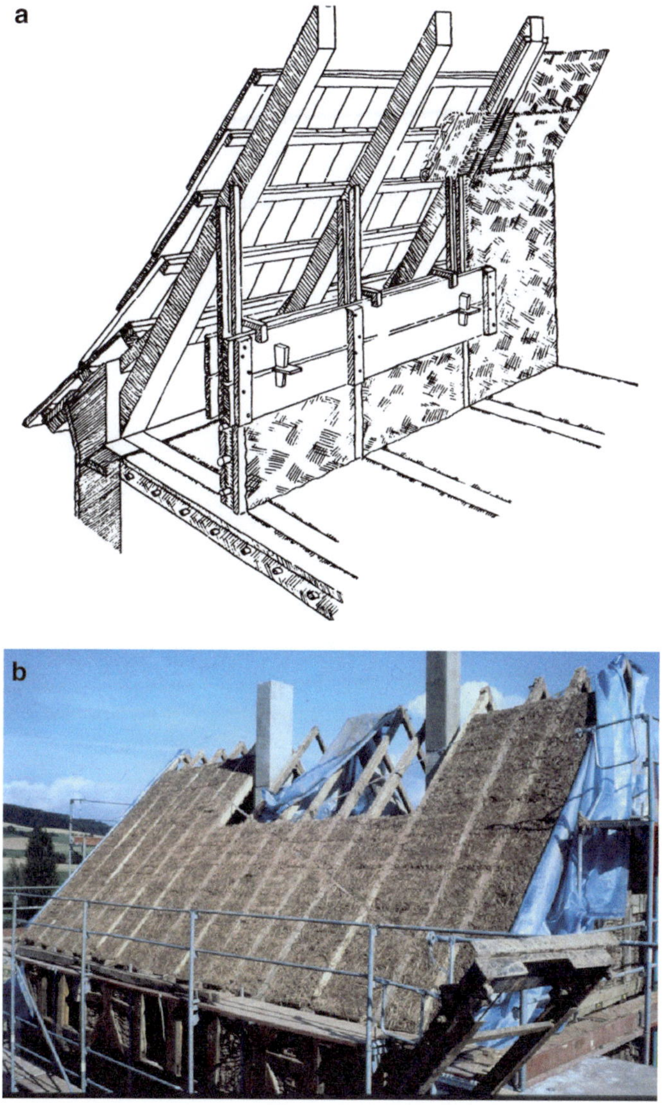

Fig. 4.59 Roof and attic finishing using light clay. (**a**) Roof insulation using inserted clay panels, partition wall made of timber framing with a light straw-clay infill. (**b**) Roof insulation with 5-cm-thick reed panels and a light straw-clay loose fill on top

Sects. 4.3.4.1–4.3.4.3. The roof insulation technique using light straw-clay panels shown in Fig. 4.59a transitions to a non-load-bearing partition wall designed as a vertical timber-frame construction with a light-clay infill. Nowadays, prefabricated light straw-clay panels are available for use as rafter inserts.

Defects in the outer skin of the roof reduced the thermal insulation effect of the moisture-penetrated light-clay layer accordingly. Leaks were very difficult to repair. Cladding the rafter bearing surfaces of the eaves purlin (bond beam, top plate), which could have also been added at a later point in time, was used to create the impression of a jamb wall. However, this technique also reduced the available floor space of the attic.

The low water resistance of earth building materials limits their use as outer skins of pitched roofs which come into direct contact with rainwater. In Germany, a general shortage of ceramic building materials and concrete after World War II, however, led to the development of a number of low-cost building techniques for roof structures using earth building materials.

Light-Clay Flat Roof

The roof structure described in [90] could be constructed using two different techniques in terms of the structural framework: as a wooden purlin roof or as a "prestressed wire concrete Rüdersdorf System" (Stahlsaitenbeton, System Rüdersdorf) with a pitch of 5°–35° and 5°–18°, respectively. Both techniques used prefabricated light-clay panels as inserts between the rafters which were installed using the same method shown in Fig. 4.59. The panels were 32 × approx. 70 cm with a thickness of 12 cm. After installation, the surface of the panels was coated with a 2-cm-thick "hard earth screed" which was modified with 2.5 kg of barley chaff per square meter of outer skin of the roof. After drying, the earth screed layer was waterproofed with oil.

Earth Shingle Roof

The earth shingle roof was meant to be regulated as a building technique by DIN V 18957. Its field of application was limited to "agricultural buildings using the open-structure building technique."

The earth shingles consisted of a layer of long-stemmed straw which was spread out on a 60-cm-wide worktable with lateral boundaries. The straw layer was approx. 7 cm thick and protruded approx. 40 cm beyond the front edge of the table. Parallel to the edge of the table, a so-called shingle stick was clamped into notches which had been cut into the lateral boundaries of the worktable. Starting at the shingle stick, a 1.5-cm-thick layer of earth mortar was spread onto the straw layer over a length of 25 cm. The protruding straw was then folded over the shingle stick and pressed into the mortar bed. Next, a second shingle stick was pushed into two notches which were located opposite each other in the lateral boundaries of the worktable. This was used to keep the straw in position before a second layer of earth mortar could be applied over a length of approx. 70 cm. This layer's function was to sufficiently "glue" the straw together. The integrated shingle stick was used to attach the shingle to the structural frame of the roof.

After drying, the earth shingles (60 × 100 cm) were laid onto the structural frame in three overlapping layers and attached. The minimum thickness of an earth shingle roof was 20 cm. From the exterior, it functioned like a thatched roof. The interior "adhesive layers," however, formed a continuous surface of earth mortar which, upon water penetration, expanded and inhibited further water transport.

4.3.5.2 Flat Roofs

Flat roofs made of earth building materials generally act as stiffening diaphragms within the structure, in the same manner as ceilings. As building elements they serve two functions:

- As *roofs* they form the upper boundary of the building. In addition to meeting structural and thermal requirements, they need to be weather resistant.
- As *ceilings* they form the upper boundaries of the rooms which they cover.

Today, flat roofs made of earth building materials are particularly common in the traditional architecture of hot and dry regions in North Africa, Arabia, the Middle East, and Central Asia but also in India. As roofs, these structures can generally be walked on. As ceilings they have always been the subject of artistic design.

The support structure of earthen flat roofs can be compared to that of ceiling structures (Sect. 4.3.4). The load-bearing frame consists of ceiling joists made of round or squared timbers or thick boards (Fig. 4.60 [74]). For longer spans, the ceiling joists rest on a load-distributing crossbeam with additional support posts.

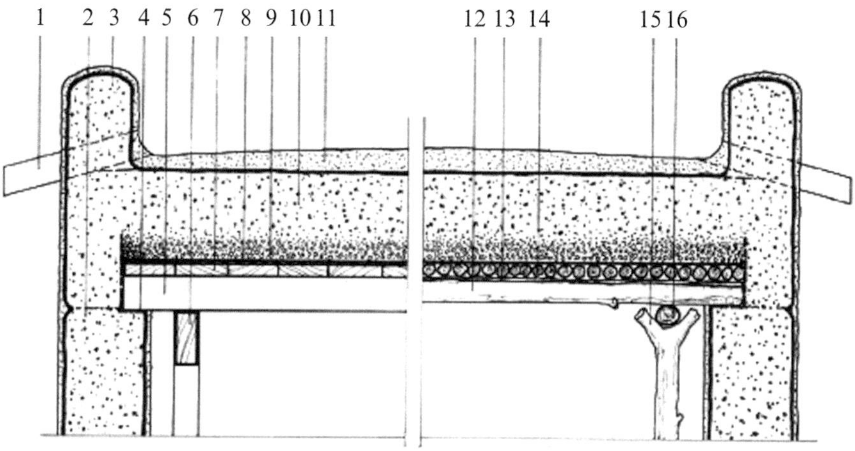

Fig. 4.60 Traditional flat roofs using earth building materials, supporting structure, and basic design [74]. (*1*) Drain pipe made from locally available material, diameter approx. 75 mm, l = 400–500 mm with sufficient wall clearance. (*2*) 250–300-mm-thick rammed earth wall plastered with stabilized soil, finished with limewash. (*3*) 200-mm-thick parapet wall, 300 mm high plastered with stabilized soil, finished with limewash. (*4*) Joist bearing surface with bituminous roofing felt. (*5*) 100 × 50 joists at 750-mm centers. (*6*) 150 × 75 crossbeams supported by 75 × 75 posts at 750-mm centers. (*7*) Ceiling boards. (*8*) Double layer of bituminous roofing felt/burlap mats dipped in bitumen. (*9*) Gravel-soil mix 50 mm. (*10*) Sand–soil mix up to 200 mm. (*11*) Stabilized soil screed 50 mm, 2 % sloped for drainage. (*12–16*) Log version

Fig. 4.61 Traditional flat roofs using earth building materials, *bottom view* of ceiling

A continuous sheathing of boards or sticks (also bamboo or reeds) is laid on top of the ceiling joists (Fig. 4.61). Then, coarse-grained earth material is added and compacted. In order to achieve a better sealing effect, the fine-grained content of the earth material should increase toward the top. As a final "outer skin," a stabilized fine earth screed (fresh cow dung, bitumen, etc.) is applied and smoothed. In newer roof systems, a moisture barrier (plastic membrane) is placed between the sheathing and the puddled earth.

A parapet wall is built on top of the exterior walls to form the lateral boundary of the flat roof. Gaps are created in the parapet walls at the top level of the screed for integrating rainwater spouts with sufficient overhang (Fig. 4.62). The screed layer is sloped toward these drains. Although rainfall is rare, the screed and drains need to be maintained regularly. Flat roofs are prone to leaking and must be repaired frequently (Sect. 5.3.3.3).

In hot and dry regions, traditional earthen flat roofs are optimal in terms of indoor climate: the high mass of the roof stores the day's heat and later releases it into the interior of the building during the cool morning hours.

In seismic areas, however, these heavy roof structures (with their high mass on top of the building) pose a potential risk to the occupants. As with ceilings, the structural elements of the flat roofs generally rest directly on the wall copings without the use of anchoring or a load-distributing bond beam.

In order to reduce mass, stick webbing based on the design principle shown in Fig. 4.61 can be integrated into the earthen flat roof system to create "false domes"

Fig. 4.62 Traditional flat roofs using earth building materials, drainage

(Sect. 4.3.5.3): Starting from a roughly square room layout, the sticks are laid, beginning in the corners, as a hypotenuse to the perpendicular sides of the room. Each new top layer is turned by 90°, respectively, and corbelled toward the center of the room layout until it is completely covered. The puddled earth material on top stabilizes the stick webbing construction.

In windowless interior spaces, a glass window is used as the "keystone" of the webbed dome to allow natural light into the room.

4.3.5.3 Curved Roofs

Similar to flat roofs, dome and vault structures made of earth blocks commonly combine the functions of "roof" and "ceiling," and often even "wall," in a single building element. They can also be designed as intermediate ceilings. *Vaults* are structures which curve along one axis (lower-height versions are called caps), while *domes* curve along two axes (Fig. 4.63a [91]).

Basic Types

In traditional earthen architecture of the Arab-Islamic cultures, construction methods using earth block masonry prevail. The spatial arrangement of the earth blocks used in domes and vaults can be divided into radial and corbelled coursing. For the first technique, the bedding joints are arranged *radially*, or in relation to the center point of a parabola. For the second technique, the earth blocks of each new course are laid in such a manner that they protrude inward a few centimeters (corbelling) until the intended height of the building has been reached and the space is enclosed. The bedding joints are *horizontal* (Fig. 4.63b), [91–93].

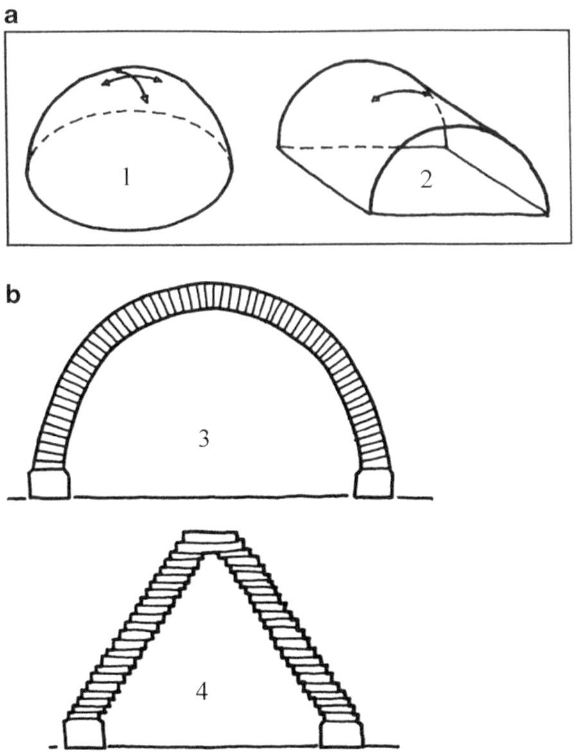

Fig. 4.63 Curved roof construction, principles [91]. (**a**) Curving, (*1*) double curved, dome; (*2*) single curved, vault. (**b**) Layout of earth blocks: (*3*) inclined courses. (*4*) Corbelled courses

Figure 4.64 shows some examples of the great variety of shapes of traditional curved roofs using the two basic types—vaults and domes—as well as combinations of the two techniques.

Figure 4.65 depicts traditional "beehive" houses in North Syria as an example of using a paraboloid to enclose space [94]. The earth blocks are corbelled and have horizontal bedding joints.

There is evidence that the construction techniques used for single- and double-curved structures made of sun-dried earth blocks were already known in present-day Egypt and Sudan more than 4000 years ago (Fig. 1.9) [95]. They are therefore also referred to as "Nubian domes and vaults." In Central Asia as well, the tradition of curved earth block structures goes back thousands of years. Baimatova [96] has analyzed the wealth of forms and shapes used in historical dome and vault construction in this region.

Today, domes and vaults made of earth blocks are particularly common as traditional building techniques in the arid climates of Asia and Africa. In this climate, they create better indoor conditions than box-like buildings:

– Vaults and domes create a greater ceiling height in the middle of the building which allows the warm, rising air to collect under the ceiling and escape to the outside through openings.

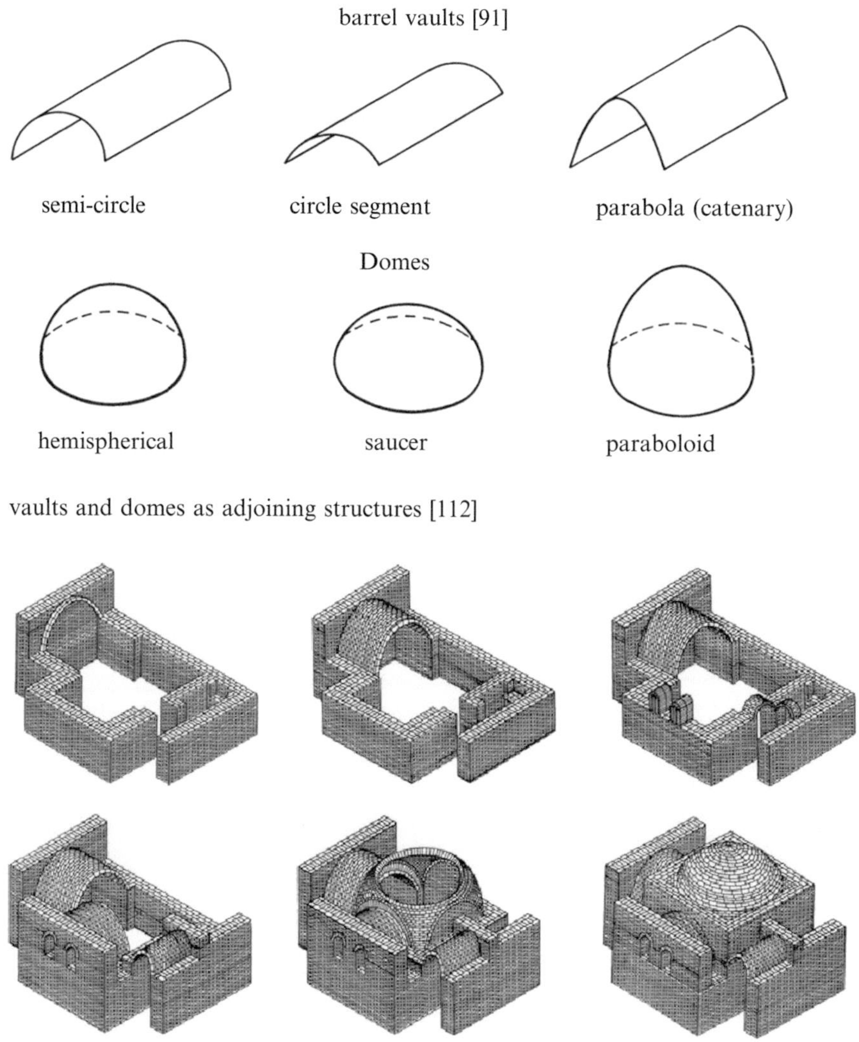

Fig. 4.64 Domes and vaults, example shapes

- Compared to flat roof surfaces, the maximum incidence of sun radiation on single-curved vaults is linear, while only point-like on double-curved domes. This significantly reduces the thermal load applied to the roof compared to flat roof surfaces.

Additionally, vaults and domes can be constructed without wooden formwork as long as skilled craftsmen are available. This is a particular advantage in arid climates where wood is scarce.

basic shape of a paraboloid with corbelling layers of earth blocks and horizontal bedding joints

Fig. 4.65 Earth block domes and vaults, "false" dome, traditional "beehive" houses in Syria [94]

Today, dome and vault masonry construction plays no role in the day-to-day construction work in Germany and Central Europe. In recent years, however, some impressive projects have been executed as "earth block domes" (Minke [75], Fig. 4.10).

Dimensioning

In structural terms, vaults, and domes are load-bearing structures with a curved plane. In contrast to shells, they almost exclusively transfer compressive stresses. Earth blocks can only absorb tensile stresses to a limited extent. Earth block dome and vault structures therefore need to be designed in a manner which guarantees that only compressive stresses occur. For a vault carrying only its dead weight, this is achieved if the cross section is shaped like an "inverted catenary" (Fig. 4.66). A hanging chain exposed only to its own weight creates the ideal shape at which only tensile forces occur. When the line is inverted 180° to form a "standing" curve, a "thrust line" is created which is the ideal cross section of a vault. Only compressive forces occur under the influence of its dead weight.

The splitting of the resultant force *R*, consisting of vertical compressive and horizontal thrust components at the intersecting point of the foundation bottom plane, shows: The steeper the angle used to transfer the resultant force to the foundation, the smaller the horizontal thrust portion and the simpler the foundation can be designed. The resultant force of the vault thrust and the wall load must lie in the core area of the foundation (measured at 1/6 the length or 1/6 the width from the center). If two of the same vaults meet on top of a strip foundation, the horizontal components of the vault thrust caused by the dead weight of the structures cancel each other out.

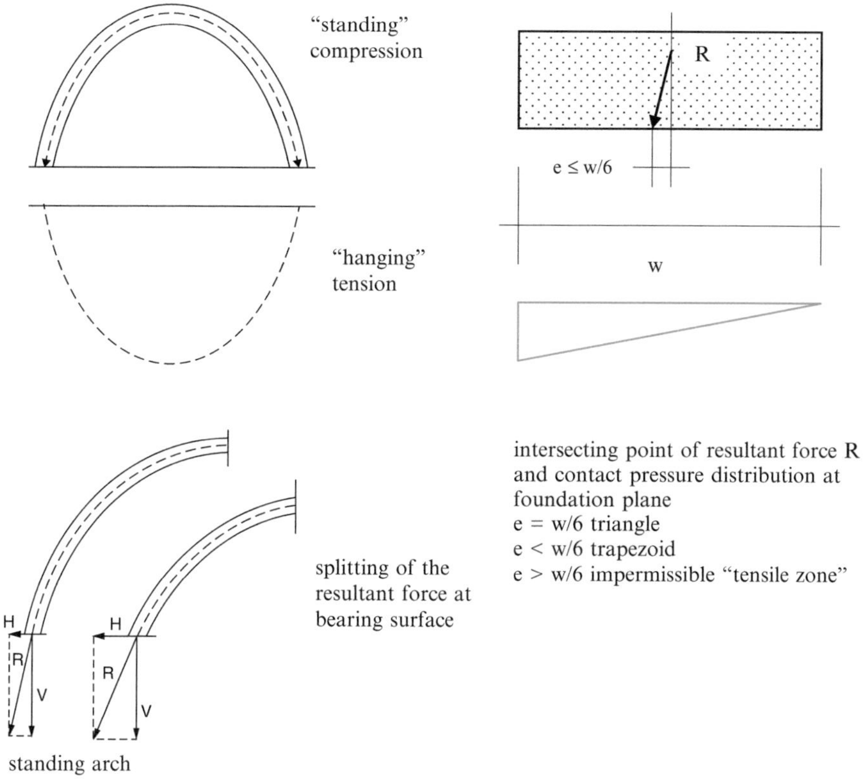

Fig. 4.66 Earth block domes and vaults, flow of forces within the "vault"

Construction Technique

In the "Nubian vault technique," thin earth blocks are laid to form inclined arches without the use of a support system or formwork during construction. The courses are laid at an angle and require one or more stable "end walls" or "arches" to lean on and absorb the horizontal thrust. The inclination of the masonry arches is approx. 20° off the vertical. To prevent sliding and tipping of the blocks in the upper section of the courses, the dimensions of the earth blocks consist of a relatively low height (5–6 cm) and a comparatively large bedding side (15 × 25 cm). Before construction commences, the shape of the arch is transferred to the end wall. The inclination of the full arch courses is established by partial courses which lean against the base forming pendentives (Fig. 4.67) [97].

Earth block domes and vaults can survive major earthquakes. Figure 5.33 shows damage patters on vaults in the Bam Citadel after the December 2003 earthquake.

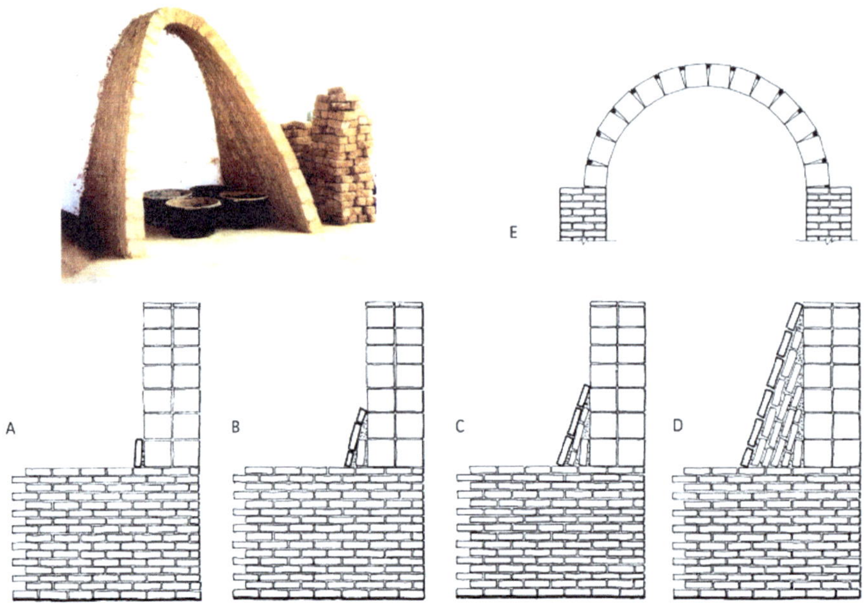

Fig. 4.67 Construction of an earth block vault without the use of formwork [97]

4.3.6 *Plaster*

4.3.6.1 Applications and Requirements

Plasters are level coatings made of plaster mortars according to Sect. 3.5.6.2. They are applied to building elements as thin layers which cover the entire surface and protect the covered building elements from various impacts which differ for interior and exterior surfaces. Plasters obtain their final properties after they have hardened on the building element.

DIN EN 998-1, in connection with DIN 18550, describes the general requirements placed on plaster mortars.

The different applications of plasters in earth building are divided into [8]:

- Earth plaster applied to earth substrates or other substrates
- Plasters using other binders applied to earth substrates

They can be applied as interior or exterior plasters and are subject to a variety of stresses:

Interior plasters (DIN 18550-2) are exposed to stresses as support surfaces for paints and wallpaper and are subject to mechanical stresses if used in public spaces. Today, specific requirements are placed on interior plasters in terms of their aesthetic and physical qualities, particularly in their capacity to quickly absorb increased levels of indoor humidity. In addition, interior plaster can improve the fire performance and sound insulation of building elements.

For the first time, the new draft of DIN 18550-2:2014-10 includes earth plaster mortars as a type of interior plaster.

Exterior plasters (DIN 18550-1) are exposed to mechanical stresses and weather conditions. Particular stress is placed on the building's weather side facing the prevailing wind direction. Here, the exterior plaster must prevent sudden water penetration of the exterior surfaces during driving rains. Water should be absorbed slowly and released again with the help of sufficient ventilation. In addition, exterior plasters can improve thermal insulation and contribute to the visual appearance of the exterior facade of a building. In large urban areas, exterior plasters are also exposed to airborne pollutants.

As is the case with other earth building materials, *earth plasters* (DIN 18947) are not weather resistant. They are therefore mostly suited for interior plaster applications or for exterior surfaces which are not exposed to driving rain. When used indoors, their limited mechanical strength needs to be taken into consideration. Compared to other mineral plaster finishes, earth plasters have a considerably higher water vapor adsorption capacity due to the retained surface activity of their clay minerals. This creates a healthy indoor climate with balanced humidity levels.

Due to their improved weather resistance, *lime plasters or stabilized earth plasters* are generally also suitable for use on surfaces of building elements which are exposed to the weather. Diffusion-open paints can further improve the plasters' weather resistance.

In addition, earth plasters and stabilized earth plasters offer a variety of design options for surfaces in terms of shape (even or organic surfaces) as well as texture, color, wall decorations, and wall paintings. In this context, a wealth of traditions can still be found in various cultures (Fig. 4.68 [98]).

4.3.6.2 Substrates

The substrate serves as the plane which the plaster must be firmly attached to while the building is in use.

Checking the Substrate

The substrate must be sufficiently firm, clean, free of dust, and dry. Loose sections should be removed or stabilized. The substrate needs to be level to ensure that the plaster mortar can be applied in even layers. It should also be free of multilayer paint coats and must not be contaminated with oils and salts (which are most likely to collect in the wall sections above stem walls which experience moisture penetration). For plastering, the temperature of the building element should be at least +5 °C.

Fig. 4.68 Earth plaster, wall decoration in Sirigu, Ghana [98]

Before the plaster work commences, the contractor needs to test the substrate for the abovementioned properties and document the test results. The following methods are suitable for testing the substrate and surrounding areas as defined in DIN 18550:

- Visual assessment of the substrate to check for impurities and foreign matter, loose and brittle sections, efflorescence
- Manual test by passing a hand across the substrate (to check for dust and dirt)
- Scratch test with a hard object (to check for flaking, spalling, and loose sand)
- Wetting (to check absorbency and for remains of formwork release agents)
- Temperature measurement (substrate, ambient temperature)

Preparation of the Substrate

After the substrate has been checked, dust and other loose particles as well as foreign matter adhering to the surface must be removed. Uneven surfaces must be leveled. Depending on the capillarity of the substrate material, it might be necessary to pretreat the substrate. As a general rule, very absorbent substrates need to be

Fig. 4.69 Earth plaster, scratch and puncture patterns to roughen the substrate [51]

pre-wetted before the plaster is applied. Smooth substrates should be roughened and pretreated with a sprayed or brushed-on clay slurry. In historical earth building, the wet substrate was scratched or punctured to facilitate better adhesion of the plaster (Fig. 4.69).

Changes in substrate materials, particularly wooden elements, as well as recesses and protrusions need to be covered or evened out using standard lath materials. Joints between wood and infill in half-timber construction need to be covered with 10 cm of lath overlap.

In contrast to lime or cement plasters, earth plasters do not set chemically. After air-drying and hardening, they only adhere to the substrate mechanically. The quality of the substrate also determines if an earth plaster should be applied in one or multiple layers.

Substrates for Earth Plaster

After appropriate pretreatment, earth plasters can be applied to most standard substrates made of mineral building materials (overview Table 4.11).

Building Elements Made of Mineral Building Materials

Standard masonry made of fired bricks, natural stone, sand–lime blocks, and earth blocks (Fig. 4.38): The substrate needs to be pretreated depending on the roughness of the masonry. Smooth surfaces made of clinker bricks require a spray coat or brushed-on pretreatment of clay slurry. Earth plaster can generally be applied directly to solid or porous bricks as well as sand–lime blocks (after pre-wetting if needed). Cutting channels into the mortar joints to a depth of approx. 1 cm improves the mechanical adhesion of the dried earth plaster. Earth blocks and light-clay blocks containing organic fibers, as well as "thick" clay panels, are generally also good substrates. They can be roughened in order to bring the fiber ends out of the substrate, which further improves the mechanical adhesion of the plaster.

Table 4.11 Earth plaster, overview of substrates

Nr.	Substrate	Single coat	Two coat	Primer coat	Spray or slurry coat	Pre-wetting	Reinforcement	Comments
1	*Masonry*							
1.1	Earth blocks	□				■		Cut channels into mortar joints 1 cm deep
1.2	"Green," unfired bricks		■		□	■		Pay attention to low water stability, cut channels into mortar joints 1 cm deep
2	*Bricks/perforated bricks*	□				■		
2.1	Clinker bricks	□			■			
2.2	Sand–lime blocks	□				■		
3	*Concrete/natural stone*							
3.1	Smooth	□			■			Difficult substrate to work with, pay attention to formwork release agent residues
3.2	Aerated concrete	□		□		■		
3.3	Natural stone	□			■			Difficult substrate to work with
4	*Earth*							
4.1	Rammed earth		◙			■		
4.2	Light-clay/half-timber framing		■			□	□, ■*	*Do not pre-wet wooden elements or reed mat
5	*Plaster*							
5.1	Earth plaster, old		■			■		Rework entire surface
5.2	Lime and gypsum plaster, old	□		■	□			
6	*Panels*							
6.1	Clay panels for dry construction	□				■	■	
6.2	Wood wool panels, reed panels		■		◙		■	Do not pre-wet
6.3	Softwood fiberboard		■				■	Roughen surface, do not pre-wet
6.4	Gypsum board	□		■			■	Difficult substrate to work with, do not pre-wet
6.5	Hard composite wood panels							Unsuitable substrate

■ mandatory, ◙ recommended, □ optional

Extrusion-shaped unfired bricks have a smooth surface and are quite sensitive to moisture. Pre-wetting should therefore be carried out very carefully. Earth plaster can generally be applied to "green" unfired bricks but pretreatment with a spray coat is sometimes necessary. In modern earth building, "green" unfired bricks with very smooth surfaces are often left as exposed masonry for aesthetic reasons.

Concrete is generally a difficult substrate. Smooth concrete surfaces should be pretreated with a spray coat of cement slurry with coarse sand or fine gravel (2–4 mm) or with a commercially available primer. The surface should first be checked for possible release agent residue which needs to be removed. Aerated concrete is highly absorbent and should therefore be pre-wetted or pretreated with a primer.

Rammed earth/light clay: Before an earth plaster can be applied to newly constructed building elements made of rammed earth or light clay, the drying process and the accompanying shrinkage deformations or settling of the wall needs to be completed. In order to improve the mechanical adhesion of the earth plaster on the rammed earth substrate, horizontal bands of roofing tiles or chunks of broken brick can be integrated into the building element during construction (Fig. 4.28). Due to the special surface structure of rammed earth walls and their design possibilities, rammed earth structures have mostly been left unplastered in recent years.

Earth plasters generally adhere better to light-clay substrates due to the (fiber) aggregates contained in them, particularly after additional roughening of the surface. If the light clay is installed with the help of permanent formwork, the earth plaster adheres to the formwork (e.g., a reed mat) and not to the light clay itself.

Existing lime, cement, gypsum, or earth plasters: Substrates made of existing and sufficiently stable lime, gypsum, and cement plasters can be covered with earth plasters. Holes need to be filled in beforehand which often creates different absorption levels of the substrate along the surface. Pretreatment of the surface with a primer is therefore recommended before an additional plaster coat is applied.

Old earth plasters which are sufficiently stable can be covered with new earth plasters, e.g., colored fine-finish plasters. Holes and loose sections need to be repaired first. Old earth plasters were often repaired with gypsum or sometimes covered with multiple layers of paints, including water vapor impermeable oil-based paints in kitchen areas. These materials have to be removed. Frequently, mold has formed between the individual layers of the oil-based paint. Before a new earth plaster is applied, the adhesive strength of the clay minerals in the existing earth plaster can be activated by wetting the surface and working it with a coarse brush.

Drywall Panels

Thin clay panels: The panel joints should be covered with strips of wide-meshed burlap. The strips are placed flat over the joint and attached with the help of a clay slurry or fine-finish earth plaster (Fig. 4.70).

Gypsum boards are generally difficult substrates to work with and are vulnerable to damage. The panel joints need to be filled with compound, sanded according to the manufacturer's instructions and covered with reinforcement mesh.

thin clay panels attached to substructure — panel joint, covered with reinforcement mesh

Fig. 4.70 Substrates for earth plaster, preparation of clay panel joints [51]

Applying reinforcement across the entire surface is recommended. The panel is then treated with a primer in order to protect it from moisture penetration from the wet fine-finish earth plaster. Plaster mortar should only be applied after the primer has dried completely.

Wood Composite Panels

Wood particle boards or oriented strand boards (*OSB*) are not intended for direct plaster application. They are difficult substrates to work with and should only be used with earth plasters in exceptional cases. If earth plaster needs to be applied to them, expert advice should be sought.

Softwood fiber boards are typically plastered with two coats. First, the boards are roughened so that the ends of the wood fibers stick out which creates better adhesion to the base coat. The substrate should not be pre-wetted. Reinforcement mesh should be embedded into the entire surface of the base coat. Next, the fine-finish plaster is applied.

Cement- or lime-bonded wood wool panels make good substrates. The panel joints should be reinforced with standard lath materials. Alternatively, reinforcement mesh can be embedded into the base coat across the entire surface. The substrate should not be pre-wetted.

Lath Materials/Reinforcement Mesh

Reed lath is the most common lath used in earth building. It should not be pre-wetted before plastering and the plaster should be applied in two coats. Embedding reinforcement mesh across the entire surface of the base coat is recommended.

Reed mats can be used to cover building elements made of wood or wooden composites, or as permanent formwork (Sect. 3.2.2.1). The individual reeds are bound together on both ends with a 1-mm-thick binding wire. This keeps the reeds fixed in place, allowing them to be rolled up as a mat (Fig. 4.45).

Fig. 4.71 Earth plaster, embedding reinforcement mesh into the wet base coat [51]

In earth building, *reed panels* serve as both plaster lath and thermal insulation. Standard panels are 20 or 50 mm thick and their dimensions are 2 × 1 m. The panels have binding wires across the reeds spaced approx. every 20 cm which are stapled every 5 cm.

Other lath materials such as *brick-wire mesh* or *metal lath* can also be used to improve the earth plaster's adhesion to the substrate.

Reinforcement mesh is generally more finely woven. It is used for reinforcing thin fine-finish plasters in multilayer plaster systems and is usually embedded into the wet base coat across the entire surface (Fig. 4.71).

Due to their smooth surface, *fiberglass meshes and synthetic fiber meshes* are only suitable as reinforcement for earth plaster mortars to a limited extent. Lime- and cement-bound plaster mortars, on the other hand, set through a chemical reaction and lock together with the mesh. Natural fiber meshes are more suitable for the reinforcement of earth plasters (e.g., hemp, burlap) [99].

4.3.6.3 Plaster Application and Drying

Processing

Industrially produced earth plaster mortars according to Sect. 3.5.6.2 are prepared on the building site following the manufacturer's instructions. The mortars can be prepared manually or using electric or conventional concrete mixers (barrel mixers). The consistency of the plaster mortar should not be too stiff but also not too wet. Information about the optimal mixing ratio can generally be obtained from the manufacturer. When in doubt, test patches of plaster mortar can always be applied first.

Fig. 4.72 Earth plaster with cracks after application [100]

Colored fine-finish earth plasters need to rest after they have been mixed with water in order to allow the binder activity to fully develop. The mix needs to be stirred again directly before application.

Test Patches

Earth plaster mortars produced on site (e.g., using soil from the building site excavation) need to be tested using a test patch before their application. Qualities to check for are sufficient adhesion, possible amendments, and optimal application consistency. The test patch should be applied to an area of at least 1 m^2 with the same substrate characteristics as the actual surface to be plastered. The test earth plaster should be applied at the intended thickness and consistency. After 2–3 days of drying, the building element should be checked for the following signs:

- Cracks in the plaster which become visible after only 1–2 days (Fig. 4.72 [100]) and cracked plaster chunks which curve outward and are separated from the substrate: the mix is too rich and the sand portion needs to be increased.
- Cracks in the plaster, but good adhesion to the substrate: the plaster mix can be used as a base coat; sand needs to be added if the mix is intended for use as a fine-finish plaster.
- The test earth plaster remains crack-free, even after 3 days of drying: the mix can be used as a fine-finish plaster.
- The earth plaster falls off the wall while being applied, or it separates from the substrate in small sections upon impact even though it is not cracked: the mix is too lean; the clay portion needs to be increased.
- The final fine-finish plaster sands off considerably: the mix is too lean; the clay or fiber portion needs to be increased.

Coat Structure

Depending on the intended thickness (but also for design purposes), earth plasters can be applied in a single coat or in multiple coats. A single-coat plaster is typically more cost-effective.

Plasters more than 1 cm thick should consist of a thicker base coat containing fiber material, followed by a thinner fine-finish plaster mortar (Sect. 3.5.6.2). Before the fine-finish plaster is applied, horizontal or diagonal grooves or embedded reinforcement mesh can be added to the wet base coat to improve adhesion. The fine-finish plaster coat is then applied to the hardened base coat which should be slightly pre-wetted.

Plaster Application

In general, earth plasters are applied in the same manner as lime plasters: throw on, trowel level, and float. The required consistency is also the same as well as the tools used for plaster application and finishing.

Earth plasters can be applied manually or with standard plaster machines as long as the fibers contained in the plaster allow for their use.

Plaster reinforcement (which increases the tensile strength of the plaster during drying) should be embedded into the upper half of the plaster coat. It needs to be taut and crease-free and should overlap by a minimum of 10 cm.

Special attention should be paid to the edges of window and door recesses. They can be rounded to improve the light incidence of windows (Fig. 4.73), or corner beading or edging profiles can be embedded into the earth plaster.

Due to their water solubility, earth plasters do not harden inside the hoses of plaster machines when not used for a longer period of time. Earth plasters do not have an irritating effect on the skin which is another advantage of working with them compared to lime or cement plasters.

Drying

As is the case with all earth building elements, earth plasters attain their full usability after completely drying. The drying time mainly depends on the drying conditions in the room but also on the substrate. Cross ventilation quickly removes moisture from the room released from the evaporating mixing water and decreases drying times. An earth plaster with a thickness of 1 cm is generally completely dry after approx. 7 days.

The risk of the formation of shrinkage cracks has already been mentioned in this context. Shrinkage cracks are unproblematic if the base coat is applied in multiple layers and cracked chunks remain firmly adhered to the substrate. Shrinkage cracks in the final top coat plaster are generally not desirable. In contrast to lime and cement mortars, they can be easily repaired due to the water solubility of earth plaster mortars. However, when using colored fine-finish earth plasters, this can lead to differences in color intensity which can only be removed by reworking the entire surface.

Fig. 4.73 Earth plaster, curving of window recesses [51]

Mold Development

Under certain conditions (high humidity with condensation on cooler surfaces), mold can form on finished building elements when the building is in use. This problem is not specific to earth plasters. However, many earth plasters contain organic fibers which can promote the growth of mold. The open-pore quality and capillarity of earth plaster guarantees that any condensation is distributed immediately, thereby preventing the development of favorable conditions for mold growth as long as the building elements are dry. This makes earth plasters a good choice for health-conscious construction.

Fresh earth plasters need to dry quickly. If fast drying cannot be ensured, the plaster needs to be dried artificially (Sect. 3.3.3). The drying process should be monitored and documented. The Technical Information Sheet TM 01 DVL [15] recommends the use of a drying protocol. Mold can form on wet plaster surfaces if the drying process is delayed.

The mold disappears from the affected surfaces after the earth plaster has dried completely.

4.3.6.4 Surface Design and Finishing

Depending on the time of finishing and the plaster method chosen, different tools can be used to smooth, float, or work the earth plaster surface—trowels, putty knives, floats, sponges, and felt. In contrast to lime and cement plasters, the finishing stage of earth plasters can be extended by keeping the surface wet.

surface smoothing using a plastic trowel

and a sponge float

smoothing and polishing of a fine-finish earth plaster surface using soap stones

colored earth plaster finish

Fig. 4.74 Design of earth plaster surfaces, surface finishing, and coloring [51, 108]

The design of the surface—perfectly smooth and flat, organic and manually worked, colored or in the natural earth plaster color—depends on the individual tastes of the client (Fig. 4.74).

Paints

Paints applied to interior earth plaster surfaces should not impede the plaster's positive effects on the indoor climate described above. Transparent and diffusion-open paints should therefore be chosen over thick paints which form layers. Suitable paints for earth plaster surfaces are brush-on earth plasters and earth paints (Sect. 3.5.6.2), as well as paints based on lime, chalk, casein, and marble dust.

In traditional earth building, earth plasters have always been used on exterior surfaces. Professionally applied paints can extend the lifetime of such earth plasters as long as they are regularly maintained.

The following aspects need to be considered when using *lime paints*, which were commonly applied to earth and lime plasters in Germany:

- Slaked lime (calcium hydroxide $Ca(OH)_2$) is the most suitable lime for lime paint. It should soak as long as possible and must be handled carefully due to the risk of skin burns.
- Using suitable additives (casein, glue), the lime putty is prepared to a thin, milky consistency. Color pigments (preferably mineral-based pigments) can be added to the paint.
- Initially, the first layer is applied to the wet plaster in a thin, transparent coat. This process is repeated each day for multiple days.
- Exterior paint coats should be applied in cool and wet, frost-free weather. Depending on their level of exposure to the elements, exterior lime paints should be reapplied in regular intervals of every few years.

In Germany, lime paints are mainly limited to exterior applications where they are used to improve the weather resistance of the plaster. In other cultures, for example, in Africa and India, the artistic treatment of these plaster surfaces is of at least equal importance. Particularly in rural areas, traditional symbolism finds its expression in graphic and sculptural designs. Figure 4.68 shows an example in Ghana [98]. Changing lifestyles, however, have put this art at risk of being lost. Other examples of artistic surface finishing can be found in Japan, where so-called polished plasters have been part of traditional architecture for centuries. The application techniques of these plasters require extensive training and great mastery of the materials used.

Plasters with Scratch and Puncture Patterns

Scratch and puncture patterns represent a special form of the sculptural design of exterior plaster surfaces, particularly in historical half-timber construction. The ornaments were pushed, cut, or scratched into the wet plaster surface by bare hand or with the help of twig brooms, nail boards, or specially cut wooden tools. In the eighteenth and nineteenth centuries, this technique was particularly common in the German regions of Baden-Wuerttemberg, Hesse, Thuringia, and Saxony. Designs included wavy lines, as well as popular floral motifs and depictions of animals [101].

Scratch and puncture patterns were used in both lime and earth plasters. In the latter, they were restricted to interior plasters or exterior plasters on surfaces which were not exposed to the weather.

Other Finishes

Generally speaking, interior earth plasters can also be finished with wallpaper. By doing so, however, the typical design elements of earth plaster (such as surface texture and color) are lost. Wallpaper also impedes the diffusion-open characteristics of the building element surfaces which represent an important aspect of health-conscious building. The use of plastic wallpaper completely eliminates these advantages.

For renovation work in older buildings with earth plasters, existing wallpaper needs to be wetted and then carefully removed from the wall. Holes in the surface should be filled with putty before new wallpaper is attached. The surfaces should then be treated with a primer to ensure that the new wallpaper can be removed more easily during future renovations.

Fiber-reinforced earth plasters can also be used in kitchens and bathrooms but should not be used in areas directly exposed to splashing water. This is an area where tiles are typically used which requires a substrate of standard water-resistant building materials. Earth plasters are not a suitable substrate for tiles.

4.3.6.5 Plaster on Exterior Wall Surfaces

The following aspects need to be considered for the exterior application of *earth plasters*:

- Earth plasters are not weatherproof. When applied to the weather side of the building, they erode over time. In such situations, lime plasters or sidings made of suitable materials (wood, slate, etc.) are preferred.
- A stem wall with a minimum height of 50 cm and sufficient roof overhangs with functioning roof drainage are required.
- The weather resistance of exterior earth plasters can be improved with organic additives (fresh cow dung, lime–casein glue). Additional protection can be achieved by applying multiple layers of a lime–casein paint. The first layer should be applied to the wet fine-finish plaster, and all subsequent layers should be applied after the previous layer has dried (Sect. 4.3.6.4).

For exterior *lime plasters* applied to an earthen substrate on the weather side of the building, the following should be kept in mind:

The lime plaster itself hardens through a chemical process but it does not bond with the earthen surface. It only adheres mechanically to the substrate which means the substrate needs to be roughened before the lime plaster is applied. This can be achieved, for example, by pressing diagonal grooves or puncture holes (diagonally downward) into the wet base coat which the lime plaster can key into.

As in the case of lime paints, calcium hydroxide $Ca(OH)_2$ is suitable for use as a plaster, in the form of quicklime which has been hydrated and slaked for several years and prepared as a lime putty. For health and safety reasons, this type of lime is difficult to find nowadays, but it is superior to the commercially available hydrated lime which comes in powder form and should soak overnight or longer. Trass limes are hydraulic limes which also harden underwater (like cement) in the absence of CO_2. They are harder than non-hydraulic limes and should be kept separate from the earthen substrate using suitable lath material. It is also possible to add clay-rich soil to the lime in order to improve its processing qualities. Such "reduced" lime plaster mixes were popular in the past for economic reasons. Lime plasters with the addition of cement should not be used on earthen substrates because they are too rigid.

4.3.6.6 Earth Plaster Requirements

The requirements placed on earth plaster as a building material are defined in DIN 18947 (Sect. 3.5.6.2). The Lehmbau Regeln [8] and the accompanying Technical Information Sheet TM 01 DVL [15], published by the German Association for Building with Earth e. V., specify additional requirements on the usability of earth plaster as a building element.

Mechanical Strength

Earth plaster must sufficiently and evenly adhere to the substrate or lath. This is also true for adhesion between the individual plaster layers of multilayer plasters. Local separations restricted to small areas do not limit the plaster's usability if the plaster's overall stability as a building element is guaranteed. Sufficient mechanical stability of earth plaster can only be achieved if the strength properties of the plaster and the substrate match.

According to DIN 18947, the mechanical strength of earth plaster mortars is generally described with the help of strength classes which include compressive, flexural, and tensile adhesive strength. Adherence to the minimum values given in Table 3.10 must be verified.

Based on [8, 15], the following parameters must be verified for the application of earth plaster according to Sect. 3.6.2.2:

Dry Compressive Strength

Earth plaster mortar used as a base for fine-finish mortars, paints, and wallpaper must have a dry compressive strength of $\beta_D \geq 1.5$ N/mm^2. This corresponds to class CSII ($\beta_D = 1.5$–5.0 N/mm^2) according to DIN EN 998-1. In [16], a compressive strength of $\beta_D \geq 1.0$ N/mm^2 is required after 28 days of hardening.

In [15], the following minimum requirements are given for additional areas of application (Table 4.12).

Adhesive Strength

A minimum value of 0.05 N/mm^2 is required for the adhesive strength of earth plaster mortars on building element surfaces achieved in a laboratory setting according to DIN EN 1015-12 and DIN 18947. Additional permissible values for adhesive strength, depending on the function of the building element, are defined in [15] (Table 4.13). The values apply to the plaster's adhesion to the substrate as well as between the individual plaster coats in a multilayer plaster system.

Table 4.12 Earth plaster mortars, dry compressive strength based on use

No.	Building material property in terms of intended use of earth plaster mortar	Dry compressive strength [N/mm^2]
1	Ancillary rooms	≥0.5
2	Retroactive surface consolidation in rooms with normal levels of use, e.g., living and work spaces in single-family homes or apartment buildings	≥1.0
3	Base for fine-finish plasters, paints, and wallpaper	≥1.5
4	Exposed, unconsolidated surfaces in rooms with normal levels of use, e.g., living and work spaces in single-family homes or apartment buildings	≥1.5
5	Exposed surfaces in high-traffic areas, e.g., public buildings (apply test patches before plaster application)	≥1.5

Table 4.13 Earth plaster surfaces, adhesive strength based on use

No.	Building material property in terms of intended use of earth plaster mortar	Adhesive strength [N/mm^2][a]	Manual assessment
1	Earth plaster surfaces in ancillary rooms	–	No visible flaking or breaking after repeated application of light pressure to the plaster surface
2	Earth plaster after surface consolidation as a base for fine-finish plasters, paints, and wallpaper	≥0.05	No visible flaking or breaking after repeated application of medium pressure to the plaster surface
3	Exposed earth plaster after surface consolidation in rooms with normal levels of use, e.g., living and work spaces in single-family homes or apartment buildings	≥0.05	No visible flaking or breaking after repeated application of medium pressure to the plaster surface
4	Exposed earth plaster after surface consolidation in high-traffic areas, e.g., public buildings	≥0.08	No visible flaking or breaking after repeated application of medium pressure to the plaster surface

[a]Adhesion to the substrate as well as between and within individual plaster coats

Abrasion Resistance

Abrasion resistance is included in DIN 18947 and TM 01 DVL [15] as an optional/supplementary test method. The permissible abrasion dust quantity can be determined according to DIN 18947 using the strength classes given in Table 3.10. Abrasion resistance can also be determined based on building element properties. The abrasion dust quantities given in Table 4.14 can be supplemented with information about the tensile adhesive strength listed in Table 4.13.

The surface should be assessed after the final steps of compressing, sponging, and brushing off have been completed.

Table 4.14 Earth plaster, permissible abrasion dust quantities

No.	Building element property in terms of intended use of earth plaster mortar	Abrasion dust quantity [g]	Manual assessment
1	Base for fine-finish plasters, paints, and wallpaper after surface consolidation	≤1.5	Color rub-off and medium dusting/sanding permissible
2	Exposed surfaces in rooms with normal levels of use, e.g., living and work spaces in single-family homes or apartment buildings	≤1	Color rub-off, loosening of only individual grains of sand permissible
3	Exposed surfaces in high-traffic areas, e.g., public buildings	≤0.7	Only minimal color rub-off and virtually no dusting/sanding permissible

Visual Appearance

Clients often have only vague ideas about the final look of an earth plaster surface. It is therefore recommended to put the client's wishes in writing (during the planning stages) in terms of visual appearance of the final earth plaster. A plaster test patch should also be applied to the actual surface before work commences.

The visual appearance of a plaster surface is determined by:

- Plaster method
- Irregularities
- Cracking
- Color

Plaster Method

The plaster method defines the type of plaster work and surface treatment which is applied in order to create a specific surface finish (see DIN EN 998-1 and DIN 18550-2). Common surface treatments for earth plasters are (Fig. 4.74 [51]):

- Smooth plaster: finished with a smooth plastic trowel
- Floated plaster: finished with a felt board, sponge or wooden float

Irregularities

The German Association of Stucco Trades (Deutscher Stuckgewerbebund) has defined quality levels for visual requirements placed on plaster surfaces which are to receive further finishing with varnishes, paints, thin-layer finishes, or wallpaper [102]. These requirements can also be applied to earth plasters and have therefore been included in TM 01 DVL [15].

According to the requirements, irregularities in standard plaster applications are acceptable within certain limits. The production of "flawless" plaster surfaces requires an unjustifiable level of effort.

There are four quality levels (Q1–Q4) with the highest quality requirements at level Q4. The different plaster methods of "troweled," "smoothed," and "floated" have to be indicated for quality levels Q2–Q4. The specifications apply to dimensional tolerances in terms of standard and increased requirements placed on the evenness of the plaster. According to these specifications, even the highest level Q4 does not exclude visible dimensional tolerances of the plaster when examined in grazing light. Hairline cracks are permissible at level Q2 because they are filled or covered up by paints or wallpaper. Level Q2 is therefore required for substrates which are to be covered with a fine-finish plaster. Level Q3 applies to substrates which will be finished with clay paints (Sect. 3.5.6.2).

Cracking

The finished plaster surface should be mostly crack-free. It is, however, virtually impossible to produce plaster surfaces which are entirely free of cracks, or their production is very difficult and time consuming. Due to differences in material properties, cracks in earth plasters can generally not be prevented in corners, sections abutting timber construction, or light partition walls. According to DIN 18550-2, plasters with small amounts of hairline cracks are not considered to be faulty. Cracks can be referred to as hairline cracks if the openings are ≤ 0.2 mm.

By determining the linear degree of shrinkage according to Sect. 3.6.2.1, the extent of cracking in earth plasters can be estimated. Cracking should not considerably restrict the usability or the visual qualities of the plaster.

Thick plaster layers dry more unevenly and are therefore more prone to cracking than thin layers. Small shrinkage cracks in base coat layers are permissible as long as they are completely filled by the subsequent fine-finish layer.

Color

Particularly in the case of colored earth plasters, repair work on cracks can lead to differences in color which might require a reworking of the entire surface.

Before using a colored fine-finish earth plaster, the plaster should be applied as a test patch together with the intended base coat. This will give an indication of the final color of both plasters used together.

Moisture Sorption Capacity

Rapid changes in room humidity can be balanced by the enclosing surfaces (to a greater or lesser extent), depending on the building elements' sorption capacity (Sects. 5.1.2.1 and 5.1.2.5). Untreated earth plasters possess a high sorption capacity.

The sorption capacity of earth plaster mortars is determined according to DIN 18947 and Sect. 3.6.3.1.

4.3.7 *Technical Installations*

Technical installations are carried out after the construction of the building shell is finished. Technical installations consist of all structural components required for the supply of energy, heating, water and wastewater, as well as ventilation. With regard to water-sensitive earth building materials, it is particularly important to pay attention to areas which are prone to accidental water damage (water pipes). All contact between the earth building elements and standing or running water needs to be prevented throughout the building's lifetime (Sect. 4.3).

4.3.7.1 Running of Cables and Pipes

Channels and recesses for installations and anchors for fastener can be created by:

- Cutting into the wet earth building material
- Inserting U-profiles, squared timber, or empty conduit
- Inserting anchors for attaching heavy objects

Pipes for water and wastewater must be easily accessible in case of emergency and need to be sufficiently insulated.

Electrical cables can be laid in channels or pulled through conduit. Channels can be created during the construction of the earth building elements (Sect. 4.3.3.1). In this context, it is important to adhere to permissible depths. It is also possible to embed electrical cables directly into the plaster as long as the plaster thickness is sufficient.

4.3.7.2 Wall Fasteners

Based on the strength of the earth building material, standard fasteners can be used to attach lighter-weight objects. These fasteners include all types of anchors, wooden slats, nails, screws, and ties (Sect. 4.3.3.1).

Edge protectors (aluminum corner rails, wood molding) which are attached to window and door reveals at a later point in time can be glued to the earth plaster. A primer needs to be applied to those areas which will be in direct contact with the glue.

4.3.7.3 Heating Systems

Heating systems or fireplaces have been an integral component of technical installations in the architectural traditions of all cultures. Depending on the climatic conditions, these systems were located either indoors or outside. Fireplaces and stoves inside buildings required suitable systems for releasing smoke to the outside. Most fires were fueled by wood.

Because earth building materials were considered fireproof, the construction of fireplaces and stoves frequently included the use of earth blocks with earth masonry mortar and earth plaster mortar, sometimes in combination with natural stone and bricks (and later firebricks).

Traditional Fireplaces and Stoves

The energy produced by traditional fireplaces and stoves had to fulfill two main tasks: to facilitate the preparation of food and to provide heat in the winter months.

Arched "Schwibbogen" Stove

At the end of the sixteenth century, open fireplaces gave way to masonry hearths in the traditional "hall houses" of Northern Germany. These hearths were constructed along the wall of the entrance hall. From here, the adjacent living quarters were heated by a stove ("wall chamber") (Fig. 4.75 [103]). The hearth itself consisted of two masonry side walls which supported a masonry arch. In German, this arch was called a "Schwibbogen." Usually, the masonry work was carried out using earth building materials. Smoke collected under the arch and was released through a smoke hole in the kitchen ceiling. Here, the smoke could be used to preserve (smoke) meat, before escaping to the outside via small roof openings ("owl holes") in the gables.

Bell-Shaped Stove

The eighteenth/nineteenth century brought a transition from the arched to the bell-shaped stove which used a separate chimney to release the smoke. In place of the massive arch, a "bell hood" with a rectangular base now rose above the fireplace. The bell was constructed of earth blocks and rested on a wooden beam which was wrapped with straw clay. Above the kitchen ceiling, the flue of the bell led into an earth masonry chimney covered with earth plaster. In the thatched roof houses common in Northern Germany, the chimney always followed the ridge line due to the risk of fires. Often, the stove was not located directly in the middle of the house, and the chimney was therefore "extended" through the attic to reach the ridge of the roof (Fig. 4.76 [104]).

The bell hood's role diminished with the introduction of improved "closed" cooking stoves. Here, the smoke was released directly into the chimney or via metal stove flue pipes leading to the chimney. The original bell hoods were closed up and later torn down. In northwestern Mecklenburg, bell hoods could still be found into the 1950s. Exhaust hoods installed in modern kitchens have the same function and are reminiscent of these almost forgotten structures.

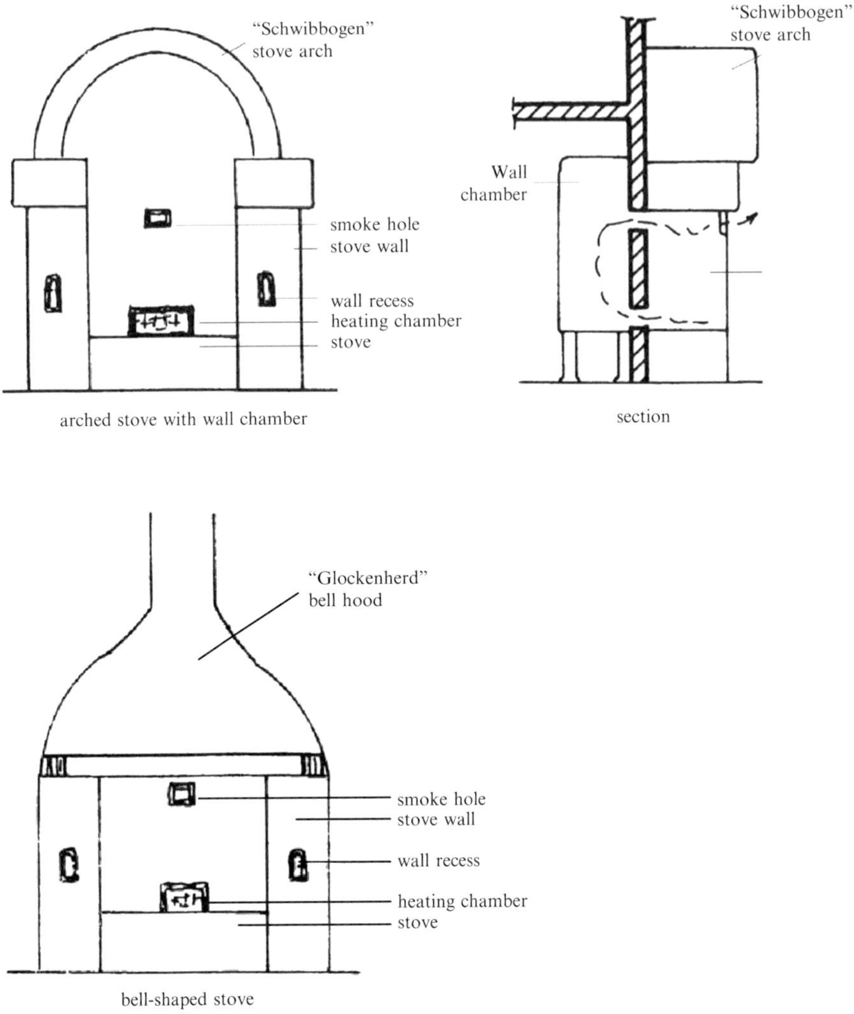

Fig. 4.75 Traditional fireplaces, hearths [103]

Earth Ovens

Traditional fireplaces have always included ovens for baking, often made of earthen materials. Because these ovens posed a potential fire hazard, they were often built outdoors as a separate roofed structure and were referred to as "bakehouses." Baking was a communal activity in many regions, and the bakehouse was available for general use at specified times.

The oven itself consisted of a brick masonry base which supported the oven cavity. The cavity was shaped as a dome or vault built of firebrick and earth masonry mortar. It included an opening with a door for firing the oven with dry brushwood. The dome or vault was covered by a 10-cm-thick layer of straw clay (for thermal insulation) which was finished with a lean earth plaster after drying.

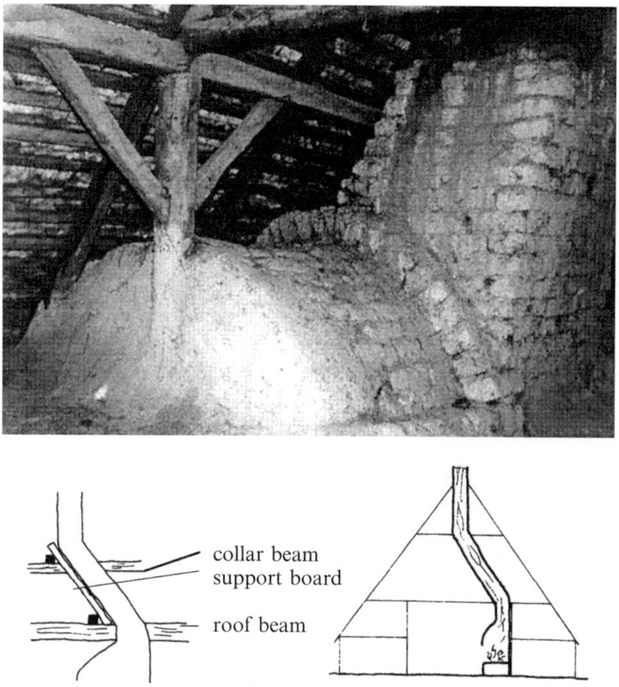

attic with upper part of "Schwibbogen" arch and extended earth block chimney [104]

extended chimney, detail

complete system with stove [103]

Fig. 4.76 Traditional fireplaces, extended chimneys

For baking, the entire oven cavity was first filled with brushwood. After the wood had completely burned down (after approx. 1.5 h), the ashes were removed. The oven had reached the correct temperature when all soot was burned off the interior surface of the vault cavity. Now the oven could be loaded with the baked goods.

In many parts of the world, traditional earth ovens are still in use today, for example, in the Middle East and Central Asia as well as in rural areas of Latin America where the oven is called "horno de barro" (oven made of earth blocks).

Hypocaust Heating System

A hypocaust heating system uses hot air flowing through a solid body but has a lower surface temperature than a conventional heater or radiator. Floors and walls but also solid benches can serve as heat transfer mediums. This type of hot air heating system dates back to ancient Rome.

The same principle is used in the traditional Korean heating system called "Ondol" found in residential buildings (see Fig. 4.2, traditional Hanok house): The floors are made of slabs of stone which rest on brick pillars and wedges (Fig. 4.77a [1]). The cavity between the ground and the underside of the floor is enclosed by the

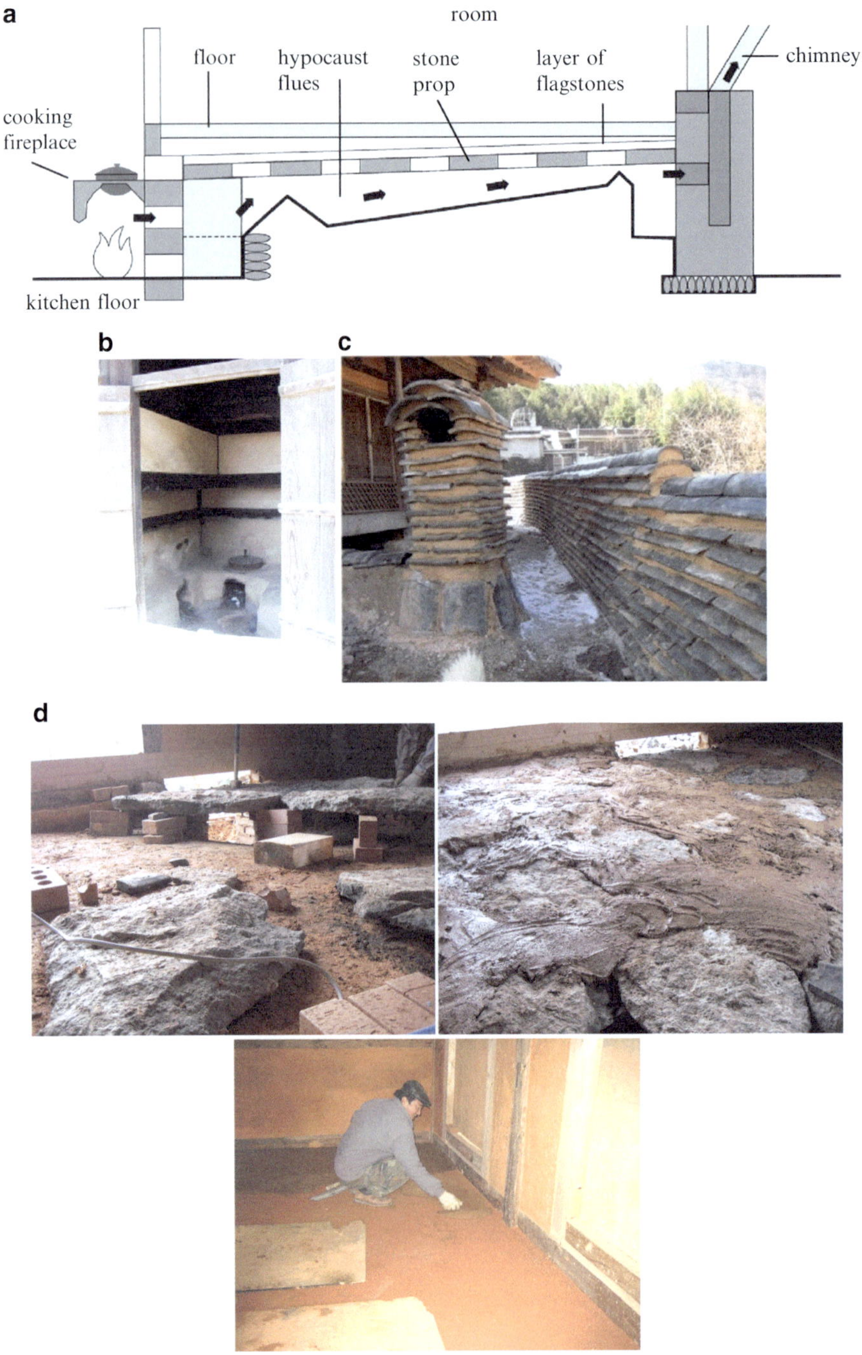

Fig. 4.77 Fireplaces, traditional Korean Ondol fireplace (cooking area combined with floor heating system) [1]. (**a**) Cross section of system, floor with basement and hypocaust heating system. (**b**) Cooking area outside the house. (**c**) Separate chimney. (**d**) Installation of "supported" flagstone floor followed by earth screed layer

exterior walls of the house and forms a hollow space. The joints between the flagstone are filled with quarry stones and sealed with earth mortar. This structure supports an earthen floor which is applied in multiple layers. The natural stone slabs considerably increase the thermal mass of the floor (Fig. 4.77d).

In winter, the hot smoke and gases pass through the space under the floor toward a flue pipe which ends in a freestanding chimney next to the house. The chimney is made of earth blocks and ceramic roof tiles for erosion protection (Fig. 4.77c). Two openings at the top of the chimney allow the smoke and gases to escape. The required heat is produced in a stove which is also used for cooking. The stove is made of earth block masonry and is located in an alcove outside the house (Fig. 4.77b).

Chimneys

Due to the risk of fires, the construction of chimneys for residential buildings was regulated in German building codes as early as the nineteenth century. Before 1850, flue pipes were designed to be accessible ("climbable") from the interior of the building [53]. Later, their dimensions were decreased in order to improve their draw characteristics. The higher temperatures of the smoke and gases, however, put more strain on the building materials of the chimney.

The function of the chimney is based on the stack effect: it creates air buoyancy (draw) in the smoke and gases which are warmer than the surrounding air. A chimney "draws" well when its height and the diameter of its flue pipe are optimally designed for the specific amount and temperature of the smoke and gases it is transporting.

During the "lighting" phase, the temperature of the smoke and gases is comparatively low. If it falls below the dew point, condensation of the accompanying water vapor can form along the interior of the flue pipe. With the condensation comes the settling of tar and sulfur (soot) which, in the worst case, permeate the chimney and become visible on its exterior as brown spots which smell of sulfur. This process is called "sooting" and typically weakens the structural strength of the chimney.

Modern flues are ceramic or stainless steel pipes made of waterproof, temperature-resistant materials. The use of earth blocks is currently not permitted. Tests on building elements need to determine if earth blocks can again be used (as they were before 1850) in the construction of chimney components which are in direct contact with smoke and gases. DIN 18945 could be used as an instrument to define the characteristic properties of earth blocks for use as a product for this specific application.

Modern Wall Heating

Wall heating is used to heat building elements through a system of heating coils integrated into the wall. They produce radiant heat which spreads evenly across the entire surface of the building element and, compared to conventional central heating systems, creates a better layering of the air temperature in the room in terms of thermal physiology. Radiant heat is also perceived as more comfortable than convection heat produced by central heating systems (Sect. 5.1.1.1) which heat the air and circulate it in the room. This makes it possible to lower the supply temperature of the water by up to 2 K without a change in the physiological perception of the heat [52].

In order to facilitate effective radiation of the heat, the "heating" walls should not be blocked by furniture. In addition, wall heating systems should be installed in interior walls. Exterior walls would require additional insulation in order to avoid losing the produced heat to the outside through thermal conduction.

There are two different heating systems which are currently available for earth building materials [52]:

- Wall heating using water-filled heating coils
- Wall heating in the form of hot air systems (hypocaust)

Hot Water Heating Systems

Heating coils made of suitable piping material are embedded in the earth base coat across the entire wall surface and covered with a second plaster coat. Once in use, water (as the heat transport medium) passes through the pipe coils. Heat is released into the earth plaster which, in turn, radiates it into the room (Fig. 4.78). Clay panels with integrated coils have recently become available from different manufactures.

Water-filled pipe coils in combination with earth building materials always pose a risk of water damage. A wall scanner should therefore be used before fasteners are attached to the wall.

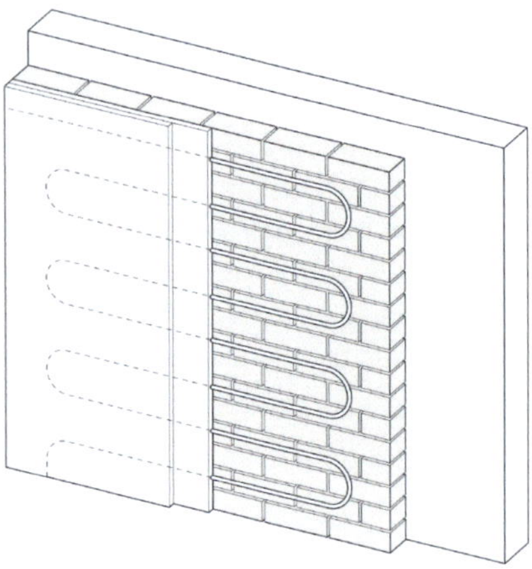

coils embedded in earth plaster [52]

clay panel with integrated pipe coils (WEM)

Fig. 4.78 Technical installations, wall heating in the form of water-filled heating coils

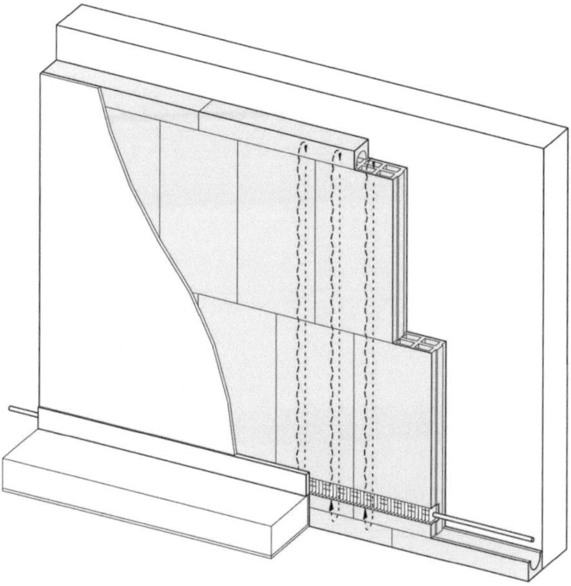

Fig. 4.79 Technical installations, wall heating in the form of a hot air system with hollow-core clay panels (hypocaust) [52]

Baseboard Heating

Baseboard heating uses air instead of water as the heat transport medium. This largely eliminates the risk of water damage.

The wall heating system consists of hollow-core clay panels which are placed as wall linings in front of the "heating wall" and covered with a U-shaped cap. A conventional baseboard heating system integrated into the wall lining panel heats up the air which, in turn, heats the wall by circulating through the hollow chambers (Fig. 4.79). The heated wall radiates heat into the room over its entire surface.

The principle of the hypocaust heating system was applied in the buildings of a printing plant in Austria [54]. The load-bearing walls consist of prefabricated rammed earth elements (Sect. 4.3.3.1) with integrated recesses for a hypocaust heating system (Figs. 3.18 and 4.26).

Earth Masonry Heaters

Earth masonry heaters are becoming increasingly popular. They consist of standard commercially available wood stoves which are wrapped with a thermal-mass layer of earth which can also include recycled earth blocks. The thermal-mass layer can be customized, for example, by integrating multiple functions (stove bench, cook stove, stove pipes, the heating of several rooms).

An earth masonry heater produces comfortable radiant heat which is retained over many hours due to its high thermal mass (Fig. 4.80). The construction of an earth masonry heater should be planned and carried out by an experienced specialist.

Fig. 4.80 Technical installations, earth masonry heater in the form of a standard wood-burning stove wrapped with a coat of high-mass earth [87]

By using of wood as a renewable firing resource, earth masonry heaters are considered to be CO_2-neutral. They contribute to the reduction of fossil fuel consumption in the production of heat for residential buildings and to sustainable building in general.

4.3.7.4 Traditional Cooling Systems

Ventilation Systems

So-called wind catchers are building elements used for the natural ventilation of buildings in the hot and dry climates of the Arabic-Islamic culture. They are an integral part of the region's traditional architecture. At unbearably high outside temperatures, fully functional wind catchers create a comfortable indoor climate using an ingenious system of fresh air supply and used air removal.

A wind catcher (called "malkaf" in Egypt and "badgir" in Iran) can be compared to a covered chimney with vertical flues for the fresh air supply and the outlet of used air. The direction of the air movement in the flues changes depending on the time of day and the direction of the wind. The diagram in Fig. 4.81b [105] shows the operating principle of a wind tower: During the day, the south side of the wind tower

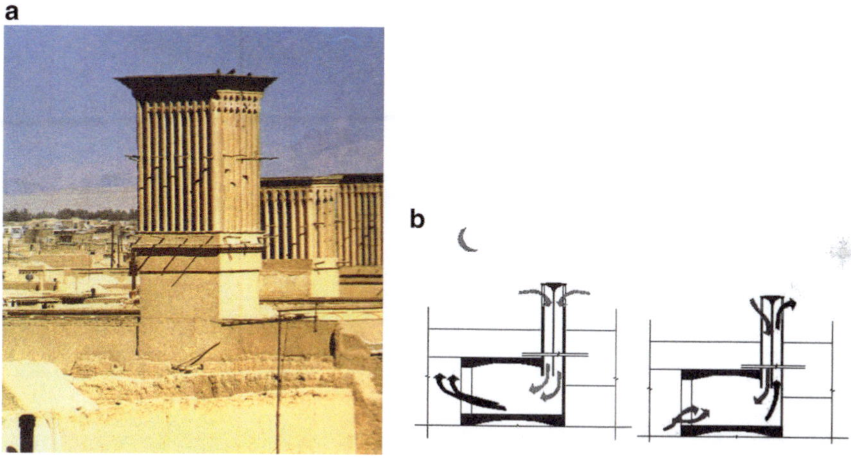

Fig. 4.81 Wind towers used for the natural climate control of interior spaces in hot and dry climates [105]. (**a**) Rooftops with wind towers in Yazd, Iran. (**b**) Operating principle of a wind tower showing changes between day and night

heats up. The warm air rises and extracts the air from the interior. This creates low pressure in the interior which, in turn, pulls fresh air from the cool interior courtyard into the interior space. At night, the cool air sinks through the wind tower into the interior pushing the day's warm air out into the interior courtyard where it rises.

Traditionally, wind towers were made of earth building materials, often in connection with timbers which were used to create the openings. Figure 4.81a [105] shows rooftops with wind towers in the Iranian oasis town of Yazd. Today, wind towers are the focus of many historical restoration projects. The perfect technical design of these often century-old structures continues to amaze. The planners who, so long ago, designed and carried out the construction of these systems using relatively simple methods deserve our respect.

Ice Houses

Before electrically operated refrigeration systems were invented, perishables were kept fresh during the summer months in special structures, so-called ice cellars or ice houses. Shallow ponds, built in the vicinity of these structures, were protected from the sunlight by walls. In winter, water in the ponds froze, and the ice was "harvested," cut into blocks, and stored in the ice house. This building consisted of a lined ice chamber and an aboveground structure made of a cold-insulating building material.

In many oasis towns in the desert regions of Iran, for example, in Kerman, Naen, and Bam, the technique of ice making and storing was highly developed. The ice was stored in circular ice chambers (yakhchal) below the ground which were up to

cross-section of an ice house built of earth block masonry with an underground ice chamber, Kerman, Iran

Fig. 4.82 Ice houses in Iran [106]

8 m deep and more than 10 m in diameter [97]. The aboveground dome structure was made of earth blocks, in the form of inward-tapering wall rings placed on top of each other. Figure 4.82 shows the ice house of the city of Kerman [106]. The cross section shows that the dome structure had a thickness of up to 2 m at its base. This tremendous thermal mass was able to prevent the ice from melting, even during extremely hot summers.

References

1. http://dev.lehmbau-atlas.de
2. Bochow, K.H., Stein, L.: Hadramaut—Geschichte und Gegenwart einer südarabischen Landschaft. VEB Brockhaus Verlag, Leipzig (1986)
3. Grigutsch, A., Keller, B., Fährmann, H.: Erfahrungen beim zweigeschossigen Lehmbau in Gotha. Bauplanung/Bautechnik. **7**(6), 128 (1952)
4. Heringer, A., Roswag, E., Ziegert, C.: Traditionelle Lehmbautechniken in der modernen Architektur: "School handmade in Bangladesh". In: LEHM 2008 Koblenz, Proceedings of the 5th International Conference on Earth Building, pp. 246–249. Dachverband Lehm e.V., Weimar (2008)
5. Harris Huges, E.: Holy Adobe. Hughes, El Paso (1982)
6. Aleyadah, B.: Prince Ahmad Bin Salman Mosque, Riyadh, Saudi Arabia. In: LEHM 2012 Weimar, Beiträge zur 6. Int. Fachtagung für Lehmbau, pp. 22–27. Dachverband Lehm e.V., Weimar (2012)
7. Fischer, F., Mönnig, H.-U., Mücke, F., Schroeder, H., Wagner, B: Ein Jugendklub aus Lehm—Baureport Herbsleben. Weimar, Wiss. Z. Hochsch. Archit. Bauwes.—A. **35**(3/4), 162–168 (1989)
8. Dachverband Lehm e.V. (Hrsg.): Lehmbau Regeln—Begriffe, Baustoffe, Bauteile, 3., überarbeitete Aufl. Vieweg+Teubner/GWV Fachverlage, Wiesbaden (2009)
9. Europäisches Parlament; Rat Der Europäischen Union: Verordnung Nr. 305/2011 vom 9.März 2011 zur Festlegung harmonisierter Bedingungen für die Vermarktung von Bauprodukten und zur Aufhebung der Richtlinie 89/106/EWG des Rates, L 88/5 v. Amtsblatt der Europäischen Union, Brüssel (2011)

10. Schroeder, H., Volhard, F., Röhlen, U., Ziegert, C.: Die Lehmbau Regeln 2008—10 Jahre Erfahrungen mit der praktischen Anwendung. In: LEHM 2008 Koblenz, Beiträge zur 5. Int. Fachtagung für Lehmbau, pp. 12–21. Dachverband Lehm e.V., Weimar (2008)
11. Dachverband Lehm e.V. (Hrsg.): Lehmsteine—Begriffe, Baustoffe, Anforderungen, Prüfverfahren. Technische Merkblätter Lehmbau, TM 02. Dachverband Lehm e.V., Weimar (2011)
12. Dachverband Lehm e.V. (Hrsg.): Lehm-Mauermörtel—Begriffe, Baustoffe, Anforderungen, Prüfverfahren. Technische Merkblätter Lehmbau, TM 03. Dachverband Lehm e.V., Weimar (2011)
13. Dachverband Lehm e.V. (Hrsg.): Lehm-Putzmörtel—Begriffe, Baustoffe, Anforderungen, Prüfverfahren. Technische Merkblätter Lehmbau, TM 04. Dachverband Lehm e.V., Weimar (2011)
14. Dachverband Lehm e.V. (Hrsg.): Qualitätsüberwachung von Baulehm als Ausgangsstoff für industriell hergestellte Lehmbaustoffe—Richtlinie. Technische Merkblätter Lehmbau, TM 05. Dachverband Lehm e.V., Weimar (2011)
15. Dachverband Lehm e.V. (Hrsg.): Anforderungen an Lehmputz als Bauteil. Technische Merkblätter Lehmbau, TM 01. Dachverband Lehm e.V., Weimar (2014)
16. Natureplus e.V. (Hrsg.): Richtlinien zur Vergabe des Qualitätszeichens "natureplus". RL 0607 Lehmanstriche und Lehmdünnlagenbeschichtungen (September 2010); RL 0803 Lehmputzmörtel (September 2010); RL 0804 Stabilisierte Lehmputzmörtel (proposed); RL 1006 Lehmbauplatten (September 2010); RL 1101 Lehmsteine (proposed); RL 0000 Basiskriterien (May 2011); Neckargemünd (2010)
17. Schroeder, H.: Modern earth building codes, standards and normative development. In: Hall, M.R., Lindsay, R., Krayenhoff, M. (eds.) Modern Earth Buildings—Materials, Engineering, Construction and Applications. Woodhead Publishing Series in Energy: Nr 33, pp. 72–109. Woodhead Publishing, Oxford (2012)
18. Cid, J., Mazarrón, F.R., Cañas, I.: Las normativas de construcción con tierra en el mundo. Informes de la Construcción **63**(523), 159–169 (2011)
19. Italia. Regione Piemonte L.R. 2/06, Norme per la valorizziazione delle costruzioni in terra cruda. B.U.R. Piemonte, n. 13 (2006)
20. Instituto Nacional de Normalización INN: Estructuras—Intervencióde construcciones patrimoniales de tierra cruda—Requisitos del proyecto structural, NCh 3332.c2012, Santiago de Chile (2012)
21. Royaume du Maroc: Règlement parasismique des constructions en terre. RPCTerre, Ministère de l'interieur, Ministère de l'equipment et du transport, Ministère de l'habitat, de l'urbanisme et de la politique de la ville, Rabat (2001)
22. AFNOR: XP P13-901: Compressed Earth Blocks for Walls and Partitions: Definitions—Specifications—Test Methods—Delivery Acceptance Conditions. AFNOR, St. Denis la Plaine CEDEX (2001)
23. MOPT: Bases para el Diseño y Construcción con Tapial. Centro de Publicationes, Secretaría General Técnica, Ministerio de Obras Públicas y Transportes, Madrid (1992)
24. Jiménez Delgado, M.C., Cañas Guerrero, I.: The selection of soils for unstabilized earth building: a normative review. Construction and Building Materials **21**, 237–251 (2007)
25. Asociación Española de Normalisación y Certificación AENOR: Bloques de tierra comprimada para muros y tabiques. Definiciones, especificaciones y métodos de ensayo. UNE 41410, Madrid (2008)
26. Achenza, M.: A national law for earthen architecture in Italy. Difficulties and expectances. In: "Living in Earthen Cities—Kerpic '05", Proceedings of the 1st International Conference, pp. 191–194, Istanbul Technical University ITU, Istanbul (2005)
27. Schweizer Ingenieur-und Architekten-Verein SIA (Hrsg.): SIA Dokumentationen D 077, Bauen mit Lehm (1991); Schaerer, A., Huber, A., Kleespies, T.: D 0111, Regeln zum Bauen mit Lehm (1994); Hugi, H., Huber, A., Kleespies, T.: D 0112, Lehmbauatlas (1994); Hugi, H., Huber, A., Kleespies, T.: Zürich: SIA 1991 u. (1994)

28. Turkish Standard Institution TSE (1995–1997): Cement Treated Adobe Bricks. TS 537 (1985); Adobe Blocks and Production Methods TM 2514 (1997); Adobe Buildings and Construction Methods TM 2515 (1985); Ankara (1997)
29. State Building Committee of the Republic of Kyrgyzstan/Gosstroi of Kyrgyzstan: Возведение малоэтжных зданий и сооружений из грунтоцементобетона РСН-2-87 (Building of low-storied houses with stabilized rammed earth), Republic Building Norms RBN-2-87, Frunse (Bischkek) Republic of Kyrgyzstan (1988)
30. Center for the Development of Industry ACP-EU/CRATerre-BASIN: Compressed Earth Blocks, Series Technologies. Nr. 5 Production equipment (1996); Nr. 11: Standards (1998); Nr. 16 Testing procedures (1998). Brussels (1996–1998)
31. Institut National de la Normalisation et de la Propriété d'Industrielle INNOPRI: Blocs de terre comprimeé—Spécifications techniques. NT 21.33 (1996); Blocs de terre comprimeé—Définition, classification et désignation. NT 21.35 (1996), Tunis (1998)
32. Standards Association of Zimbabwe: Rammed Earth Structures. Zimbabwe Standard Code of Practice SAZS 724, Harare (2001)
33. Shittu, T.A.: Lehmbau Normen und Regeln—ein Überblick über die nigerianischen Baugesetze. In: LEHM 2008 Koblenz, Beiträge zur 5. Int. Fachtagung für Lehmbau, pp. 40–47. Dachverband Lehm e.V., Weimar (2008)
34. Federal Republic of Nigeria: National Building Code, 1st edn. LexisNexis Butterworths, Capetown (2006)
35. Kenya Bureau of Standards KEBS: Specifications for Stabilised Soil Blocks. KS02-1070:1993 (1999), Nairobi (1999)
36. Bureau of Indian Standards BIS (ed.): Improving Earthquake Resistance of Earthen Buildings—Guidelines. Indian Standard IS 13827, New Delhi (1993)
37. Bureau of Indian Standards: Indian Standard 2110-1980: Code of Practice for In Situ Construction of Walls in Buildings with Soil-Cement (1991 1st rev., 1998 reaffirmed). New Dehli (1981)
38. Bureau of Indian Standards BIS (ed.): Stabilized Soil Blocks Used in General Building Construction—Specification. Indian Standards IS 1725, (Second revision). New Delhi (2013)
39. Sri Lanka Standard Institution: Specification for Compressed Stabilized Earth Blocks. Sri Lanka Standard SLS 1382; Part 1: Requirements; Part 2: Test methods; Part 3: Guidelines on production, design and construction. Colombo (2009)
40. Regulation & Licensing Dept., Construction Industries Div., General Constr. Bureau: 2006 New Mexico Earthen Building Materials Code. CID-GCB-NMBC-14.7.4. Santa Fe (2006)
41. ASTM International, Standard Guide for Design of Earthen Wall Building Systems: ASTM E2392/E2392—10, West Conshohocken (2010)
42. National Building Standards of Peru (ed.): Adobe. Technical Building Standard NTE E. 080 (engl.). Lima (2000)
43. Associação Brasileira de Normas Técnicas ABNT, Rio de Janeiro (1984–1996); NBR 8491 EB1481, Tijolo maciço de solo-cimento (1984); NBR 8492 MB1960, Tijolo maciço de solo-cimento—Determinação da resistência à compressão e de absorção d'água (1984); NBR 10832 NB1221, Fabricação de tijolo maciço de solo-cimento com a utilização de prensa manual (1989); NBR 10833 NB1222, Fabricação de tijolo maciço e bloco vazado de solo-cimento com a utilização de prensa hidráulica (1989); NBR 10834 EB 1969, Bloco vazado de solo-cimento sem função strutural (1994); NBR 10835 PB 1391, Bloco vazado de solo-cimento sem função estrutural (1994); NBR 10836 MB3072, Bloco vazado de solo-cimento sem função estrutural—Determinação da resistência à compressão e de absorção d'água (1994); NBR 12023 MB 3359, Solo-cimento—Ensaio de compactação (1992); NBR 12024 MB3360, Solo-cimento—Moldagem e cura de corpos-de-prova cilíndricos (1992); NBR 12025 MB3361, Solo-cimento—Ensaio de compressção simples de corpos-de-prova cilíndricos (1990); NBR 13554, Solo-cimento—Ensaio de duribilidade por moldagem e secagem (1996); NBR 13555, Solo-cimento—Determinação da absorção d'água (1996); NBR 13553, Materiais para emprego em parede monolítica de solo-cimento sem função estrutural (1996)

44. Instituto Colombiano de Normas Técnicas y Certificación ICONTEC: Bloques de suelo cemento para muros y divisiones. Definiciones. Especificaciones. Métodos de ensayo. Condiciones de entrega. NTC 5324. Bogotá (2004)
45. Standards New Zealand: NZS 4297: 1998 Engineering Design of Earth Buildings; NZS 4298: 1998 Materials and Workmanship for Earth Buildings; NZS 4299: 1998 Earth Building Not Requiring Specific Design. Standards New Zealand, Private Bag 2439, Wellington (1998)
46. CSIRO Australia, Division of Building, Construction and Engineering, Bull. 5: Earth-Wall Construction, 4th edn. North Ryde (1995)
47. Earth Building Association of Australia EBAA (ed.): Building with Earth Bricks & Rammed Earth in Australia. EBAA, Wangaratta (2004)
48. Walker, P., Standards Australia: The Australian Earth Building Handbook. Standards Australia International, Sydney (2002)
49. Dahlhaus, U., Kortlepel, U., et al.: Lehmbau 2004—aktuelles Planungshandbuch für den Lehmbau (Ergänzungen 2001 u. 2004). Manudom Verlag, Aachen (1997)
50. Schmitz, H., Krings, E., Dahlhaus, U., Meisel, U.: Baukosten 2012/13; Band 1: Instandsetzung, Sanierung, Modernisierung, Umnutzung; Band 2: Preiswerter Neubau von Ein-u. Mehrfamilienhäusern, 21. Aufl. Verlag Hubert Wingen, Essen (2013)
51. Dachverband Lehm e.V. (Hrsg.): Kurslehrbuch Fachkraft Lehmbau. Dachverband Lehm e.V., Weimar (2005)
52. Dachverband Lehm e.V. (Hrsg.): Lehmbau Verbraucherinformation. Dachverband Lehm e.V., Weimar (2004)
53. Ahnert, R., Krause, K.H.: Typische Baukonstruktionen von 1860 bis 1960 zur Beurteilung der vorhandenen Bausubstanz; Bd. I: Gründungen, Wände, Decken, Dachtragwerke; Bd. II: Stützen, Treppen, Erker und Balkone, Bogen, Fußböden, Dachdeckungen. VEB Verlag f. Bauwesen, Berlin 1985 (I), 1988 (II)
54. Kapfinger, O., Rauch, M.: Rammed Earth—Lehm und Architektur—Terra cruda. Birkhäuser, Basel (2001)
55. Schroeder, H., Bieber, A.: Neue Stampflehmprojekte in Thüringen. In: LEHM 2004 Leipzig, Beiträge zur 4. Int. Fachtagung für Lehmbau, pp. 190–201. Dachverband Lehm e.V., Weimar (2004)
56. Schroeder, H.: Stampflehmböden in historischen Orangeriegebäuden in Deutschland. In: LEHM 2012 Weimar, Beiträge zur 6. Int. Fachtagung für Lehmbau, pp. 354–355. Dachverband Lehm e.V., Weimar (2012)
57. Walker, P., Keable, R., Martin, J., Maniatidis, V.: Rammed Earth—Design and Construction Guidelines. BRE Bookshop, Watford (2005)
58. Rischanek, A.: Sicherheitskonzept für den Lehmsteinbau. Technische Universität, Fak. Bauingenieurwesen, Wien (2009)
59. Dierks, K., Stein, R.: Ein Bemessungskonzept für tragende Stampflehmwände. In: Moderner Lehmbau 2002; Beiträge, pp. 37–48. Fraunhofer IRB Verlag, Stuttgart (2002)
60. Fischer, F., Mönnig, H.-U., Schroeder, H., Wagner, B.: Erarbeitung einer technologischen Konzeption zur Errichtung von Gebäuden aus Erdstoff mit Erprobung unter den Bedingungen der Bauausführung, Teil I: 1988; Teil II: 1989. Unpublished research report. Hochschule f. Architektur u. Bauwesen, WBI, Weimar (1989)
61. Miller, T., Grigutsch, A., Schultze, K.W.: Lehmbaufibel—Darstellung der reinen Lehmbauweise. Schriftenreihe d. Forschungsgemeinschaften Hochschule, Heft 3, Reprint der Originalausgabe von 1947. Verlag Bauhaus-Universität, Weimar (1999)
62. Krüger, S.: Stampflehm—Renaissance einer alten Technik. manudom Verlag, Aachen (2004)
63. www.rammedearth.blogspot.com
64. Hetzl, M.: Der Ingenieur-Lehmbau. Die Entwicklung des ingenieurmäßigen Lehmbaus seit der Französischen Revolution und der Entwurf neuer Lehmbautechnologien. Dissertation, Technische Universität, Fak. Architektur, Wien (1994)
65. Williams-Ellis, C.: Building in Cob, Pisé and Stabilized Earth, Reprinted edition. Donhead, Shaftesbury (1999)

66. Keefe, L.: Earth Building—Methods and Materials, Repair and Conservation. Taylor & Francis, London (2005)
67. McDonald, F., Doyle, P., MacConville, H.: Ireland's Earthen Houses. A. & A. Farmar, Dublin (1997)
68. Bertagnin, M.: Architetture di Terra in Italia—Tipologie, tecnologie e culture costruttive. Edicom Edizioni—Culture costruttive, Monfalcone (1999)
69. Easton, D.: Industrielle Formgebung von Stampflehm. In: LEHM 2008 Koblenz, Beiträge zur 5. Int. Fachtagung für Lehmbau, pp. 90–97. Dachverband Lehm e.V., Weimar (2008)
70. Fontaine, L., Anger, R.: Bâtir en terre—Du grain de sable à l'architecture. Édition Belin/Citè des sciences et de l'industrie, Paris (2009)
71. Kafescioglu, R., Isik, B.: The relevance of earth construction for the contemporary world and Alker—gypsum stabilized earth. In: Kerpic '05, Proceedings of the 1st International Conference "Living in Earthen Cities", pp. 166–173. Technical University ITU, Istanbul (2005)
72. v. Bodelschwingh, G.: Ein alter Baumeister und was wir von ihm gelernt haben—Der Dünner Lehmbrotebau, 3. Aufl. Heimstätte Dünne, Bünde (1990)
73. Žabičková, I.: Hliněné stavby. ERA vyd., Brno (2002)
74. Schreckenbach, H., Abankwa, J.G.K.: Construction Technology for a Tropical Developing Country. Gesellschaft für Technische Zusammenarbeit GTZ, Eschborn (1982)
75. Minke, G.: Handbuch Lehmbau. Baustoffkunde, Techniken, Lehmarchitektur, 7. erw. Aufl. Ökobuch Verlag, Staufen (2009)
76. Tulaganov, B.A., Schroeder, H., Schwarz, J.: Lehmbau in Usbekistan. In: Zukunft Lehmbau 2002—10 Jahre Dachverband Lehm e.V., pp. 91–101. Dachverband Lehm e.V./Bauhaus-Universität, Weimar (2002)
77. Brändle, E.: Sanierung Alter Häuser. BLV Verlagsgesellschaft, München (1991)
78. Schrammel-Schäl, N.: Gefiederte Wände—Wandbehang mit Stroh, eingebettet in Lehm. In: LEHM 2012 Weimar, Beiträge zur 6. Int. Fachtagung für Lehmbau, pp. 314–319. Dachverband Lehm e.V., Weimar (2012)
79. Kraft, J.: Was wie machen? Instandsetzen u. Erhalten alter Bausubstanz. Interessengemeinschaft Bauernhaus IGB, Grafischer Betrieb, Weyhe (1992)
80. Naasner, C., Paulick, R., Stendel, G., Wunderlich, A.: Lehmbaupraktikum Lindig 2004. Unpublished internship report. Bauhaus-Universität, Fak. Architektur, Weimar (2004)
81. Schüler, G., Schroeder, H: Naturbelassene Baustoffe—Beispiel Kalktuff Magdala. Wiss. Z. Hochsch. Archit. Bauwes. – A/B – Weimar **39**(4), 337–341 (1993)
82. Zweite Verordnung zur Änderung der Energieeinsparverordnung (EnEV 2009) v. 18.11.2013. Bundesgesetzblatt I, Nr. 61, Berlin (2013)
83. Pollack, E., Richter, E.: Technik des Lehmbaus. Verlag Technik, Berlin (1952)
84. Richter, A.: Analyse und Bewertung der aktuellen Lehmbauweisen im Bereich des ein-und mehr-geschossigen Wohnungsbaus. Unpublished diploma thesis. Bauhaus-Universität, Fak. Architektur, Weimar (1996)
85. Nordische Naturbau GmbH: Lehmoment Strohlehm-Elemente. o.J.—Company Brochure, Havetoftloit
86. Minke, G., Mahlke, F.: Der Strohballenbau—ein Konstruktionshandbuch. Ökobuch-Verlag, Staufen (2002)
87. Schroeder, H.: Ökologischer Wohnhausneubau mit Lehmbaustoffen in Weimar-Taubach. In: Zdravé domy 2006—Přírodní materiály ve stavbách, Conference Proceedings, pp. 33–40. Technical University, Brno (2006)
88. Morton, T., et al.: Low Cost Earth Brick Construction, Dalguise: Monitoring & Evaluation. Arc, Chartered Architects, Auchtermuchty (2005)
89. Karphosit: Karphosit Lehmbauplatten. o.J. Company Brochure, Peißen
90. Richter, E.: Decken und Flachdächer aus vorgefertigten Lehmplatten. SR zur Förderung der Naturbauweisen, Nr. 3. Beratungs-u. Lehrstelle f. Naturbauweisen, Wallwitz (1951)
91. Minke, G., Mukerji, K.: Structurally Optimised Domes—A Manual of Design and Construction. GTZ/GATE/Vieweg-Verlag, Braunschweig (1995)

92. Bendakir, M.: Architectures de terre en Syrie—une tradition de onze millénaires. ENSAG-CRATerre, Grenoble (2008)
93. Joffroy, T., Guillaud, H.: The Basics of Building with Arches, Vaults and Cupolas. SKAT/CRATerre-EAG, St. Gallen (1994)
94. Orgel, R.: Traditioneller Wohnungsbau in Syrien. Unpublished student research paper. Bauhaus-Universität, Fak. Architektur, Weimar (2004)
95. Fathy, H.: Architecture for the Poor. University of Chicago Press, Chicago (1973)
96. Baimatova, N.S.: 5000 Jahre Architektur in Mittelasien—Lehmziegelgewölbe vom 4./3. Jahrtausend. v. Chr. bis zum Ende des 8. Jahrhunderts. n. Chr. Archäologie in Iran und Turan, 7, Deutsches Archäologisches Institut, Eurasien-Abteilung, Außenstelle Teheran. Verlag Philipp v. Zabern, Mainz (2008)
97. Khalili, N.: Ceramic Houses & Earth Architecture—How to Build Your Own, 2nd edn. Burning Gate Press, Los Angeles (1994)
98. Navrongo-Bolgatanga Diocese, Ghana Museums and Monuments Board, CRATerre-EAG: Navrongo Cathedral—The Merge of Two Cultures. CRATerre Editions, Villefontaine (2004)
99. Figgemeier, M.: Die multifunktionalen Eigenschaften—keine Überraschungen beim Lehmputz-Einsatz. Der Maler- und Lackiermeister **11**, 37–41 (2009)
100. Röhlen, U., Ziegert, C.: Lehmbau-Praxis, Planung und Ausführung. Bauwerk Verlag, Berlin (2010)
101. Stappel, M.: Kratzputz, Stippputz—historische Verzierungen von Gefachen. Informationsblatt 36. Aus den Arbeiten des Freilichtmuseums Hessenpark, Neu-Anspach (2005)
102. Deutscher Stuckgewerbebund im ZV des Deutschen Baugewerbes u. a.: Putzoberflächen im Innenbereich. Berlin (1999)
103. Frimodig, H., Schulz, C., Molzen, C., Parschau, J., Stutz, H.: Bauernhäuser in Nordwestmecklenburg. Einblicke zwischen Schaalsee und Salzhaff, H. 5. NWM-Verlag, Grevesmühlen (1998)
104. Goer, M.: Ungarndeutsche Bauernhäuser in der Baranya. Denkmalpflege in Baden-Württemberg **4**, 9–21 (1996)
105. Farahza, N., Khajehrezaei, I.: Badgir, earthen resistant structure. In: TerrAsia 2011, Proceedings of the International Conference on Earthen Architecture in Asia, pp. 247–258. Mokpo National University, Dept. of Architecture, Mokpo (2011)
106. Fakhar Tehrani, F., Koosheshgaran, M.: The role of soil and land in the regional architecture of the central plateau of Iran. In: Terra 2003—Proceedings of the 9th International Conference on the Study and Conservation of Earthen Architecture, pp. 582–592. Yazd (2003)
107. Evangelische versöhnungsgemeinde Berlin (Hrsg.): Kapelle der Versöhnung—Ausstellung. Broschüre, Berlin (2000)
108. Claytec e.K.: Die Anleitung zum modernen Lehmbau—Architektenmappe 2008. Company Publication, Viersen (2008)
109. Dachverband Lehm e.V. (Hrsg.): Lehmdünnlagenbeschichtungen—Begriffe, Anforderungen, Prüfverfahren, Deklaration. Technische Merkblätter Lehmbau, TM 06. Dachverband Lehm e.V., Weimar (2015)
110. Standards Organisation of Nigeria SON: Standard for Stabilized Earth Bricks. NIS 369:1997, Lagos (1997)
111. International Conference of Building Officials: Uniform Building Code Standards, Section 2405, Unburned Clay Masonry, Whittier (1982)
112. Norton, J.: Building with Earth—A Handbook. Development Workshop. Intermediate Technology Publications, Rugby (1986)

5

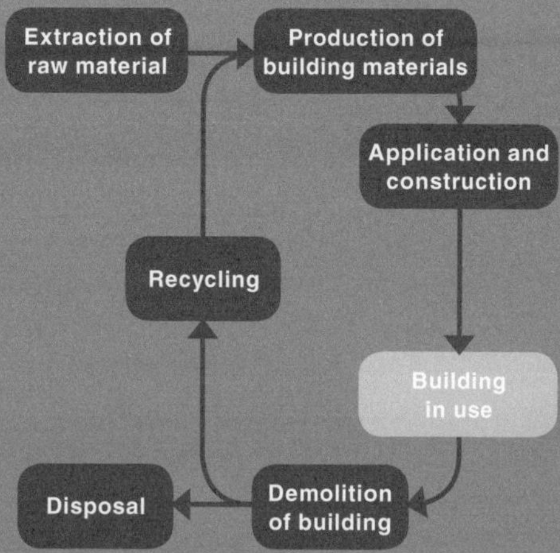

Structures Built of Earth Building Materials – Impacts, Structural Damage and Preservation

During its lifetime, a building is exposed to a multitude of exterior impacts, natural aging of its building materials, as well as stresses as a result of the building's use (Fig. 5.1).

In order to guarantee the unrestricted use of the building, all general requirements placed on the structure must be met during its entire useful life, independent of the building materials used.

© Springer International Publishing Switzerland 2016
H. Schroeder, *Sustainable Building with Earth*,
DOI 10.1007/978-3-319-19491-2_5

5.1 Performance of Building Elements and Structures While in Use

The impacts on structures during the building's useful life are recorded by DIN EN 1991-1-1. These actions primarily consist of:

- The prevailing climate: temperature, precipitation, freeze–thaw cycles, insolation, and wind
- The building ground: deformations and moisture (with high salt levels)
- The location and traffic: noise, vibrations, and air pollution
- User activity: mechanical and dynamic stresses, moisture accumulation and tobacco smoke in interior spaces, and accidental damage

Exceptional natural impacts, such as earthquakes (DIN EN 1998-1), tornadoes, and flooding, require special structural design features.

These impacts are grouped together as partial concepts for thermal insulation, moisture protection, fire protection, and sound insulation. They are used to derive the respective requirements in terms of a structure's usability. Adherence to these requirements must be verified together with structural dimensioning. In the planning phase, these verifications form the *complex dimensioning* or construction design of a building.

A building's response to these impacts is described with the help of appropriate parameters which must be determined using standardized test methods (Table 1.1). Compared to other mineral building materials, structures made of earth building materials exhibit special characteristic features.

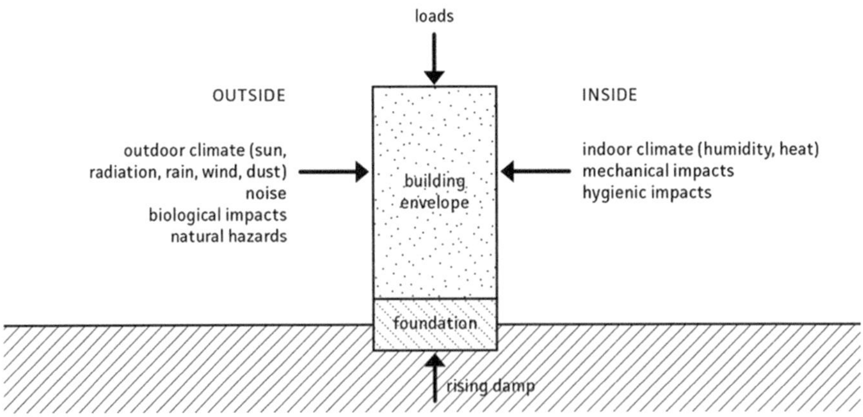

Fig. 5.1 Impacts on the building envelope during its useful life

5.1.1 Thermal Insulation Parameters

The mechanisms of thermal transfer and temperature distribution in building elements and structures made of earth building materials can be described and illustrated with the help of thermal insulation parameters.

For the climate in Central Europe, the following general requirements must be met in terms of thermal insulation, particularly in structures made of earth building materials, for the building's useful life:

- Energy conservation through thermal insulation, especially in order to decrease heat loss during the heating period
- Prevention of condensation on the interior surfaces of building elements caused by insufficient thermal insulation
- Prevention of the overheating of interior spaces during the summer months

5.1.1.1 Mechanisms of Heat Transfer

In physical terms, heat is a form of energy just like mechanical, electrical, or magnetic energy. It represents the kinetic energy of the elementary particles of matter. Its intensity is expressed in temperature (Table 5.1).

Heat can be transferred from one body to another by *radiation* or *conduction*. In liquids or gases, heat is also transferred by *convection*. All three mechanisms occur more or less simultaneously.

Thermal Radiation

Heat transfer by radiation is the transfer from a warmer (+) to a cooler (−) surface through contact (Fig. 5.2). This process uses electromagnetic waves, which are mainly infrared, and does not require a transfer medium such as air.

The warmer surface emits thermal energy in the form of radiant heat which is passed on to the cooler surface. The emitted quantity of heat is a function of the temperature difference between the surfaces and the material properties of the warmer surface (emissivity). The intensity of thermal radiation increases substantially as the temperature rises.

The emitted heat is partially reflected (r) and partially absorbed (a) by the cooler surface. The material properties of the cooler surface determine the extent of this process: smooth and light-colored surfaces are good reflectors, whereas dark and rough surfaces absorb well. These characteristics result in heat gains which can have a positive effect on a building's energy balance.

The total sum of *absorption a* and *reflection r* always equals the quantity of radiated heat.

The power radiated by the total energy E_s per unit of surface area is determined with the help of the Stefan–Boltzmann law. It is proportional to the fourth power of the absolute temperature T of the radiating body:

$$E_s = \sigma' \cdot \Delta T^4 \left[\mathrm{W/m^2}\right]$$

Table 5.1 Thermal parameters for calculating temperature distribution in earth building elements

No.	Parameter	Symbol	Unit	Formula	Meaning
1	Thickness	d	m		Thickness of the building element or building element layer
2	Area	A	m^2		
3	Time	t	s, h		
4	Temperature	T, Θ	K, °C 1 K = 1 °C		Material temperatures are expressed in °C, temperature differences in K
5	Heat, quantity of heat	Q	J, Ws 1 J = 1 Ws		Type of energy, heat quantity/time unit = heat flow
6	Heat flow	Φ	W	$\Phi = dQ/dt$	Heat quantity dQ which is transferred in time dt
7	Heat flow density	q	W/m^2	$q = \Phi/A$	Transferred heat flow per area unit
8	Thermal conductivity	Λ (k)	W/m K		Heat quantity Q which flows through 1 m^2 of a 1-m-thick layer of a material in one second at a constant temperature difference of 1 K between both surfaces (Sect. 3.6.3.2, Table 3.33)
9	Heat transfer coefficient	Λ	W/m^2 K	$\Lambda = \lambda/d$	Ratio of the thermal conductivity λ to the layer thickness d of a building element. It describes the heat flow Φ in W which passes through 1 m^2 of a building element layer with a thickness d in m at a constant temperature difference of 1 K between both surfaces
10	Heat transfer resistance	R	m^2 K/W	$R = d/\lambda = 1/\Lambda$	Ratio of the layer thickness d of a building element to the thermal conductivity, which means that it is the reciprocal value of the heat transfer coefficient Λ. It describes the resistance to the heat flow Φ in W which passes through 1 m^2 of a building element layer with a thickness d in m at a constant temperature difference of 1 K between both surfaces
11	Surface heat transfer coefficient, interior (i) and exterior (e)	$1/R_{si}$ $1/R_{se}$	W/m^2 K		Heat flow which, taking wind conditions (surface color and texture) into consideration, is transferred between 1 m^2 of building element surface and the air at a temperature difference of 1 K between air and surface

12	Surface heat transfer resistance, interior (i) and exterior (e)	R_{si}, R_{se}	m² K/W		Reciprocal value of the surface heat transfer coefficient, required for determining heat loss
13	Overall heat transfer coefficient (*U*-value)	U	W/m² K	$U = 1/R_T = 1/(R_{si} + \Sigma d/\lambda + R_{se})$	Reciprocal value of the overall heat transfer resistance which describes the heat flow in W which passes through a building element per m² at a temperature difference of 1 K
14	Overall heat transfer resistance	R_T	m² K/W	$R_T = 1/U$	Total resistance to heat flow consisting of the heat transfer resistances of the individual building element layers and the surface heat transfer resistances of the air layers
15	Specific heat capacity	c_p	W s/kg K W h/kg K kJ/kg K		Heat quantity in Ws which is needed to change the temperature of 1 kg of a specific material by 1 K (Sect. 3.6.3.2, Table 3.34)
16	Heat storage capacity	Q_s	Ws/m²K	$Q_s = c \bullet \rho \bullet d$	Heat quantity Q stored in 1 m² of a panel-shaped building element with a thickness d in m and made of a material with a bulk density ρ in kg/m³ at an excess temperature of 1 K
17	Cooling behavior	t_A	h	$Q_s/\Lambda = \lambda \bullet \rho \bullet c/\Lambda = \lambda \bullet \rho \bullet c^2/\lambda$	Ratio of the stored heat quantity Q_s in 1 m² of wall to the heat transfer coefficient Λ at an excess temperature of 1 K
18	Thermal effusivity coefficient	b	Ws/s$^{0.5}$m² K	$b = (\lambda \bullet \rho \bullet c)^{0.5}$	Measured value for the speed of thermal absorption and thermal emission of a material which forms the space-enclosing surfaces. The smaller b, the faster the heating of the surfaces ("warm to the touch")
19	Temperature amplitude	$\Delta\Theta$	K	$\Delta\Theta = \Theta_{si} - \Theta_{se}$	Difference between the maximum temperatures on the interior and exterior surfaces of a building element within 24 h
20	Temperature amplitude attenuation	$\Delta\Theta_{se}/\Delta\Theta_{si}$	–		Ratio between the temperature amplitudes on the exterior and interior surfaces of the building element (Fig. 5.4)
21	Temperature amplitude ratio	$\Delta\Theta_{si}/\Delta\Theta_{se}$	–		Reciprocal value (Fig. 5.4)
22	Phase displacement	φ	h		Time difference between reaching the maximum temperature on the exterior and interior surfaces of the building element (Fig. 5.4)

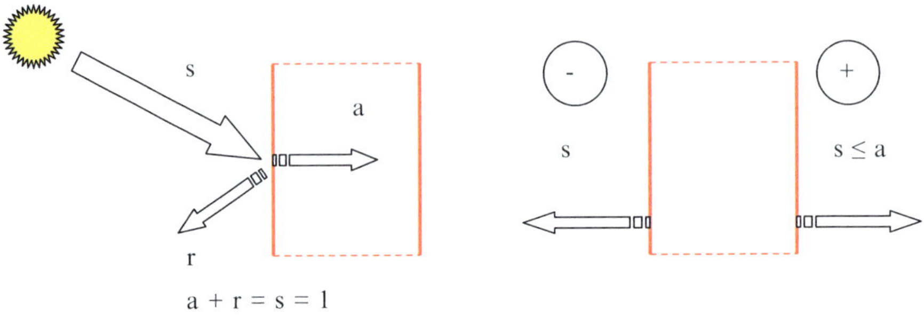

Fig. 5.2 Thermal radiation s, adsorption a, and reflection r

with $\sigma' = 5.67 \times 10^{-8}$ [W/m² K⁴] (Stefan–Boltzmann constant).

The proportion of individual wavelengths in relation to the total radiation varies. The wavelength λ_{max} is of particular interest because it describes the largest portion of emitted radiation. According to Wien's displacement law, this portion is inversely proportional to the absolute temperature of the radiating body:

$$\lambda_{max} = c'/T$$

with $c' = 2898$ [μm K] (third radiation constant).

A change in temperature not only results in a change in the intensity of the radiation but also a change in its spectral composition. The radiation maximum shifts to shorter wavelengths when a rise in temperature occurs.

Strictly speaking, the Stefan–Boltzmann law only applies to a "black body" with regard to the specified constants. The term "black body" refers to an ideal body which absorbs all incident radiation and emits the most energy compared to all other bodies at the same temperature. In contrast, real bodies only emit a portion of the incident thermal radiation.

The thermal radiation of a "nonblack" body (building elements made of the examined material) is determined as follows:

$$E_s = \varepsilon' \cdot \Delta T^4 \times 10^{-8} \left[W/m^2 \right] \text{with}$$

$\varepsilon' = C/C_s$, emissivity

$C_s = 4.96$ kcal/hm² K⁴ = 5.78 W/m² K⁴, radiant coefficient for a "black body"

C radiation coefficient for a "nonblack" body

Accordingly, the radiation coefficients C of "nonblack" bodies (building elements) are smaller.

Using a radiation pyrometer (infrared camera), Reincke [1] determined an emissivity of $\varepsilon' \sim 0.93$ for earth building surfaces. This means that earth as a building material has a similarly high level of emissivity as all other common nonmetallic building materials. This result corresponds to the abovementioned theory of electromagnetic waves which predicts high emissivity for non-conducting materials.

For light-colored, shiny metal building materials, C is in the range of <1 kcal/hm² K⁴. Their radiation is particularly low, but they reflect especially well.

Convection

Heat transfer by convection in liquids or gases is always linked to a transport of materials. The increase in volume connected to the rising temperature results in buoyancy which can, for example, be observed in warm air rising in the atmosphere.

Thermal Conduction

In the event of direct contact between a cold and a warm body, or within a body, thermal conduction occurs through a balancing of the motion intensity of matter particles. In other words, energy transport takes place from the higher temperature body to the lower temperature body. The quantity of heat which passes through a defined cross section of a conducting body per time unit is referred to as *heat flow*. This process is dependent on the molecular composition of a substance as well as its structure and air void content. It is thus defined as a *specific material property* which is described by the coefficient of thermal conductivity λ (Sect. 3.6.3.2, Table 3.33).

For the calculation of heat transfer in building materials and building elements, the three mechanisms of heat transfer (heat conduction, radiation, and convection) are generally grouped together into the total energy flow. The thermal conductivity *specific to the building element* is then determined.

5.1.1.2 Temperature Distribution in Earth Building Elements

Limiting Conditions

In general, the calculational procedure for recording temperature distribution in building elements is a time-consuming, geometry-dependent, three-dimensional problem. For the thermal assessment of a building element, it is generally sufficient to reduce the process to a one-dimensional, *steady-state* problem (Fig. 5.3 [2]): it is assumed that the temperatures on both sides of the building element remain constant over time and that a varying heat transfer only occurs toward the building element thickness. This results in a linear temperature profile in the building element. For multilayer wall systems, the temperature profile is bent due to the different degrees of thermal conductivity in the individual layers.

There are, however, a number of thermal situations which cannot be sufficiently described using steady-state, time-independent conditions. For example, fast and significant temperature changes on the surface of building elements lead to heat flow changes within the building element. In this context, the heat storage capacity of the building materials and building elements, as a function of time, is of significance. This situation is referred to as an *unsteady-state* problem.

Furthermore, heat flows in building elements made of porous building materials, including earth building materials, are always linked to moisture transport processes (Sect. 5.1.2.1).

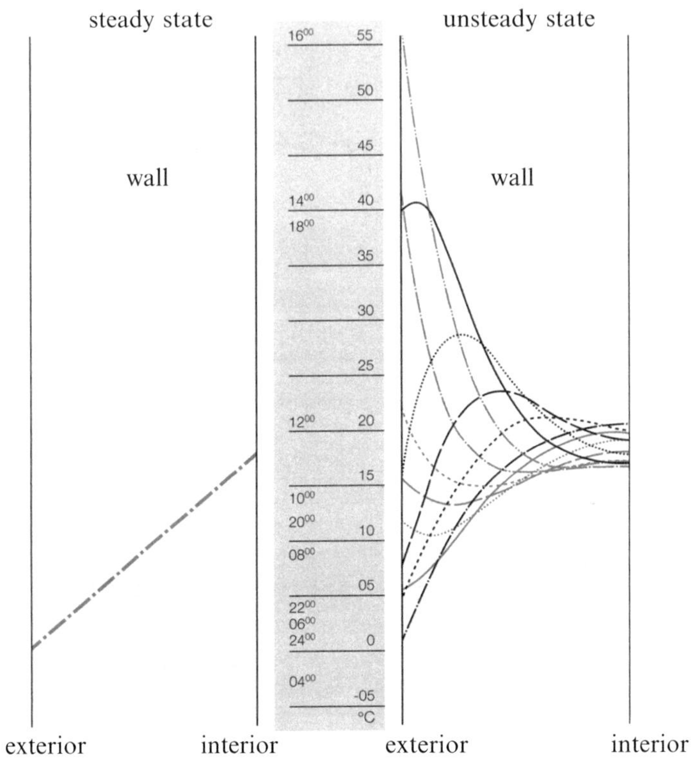

Fig. 5.3 Temperature distribution in earth building elements, according to [2]

Thermal Parameters

The respective parameters for thermal calculations of temperature distribution in earth building elements are listed in Table 5.1 including (preliminary) symbols and units based on DIN EN ISO 7345 and 6946.

Heat Transfer Coefficient and Heat Transfer Resistance

(9 and 10): building elements generally consist of multiple layers. This means that in order to determine heat loss, the sum of the resistances of each building element layer to the heat flow Φ must be used. For multilayer systems, the total heat transfer resistance is determined with the help of the sum of the partial resistances.

In practical building terms, the heat transfer resistance R is the measure used to indicate the thermal insulation properties of a building element: high heat transfer resistance means that little heat flows from the warm side to the cold side and is an indicator of good thermal insulation properties.

Surface Heat Transfer Coefficient and Surface Heat Transfer Resistance

(11 and 12): in addition to building element layers, air layers also contribute to thermal insulation. Friction on both sides of a building element "slows down" the air movement and creates "heat transfer layers." The color and texture of the surface, and particularly the wind conditions, have an impact on the heat transfer process. There are two different surface heat transfer resistances R_{si} and R_{se} which apply to the interior and exterior building element surfaces, respectively. DIN EN ISO 6946 defines consistent limiting and transitional conditions for thermal calculations.

Overall Heat Transfer Coefficient and Overall Heat Transfer Resistance

(13 and 14): in terms of building practice, the *U*-value indicates the heat loss of a building element or exterior surfaces of the building. Therefore, it represents the most important building element parameter for the thermal insulation of buildings. A low *U*-value is an indicator of little heat loss and good thermal insulation of the building element or structure.

Heat loss in winter should be kept as low as possible. The same applies to the heating energy demands in new buildings. In existing buildings, thermal insulation should be improved in order to reduce heating energy demands and pollution. The minimum requirements for thermal insulation of the building envelope and related weather-dependent moisture protection are specified in DIN 4108-2. Since 2002 and 2007, proof of compliance is required according to the regulations set forth by the German Energy Conservation Regulation (EnEV) [3] which stipulates significantly higher standards.

The "reference building process" is a continuation of the EnEV as the EnEV 2009 (2014) for residential buildings (DIN 18599 (D)). Here, the maximum permissible annual primary energy demand for a planned building is determined with the help of an identical reference building with standardized building elements and required systems engineering.

The *U*-value cannot be used to analyze specific topographical conditions, different layouts, or building densities. In addition, when calculating the thermal energy demand, the *U*-value does not account for heat gain resulting from the high heat storage capacity of buildings made of high-mass building materials. The same is true for solar gain. For the verification procedure of the energy balance of buildings according to DIN 4108-6 (D), the abovementioned effects as well as additional effects can be accounted for with the help of correction factors.

As the building elements or the insulation layers get thicker, the insulation effect improves only slightly. The ratio of insulation expenses to energy savings thus becomes increasingly unfavorable.

Heat Storage Capacity and Cooling Behavior

(16 and 17): in transitional periods with rapid fluctuations in outside temperatures, building elements which can store heat prevent the building's interior from cooling down or heating up too quickly. The higher a building element's heat storage

capacity Q_s and the lower its heat transfer coefficient Λ, the better its effectiveness in terms of heat storage and cooling. The quotient of both values is referred to as the thermal inertia or cooling behavior of a building element. The higher the Q_s/Λ value, the slower the building element cools down.

Thermal Effusivity Coefficient

(18): the thermal effusivity coefficient b of a material describes how much heat is removed from the human body when a person touches the material by hand or foot. The higher the thermal effusivity coefficient *b*, the more heat that is removed and the cooler the material feels. In rooms with high air temperatures, building elements with surface layers made of materials with a high thermal effusivity coefficient stay "cool" for a longer period of time. Heavy building materials such as natural stone, concrete, and rammed earth have high *b* values, whereas the values for light building materials such as wood, cork, or foamed materials are correspondingly lower. Earth plasters with a high organic fiber aggregate content therefore seem "warmer to the touch."

Temperature Amplitude Attenuation and Phase Displacement

(19–22): the temperature amplitudes which form on the exterior surface of the building element continue as oscillations within the building element and reach the interior surface by means of thermal conduction. When passing through the building element, the amplitudes become weakened (attenuated) (Fig. 5.4, according to [4]). In order to maintain a pleasant indoor climate, the large temperature fluctuations of the outdoor air need to be reduced to a comfortable level in the interior of the building. This is based on the assumption that the temperature of the indoor air fluctuates to the same degree as the temperature on the interior surface of the building element. This means that the thermal behavior of the adjoining indoor space (e.g., a thermal storage wall) is not taken into account.

An attenuation of the temperature amplitudes on the interior surface of the building element can be described as the cooling behavior Q_s/Λ of the building element and can be achieved through the use of a building material with a high thermal storage capacity or through adequate thermal insulation. High-mass earth building materials, such as rammed earth, store more heat than light clay but are also better thermal conductors. This means that their thermal insulation properties are not as high and they should therefore be used for interior walls. Light clays are better insulators for exterior walls.

Depending on the climate type, the temperature amplitude attenuation has different target functions:

- In climates with high annual temperature amplitudes (polar regions, temperate climates), thermal insulation is most important. For moderate climates, protection against the summer heat also needs to be considered.
- In hot and dry climates with high daily temperature amplitudes, amplitude attenuation through the use of high-mass building materials and greater building ele-

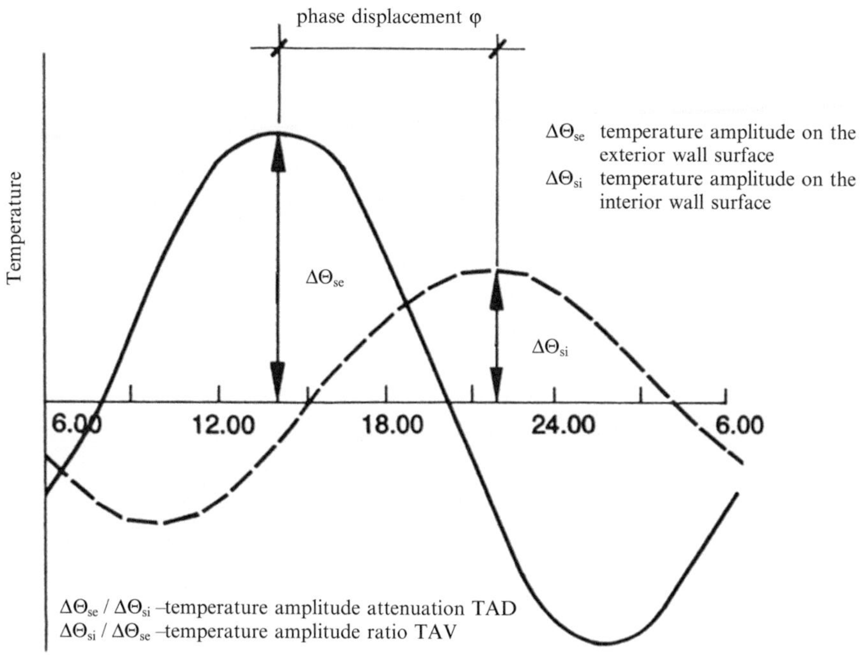

Fig. 5.4 Temperature amplitude attenuation and phase displacement

ment thicknesses is recommended: as a result of a time lag or phase displacement, the cooler nighttime temperatures reach the indoor surfaces during the midday heat of the following day. Conversely, the stored midday heat from the previous day is released into the interior during the cool morning hours ("storage discharge") (Fig. 5.4, according to [4]).

- In hot and humid climates with low annual and daily temperature amplitudes, only building materials with a low heat storage capacity are recommended.

Phase displacement is a function of the building element thickness and the thermal properties of the building materials used (thermal conductivity, specific heat, heat storage capacity). It should only be noticeable on the interior surface after approx. 8–10 h. The general rule is that the smaller the temperature amplitude ratio the larger the phase displacement.

5.1.2 *Hygric Parameters*

The hygric parameters which are used to describe moisture transport in building elements are defined in DIN EN ISO 9346.

Earth building materials are more sensitive to moisture than all other mineral building materials. It is therefore one of the most important tasks in earth building to effectively protect the building from moisture during its entire useful life and to repair buildings which have been damaged by moisture.

5.1.2.1 Mechanisms of Moisture Transport

Moisture can enter an earth building element as a liquid or gas due to the material's hygroscopic properties and the open pore structure of interconnected capillaries. Depending on the differences in moisture, temperature, and vapor pressure between the building element and the adjacent medium, moisture can be transported through the building element, stored in it and released again. The transport mechanisms are influenced by the material composition and the pore structure of the earth building materials.

The absorption and transport of moisture as a liquid or gas (water vapor) takes place via different transport mechanisms depending on the pore classes of different pore sizes (Table 5.2, according to [5]):

Weather and groundwater act on the building element from the exterior as *liquid* water which is absorbed by *capillary* action and distributed (Sect. 3.6.3.1). After extended exposure, the macropores also fill with water which can result in a loss of structural integrity of the earth building element. This is the case in the event of accidental water damage or natural disasters which result in standing or running water (Sect. 5.2.4.1). A similar effect occurs when liquid water freezes and thaws in the pore spaces.

On the interior of the building, water vapor from the air is absorbed by the space-enclosing building elements *hygroscopically* in its *gaseous state* and then distributed. The following processes can be distinguished between in connection with the equalization of water vapor pressure within and outside the building element [5]:

Sorption: equalization between the moisture contained in the air and in materials
Absorption: uptake of water due to an increase in relative humidity
Adsorption: collecting of water on the surface (of the capillaries) due to an increase in relative humidity, requires pore diameters of ≤0.1 μm
Desorption: release of water due to a decrease in relative humidity

Table 5.2 Moisture transport mechanisms and pore sizes

No.	Pore class	Size range	Transport mechanism
1.	Micropores	<0.1 μm	Hygroscopic water absorption (adsorption), capillary condensation, vapor diffusion
2.	Capillary pores	0.1 μm–1 mm	Capillary action as liquid water against the influence of gravity, vapor diffusion, capillary condensation
3.	Macropores	>1 mm	Saturated or unsaturated flow based on the fill level under the influence of gravity; due to a decrease in capillary pressure, water no longer enters the pores through the capillaries

The van der Waals force creates an adsorptive bond between the water molecules and the pore walls of the earth building material in its micropores. In addition, vapor pressure decreases via the menisci of the water surfaces so that water is released from the air (capillary condensation) even before the saturation vapor pressure is reached. Here, the preferred mechanism of water transport is vapor diffusion. Liquid water is no longer transported. Therefore, these pores contain water even in relatively dry surroundings which makes up the main portion of equilibrium moisture (Sect. 5.1.2.4).

During moisture migration in earth building elements due to differences in temperature between exterior and indoor air, different transport mechanisms occur parallel to each other (Table 5.2). These mechanisms can also work in opposite directions. In winter, water vapor diffusion generally occurs from the warm side (interior) to the cold side (exterior). If the adsorptive moisture on the outside is high enough (>50 % relative humidity), capillary water transport takes place from moist (outside) to dry (inside) independent of the temperature. This is an example of transport in the opposite direction. Vapor-proof insulation materials and vapor barrier sheets which are integrated into the wall (as well as waterproofing agents) limit or prevent these transport mechanisms.

5.1.2.2 Water Vapor Diffusion Resistance Factor

Depending on the temperature, water has a specific vapor pressure which causes water saturation of the air (partial pressure of water vapor). If there is a difference in pressure between the indoor air and the pores of the surrounding surfaces of the building elements, water vapor flows toward the low-pressure side until an equilibrium has been reached. The water vapor diffuses through the porous building material. This property is indicated by the water vapor diffusion resistance factor μ. The value μ is a dimensionless ratio which compares the diffusion density of the water vapor flow in a building material with that found in an air layer of an equivalent thickness s_d. Still air has a μ value of 1.

The s_d value indicates the required thickness of the air layer in order to possess the same diffusion resistance as the specified building material with a layer thickness d. This makes it possible to compare building element layers of different thicknesses:

$$s_{\mathrm{d}} = \mu \cdot d.$$

Pore structure, material bulk density, and temperature influence water vapor diffusion. An increase in bulk density (and the associated decrease in pore space) leads to a higher μ value and an increase in vapor resistance of the building material.

For finish plasters on interior building element surfaces, the water vapor permeability should be as high as possible [6]. Earth plasters have relatively low μ values and high diffusivity. This means that equilibrium moisture is easily restored.

Table 5.3 Comparison of the water vapor diffusion resistance of different building materials

No.	Building material	μ range	Source
1	Masonry made of		
	– Fired bricks	5–10	DIN 4108-4
	– Clinker bricks	50–100	DIN 4108-4
	– Sand-lime bricks	5–25	DIN 4108-4
	– Lightweight concrete blocks	5–15	DIN 4108-4
	– Tuff bricks	20–50	[6]
2	Earth building materials	5–10	[15]
	– Light clay	2–5	[87]
3	Standard concrete	70–150	DIN 4108-4
4	Plasters	10–35	DIN 4108-4
	– Cement plasters	50–100	[6]
	– Lime-cement plasters	10–20	[6]
	– Lime plasters	9–15	[6]
	– Thermal insulation plasters	5–10	[6]
5	Wood		
	– Wet	20–80	DIN 4108-4
	– Dry	100–500	DIN 4108-4
	– Wood composites, depending on bulk density	1–400	DIN 4108-4
6	Insulation materials		
	– Mineral and plant fibers	1–10	DIN 4108-4
	– Synthetic materials	1–300	DIN 4108-4
	– Foam glass	Virtually vapor-proof	DIN 4108-4
7	Vapor barriers		
	– Bitumized felt, smooth	2,000–20,000	DIN 4108-4
	– Roofing felt, plastic membrane	10,000–100,000	DIN 4108-4
	– Aluminum foil ≥125 g/m^2	Virtually vapor-proof	DIN 4108-4

The ability of the space-enclosing surfaces to absorb and release excess water vapor from and into the indoor air during the building's useful life helps establish a balanced and comfortable indoor climate (Sect. 5.1.1.2). In situations with rapidly changing vapor pressures, it is therefore important to keep the μ values low.

The μ value of hardened mortar for plaster mortar is determined according to DIN EN 1015-19. Table 5.3 shows a comparison of μ values of different building materials.

5.1.2.3 Condensation

Condensation of water vapor on interior building element surfaces occurs when the temperature of the indoor air falls below the dew point. The *dew point* is the air temperature at 100 % water vapor saturation. According to DIN 4108-3, the calculated condensate forming on the interior of a building element should generally not exceed a value of 0.5 kg/m^2 during a period of frost. This problem is becoming increasingly important in light of highly insulated and wind-proof exterior building elements and has prompted discussions about mold development and "proper ventilation."

Practical experiences with residential buildings under normal conditions of use have shown that dry earth building materials possess a sufficient level of capillarity which properly distributes condensate forming on the building element surfaces through capillary action. Moisture penetration can therefore be prevented as long as the minimum thermal insulation requirements according to DIN 4108-2 are met. This eliminates the conditions required for mold formation.

Under unfavorable conditions (mostly in winter), the water vapor saturation pressure might also be reached in the interior of the earth building element causing the vapor to change over to the liquid phase. Here, practical earth building experiences have also shown that the calculated condensate (e.g., determined using the Glaser method) can be distributed via capillary action without any harm to the earth building element.

In order to prevent the development of condensate inside the building element, the building element layers should generally be structured from the inside to the outside as follows: the heat transfer resistance R should increase, while the water vapor diffusion resistance factors μ should decrease (e.g., a light-clay layer on the exterior, high-mass earth block wall lining on the interior). Within the context of planning the thermal retrofit of historical earth buildings with the help of insulation layers, interior insulation is increasingly playing a larger role (Sect. 5.3.3.2).

5.1.2.4 Equilibrium Moisture

Equilibrium moisture refers to the moisture content of the building element which gradually sets in as the average value for the building's useful life. This moisture content is first reached after the drying process of the finished building element is complete (Sect. 3.3.3). After that, this level is rarely exceeded or undercut under standard conditions in indoor spaces (40–70 % RH, +20 °C). It does not interfere with the performance of the building element and is taken into account when determining the coefficient of thermal conductivity and load assumptions. Fluctuations in indoor humidity can only be compensated for within the fluctuation limits of the equilibrium moisture content. Various sources define values between 2 and 3 mass in % for earth building materials. Niemeyer [7] specifies values between 2.5 and 4.5 % "depending on the type, location, and age" of the building element.

The equilibrium moisture content also depends on the clay mineral structure of the earth building element. Three-layer clay minerals (montmorillonite) have a larger specific surface area and, accordingly, a higher percentage of capillary pores and micropores than two-layer clay minerals (kaolinite). They therefore also possess higher potential for accumulating water molecules from the air (Sect. 2.2.3.4). In such cases, the equilibrium moisture content can be significantly higher than 3 %.

Balancing the equilibrium moisture content in different building materials, for example, in half-timber or timber-frame construction, has practical significance for earth building. In these building methods, earth and wood are used as composite building materials. Dry wood has a higher equilibrium moisture content than dry earth, approx. 10 % compared to 3 %. If the earthen material in the building element is always kept dry and if both building materials are permanently bonded allowing for effective diffusion, a diffusion gradient of the equilibrium moisture content from the wood toward the earthen material develops: the earthen material keeps the wood dry and has a preserving effect. In the long run, mold and wood pests cannot find suitably moist living conditions, and the wood and earthen material can maintain their functions over centuries (Fig. 1.19).

5.1.2.5 Air Humidity Sorption

The equilibrium moisture content can be described as a function of the relative humidity (RH) and the pore sizes of the building material in the form of a *sorption isotherm*. Porous mineral building materials, including earth building materials, have a very large pore surface. At a relative humidity of <50 %, water vapor is taken up mainly by adsorption. At a higher humidity level, take-up increasingly occurs through capillary condensation (Sect. 5.1.2.1 [5]).

The profile of the sorption isotherm can be used to derive the quantity of moisture taken up and released by the building material under certain limiting conditions. The effect is not only dependent on the "water vapor storage capacity" of the building element but also on the temperature fluctuations in the interior space. These fluctuations are low in well-insulated, centrally heated homes.

Only the first 1–2 cm of the interior building element surface (usually made up of the plaster layers) have an effect on the indoor climate in terms of balancing rapidly changing vapor pressures in the indoor air. As the exposure increases, the take-up and release of moisture reaches further into the building element and the transport mechanisms described in Sect. 5.1.2.1 begin to occur.

Minke has performed extensive sorption behavior measurements (Sect. 4.3.6.6 [8]) of earth building materials compared to other building materials. These measurements, as well as the measurements of water vapor sorption of different earth plasters carried out by Holl/Ziegert according to DIN EN ISO 12571 [9], show that earth building materials perform considerably better (under normal indoor humidity levels of 40–70 % RH) than conventional building materials which use lime, gypsum, or cement as a binder (Fig. 5.5). This confirms that the clay mineral structure of the construction soil influences the sorption quantity: for three-layer minerals it is higher than for construction soils containing two-layer minerals (Sect. 2.2.3.4).

When comparing the sorption isotherms of different materials, it becomes apparent that the equilibrium moisture content of textiles, paper, and wood is considerably higher than that of mineral building materials, including earth (Fig. 5.6) [5]. In this context, the following aspects need to be considered: during the useful life of an earth building, interior spaces are furnished with objects made of the abovementioned materials. In addition, a healthy indoor climate requires air exchange via ventilation through windows, mechanical ventilation systems, or infiltration air exchange through air leaks in the exterior building elements (Sect. 5.1.5). This leads

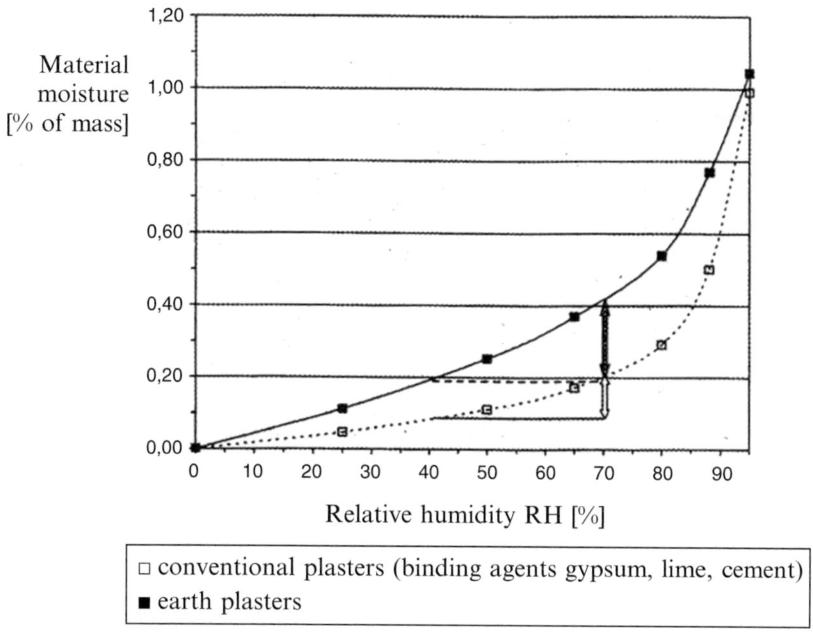

Fig. 5.5 Water vapor sorption of examined earth plasters compared to conventional plasters [9]

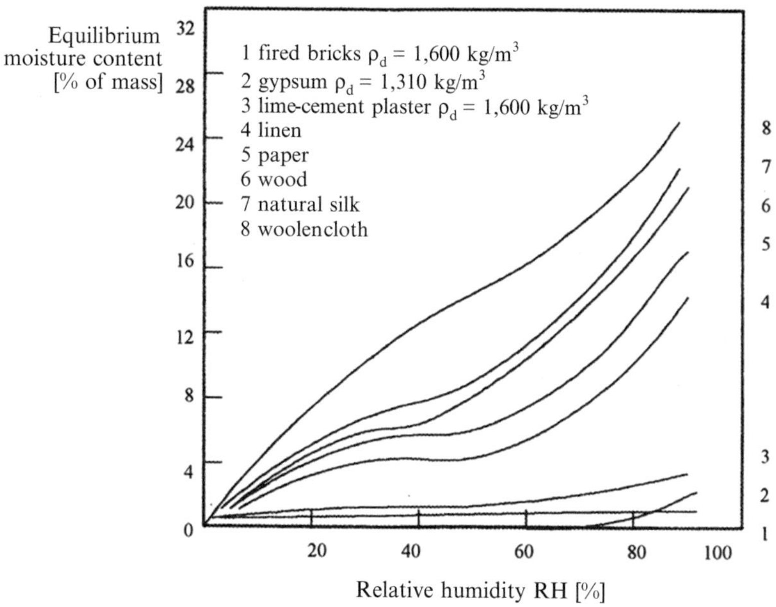

Fig. 5.6 Sorption isotherms for different materials at room temperature [5]

to the fundamental question of the effectiveness of the "buffer" function of space-enclosing earth building elements within the context of normally furnished rooms exposed to the respective ventilation cycles.

5.1.2.6 Erosion Resistance

In the Central European climate, erosion of earth building structures due to wind-driven rain generally does not create structural problems as long as the building has a fully functional roof and foundation (Sect. 5.2.1.2). The well-known phrase that an earth building needs "a good pair of boots and a good hat" refers to an extended roof overhang and a stable foundation with a high stem wall. A loss of building material due to natural erosion during the building's useful life has been "accounted for" in the great wall thicknesses used in traditional load-bearing structures and is only an aesthetic problem (Fig. 5.21).

Under calm conditions, rain falls to the ground vertically under the influence of gravity. Under windy conditions, it hits the wall surface at an angle: the stronger the wind, the steeper the inclination. This is referred to as driving rain or wind-driven rain. The risk of wind-driven rain varies by region and, according to DIN 4108-3, is divided into stress groups according to annual rainfall. Often, local terrains have a greater impact on the risk, for example, whether the building is sheltered from the wind or in an exposed location on top of a hill.

Exposure to wind-driven rain also varies for the individual sections of the exterior walls: corners of buildings are under the greatest stress due to higher wind speeds and dynamic pressure. Surfaces facing the wind (which in Central Europe is southwest to west) as well as higher wall sections are at a greater risk.

In climates with large amounts of annual precipitation and high precipitation intensity, fear of erosion and suspected accompanying structural losses in earth building elements leads to the use of mostly artificial stabilizers (generally cement) in modern earth building.

This fear might not be justified if the building has a well-functioning roof, foundation, and stem walls as well as sufficient wall thicknesses (rammed earth or cob, approx. 50-cm thick). Figure 5.7 shows residential buildings constructed of (non-stabilized) cob in the Bangalore region of India which are approx. 200 years old. These buildings have survived large amounts of precipitation during the annual monsoon (approx. 1000 mm in roughly 4 months) without damage because they are regularly maintained. Edwards [10] describes Cyclone Winifred which devastated the northern coast of Queensland (Australia) and the city of Cairns in February 1986 with wind speeds exceeding 200 km/h and heavy rainfall. Rammed earth buildings which had just been completed as well as older rammed earth structures did not suffer any damage.

The Australian and New Zealand earth building regulations [11, 12] describe a scale model test for analyzing the stress on earthen building element surfaces through exposure to wind-driven rain. In this test, the building element surface

Fig. 5.7 Erosion resistance of cob buildings in a monsoon climate (India)

(building material) is sprayed with a water jet at a pressure of 70 kN/m^2 using a nozzle with a perforated disk (Fig. 5.8). The exposure time t_E of the water jet is predefined based on the annual amount of rainfall and the wind force of the specific site according to the following rule:

$$t_E\,[\text{min}] = \frac{\text{annual rain fall}\,[\text{mm}]}{10 \times \text{wind factor}}$$

Wind factors:

0.5 at an average wind speed with rain = 4 m/s
1.0 at an average wind speed with rain = 7 m/s
2.0 at an average wind speed with rain = 10 m/s

After the respective exposure time, the erosion depth is measured and the average value expected under actual conditions is determined. For this, an empirically determined reference value has been defined which corresponds to the average wind speed of rain under natural conditions (around 7 m/s). This is based on an estimated useful life of the building of 50 years.

In addition, the measurement is multiplied by a safety factor of 2. It is also assumed that moisture penetration in some sections could be 50 % higher than the average erosion depth. Therefore, the erosion depth of these sections is increased by the additional 50 %.

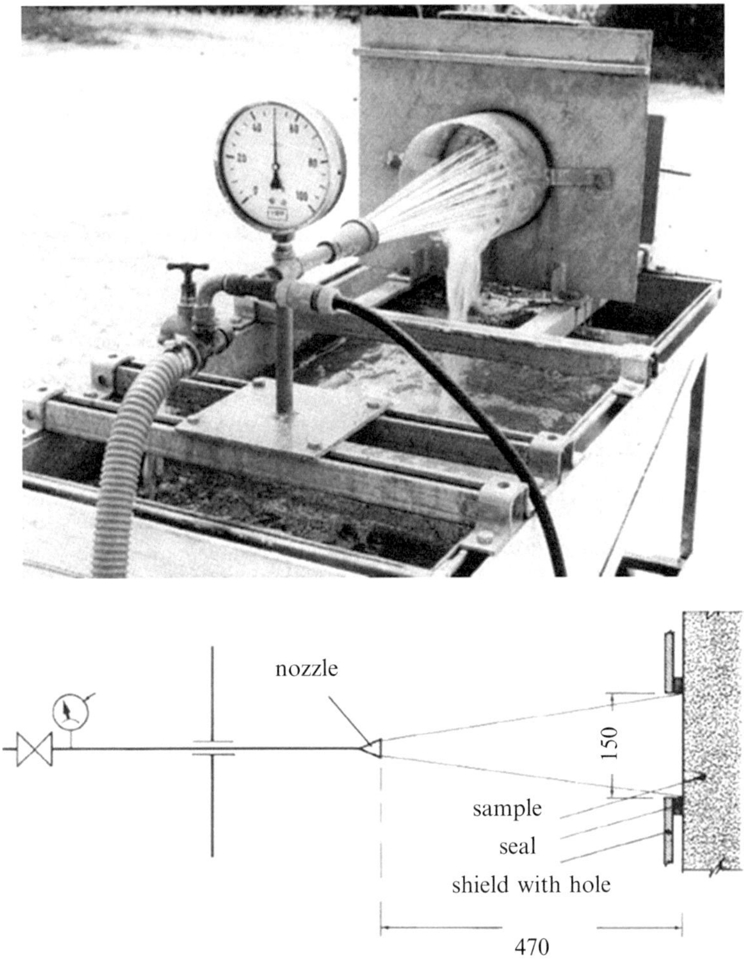

Fig. 5.8 Testing the effects of wind-driven rain on earth building materials according to [11]

5.1.3 *Indoor Climate*

Healthy conditions must be guaranteed in the interior spaces enclosed by the building envelope for the entire useful life of the building. A climate is considered physiologically ideal and comfortable when the human body is in thermal equilibrium with its surroundings.

The way a person perceives climate conditions differs depending on age, constitution, gender, diet, as well as the ability to adapt to the climate. The indoor climate is especially influenced by the following factors:

- Air temperature/radiation from surrounding surfaces
- Relative humidity of the air
- Air movement

5.1.3.1 Comfort Diagrams

The relationships between these factors can be combined in bioclimatic maps, so-called comfort diagrams. In these diagrams, the air temperature and relative humidity are displayed within an optimal human comfort range. The relative humidity of the air describes the ratio of the existing moisture content to the possible moisture content of the air or the corresponding water vapor pressure p. The saturation moisture content of the air corresponds to the saturation vapor pressure ($p_S = 100$ %).

In the Central European climate, indoor temperatures between 20 and 26 °C and RH levels between 40 and 70 % are generally considered to be "comfortable." The difference between the indoor air temperature and the temperature of the surrounding surfaces should not exceed 2 K, while, in order to prevent drafts, the average air velocity should be lower than 0.3 m/s. In tropical climates, the generally perceived "comfortable" temperature level is approx. 2–3 K higher (Lippsmeier [13] Fig. 5.9).

The design of a building always needs to take the most unbearable level of external influences on a person into account. In the Central European climate, this condition is represented by the winter cold which generally requires buildings to be heated. The exterior building envelope must be designed in such a manner that

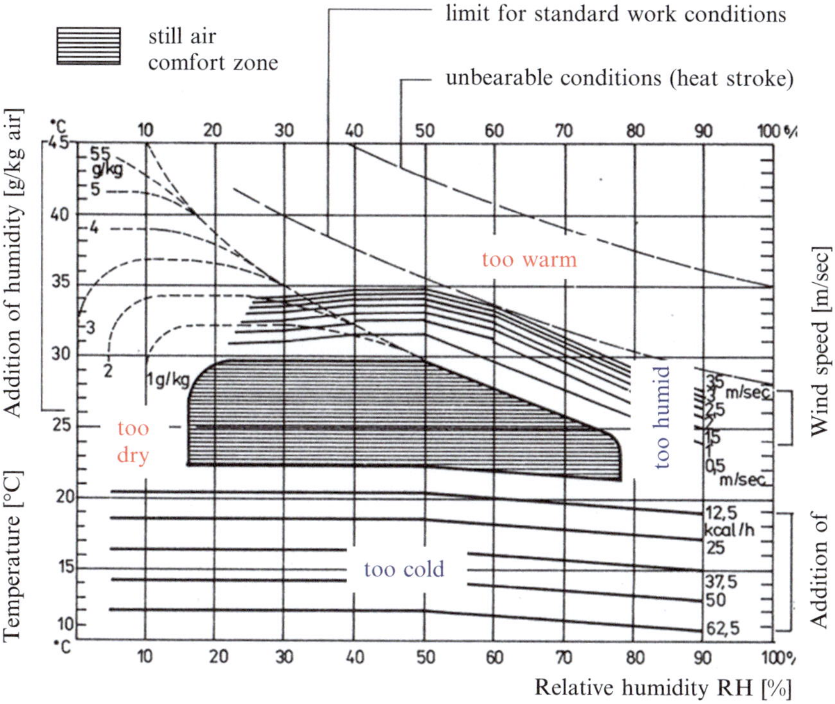

Fig. 5.9 Comfort diagram according to [13]

it primarily protects from the cold and guarantees that the indoor temperature in winter does not fall below the perceived comfort level.

In addition, the building envelope in the Central European climate must provide protection against heat in the summer months. This is mainly achieved with the help of shading devices but also with an adequate ratio of window area to total exterior wall area and suitable air conditioning.

When selecting building materials, it is important to choose materials which can reduce the temperature extremes of the outdoor air in the interior space. These are materials which offer a combination of thermal insulation and heat storage. With earth building materials, this can be achieved by using insulating light clay on the exterior wall and heat-storing high-mass earthen materials on interior walls and in ceilings.

5.1.3.2 Indoor Air Quality

The general requirements placed on the usability of building materials and building elements specified in Regulation EU No. 305/2011 [14] (Sect. 4.2.1.2) are explained in detail in supplemental interpretative documents. Document No. 3 "Hygiene, health and environment" deals, among other things, with the "interior environment of buildings." According to this document, the suitability of a building must also ensure that the structure does not have any harmful impacts on the occupants' hygiene or health. Such impacts particularly include the following:

- The release of toxic gases and hazardous particles into the indoor air
- The emission of harmful radiation
- Moisture accumulation in building elements and on building element surfaces in interior spaces (mold growth)

The following harmful substances are listed in this context:

- Metabolites (excess water vapor, body odor, CO_2) (Sect. 5.2.3)
- Products of combustion (excess water vapor, CO, NO_x, CO_2, C_mH_n, etc.) and tobacco smoke
- Volatile organic compounds (formaldehyde, solvents, etc.) (Sect. 6.2.2.1)
- Inorganic particles (respirable and non-respirable airborne particles and fibers)
- Organic particles and microorganisms (fungi, bacteria, viruses, also including small insects such as bed and house bugs) (Sect. 5.2.3)
- Electrical and electronic device emissions (ozone) as well as radon and radioactive materials (gamma rays) (Sect. 5.1.6)

Healthy indoor air is high in oxygen, odorless, and low in harmful substances. In this context, exhaled air containing CO_2 plays an important role because a CO_2 concentration as low as 0.1–0.15 volume in % is perceived as "stale" air. For this reason, indoor air needs to be regularly replaced by fresh outside air. This process is described using the term *air changes per hour n*. It specifies how often the room's net volume is replaced by fresh outside air per hour.

For health reasons, the air change rate for balancing CO_2 and for removing water vapor should be 0.3–0.5 h^{-1} in residential spaces and 1.0–2.0 h^{-1} in offices. This air change is accomplished via window ventilation and mechanical ventilation systems. For residential spaces, DIN 4108-2 specifies a minimum air change rate of 30 $m^3/(pers\ h)$ based on the number of occupants.

Another, uncontrolled, portion of air changes is the infiltration rate caused by leaks in the building envelope and building element joints. It can range from 0.1 h^{-1} in very airtight buildings to 0.3 h^{-1} in buildings which are less airtight. A very small amount of excess water vapor is transported to the outside by means of diffusion via space-enclosing building elements (Sect. 5.1.2.1).

Unhealthy indoor air is mainly caused by building materials which emit harmful substances. These include materials for flooring and their coatings, room dividers and furniture, walls and wall coverings, insulation materials, paints, varnishes, putties and adhesives, vapor barriers, wood preservatives, technical installations, and masonry materials containing aggregates and additives.

Air change rates have been reduced by an increase in the use of mechanical ventilation systems and a subsequent decrease in the use of window ventilation. This exacerbates the problem of harmful substances contained in the indoor air. The limits which had previously been defined for harmful substances in building products need to be reevaluated if the actual air change rates of rooms are below those of the conditions used in the building material tests.

Naturally formed construction soils are considered to be free of harmful substances and are "recommended based on the principles of building biology" (Sect. 6.2.2.1). However, substances can enter the earth building materials through the addition of aggregates and additives. These substances could be harmful if they exceed a certain concentration or level of exposure. Such effects might, for example, occur in structures made of asphalt emulsion-stabilized earth blocks (Sect. 3.4.2), although no studies have been conducted so far.

Gaseous radioactive materials (radon and its decay products, Sect. 5.1.6) present a particular problem. They can escape from building elements which are in direct contact with the ground or from mineral building products. For earth blocks and earth mortars, DIN 18945–47 recommend adherence to the activity concentration index for natural radionuclide activity.

For earth plasters and clay panels, recommendations have been specified for quantitative limits of harmful substances (Sects. 3.5.6.2 and 3.5.8). Methods for determining different types of indoor air pollution are defined in the DIN EN ISO 16000 series of standards.

5.1.3.3 Wind Tightness

A general requirement for lowering heating energy is the wind tightness of the exterior building envelope. Buildings which are not windtight feel uncomfortable from a physiological perspective: they feel "drafty." Except in summer months, a permanent air velocity which exceeds 0.3 m/s is perceived as unpleasant (Sect. 5.1.3.1).

The sufficient degree of air tightness is defined in DIN 4108-7. According to the standard, buildings with mechanical ventilation systems need to be more airtight than those using window ventilation. The air tightness of a building is determined with a blower door test. Using a fan, a positive or negative pressure of 50 Pa is created between the interior of the building and the outdoor air. The resulting volume flow which is forced through gaps and seams in the building envelope is described by the n_{50} value. This value should not exceed 3.0 h^{-1} for window-ventilated buildings and 1.5 h^{-1} for buildings with mechanical ventilation systems.

With regard to earthen structures, the problem of wind tightness primarily applies to half-timber construction and not to earth building materials in particular. Regardless of the infill material used, half-timber structures can only be windtight to a limited extent. The joint between the wooden framework and the panel infill is an expansion joint which absorbs shrinkage and swelling deformations. From a structural perspective, this joint is the weak link in half-timber construction.

In older half-timber buildings, a 5–10-cm-thick lining of light clay was typically applied to the interior surface of exterior walls of heated rooms. This lining generally provided sufficient wind tightness.

According to the Lehmbau Regeln [15], earth building materials with a density of 900 kg/m^3 or higher are considered windtight. At lower densities, the building element needs to be finished with a plaster. A plaster applied to one side of the building element is considered to provide sufficient wind tightness.

5.1.3.4 Subjective Perceptions

In addition to considerations in terms of the general principles of structure, design, and economics, one aspect of building design which is often neglected is the subjective perception of the indoor climate.

The standard DIN EN ISO 7730 "Ergonomics of the thermal environment" describes and optimizes comfort levels in indoor spaces and their subjectively perceived effects on the user. Individual user perceptions are statistically recorded using the so-called PMV index (predicted mean vote), followed by an evaluation and implementation into the building design.

In a survey conducted for a research project at the ETH Zurich [16], occupants of 22 existing and newly constructed earth buildings in Switzerland and Southern Germany unanimously agreed that their homes felt "comfortable."

5.1.4 Fire Protection Parameters

In general, a distinction is made between the *fire performance of a building material* and the *fire resistance of a building element*. In the event of a fire, both are important and are therefore applied together in the respective building regulations. Whereas the performance of a building material is important in the starting phase of a fire, the term fire resistance describes a building element's behavior during a fully developed fire.

5.1.4.1 Fire Performance of Earth Building Materials

See Sect. 3.4.8.

5.1.4.2 Fire Resistance of Earth Building Elements

The fire resistance of a building element is defined by the time period in which the building element meets the requirements of a fire resistance test specified in DIN 4102-2. These requirements include criteria which are crucial during a fire such as "passage of flame," "increase in surface temperature on the unexposed side," and "guarantee of structural stability."

DIN 4102-2 identifies different fire resistance classes (FKW) for defining the fire resistance of building elements: F 30, F 60, F 90, F 120, and F 180. The numbers indicate the duration in minutes which a building element can withstand fire. F stands for the building element category, in this case: walls, ceilings, pillars and joists, and stairs.

For determining the requirements which are placed on an individual building element, the following terms are used as a combination of FKW and the building material class (DIN 4102-1):

Fire retardant: F 30–B (FWK 30, made of combustible building materials)
Fire resistant: F 90–A (FWK 90, made of noncombustible building materials)

The DIBt has invalidated "Sect. 5.1.4 Fire Performance" of the Lehmbau Regeln [15]. The corresponding Tables 5.7 and 5.8 on FKW of building elements containing earth building materials can therefore no longer be applied. Table 5.4 provides an overview which can be used as a guideline.

These specifications apply to traditional structures (half-timber and timber-frame construction) and have been compiled from older standards, some of which have been withdrawn. No information for modern construction with earth building materials exists to date. Such specifications need to be determined through systematically planned fire tests carried out by approved testing facilities.

As is the case with the fire performance of building materials, the classification of the fire resistance of building elements is currently based on DIN as well as European standards: DIN 4102-2 and DIN EN 13501-2. Compared to the present DIN system, the European classification system is much more detailed allowing for a large number of possible combinations.

5.1.5 Sound Insulation Parameters

Fire protection and thermal insulation are two essential characteristics of a structure which cannot be tested directly using quantitative methods. This is not the case with sound: people inside a building perceive ambient noise from the outside or from

Table 5.4 Fire resistance of building elements containing earth building materials, guideline based on older DIN standards, according to [15]

No.	Building element	Description	Classification	Source
1.	Walls	Solid masonry or rammed earth walls with a thickness of 24 cm	F 90 A	DIN V 18954
2.		Half-timber walls with infill: a minimum timber cross section of 100 × 100 mm when exposed to fire from one side, 120 × 120 mm when exposed to fire from both sides, straw-clay infill, finished with plaster on at least one side	F 30 B	DIN 4102-4:1994
3.	Timber beam ceilings	With fully exposed timber beams, exposed to fire on three sides: ceiling overlay, for example, made of earth building materials of any thickness, depending on the distance between beams and the beams' cross section, sheathing, and floor system	F 30 B to F 60 B	DIN 4102-4:1994
4.		With covered beams: ceiling infill or slatted timbers with earthen loose fill of ≥60 mm depending on distance between beams, sheathing above, and cladding below	F 30 B to F 60 B	DIN 4102-4:1994
5.		Ceiling overlays: only applies to fire exposure from the top, earthen loose fill ≥ 50 mm	F 30	DIN 4102-4:1970

within the structure at all times. The well-being of the occupants is affected if noise inside the building exceeds a certain level and is perceived as disturbing.

An assessment of a structure's sound insulation properties does not analyze material-specific parameters. Instead, it examines the area density of a building element (the mass of the materials in *g* based on 1 m^2 of the building element), as well as its flexural rigidity and tightness. Adjacent building elements (such as walls and cable and pipe ducts) also have an influence on sound transmission.

Values for the sound parameters of building elements can be determined in tests based on DIN 4109-1. For additional structures, these values can serve as the basis for mathematical calculations through extrapolation. Sounds can be distinguished by their transmission method:

- Airborne sounds: sounds which spread through the air, for example, caused by human speech
- Structure-born sounds: sounds which spread through solid materials, for example, impact sounds

Sound insulation played no role in the field of earth building in Germany after World War II. The main focus at that time was on selecting suitable construction soils and determining strength properties for load-bearing construction. Therefore, the DIN standards for earth building from the 1950s do not contain sound insulation parameters derived from tests.

Even today, no systematically planned and executed tests for determining sound insulation parameters for earth building exist. With the development and use of earth building materials in dry construction (starting around 1997), however, sound insulation properties have become increasingly important. Individual German manufacturers of earth building materials have therefore initiated product-specific tests to determine these parameters [17].

As is the case with fire protection, national standards (DIN 4109-1) and European standards continue to apply in the area of sound insulation concurrently. The respective standards are DIN EN 12354-1 for airborne sound insulation and DIN EN 12354-2 for impact sound insulation between rooms.

5.1.5.1 Airborne Sound Insulation of Walls

Terminology

The term *airborne sound excitation* describes the process in which airborne sound in a source room generates vibration in a building element that divides two rooms. This vibration, in turn, creates airborne sound in the receiving room. The building element's resistance to sound transmission is called *airborne sound insulation* [18]. The *sound reduction index R* describes the airborne sound insulation of building elements. It is calculated with the help of the difference in sound levels between two rooms: typically, the source room and the receiving room. The most important reference value for assessing airborne sound insulation is the *weighted sound reduction index* R_w [dB] as an individual value for the simplified designation of building elements.

If sound transmission through adjacent building elements is taken into account, the *weighted sound reduction index* R_w' is given. This flanking transmission does not play a role in sound reduction indices below $R_w' < 48$ dB because $R_w' = R_w$. However, it must be included for higher sound insulation values.

Wall Structure Requirements

DIN 4109-1 specifies the following required sound reduction indices R_w' [dB] for exterior building elements of residential buildings as well as for sound transmission from adjacent living or work areas depending on the exterior noise level (Table 5.5):

Table 5.5 Required sound reduction indices for exterior building elements

No.	Exterior noise level [dB]	Required R'_w for apartments, bedrooms, and classrooms [dB]	Required R'_w for office spaces [dB]
1	up to 55	30	–
2	56–60	30	30
3	61–65	35	30
4	66–70	40	35
5	71–75	45	40
6	76–80	50	45
7	>80	[a]	50

[a]Depending on local conditions

Required R_w' [dB] for interior building elements:

- Partition walls between apartments or offices: 53
- Staircase or corridor walls: 52
- Walls between single family houses, duplexes, or row houses: 57

Values Determined Through Testing and Calculated Values

Sound insulation tests initiated by a German manufacturer of earth building materials [17] to test the company's clay panels show the following results: in a comparison between different traditional wall construction systems using rammed earth, light-clay blocks, earth blocks, green "unfired" bricks, and wood-chip light clay, the building materials with the highest dry bulk density (green "unfired" bricks) have the highest sound reduction index.

The measured values for partition walls consisting of a timber frame covered on both sides with clay panels and a finish earth plaster, with or without a filled wall cavity, are only slightly below the calculated sound reduction indices R_w for traditionally constructed walls. The partition wall shown in Fig. 4.50 has a value $R_w = 56$ dB (cavity $d = 8$ cm insulated using wool). This value corresponds to the sound insulation reduction index of a wall made of light-clay blocks ($d = 36.5$ cm, $\rho_d = 1200$ kg/m^3, $R_w = 55$ dB). This makes this type of partition wall a very suitable design in terms of sound insulation.

5.1.5.2 Sound Insulation of Timber Beam Ceilings

Terminology

The term *impact sound* refers to any structure-born sound which is created on floors, ceilings, stairs, etc. through walking or other activities. It radiates partially as airborne sound directly into the room located below or beyond or propagates as structure-born sound waves. The *impact sound level* L_n is the sound level which is measured in a receiving room when the examined building element—typically a ceiling or staircase—is excited by a standardized tapping machine.

For airborne sound, the insulation properties are described using the airborne sound insulation index. In contrast, the impact sound insulation properties of a ceiling are described using the impact sound pressure level in the receiving room. This means that high-impact sound pressure levels indicate a low degree of soundproofing. The weighted normalized impact sound pressure level $L_{n,w}$ is determined in the same manner as the sound reduction index.

Ceiling Structure Requirements

According to DIN 4109-1, a value of $L_{n,w} = 53$ dB is required for ceilings which separate different units in multistory apartment and office buildings.

Table 5.6 Calculated sound reduction indices and impact sound pressure levels for timber beam ceilings

No.	Ceiling system	Sound reduction index R_w [dB]	Impact sound pressure level $L_{n,w}$ [dB]
1	Slatted timber ceiling with straw clay, approx. 8 cm	Approximately 45	Approximately 72
2	Insert ceiling using earthen loose fill >200 kg/m^2	>54	<60
3	Ceiling overlay using "green" unfired bricks laid with open joints of approx. 2 mm	>51	<53

Calculated Values

In [17], weighted impact sound pressure levels $L_{n,w}$ [dB] are determined for traditional timber beam ceilings which incorporate various earth building materials (Table 5.6):

Additional example calculations are given in [19].

5.1.6 Exposure to Radiation

Media reports about "radiating building materials" often lead to uncertainty among consumers. This uncertainty is largely the result of ignorance with regard to the causes of radioactivity and high-frequency radiation and their health risks.

5.1.6.1 Radioactive Radiation

All living beings on earth, including humans, are exposed to natural high-energy (ionizing) rays. This natural exposure to radiation is divided into unmodified natural exposure and artificial or man-made exposure.

Unmodified natural exposure consists of cosmic and terrestrial radiation as well as the incorporation of radioactive material (ingested with food). It also includes natural exposure modified by humans in the form of building materials and through inhalation of radon within buildings. *Man-made exposure* includes artificial radiation in the field of medical diagnostics and therapy or as a result of the Chernobyl disaster [20].

Parameters

The parameters listed in Table 5.7 [20] are of significance within the context of radioactive radiation of building materials and structures:

In general, the production and use of building materials represent a form of modified natural exposure to radiation. Virtually all standard building materials

Table 5.7 Parameters for radioactive radiation of building materials

No.	Parameter	Measuring unit	Description
1	Activity *A*	Becquerel Bq; 1 Bq = 1 decay/s	Number of decays of a radioactive substance/unit of time
2	Specific activity *a*	Bq/kg	The activity based on the mass unit of a radioactive substance
3	Dose equivalent *H*	Sievert Sv; 1 Sv = 1 J/kg	Assesses the radiation risk for a biological tissue and is calculated as the absorbed dose D with a dimensionless assessment factor *q*: $H = q \cdot D$
4	Absorbed dose *D*	Gray Gy; 1 Gy = 1 J/kg	Energy transmitted by high-energy radiation to matter with a specific mass
5	Dose equivalent rate *h*	Sv/a or mSv/a	A measure for the biological effect of radioactive radiation on humans at an exposure to the dose equivalent *H* of 1 Sv over 1 year

used today contain radioactive substances and thus lead to an increased natural exposure to radiation in buildings. This includes all building materials produced from natural stone and earth as well as certain industrial waste products which are used as building materials or aggregates. This raises the question if earth building materials should also be seen as possible "sources of radiation."

The radionuclides responsible for increasing terrestrial ambient radiation (γ radiation) are potassium-40, radium-226, and thorium-232. In addition, the radioactive noble gas radon-222 develops as a decay product of radium. This gas can leak into the breathable air but, in chemical terms, does not pose a health risk to humans. At certain concentration levels, however, the radioactivity of the inhaled gas and its decay products (which are deposited in the bronchi) can lead to health risks (lung cancer) for humans.

Average radon exhalation (outgassing) from the open ground occurs in the range of 20–80 Bq/m^2 h. It is especially high in areas with volcanic rock (granite, porphyry, etc.) and its weathering products and sediments. Radon outgassing is also higher when an increase in gas permeability (porosity) of the soil is present.

The degree of radon exhalation of building materials and building elements is dependent on the same causes: the radium concentration of the raw mineral material used and the porosity and moisture content of the building material. In connection with architectural structures, this means that the indoor radon concentration level increases as more radon is able to enter the building through cracks and joints in those building elements which are in direct contact with the ground. Small rooms and a high exhalation rate of the space-enclosing building elements can also lead to increased radon levels in the room. Higher air change rates can lower this concentration considerably.

Requirements Placed on Building Materials

According to a recommendation by the German Commission on Radiological Protection, radon levels in indoor air should not exceed a maximum reference value (not limit) of 250 Bq/m^3. In the federal states in Western Germany, the average value of radon concentration is around 50 Bq/m^3, and only 1 % of all apartments exceed the maximum reference value of 250 Bq/m^3. In mountainous regions with radium-rich rock formations which are used as building ground, the radon concentration in indoor air can be much higher.

Research has shown [20] that the concentration level of the radionuclides which are responsible for ambient terrestrial radiation varies considerably between different building materials but also within the individual building material groups themselves. Although most tested building materials do not pose a significant risk in terms of radiation exposure, the results for synthetic gypsum and its intermediate products (bauxite, hard coal fly ash, and red mud) are above the reference value which is considered harmless.

To estimate the concentration level of γ radiation of building materials, different evaluation formulas for comparing building materials have been developed. The so-called Leningrad molecular formula [5] is among them

$$\frac{c_{\mathrm{K}}}{4810} + \frac{c_{\mathrm{Ra}}}{370} + \frac{c_{\mathrm{Th}}}{260} \leq 1\left[\mathrm{Bq}/\mathrm{kg}\right],$$

where C_{K}, c_{Ra}, and c_{Th} are the specific activities of the radionuclides potassium-40 (K), radium-226 (Ra), and thorium-232 (Th) for the tested building material. If the reference value (which is determined from the sum of the partial activities) is ≤ 1, the radiation level of the tested building material does not pose appreciable risks.

Following a recommendation by the European Commission "Radiation Protection 112" [21] from the year 1999 (which was based on the Austrian standard ÖNORM S 5200 "Radioactivity in Building Materials"), an "Activity Concentration Index I" was determined according to the principle of the Leningrad molecular formula:

$$I = \frac{c_{\mathrm{Ra}}}{300} + \frac{c_{\mathrm{Th}}}{200} + \frac{c_{\mathrm{K}}}{3000} \quad \left[\mathrm{Bq}/\mathrm{kg}\right].$$

The I values in Table 5.8 are recommended values for the different building material groups based on the dose equivalent rate h:

The index I value does not make explicit statements about potential health risks. It can only be used as a measure for comparing different building materials. The use of a specific building material in its finished state can be assessed as "not affecting indoor air quality," as "needs further testing," or as "critical" depending on the material's adherence to the defined criteria.

The value for soils and pure clays, which was determined based on the Leningrad molecular formula, is considered harmless, but it should be noted that there are

Table 5.8 Activity concentration index I for building materials based on the dose equivalent rate

Dose equivalent rate [mSv/a]	0.3	1.0
Mass building materials, e.g., concrete	$I \leq 0.5$	$I \leq 1$
Building materials for finishing surfaces, e.g., tiles, panels	$I \leq 2$	$I \leq 6$

significant variations between the individual values. This can be explained by the different levels of radionuclides contained in the parent rock material for soils and pure clays which were formed through weathering and, possibly, transport. In Germany, radium-rich rock formations can be found in Hunsrück, in the Eifel, in the Bavarian Forest, in the Fichtel Mountains, and in the Erz Mountains. Soils and pure clays from these regions might therefore contain higher levels of the abovementioned radionuclides. Their use as earth building materials and earth building elements, however, does not lead to an increased radiation risk in indoor spaces. In these regions, a higher risk can be expected from radon exhalation from the building ground and gas entering buildings through building elements which are in direct contact with the ground. On its website, Germany's Federal Office for Radiation Protection (BfS) has published a general map of radon concentration in soil air at a depth of 1 m [22] (www.bfs.de/de/ion/anthropg/radon/radon_boden/radonkarte.html). The map shows radon expectancy classes based on regional geological conditions.

Triggered by recent discussions about occurrences of radon-222 (a decay product of radon-226) in construction soils and their potential health risks, German manufacturers of earth building materials have initiated Index *I* testing for their products. Eight different earth building materials were tested and the Index *I* values ranged from 0.19 to 0.31. This means that these earth building materials can all be assigned to the "nonhazardous" group.

The DVL has supplemented its Technical Information Sheet TM DVL 05 "Quality Monitoring of Construction Soils" [23] with a recommendation for determining the Activity Concentration Index I based on the radon expectancy classes. This recommendation has been included in DIN 18945–47 as a voluntary test (appendix). It specifies an index value of <1.

5.1.6.2 Shielding from High-Frequency Electromagnetic Radiation

Until recently, structures which were designed and built according to traditional rules of construction provided sufficient protection from various external impacts. The last decade of the twentieth century, however, saw the emergence of a desire for virtually unlimited communication which has led to the development of a new technical medium—wireless data transfer and cellular technology. This form of communication uses high-frequency waves in the range of 10–100 kHz over the entire MHz range up to 150–300 GHz. Transfer is facilitated by an extensive system of transmitters. In 2012, Germany had more than 114 million cell phone users and approx. 85,000 transmitters [www.bundesnetzagentur.de]. This means that the number of cell phone users considerably exceeded the number of inhabitants.

This technological development has created a new type of impact on structures: high-frequency electromagnetic radiation which, for the most common functions of cellular communication and GPS, is in the frequency range of 890–2170 MHz.

The characteristics of HF waves are similar to those of light waves: upon impact with an object, such as a building, they are reflected or they pass through the object

(Sect. 5.1.1.1) and are absorbed. Both occurrences depend on the type and structure of the building material and the characteristics of the specific electromagnetic wave. Whereas radio and television stations used to work almost exclusively with continuous radiation of analog amplitude- and frequency-modulated waves, mobile communications today use pulsed frequencies. These frequencies consist of continuous changes between pulse on and pulse off at millisecond intervals. For example, a cell phone conversation connects 217 times per second between the phone and the cell phone tower. This allows multiple devices to use the same frequency simultaneously.

The physiological impacts of HF radiation on the human body can be divided into two effect groups: thermal and nonthermal. *Thermal effects* are caused by electromagnetic waves entering the human body. They increase the kinetic energy of molecules and lead to a rise in local temperature.

Individual research results also point to *nonthermal effects*. These include changes in behavior, neurological effects, increased cancer risks, sleep disorders, impacts on hormones and metabolism, increased stress, and impacts on genetic material (DNA) [24]. However, the results have provided no clear picture so far.

It is therefore of particular interest to examine the extent to which building materials and building elements can shield HF radiation. A study conducted at the University of the Federal Armed Forces in Munich, Germany [25], tested the shielding effects of building elements made of different building materials at different thicknesses. Independent of the specific material, the shielding effects improved with an increase in the building element thickness.

Figure 5.10 compares a selection of building elements in terms of their attenuation effects in the frequency range of 890–2170 MHz for the most common functions of cellular communication and GPS. The building elements are made of different building materials but have the same thickness of $d = 24$ cm. Compared to other mineral building materials, earth building materials tested at the respective building element thickness showed very favorable results: earth block masonry ($\rho_d = 1600$ kg/m^3) has a significantly higher attenuation effect than vertically perforated bricks ($\rho_d = 1200$ kg/m^3) and sand-lime blocks ($\rho_d = 1800$ kg/m^3) of the same thickness. Even masonry walls made of vertically perforated bricks with a thickness of $d = 36$ cm (but only $\rho_d = 800$ kg/m^3) are less favorable in terms of frequency attenuation.

5.2 Structural Damage as a Result of External Impacts

Based on the external impacts described in Sect. 5.1, structures made of earth building materials are subject to varying degrees of wear during their entire useful life depending on the intensity of stresses [26, 27].

In this context, the term *wear* describes the process or state of material deterioration of the structure or parts of the structure as a result of their use and/or environmental factors. *Structural damage* is present if a building or part of a building can

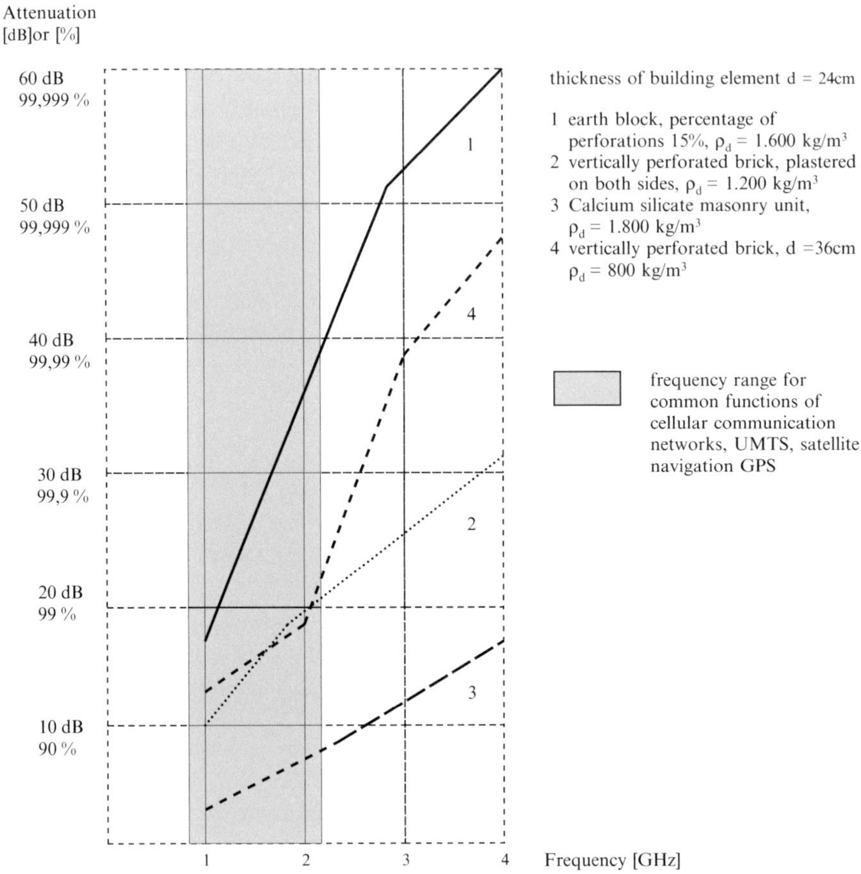

Fig. 5.10 HF transmission attenuation by building elements made of different earth building materials, according to [25]

no longer be properly used. Material wear can lead to structural damage. To repair such damage, suitable renovation measures are required with the goal of regaining full usability (Sect. 5.3). Serious damage can lead to the demolition of the building.

Regular maintenance and repair can delay wear, prevent structural damage, and extend a building's useful life. In addition, there are general design principles which can be observed in the design stages of a building in order to reduce or prevent the negative effects of external impacts on the building envelope.

Depending on their characteristics, the external impacts described in Sect. 5.1 can be divided into mechanical, chemical, and biological processes, although a combination of different mechanisms often occurs. Natural disasters and planning mistakes form separate categories.

5.2.1 *Mechanical Impacts*

Compared to other mineral building materials, earth building materials have lower strength properties and are susceptible to moisture. In terms of their use, this means that areas of activity which are subject to higher mechanical stresses experience more mechanical wear, for example, wall surfaces in high-traffic staircases or edges along door and window openings. The same is true for agricultural livestock buildings.

5.2.1.1 Mechanical Wear

The mechanical wear of earth building element surfaces is the result of exposure to various types of stresses during the building's useful life. Section 3.6.2.2 combines these stresses under the term "wear resistance" and describes respective test methods. Requirements placed on the mechanical stability of earth plasters in terms of wear are described in Sect. 4.3.6.6.

5.2.1.2 Moisture Penetration

Moisture damage to earth buildings leads to a reduction or loss of strength within the earth building element. This is the result of capillary moisture penetration of standing water and/or the erosion effects of running water. The effects of running water become visible as serious damage patterns after only a short period of time. The damaging effects of capillary moisture penetration, on the other hand, show themselves only after some time has passed in the form of gray and dark discolorations on the building element surfaces or peeling paint.

The general diagram depicted in Fig. 5.11 (based on [28]) shows the impact of moisture on structures. Depending on the type of moisture absorption, typical damage patterns for individual building elements can be differentiated between and compared based on the "cause-and-effect" principle.

Foundation and Stem Wall

Defective and moisture-penetrated foundations and stem walls form a core area in the list of structural damage. The concentration of moisture in foundations and stem walls can have numerous causes which frequently overlap and create complex damage patterns.

Building Ground

Causes: groundwater can move freely through coarse-pored soils (gravels, sands). At its upper surface boundary, the water pressure is equal to the atmospheric pressure. In soils with fine and finest pores (clay-rich soils, pure clay), water is

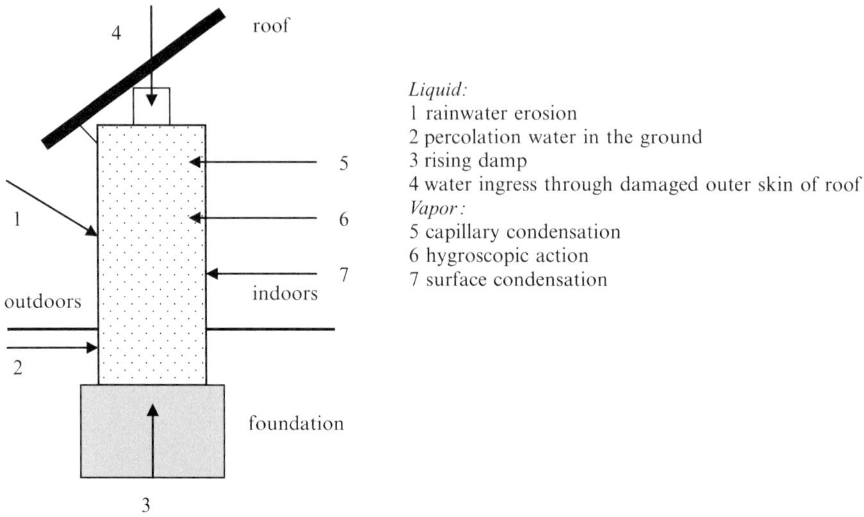

Fig. 5.11 Impact of moisture on structures made of earth building materials

transported by capillary action (Sect. 5.1.2.1). "Confined" groundwater develops when a clay-rich layer or a layer of pure clay is underlaid by an aquiferous gravel or sand layer. The water is under pressure and the upper surface boundary of the pressure-equalizing water level, the "capillary fringe," can reach heights of several meters above ground level.

A missing or defective horizontal groundwater barrier in areas which are in direct contact with the ground enables capillary rising damp to transport salts (which are dissolved in the ground or in the building materials) into the rising earth walls. Pore water evaporates on the wall surfaces leaving salts behind which collect in the pores. In this process, readily soluble salts rise higher than salts which are difficult to dissolve.

Figure 5.12 [6] provides an overview of the types and distribution of harmful salts in stem wall areas. Sulfates typically originate in the ground or in the building materials themselves. Chlorides mostly come from road salt. Nitrates occur in the vicinity of nitrogen sources, which in rural areas include livestock barns, cesspits, and manure pits. In the seventeenth and eighteenth centuries, nitrates appearing as saltpeter efflorescence on the surfaces of earth walls (sodium nitrate $NaNO_3$) even gained importance in military circles as an ingredient for gunpowder [29].

Effects: the crystallization of salts, frequently connected to an accumulation of water molecules (hydration), leads to an increase in volume. The volume increase and repeated freeze–thaw cycles destroy the structural strength of the earth building materials in the affected zone. The lower strength of earth building materials compared to brick and cement results in a faster decay of the material. Shell-like flaking of the loosened areas leads to a weakening in the structurally effective cross section of the load-bearing exterior walls (Fig. 5.13). As a result, the salt-loaded material becomes virtually pulverized on the outside. Through the chemically altered clay

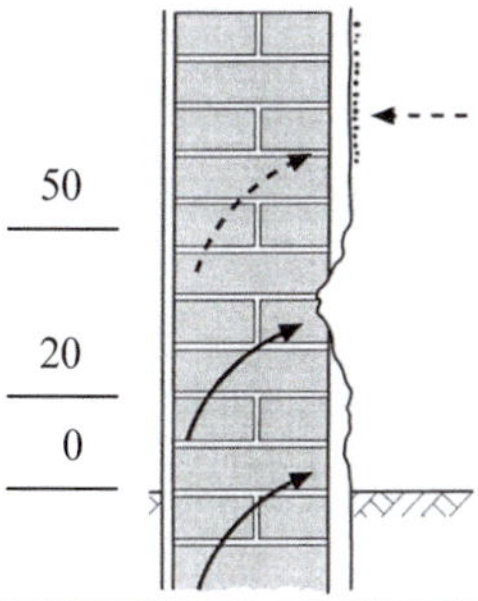

name	chemical formula	solubility per 100 ml of water
calcium nitrate	$Ca(NO_3)_2$	226
sodiumnitrate	$NaNO_3$	92
calcium chloride	$CaCl_2$	75
halite	NaCl	39
potassium chloride	KCl	24
Glauber's salt	$Na_2SO_4 \cdot 10H_2O$	92
magnesium sulfate	$MgSO_4 \cdot 7H_2O$	71
potassiumnitrate	KNO_3	13
gypsum	$CaSO4 \cdot 2H_2O$	0.3

Fig. 5.12 Types and distribution of harmful salts around the foundation and stem wall [6]

Fig. 5.13 Weakening of the cross section of a load-bearing earth wall caused by salt attack in the foundation and stem wall

minerals, the plastic properties (but also the strength of the soil) are largely lost. This limits or prevents the material's reuse as recycled earth.

Insufficient or Missing Drainage

Causes: defective or missing gutters or rainwater pipes cause stresses in the stem wall area through running water or splash back (Fig. 5.14 [30]).

Drainage around the stem wall often functions improperly or is missing altogether. Particularly in rural areas, leaking wastewater pipes and cesspits and manure

Fig. 5.14 Structural damage caused by a defective rainwater pipe and a water-penetrated half-timber wall with earth infill [30]

pits near foundations are quite common. They increase the salt load contained in the rising damp in the walls.

Effects: the same as for "building ground."

Earthen wall building materials which are exposed to these impacts become moisture penetrated and wear away after only a short period of time. The "excess" of moisture can also lead to biological impacts (Sect. 5.2.3).

Missing or Damaged Foundation and Stem Walls

Causes: oftentimes, older structures do not have stem walls or foundations made of water-impermeable materials, or they have become covered up by raised street levels. In many cases, the visible remains show washed-out masonry mortar (Fig. 5.15 [30]). Additional moisture frequently comes from water which has collected as a result of storing building waste or building materials around the stem wall or against higher sections of the earth wall. In agricultural settings, this area is also a popular location for manure piles (Fig. 5.16).

Effects: the stem walls have lost their original function of providing splash water protection. The wall sections lying within the capillary fringe exhibit gray and white films of crystallized salts or peeled-off paint (Fig. 5.17 [30]). In addition, water penetration of exterior walls made of earthen materials leads to a reduction in thermal insulation properties and provides ideal conditions for the development of mold inside the building. Other effects are the same as for "building ground."

Fig. 5.15 Structural damage in the stem wall area caused by raising of the street level [30]

Fig. 5.16 Structural damage in the stem wall area: salt attack, intensified by a manure pile

Fig. 5.17 Structural damage in the stem wall area: crystallized salts visible as a gray to white film within the capillary fringe [30]

Exterior Wall Surfaces

The extent to which exterior wall surfaces are penetrated and eroded by rain is not only determined by the abovementioned external impacts. Additional factors are the properties of the wall building materials themselves, their use, as well as the design of the facade. Dense building materials with low absorption capacity absorb less water than coarse-pored building materials. Smooth and solid surfaces allow water to run off more quickly than uneven surfaces. This can, however, also lead to more erosion. Rough surfaces and edges, which inhibit drainage and allow snow to collect, promote moisture penetration of the wall surfaces.

Which construction method is used for the exterior earthen surfaces is therefore also important: rainwater drains more slowly and less thoroughly from textured wall surfaces in exposed half-timber construction than from even, plastered wall surfaces made of rammed earth or cob. Typical damage patterns vary accordingly.

Exterior Surfaces of Half-Timber Walls with Earth Infill

Causes: tests at the ZHD Fulda have shown [31] that the moisture absorption of half-timber walls with earth infill during rain occurs primarily along the shrinkage joints between the infill and the timber frame. Water enters quickly along the edges of the earth infill. After the rain has stopped, however, these sections also dry again quickly, assisted by air movement. Watertight joint sealers or waterproof paints applied to the wooden structure in the course of restoration measures delay and inhibit this mechanism and lead to longer periods of moisture penetration along the edges of the panels. In the worst case, plastered infill panels can experience spalling of the plaster along the joints after freeze–thaw cycles. This can also negatively affect the load-bearing properties of the timber construction.

Water-repellent or watertight paints and plasters applied to the infill or timber frame can lead to problems for the same reason outlined above. Cracks in the paint or plaster would allow rainwater to enter through this "watertight" shell and dry out only very slowly or not at all.

Capillary moisture absorption via the earth infill of the panel is relatively low compared to moisture absorption via the joints. In this context, the type of infill also plays a role. In the abovementioned tests [31], the moisture absorption of an infill of earth blocks laid in lime mortar was considerably higher than that laid in earth mortar. This can be explained by the higher absorptive capacity of the lime mortar. Infills using light straw clay and straw clay over stakes displayed the lowest moisture absorption. Here, moisture penetration did not exceed more than a few millimeters. This is a result of the swelling capacity of the clay minerals in the soil which prevent a deeper penetration of water into the wall via a "sealing" effect.

Effects: a long-term effect of capillary moisture penetration and subsequent drying of exterior surfaces of half-timber walls is the weathering of the earth infill in the panels. Typical damage patterns arise which depend on the type of infill used. Soil components wash away and expose the organic fiber content. In straw-clay infills over stakes or a woven lattice, the stake work becomes visible (Fig. 5.18). In panels

Fig. 5.18 Wall surface of a half-timber structure: exposed stake work caused by erosion of the earth building material [30]

with earth block infill, the mortar gets washed out. Professionally applied plasters and paints can delay this process.

The eroding effects caused by defective gutters and rainwater pipes not only lead to the washing out of the earth building materials from the infill panels but also cause damage to the load-bearing timber frame. In the affected areas, this can result in a complete loss of the structural integrity of the construction. Dense undergrowth delays a quick drying of the water-penetrated wall surfaces and, in addition, provides a habitat for rodents which can contribute to a decrease in strength of the earth walls by impacting the structure mechanically and chemically (Sect. 5.2.3).

Exterior Surfaces of Earth Block Walls

Causes: in moderate climates, capillary moisture penetration in earth blocks due to wind-driven rain only reaches a depth of a few millimeters as the swelling of the clay minerals prevents further transport. Earth masonry mortars have a higher sand content and therefore quickly absorb a large amount of water. This makes them less erosion resistant and leads to deep erosion of the joints. Depending on the depth of the erosion level of the earth masonry mortar, the protruding earth blocks can break off up to this same level (Fig. 5.19a).

Although lime masonry mortar is more weather resistant, it also has a higher absorption capacity than earth masonry mortar. This can cause moisture to penetrate deeper into the earth blocks via the bedding and head joints, leading to more extensive loosening of the earth block structure and a subsequent loss in strength during freeze–thaw cycles.

Due to the increased amount of precipitation in humid tropical climates, the eroding effects of wind-driven rain on earth block surfaces occur considerably faster and are more severe (Fig. 5.19b). Earth blocks are therefore frequently stabilized with cement in these regions.

Effects: similar to half-timber construction, properly functioning gutters and rainwater pipes are of utmost importance. Rainwater which drains onto earth block wall surfaces at concentrated flows quickly creates deep erosion channels in the moisture-penetrated area, leading to a weakening or a complete loss of the structural integrity of the masonry wall in the affected section.

Professionally applied plasters and suitable paints can provide additional protection to exterior surfaces of earth block walls. For weather-exposed sides impacted by wind-driven rain, lime plasters offer a higher degree of mechanical protection against pounding rain. However, compared to earth plasters, they have a higher moisture absorption capacity. They also have lower shrinkage and swelling tendencies than the covered earth blocks. These differences in deformation behavior can quickly lead to the separation of the lime plaster from its substrate.

In the 1940s and 1950s, the construction of exterior weather-exposed walls using "green" unfired bricks and extrusion-shaped earth blocks (initially intended for firing) led to distinct damage patterns. Water entering the wall activated (production-related) flow planes in the earth blocks (Sect. 3.2.2.1) which led to material scaling off the blocks in the affected sections as seen in Fig. 5.20 [32]. This image clearly shows the stronger masonry mortar, most likely a lime mortar.

a

b

Fig. 5.19 Wall surface of earth block construction; erosion. (**a**) Washed-out sandy earth masonry mortar and breaking off of protruding earth blocks in a moderate climate [30] and (**b**) advanced erosion caused by wind-driven rain in a tropical monsoon climate (India)

Exterior Surfaces of Cob and Rammed Earth Walls

Causes: professionally constructed rammed earth and cob walls have smooth surfaces. They allow rainwater to drain more easily but also lead to faster erosion. Evenly draining rainwater results in even weathering of the straw-clay or cob surfaces (Fig. 5.21 [30]). In the climate of Central Europe (depending on local conditions), erosion does not exceed a depth of a few centimeters over the course of several decades and typically does not affect the structural stability of the building.

Effects: Fig. 5.22 [33] shows the erosion effects of draining rainwater. Missing gutters and an insufficient roof overhang have led to deep erosion channels in the sur-

Fig. 5.20 Structural damage to weathered exterior wall surfaces of "green" unfired bricks [32]

Fig. 5.21 Cob wall surfaces: "normal" erosion due to weathering [30]

Fig. 5.22 Cob wall surfaces: erosion channels formed by concentrated flows of draining rainwater [33]

face of the wall. The channels are cut into the cob wall below dips in the roofing panels.

In contrast to earth block wall surfaces with masonry joints, plasters applied to a rammed earth substrate have little mechanical adhesion. Similar to earth blocks, lime and cement plasters on a moisture-penetrated substrate quickly lose adhesion and brake off in chunks (Fig. 5.23 [30]).

Eaves

Causes: typical causes for moisture penetration of earth walls along the eaves include holes in the outer skin of the roof, defective or clogged gutters, and insufficient roof overhangs. Entering rainwater softens the wall coping and drains along the wall surfaces.

Eaves are also popular areas for birds to build their nests. Bird droppings can have a negative aesthetic effect on the wall surfaces as well as chemical impacts on the coping. They lead to a reduction in strength of the coping and problems around the bearing surfaces of roof and ceiling beams.

Effects: figure 5.24 [33] shows a deep erosion channel in the exterior surface of a cob wall as a result of a leak in the outer skin of the roof. In the worst case, roof beam and rafter supports are washed away, leading to a weakening of the supporting structure of the roof system (Fig. 5.25 [30]).

If buildings are located in close proximity to each other, splash back along the eaves can lead to erosion of the wall surfaces of adjacent buildings (Fig. 5.26 [33]).

Fig. 5.23 Cob wall surfaces: rigid cement plaster separated from the substrate

Fig. 5.24 Damage along the eaves: deep erosion channel caused by a hole in the outer skin of the roof [33]

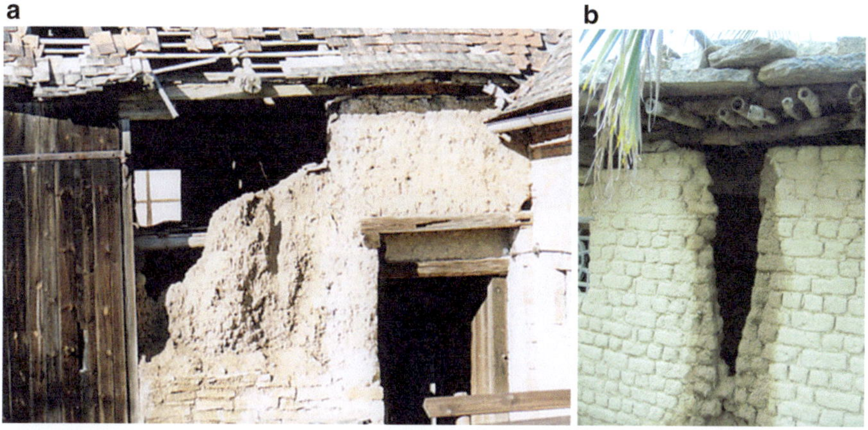

Fig. 5.25 Damage along the eaves: collapse caused by a hole in the outer skin of the roof. (a) Moderate climate [30] and (b) monsoon climate, India

Fig. 5.26 Damage along the eaves: erosion caused by splash back onto an adjacent building, insufficient distance between buildings [33]

Accidents Involving Running Water

Causes: during the entire useful life of earth buildings, all areas involving running water pose potential damage risks to load-bearing and non-load-bearing earth building elements. Clogged drains represent the main risks in kitchens and bathrooms, while frozen water can lead to pipe bursts in unheated rooms in the winter. A "secondary" accident situation can be caused by water used for fighting a fire.

Fig. 5.27 Accidental water damage: soaked earthen ceiling and subjacent half-timber wall with earth infill caused by a burst water pipe

Effects: water accidents result in moisture penetration of adjoining building elements when water accumulates on floors or ceilings or when drainage is uncontrolled. Figure 5.27 shows water damage to an earthen ceiling and the subjacent half-timber wall with a straw-clay infill. After a water pipe burst, the ceiling became completely soaked and the earth plaster washed off.

Another risk comes from pipe coils used in wall and floor heating systems which are installed directly into earth plaster or earthen ceilings. Defective pipes can lead to water leaks which remain unnoticed for longer periods of time. Example defects include holes in the material of the pipe coils as well as damage caused by attaching anchors to walls containing pipes (Sect. 4.3.7.3).

5.2.2 Chemical Impacts

Moisture transported by capillary action not only leads to a loss in mechanical strength, it can also have chemical impacts (Sect. 5.1.2.1).

In combination with the effects of wind-driven rain on exterior wall surfaces of earthen structures (which were described in Sect. 5.2.1.2), various climatic elements such as air temperature and humidity as well as solar radiation intensity and air pollutants (Sect. 1.4.3.3) lead to the chemical disintegration (aging) of the earthen material's mineral content over the course of the building's useful life [27].

The primarily silica-based raw materials split into ions of the alkaline and alkaline earth groups, into Fe and Al oxides, as well as into SiO_2 residue (Sect. 2.1.1.3). Depending on the prevailing climate, clay minerals are also broken down in the process: the more complex three-layer minerals (montmorillonite) break down into two-layer minerals (kaolinite) with a simpler structure.

In hot and humid climates, this process occurs considerably faster and more intensively (Sect. 2.1.2.6) than in moderate climates, leading to a loosening of the structure and a decrease in strength of exterior earthen surfaces. In hot and dry climates, sand storms transport salts which have a similar effect as wind-driven rain: upon impact, grains of sand and salt cause a mechanical loosening of the building element surfaces. In the presence of moisture, the remaining salts promote crystal growth which, in turn, can lead to additional structural loosening.

5.2.3 Biological Impacts

Causes: the roots of bushes or trees growing around the stem wall of a building (e.g., elder bushes, Fig. 5.28) penetrate cracks and joints of foundations, stem walls, and even earthen walls. Trees can also influence the local groundwater situation.

Fig. 5.28 Biological impacts: rampant elder bushes around a foundation [30]

In addition, shade created by trees and bushes prevents a quick drying of the stem wall after rain or snowmelt. This can promote the growth of mosses, algae, bacteria, and mushrooms.

Dry rot is the most dangerous problem for installed wood, especially pine. True dry rot also attacks dry wood and, via a network of strand-like mycelia, channels the water necessary for growth over distances of several meters (e.g., from moisture-penetrated cellar walls).

Rodents (rats, mice) and parasites often enter earthen walls via previously damaged stem walls. The droppings of nesting birds can cause additional damage to the affected areas. Walls made of earth building materials also provide suitable habitats for insects (bumble bees and termites in hot and dry climates).

Effects: water can enter the damaged wall and is distributed by capillary action.

By dissolving the cellulose, dry rot destroys the load-bearing wood frame in half-timber construction. Under favorable conditions, fungal spores can survive in infested building elements for years. Therefore, earth building materials salvaged from affected building elements should not be reused (Sect. 6.2.2.1).

Rodents and insects loosen the structure leading to a decrease in the load-bearing capacity of earth walls. Rodent droppings and urine further reduce the strength of previously damaged walls. Figure 5.29 shows a wall hollowed out by termite channels in the Bam Citadel, Iran, which was destroyed in an earthquake in 2003.

Fig. 5.29 Biological impacts: termite channels in an earth wall, Bam Citadel, Iran

In the regions of Central and South America, occupants of earth buildings face a particular health risk in connection with unplastered or cracked earth building elements. It appears that a certain type of bug which transmits the Chagas disease (a kind of sleeping sickness) finds particularly favorable living conditions in such structures. According to WHO estimates, more than ten million people are presently infected. Construction measures developed to avert this health risk include strategies for removing cracks in the building element surfaces (floors, walls, ceilings), for example, by installing dense, mesh-reinforced screeds and lime and cement-based plasters.

Animal husbandry in barns represents a special type of biological impact on structures made of earth building materials. Together with water vapor and dust, strong odors and pathogenic germs are released into the indoor air. These odors and germs are neutralized through their integration into the clay mineral structure of the earth building material in the form of volatile organic compounds. This continuously regenerates the microclimate in the barn. Bielenberg [34, 35] refers to this process as the "regeneration capacity of the building material": the ability of the material to adjust the continuously changing, pathogenic indoor conditions to the physiological outdoor conditions (thus stabilizing them for the animals). Depending on the amount of moisture in the barn, this regeneration capacity is limited to a time period of 15–30 years. Straw-clay building elements are particularly absorbent but decompose in the process. For hygienic reasons, disassembled earth building elements from barns (plaster, ceilings) should therefore not be reused (Sect. 6.2.2.1).

5.2.4 *Natural Disasters*

In the twentieth century, four million people worldwide were victims of natural disasters, primarily in the geographic regions of Asia and the Pacific, but also about 7 % in Europe [36]. Approximately half of all disasters were caused by earthquakes and one third by flooding. In rural areas of the two most populated countries, China and India, existing buildings are primarily constructed of earth. Here, but also in most developing countries in arid climates, earth is and will remain (for the unforeseeable future) the only available building material for the majority of the population. Therefore, disaster prevention needs to start in the planning stages of construction.

For industrial nations, disaster prevention will take on a more important role in the context of development work with these countries. Informing people about how safe buildings are constructed can save lives.

5.2.4.1 Flooding

Earthen structures must not be built in potential flood zones. Particularly in arid climates, river beds can be dry for long periods of time. Due to nonexistent or insufficiently enforced building regulations, buildings are repeatedly constructed in such endangered areas, with devastating effects during floods (Fig. 5.30). Due to climate

Fig. 5.30 Flooding of the Nile in Khartoum, Sudan, 1988: collapse of numerous earth buildings [UNDP, World Development, May 1989]

changes, regions which were once considered safe can now also be affected by flooding.

In the traditional earth building regions of sub-Saharan Africa, exceptionally heavy rains led to extensive flooding and the destruction of numerous existing earth buildings in August and September of 2007 [37]. In October 2008, the famous earth block tower houses of the Wadi Hadramaut in Yemen were affected by catastrophic flooding.

In Central Europe, as well, the number of extensive floods is on the rise and can affect earth buildings which were constructed in regions which traditionally appeared to be safe. Among the structures destroyed by the flooding of the Elbe River in the German state of Saxony and in the Czech Republic in 2002 and 2006 were also numerous earth buildings [38].

A practical guide for the restoration of flood-damaged earthen structures (adobe) has been published by the organization Cornerstones Community Partnerships [39].

5.2.4.2 Earthquakes

Causes

Earthquakes are primarily caused by tectonic movement in the Earth's crust. The Earth's solid rock mantle is not homogeneous or of even thickness. Instead, it is broken into large plates which "float" on a core of liquid magma. The plate boundaries

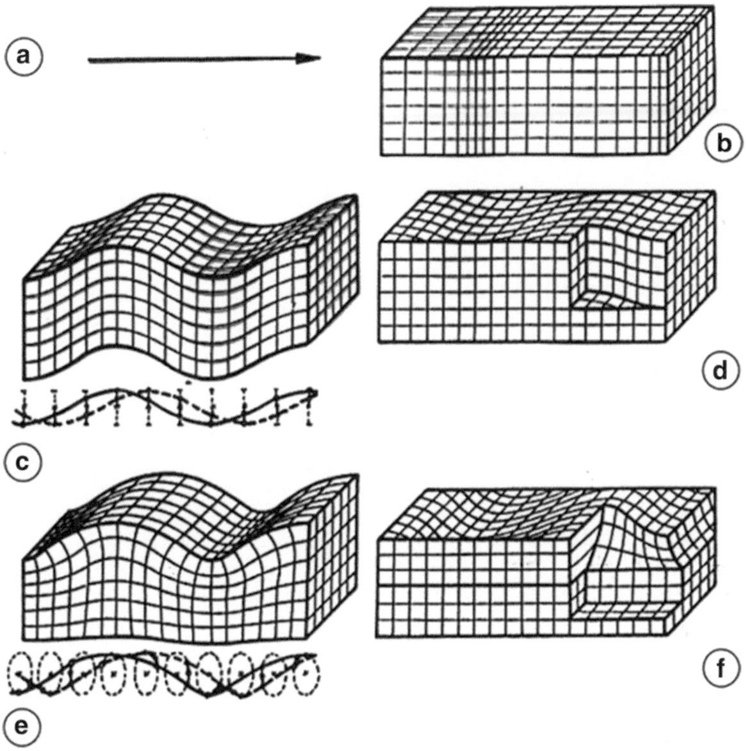

Fig. 5.31 Deformation of subsoil during passing of earthquake waves [40]. (**a**) Direction of wave, (**b**) compaction wave (P), (**c**) shear wave with vertical oscillation (S), (**d**) shear wave with horizontal oscillation (S), (**e**) "Rayleigh" wave (R), and (**f**) transverse oscillation ("love" wave, L)

slide past each other or on top of each other or lift each other up. This creates stresses in the form of friction energy in the faults between the plates. When the rock strength limits are exceeded, this energy is released as countermovements in the plate boundaries. The movements occur without warning and within seconds. After an earthquake, new stress potential can develop along the fault lines. After some time, this stress potential is released again in the form of movements which can be felt as "tremors" on the Earth's surface.

The movements of the plates propagate as earthquakes from the *hypocenter* (the center of movement) in the form of body or surface waves which move through or across the entire globe in all directions. The cubic lattice model in Fig. 5.31 [40] shows the nature of earthquake waves: when waves pass through, the rock or building ground expands and contracts within short periods of time. Waves propagating in different directions, but also waves with different duration and amplitudes, can overlap during an earthquake, creating complex stress patterns within the rock, building ground, and the structures constructed on it.

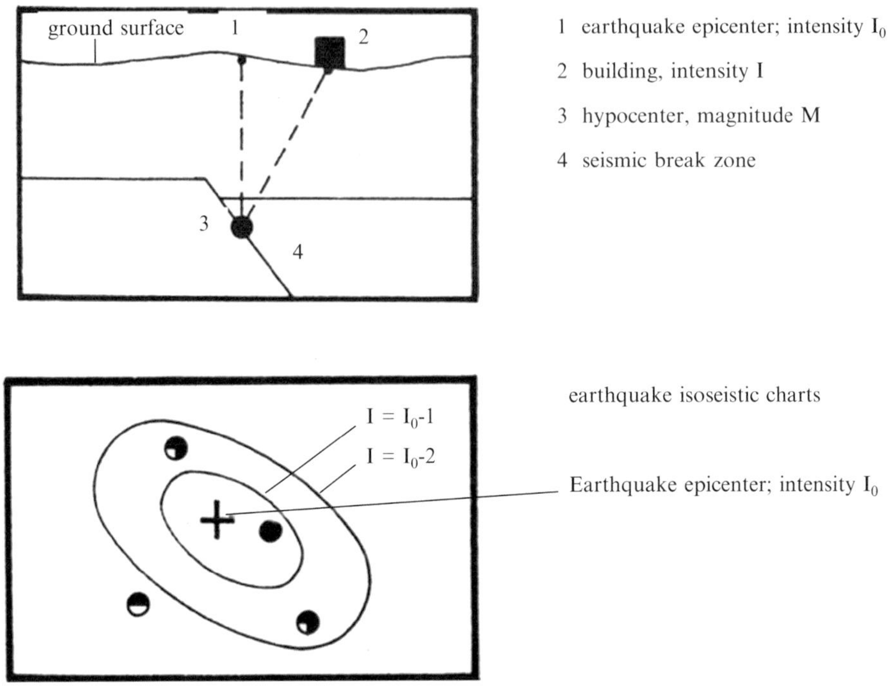

Fig. 5.32 Magnitude *M* and intensity *I* of earthquakes

Impacts on Structures Made of Earth Building Materials

The *Magnitude M* or intensity of an earthquake is measured using a seismograph and is determined with the help of an open-ended magnitude scale (Richter) as a measure of the amount of energy released. Major earthquakes have a magnitude between 7.0 and 7.9, while most earthquakes are below 3.5. A magnitude 7.0 earthquake corresponds to the amount of energy released by ten Hiroshima bombs.

The *Intensity I* quantifies the effects of an earthquake on the Earth's surface, for example, on buildings. In contrast to the magnitude, it is expressed using the Roman numerals I to XII based on a 12-degree scale of macroseismic intensity (designed in 1897 by Mercalli and later modified numerous times, e.g., MSK-64). Depending on the degree of damage on the Earth's surface, this classification differentiates between weak (I=I to IV), strong (I=V to VII), and very strong earthquakes ($I \geq$ VIII). Intensity I=XII is virtually never reached. Figure 5.32 shows the intensity and magnitude of an earthquake in a diagram: the greater the distance between the building and the epicenter, and the deeper it lies below the Earth's surface, the lower the degree of damage.

Particularly in the event of earthquake damage to earthen structures, the question of the "earthquake resistance" of these buildings arises. In this context, the media frequently blame the building material earth by making sweeping generalizations. This is especially the case after earthquakes which have led to massive destruction

Fig. 5.33 Damage patterns after earthquake impacts: intact earth block vault structures, Bam, Iran, 2003

and the loss of human life. As a result, the use of earth as a building material often becomes restricted or prohibited by the building authorities.

Damage patterns in buildings have shown that the damages incurred are not primarily a result of the building material used. Figure 5.33 shows vault structures made of earth blocks in the Bam Citadel, Iran, which are still intact after the devastating earthquake of 2003. In contrast, nearby buildings made of reinforced concrete were completely destroyed. In this particular case, fundamental principles of earthquake-resistant construction [41, 42], independent of the building materials used, were not observed in the planning stages of construction or had been neglected during subsequent remodeling. Mistakes in planning and, above all, bad workmanship coincided with the natural seismic impacts.

Assessment of Seismic Resistance

Macroseismic scales (e.g., [43]) are suitable tools for assessing the seismic resistance of existing structures and for determining appropriate measures for their seismic retrofitting. This process should be carried out in a predefined order which is described in [44]:

- First, the *seismic hazard* of the specific geographic area is assessed, independent of the structure. It describes the probability of an earthquake occurring in a given location, within a given time period, and with a given intensity. The seismic hazard is defined in respective national standards of individual countries.

– Next, the *expectation of damage* is assessed. For this, structures are evaluated in terms of their *vulnerability* to earthquakes of a specific magnitude based on their construction method. According to the European Macroseismic Scale (EMS) 98 [43], all structures of a specific construction method group can be assigned to an earthquake vulnerability class A–F. Class A represents the buildings with the lowest level of earthquake resistance (Fig. 5.34a). Structures made of earth

a

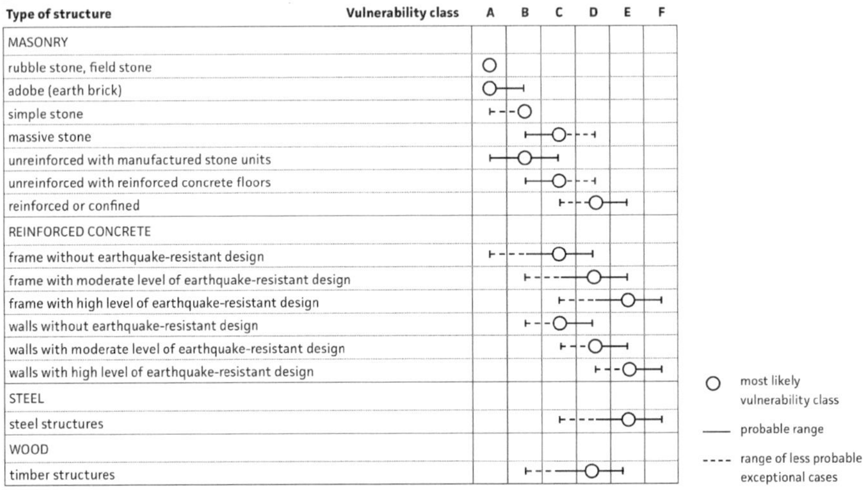

b

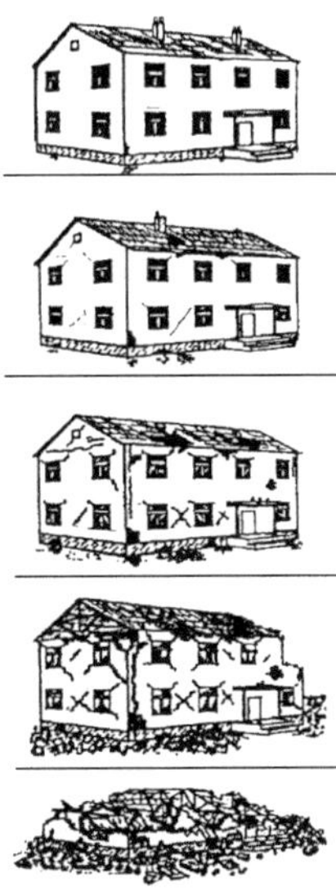

Grade 1 – Negligible to slight damage
(no structural damage, slight non-structural damage)
Hairline cracks in very few walls. Detachment of small pieces of plaster only. Falling of loose stones form upper parts of buildings in very few cases

Grade 2 – Moderate damage
(slight structural damage, moderate non-structural damage)
Cracks in many walls. Detachment of large pieces of plaster. Partial collapse of chimneys.

Grade 3 – Substantial to heavy damage
(moderate structural damage, heavy non-structural damage)
Large and extensive cracks in most walls. Roof tiles detach. Chimney's fracture at the roof line. Failure of individual non-structural elements (partitions, gable walls).

Grade 4 – Very heavy damage
(heavy structural damage, very heavy non-structural damage)
Serious failure of walls. Partial structural failure of roofs and floors.

Grade 5 – Destruction
(very heavy structural damage)
Total or near-total collapse.

Fig. 5.34 Vulnerability classes (**a**) and damage grades (**b**) according to the European Macroseismic Scale (EMS) 98 [44]. (**a**) Vulnerability classes of different types of structure and (**b**) damage degrees of masonry buildings building materials using traditional construction methods (earth blocks) are assigned to vulnerability classes A–B. Through seismic retrofitting, buildings can "move up" to a "better" vulnerability class.

- The expected damage to a structure is primarily determined by earthquake intensity and the construction method used. All structures of a specific construction method group, which are described with the help of the vulnerability classes, are therefore assigned a *damage grade* based on earthquake intensity. In Fig. 5.34b, the damage grades according to [44] are described in words in combination with the numerals 1–5: from 1, negligible damage, to 5, complete destruction.
- The EMS 98 [43] then links vulnerability classes and damage grades to earthquake intensities:
  Examples for intensities VIII and IX:
  Intensity VIII: many buildings of vulnerability class A and some of class B suffer damage of damage grade 4; some buildings of class A suffer damage of damage grade 5.
  Intensity IX: many buildings of vulnerability class B and some of class C suffer damage of damage grade 4; many building of class B suffer damage of damage grade 5.
  Damage quantities are defined as follows:
  Some: 0–15 % of all buildings
  Many: 15–55 % of all buildings
  Most: 55–100 % of all buildings
- Data collected in field studies is used to classify earthquake-related damage into damage grades according to the respective construction methods. With this information, the distribution of damage specific to the construction method, the so-called damage grade index, can be determined based on the known intensity of the earthquake. Figure 5.35 shows the damage grade index after two earthquakes of the intensities VIII and IX occurring within five weeks of each other in Gazli, Uzbekistan, in 1975. Structures using the following construction methods are included: earth block masonry, brick masonry, and half-timber construction with

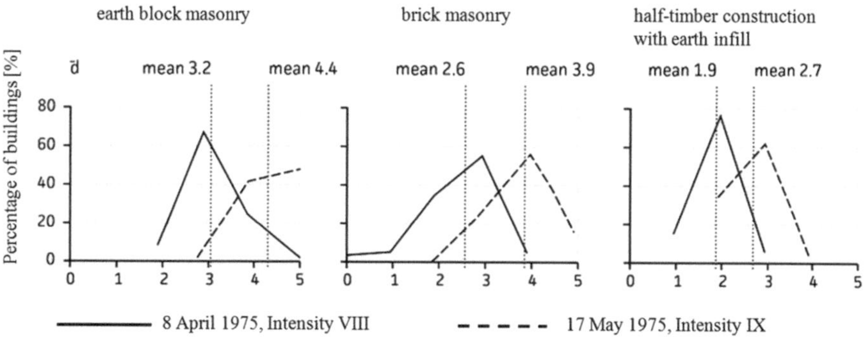

Fig. 5.35 Mean damage degrees (damageability index) for different wall construction methods after two earthquakes (intensities XIII and IX) in Gazli, Uzbekistan, 1975 [44]

earth infill. Based on the obtained data, a comparison of the construction methods showed that the earth block masonry buildings suffered the most damage, with a damage grade index of 3.2 at an intensity of VIII or 4.3 at an intensity of IX.

Using this process, the average damage grade of typical construction methods can be defined for specific regions (e.g., Uzbekistan in [44]). Regions with buildings with an average damage index >3 are at particular risk in the event of an earthquake. In these cases, the planning and execution of seismic retrofitting is of utmost importance.

Testing Facilities

Special testing facilities ("shaking tables") can be used to expose structures made of earth building materials to dynamic loads at a 1:1 scale. The level of resulting damage is examined so that suitable measures for seismic retrofitting can be derived (Fig. 5.36).

Case Studies of Damage Patterns

The following case studies illustrate the vulnerability of structures made of earth building materials during earthquakes. Particularly in rural areas of developing countries, such buildings are typically constructed by owner-builders and without

Fig. 5.36 Testing facility for determining damage patterns in earth block walls: test station with dynamic loading equipment at a scale of 1:1 ("shaking table," University of Technology Sydney, 2005)

employing necessary structural measures. Even during earthquakes of intensity VII, buildings frequently suffer irreparable damage, particularly through:

- Individual vertical walls tearing away from the construction
- Collapse of the roof structure
- Collapse of the pillars between wall openings (windows, doors)

The damage patterns described above are primarily caused by the following structural deficiencies:

- Insufficient tie-in of the perpendicular walls at the corners due to lack of adequate structural reinforcement. Faced with different vibration directions, individual walls tear away and fall over. Frequently, there is a lack of corner reinforcement in the form of exterior wall extensions in the direction of the longitudinal axes. Reinforcement of the longer exterior walls using vertical buttressing is also missing.
- Insufficient strength and quality of the earth building materials.
- Insufficient width of pillars between wall openings.
- Missing foundation grade beam and missing bond beam at ceiling bearing surface or coping levels. In traditional architecture of rural regions in Central Asia, heavyweight flat roofs with puddled earth up to a thickness of 50 cm are very common (Sect. 4.3.5.2). The load-bearing structure of the roof consists of logs of lightweight poplar. Due to their high mass, these roofs are optimal "heat buffers" in hot and dry climates. In the event of an earthquake, however, they are very unfavorable from a structural perspective. A frequently observed, potentially fatal, structural deficit can be found in the missing bond beams which would normally distribute loads. Instead, the load-bearing timber beams rest directly on the wall copings, without any anchoring.

Example 1 Residential building made of cob (pachsah) in the settlement of Oinakul, Kamashi Region, Uzbekistan, 2000 (Fig. 5.37 [45])

The unanchored ceiling beams shifted and collapsed into the building's interior with most of the puddled earth. The solid earth walls partially cracked but did not collapse.

Example 2 Earth block masonry structure in the settlement of Lugovaya, Kazakhstan, 2003 (Fig. 5.38 [45])

An exterior wall of the building tore away from the structure and fell over. There was no bond beam at the coping level. The lightweight pavilion roof system partially collapsed. The earthquake's intensity was $I = \text{VII}$, with a magnitude of $M = 5.3$.

Example 3 Residential half-timber building with earth clump infill (sintsch) in Kairakkum, Tajikistan, 1985 (Fig. 5.39 [45])

The load-bearing frame of the walls remained partially standing. The top plate still exists, and roughly half of the earth clumps (guvalja) fell out. The heavy flat roof was not anchored to the top plate and collapsed, causing serious injury to the occupants.

Fig. 5.37 Damage patterns after earthquake impact: destroyed residential building made of cob (pachsah) (Kamashi, Uzbekistan, 2000 [45])

Fig. 5.38 Damage patterns after earthquake impact: destroyed building made of earth block masonry (Lugovaya, Kazakhstan, 2003) [45]

Fig. 5.39 Damage patterns after earthquake impact: destroyed residential building using half-timber construction with earth clump infill (sintsch) (Kairakkum, Tajikistan, 1985) [45]

Fig. 5.40 Damage patterns after earthquake impact: destroyed school building made of earth block masonry (Gazli, Uzbekistan, 1976) [45]

Example 4 School building made of earth block masonry in Gazli, Uzbekistan, 1976 (Fig. 5.40 [45])

In 1976, there was an usual succession of two major earthquakes in Gazli, Uzbekistan, in the space of just 5 weeks.

Fig. 5.41 Damage patterns after earthquake impact: destroyed structures employing traditional earth block and cob techniques (tschineh), Bam, Iran, 2003

The upper image shows damage patterns after the first earthquake on April 8, 1976, with an intensity of $I = \text{VIII}$. The school building was able to resist the earthquake, apart from the typical cross-shaped cracks in the pillars between the windows. After the second, even more severe earthquake with an intensity of IX on May 17, 1976, the damaged building was completely destroyed. Figure 5.35 shows the distribution of the damage grades of both earthquakes for different wall construction methods in the region.

Example 5 Bam Citadel, Iran, 2003 (Fig. 5.41)

The earthquake on December 26, 2003 (with a magnitude of $M = 6.1$), claimed approx. 27,000 lives. It resulted in extensive destruction of the historic citadel and the new town of Bam. The buildings primarily consisted of traditional earth block and cob structures (tschineh).

An analysis of the damage patterns shows two common, overlapping planning mistakes as causes of the devastating destruction:

As in Example 1 above, the heavy flat roofs were among the main causes for the incurred damages. In many buildings, the original timber joists of the load-bearing system for the flat roofs had been replaced by much heavier steel joists. The flanges of the steel joists were laid directly on the wall copings of the earth block masonry, without the use of load-distributing bond beams or anchors in the load-bearing walls (Fig. 5.42 [46]). This led to strength failure of the earth block masonry at the bearing surface of the steel joists during the earthquake.

Fig. 5.42 Damage patterns after earthquake impact: steel ceiling joist laid on earth block wall coping without load distribution, Bam, Iran, 2003 [46]

The effects of the damage were compounded by another design element of the buildings: the floor plan. The typical regional floor plan of the load-bearing exterior earth block walls is U shaped: it consists of two long side walls which are placed directly on the property line (where they touch the side walls of the neighboring houses) and a short wall with an entrance facing the street. On the garden side, the houses are typically open and only separated from the garden by a ceiling-high gate. This is where the floor plan loses the bracing effect of the transverse wall. When the longitudinal walls fell over, a "domino effect" caused entire neighborhoods to collapse.

5.2.5 Structural Damage Caused by Planning Mistakes

Other sources of structural damage in earth buildings are structural deficiencies caused by insufficient dimensioning of load-bearing building elements in the planning stages, as well as during the construction of the building itself. Remodeling (removal of structural components, increase in live loads, etc.) can also lead to structural damage. In these cases, restoration only makes sense once the often complex causes of cracking have been identified and, if possible, eliminated.

Vertical cracks which carry through the entire wall cross section are typical damage patterns in cob and rammed earth walls. They are particularly common in

Fig. 5.43 Structural damage caused by improper execution of work: vertical crack in cob wall caused by a difference in wall height

barns which, in contrast to residential buildings, often lack the required spatial rigidity due to their comparatively slim load-bearing exterior walls. This can be traced back to the basic function of barns (large, high, and unobstructed storage areas) and the thriftiness of their owners (in terms of material).

In addition, mistakes made during the restoration of buildings can also lead to structural damage and subsequent work involving the repair of the "repair work." In this context, insufficient knowledge of structural and physical correlations coincides with a lack of or insufficient knowledge of the special characteristics of earth building materials.

Case Studies

Example 1 Barn in Baumersroda near Leipzig, Germany, load-bearing exterior cob walls (Fig. 5.43)

The top of the approx. 2-m high masonry wall or wall cladding $d \geq 24$ cm forms the bearing surface for ceiling beams and, at the same time, the base of the rising cob wall. The ground rises toward the front part of the building where the natural stone stem wall forms the base of the cob wall. This creates a difference in height of approx. 1.8 m in the cob wall which was erected as one continuous section, resulting in a vertical crack spanning the entire height of the wall.

Example 2 Barn in Saubach, Saxony-Anhalt, Germany, load-bearing exterior cob walls (Fig. 5.44 [33])

Fig. 5.44 Structural damage caused by planning mistakes: separation of an end wall (which is exposed to higher loads) from a longitudinal cob wall [33]

There are significant differences in the load types applied at the level of the wall copings of the longitudinal walls:

End wall: vertical point loads, applied off center from the gable

Load-bearing longitudinal wall: horizontal stresses caused by the roof loads as a result of missing or defective ceiling joists (tying effect) with increased load stress through the use of concrete roofing tiles installed at a later time.

The tensile stresses which formed in the corners of the building could not be absorbed by the cob material. The end wall separated from the load-bearing longitudinal wall through a crack, starting at the height of the eaves, and leaned outward. In addition, the roof loads pushed the longitudinal wall outward and away from the end wall.

A structural analysis needs to be carried out for this building which has been designed and used as a barn from the very beginning. The load-bearing exterior walls (including the gable) are too slim. Missing transverse walls, ceiling planes, and the lack of a bond beam result in insufficient rigidity of the building, making it particularly susceptible to external impacts.

In order to prevent the end wall from falling over, a buttress was installed at the vulnerable corner. Layers of corner reinforcement at intervals of approx. 0.5 m could have possibly prevented this damage.

Example 3 Barn in Ostramondra, Thuringia,Germany, load-bearing exterior cob walls (Fig. 5.45 [30])

Fig. 5.45 Structural damage caused by planning mistakes: cracks in a cob wall caused by uneven settling of the building ground [30]

Later remodeling work has probably led to overlapping causes of damage. In addition to the cause described in Example 2, a change in the conditions of the building ground occurred as a result of roof repair using heavyweight concrete roofing tiles: although roof drainage has been installed on the yard side, it is missing on the garden side. Here, concentrated seeping rainwater causes a significantly higher moisture content in the building ground than on the paved yard side. This results in different settling patterns of the building and cracking in the foundation which continues through the stem wall into the cob wall. Above the second course of cob, the cracks intersect with a further crack originating at the height of the eaves (see Example 2). In this section, the cob material has loosened and fallen out.

Example 4 Residential building in Wermelskirchen, Rhineland-Palatinate, Germany, exposed half-timber construction with straw clay over stakes (Fig. 5.46 [47])

The original exposed half-timber construction with a traditional straw-clay infill over stakes was thermally retrofitted with a wall reinforcement of wood-chip light clay. For this, an additional 10/10 frame construction with a mounting board for an inside layer of wood wool panels was added on the interior. On the exterior side of the half-timber panels, wood wool panels were attached to strips of wood mounted along the edges of the half-timber panels. The wood wool panels were offset from the flush edge by the thickness of the plaster and then finished with plaster. The panels were to serve two functions: as permanent formwork for the wet wood-chip light-clay mix and as a substrate for the exterior and interior plaster.

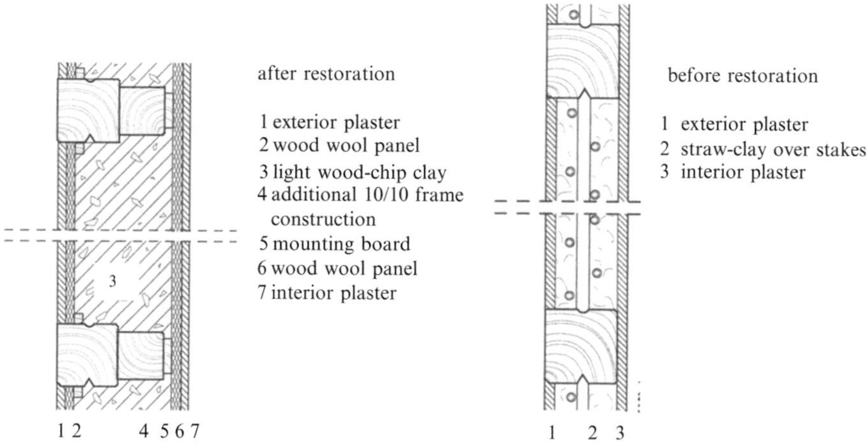

Fig. 5.46 Structural damage caused by planning mistakes: improper restoration through the use of unsuitable thermal insulation [47]

A short time after the restoration in 1992, the following damage pattern emerged: the wood wool panels with the adhering wood-chip light clay fell out. The panel mounting strips had rotted and the fasteners (which partly consisted of non-galvanized nails) were rusted. Plaster had fallen off on the interior. The mounting board and the interior frame construction were moisture penetrated and partially rotted.

The damage was caused by the joints between the exterior plaster and the half-timber construction. The narrow joints' openings were wide enough to allow rainwater to enter and drain along the timber frame and the wood wool panels. This water collected on the horizontal noggin pieces and sills causes the abovementioned damage.

5.3 Preservation

There are only rough estimates of the number of existing earthen structures in Germany and their state of preservation. Different sources, e.g., [48], specify the total number of "earth buildings" as more than two million. But what exactly is an "earth building"? Does this term include half-timber construction or does it only refer to structures built with load-bearing earthen walls? In addition, earth building materials were also used in many historical structures made of brick or stone, for example, in the form of mortars or plasters.

What can be said for sure is that the number of structures in Germany which incorporate earth as a building material in one form or another amounts to several million. It is also certain that most of these structures are in a poor state of repair.

This is especially true for agricultural buildings which have lost their original function. In these cases, the future of the buildings needs to be examined: should they be demolished or preserved? The topic of building preservation is therefore of high priority in the context of existing historical earthen structures.

5.3.1 Legal Framework

When planning and executing building preservation measures in Germany, the building regulations of the respective state building codes (LBO) must be adhered to. In addition, contractors must perform the services on their own responsibility and follow the generally accepted rules of technology (Sect. 4.2.1.1).

In Germany, older buildings (including those which were constructed using earth building materials) are "grandfathered in" as they were built in accordance with formerly valid building regulations and permits. This protection does not apply, however, to changes in the building's use, to the removal or change of load-bearing building elements (if there is the risk of collapse of individual building elements or entire buildings), or to changes in the exterior appearance of the building. In such cases, the older building is treated as new construction and all newly installed and remaining building elements must conform to current building regulations. It is not always sensible to apply these regulations to older buildings, and a case-by-case assessment by the responsible building authority can determine to which extent the preservation measures are allowed to deviate from the new construction standard. This particularly applies to requirements in terms of thermal insulation. The recommended solutions, however, still need to meet the minimum requirements of the building authorities.

Owners of *registered historic buildings* have an obligation to preserve the structure. They need to coordinate all structural changes with the responsible preservation authorities. Registered historic buildings can only be modified if the planned structural measures are necessary for the preservation of the building as a whole. According to the principles of conservation and restoration of historic structures in the "Venice Charter," the preservation of monuments is "always facilitated by making use of them for some socially useful purpose. Such use is therefore desirable but it must not change the layout or decoration of the building. It is within these limits only that modifications demanded by a change of function should be envisaged and may be permitted" (Venice Charter, 1965, Article 5).

In addition, community-level restoration and design regulations can also apply to buildings which are not registered historic buildings. In Germany, the objectives of historic preservation are also codified in the respective laws of the individual states. Owners of registered historic buildings should always resolve questions with the responsible authorities before the planning of preservation measures commences.

5.3.2 *Planning Preservation Measures*

Most older earth buildings which exist in Germany today are good examples of "common" structures and do not possess the status of a registered historic building (which would place special requirements on the preservation of the building from the point of view of the building authorities). Basically, these buildings are "grandfathered in" but are considered to have a relatively low protection status. In this context, hundreds of historical cob buildings were demolished around Leipzig, Germany, over the past decades to make room for open pit soft coal mining.

5.3.2.1 Methods

Classification into the category of "registered historic building" or "common building" largely determines which preservation methods are permissible or applicable. Preservation methods for older buildings and monuments can be distinguished as follows [26, 49–51]:

Maintenance

The term "maintenance" refers to *preventive* measures in the form of ongoing work which aims to preserve the usability of a building by removing or delaying damage caused by wear, aging, or defects. This includes such self-evident tasks as regular cleaning and repairing of gutters or replacing missing roof tiles—both of which are "life-saving" measures for earth buildings.

Repair and Restoration

The repair of structures involves the remediation of physical wear of individual building elements in order to restore the full usability of the building.

Damage in registered historic buildings is repaired by hand using the same materials as the original construction. *Repair* needs to be limited to the absolute necessities and should be given priority over renewal. *Restoration* restores the structure to its original condition with extensive use of appropriate, non-original building materials. In addition, it preserves the technical, historical, and aesthetic values of the monument. For example, panel infill which is still in good condition should be repaired with earth building materials. It should not be removed and filled with insulation material. Restoration work should be executed by employing the original historical building technique, e.g., wattle and daub. In addition, all repair measures should be reversible.

The restoration of registered historic buildings also includes special *conservation and consolidation* measures such as stabilization, hardening, waterproofing, and injection grouts or pinning. In this context and in the interest of preserving the historic building, the utilization of building materials and building techniques which have not previously been used in the structure represents an unavoidable invasion of the original building substance. This process creates a new "composite building material." Examples can be found in the repair of continuous cracks in rammed earth and cob walls and in structural stabilization measures applied to foundations and load-bearing structures. Other examples are found in the stabilization of paints, paint layers, or earth plasters on building substrates or the stabilization and hardening of surfaces in archaeological ruins of earthen structures.

Renovation

The term "renovation" can be interpreted in many different ways. The German word "Sanierung" originates from the Latin term "sanare" which means "healing." Renovation work consists of all protective measures taken to preserve older buildings which entail extensive interventions into the original substance. Often, building materials and building techniques are employed which have previously not been used in the building.

This especially includes all measures within the context of planning conversion and revitalization projects. These are primarily aimed at improving technical installations which help extend the building's lifetime and deal with more comprehensive and extensive measures than restoration and manually executed repair work. The modernization measures generally include the installation of new heating systems or sanitary improvements but also the addition of thermal insulation on exterior walls.

For the renovation of historic monuments, the following principle applies: keep interventions into the original building substance to a minimum while ensuring the future usability of the structure.

Structure Relocation (Reassembly Moves)

The relocation of a historic monument from its original to a new location is only carried out in *exceptional situations*. It involves the dis- and reassembly of the building or transport of the structure as a whole. Structures are relocated for a variety of reasons which include the development of new traffic patterns or saving the building from impending deterioration. A typical example of the relocation of buildings incorporating earth building materials is the transport of half-timber structures with earth infill to outdoor museums (Fig. 5.47). In 1994, the first cob building to be relocated as a whole structure was moved from its original location in Utzberg near Weimar to the grounds of the Thuringian Outdoor Museum in Hohenfelden, Germany (Fig. 5.48 [52]). The building combines living quarters with an animal barn and dates back to the year 1683.

Different opinions exist about the pros and cons of this method used in the preservation of historic monuments.

Fig. 5.47 Relocation of architectural monuments: half-timber construction with earth infill, outdoor museum in Bokrijk, Belgium

Reconstruction

If a monument of significant architectural and cultural importance is destroyed, it is possible to create a replica or copy of the original condition of the building on the basis of existing building documentation. This preserves the identity-forming character of the cultural site for future generations.

The *copy* of a destroyed monument can be constructed with the help of existing, detailed building documentation and can also incorporate any existing structural building substance. The *replica* of a monument is constructed without the use of original building documentation. One example is the reconstruction of Roman rammed earth buildings in the Archaeological Park in Xanten, Germany, from 2007 to 2009 [53].

The limitations of this distinction (which is based on a European interpretation of historic preservation) become clear in the more than 2000-year-old Bam Citadel in Iran, which was destroyed by a devastating earthquake in 2003 and is one of the world's largest complexes of earthen monuments (Fig. 5.41). In non-European traditions, century-old cultural landscapes and their architectural complexes (particularly those consisting of earth buildings) could only be preserved into the present day because existing building materials were continuously reused with the help of traditional building techniques. In this manner, the authenticity of the "architectural monument" was preserved [49].

Fig. 5.48 Relocation of architectural monuments: complete relocation of a cob structure, Thuringian Outdoor Museum in Hohenfelden, Germany 1994 [52]. (**a**) lifting, (**b**) transporting, and (**c**) new location

5.3.2.2 Planning Stages

The planning of building preservation measures generally consists of three stages:

- Survey of the current condition or *anamnesis*
- Assessment of the current condition or *diagnosis*
- Planning of the preservation work or *therapy*

Survey of the Current Condition

The process of taking a survey of the current condition of a structure consists of an evaluation of existing building plans, images, historical modifications, renovation work, and changes to the building's functions. With regard to earthen structures and their sensitivity to water, particular attention should be paid to roadside back filling which was carried out at a later point in time, changes in traffic patterns and drainage, as well as the condition of existing foundations.

This process also includes the recording of the structural conditions and building physics (thermal conditions) of the existing building as well as the documentation of structural damage. In connection with registered historic buildings, special requirements are placed on building inventory documentation and descriptions of damage. These requirements are also defined in the laws of the German federal states. The accuracy of the inventory plans which need to be prepared can generally be divided into four levels:

- Level 1: schematic survey drawing (*M* 1:100)
- Level 2: nearly realistic survey drawing (*M* 1:50)
- Level 3: deformation survey drawing (e.g., complex buildings with skewed lines, *M* 1:50)
- Level 4: deformation survey drawing with detailed description (as in Level 3, *M* 1:25)

The level of accuracy required for a particular case must be determined by the responsible preservation authorities. Generally speaking, the higher the building's value as a monument, the higher the required level of documentation and the higher the associated costs for the owner and the professional planners involved in the project.

The main damage is charted in the as-built plan in the form of appropriate descriptions and supplemented by documentation consisting of photographs or drawings. It is recommended to record and chart the damage in a building logbook which is used to list all cases of damage individually and classify them according to their causes.

Assessment of the Current Condition

The current condition can be described with the help of structural condition levels as a classification characteristic of the wear which has occurred (Table 5.9 [26]):

Table 5.9 Structural condition levels used in the assessment of the current condition

No.	Structural condition level	Average wear [%]	Assessment
1	1	0–10	Very good
2	2	11–25	Good
3	3	26–50	Satisfactory
4	3–4	51–80	Poor
5	4	81–100	Failing

Safe use of the building cannot be guaranteed from structural condition level 3–4 upward.

In the assessment of the current condition, the duration of use or the age of the building is an important factor. According to [26], it is limited to 90 years based on financial considerations. If a building is maintained regularly, its usability can be extended by decades, and if maintenance is neglected, it will shorten accordingly.

Earthen structures generally exhibit an unfavorable age pattern. Many of the buildings, particularly in rural areas of the former GDR, are more than 100 years old and are in a poor state of repair.

The assessment of the current condition also includes an identification of the causes of the surveyed damage which occurred during the useful life of the building (Sect. 5.2).

In this context, the remaining structural stability of damaged building elements must be evaluated based on a material analysis. Typical examples of damage in earth buildings are a weakening of the cross section around the base of the wall and structural vertical cracks in load-bearing earth walls (Sect. 5.2.5). A decision must be made to what depth the damaged (salt-loaded) earth building material needs to be removed or if the entire structure must be demolished. With the help of preliminary investigations and advice from experts from relevant fields, it should be verified if further measures are necessary and if they are compatible with the existing structure.

For functional reasons, the design of agricultural buildings often included exterior load-bearing earth walls which were too slender (Sect. 5.2.5). An assessment of the current structural stability of the earth building material based on the Lehmbau Regeln [15] could have an impact on the layout design and renovation strategies when planning the repurposing of such a structure.

Strategies

Preservation strategies can only be developed after the current condition has been surveyed and assessed. Strategies primarily include:

- Utilization concepts
- Selecting preservation methods
- Selecting (earth) building materials and respective building methods
- Conceptual design including a structural analysis
- Costs

Utilization Concepts

When formulating concepts in terms of contemporary user requirements placed on older structures, owners always need to be aware of the historic character of the building. Despite the large number of regulations which need to be adhered to, owners should not perceive their property as a "burden," but rather focus on the exciting challenges of finding unique possibilities for the contemporary use of a historical building.

For centuries, user requirements placed on residential buildings remained largely unchanged. In winter, the rooms were heated sparingly, most of the time with the help of a kitchen stove or masonry heater which stored the heat for long periods of time. To reduce heat loss at night, the windows could be covered on the outside with shutters. The coldest parts of the house were the window panes. Here, excess water vapor from the air collected as condensation. The attic remained unheated and was only used for storage. In summer, buildings stayed cool due to the relatively small size of their windows.

Over the past 50 years, user requirements placed on residential buildings have fundamentally changed. Growing prosperity has resulted in higher user demands. Attics are now often finished up to the ridge line and used as living space throughout the year. In winter, every room of the house is heated.

These changes in the requirements placed on residential construction have led to a considerable rise in heating energy demands and, accordingly, to higher emissions. As a result of this development, German legislative bodies have defined limits for heating energy demands and insulation standards in new construction (DIN 4108-2, EnEV [3]). Based on Sections 16 and 17 of the EnEV, older buildings can be exempted from these limits.

Selecting Preservation Methods

The main focus of preservation methods used for earth building structures lies in the field of repair and renovation work, supplemented by a plan for regular maintenance. Relocation and reconstruction of earth buildings are exceptional cases which require appropriately detailed planning.

Selecting Earth Building Materials and Building Methods

A decision in terms of earth building materials and respective building methods is determined by a number of factors. In addition to structural, material, and economical aspects, owners increasingly request that building physics (health-conscious building) and ecological and aesthetic parameters also be considered in the planning stages. Historic preservation requirements can also determine the selection of specific earth building materials and building methods.

Conceptual Design

In Germany, the Lehmbau Regeln [15] (Sect. 4.2.1.2)—serving as the current regulations introduced by the building authorities—as well as DIN 18945–47 for earth blocks and earth mortar (published in 2013), provide assistance with the conceptual design and possible structural analysis of earth buildings with regard to renovation measures. Other current building regulations must also be adhered to, for example, in terms of fire protection and sound insulation.

In addition to challenges regarding building materials and structural concerns, contemporary preservation of older buildings incorporating earth building materials generally includes thermal retrofitting measures using additional insulation. Depending on the building method, appropriate thermal insulation measures can significantly lower heating energy demands. According to [19], building elements of older unrenovated structures contribute to heat loss to varying degrees:

- Exterior walls approx. 30–40 %
- Roof approx. 20 %
- Attic approx. 10–20 %
- Windows approx. 10–15 %

The planning and execution of renovation work requires expert knowledge. The current condition of the building and its location always need to be taken into account. But above all, when applying the EnEV [3], the specific properties of the historical and potential modern earth building materials as well as the respective building methods need to be considered in connection with processes of building physics. With regard to structural and hygienic requirements, the minimum thermal insulation values according to DIN 4108-2 should be adhered to in all areas.

Costs

For renovation work using earth building materials, call for bid documents with field-tested typical work times and average prices are now available (Sect. 4.2.2.1 [54, 55]). If specific local building methods are requested (e.g., based on historic preservation requirements), they need to be described in detail in the call for bid documentation in order to enable contractors to formulate their offers accordingly.

5.3.3 Execution of Repair and Restoration Work

5.3.3.1 Foundations

Repair and restoration work in rising walls using earth building materials can only begin after all necessary repair work on foundations and stem walls has been professionally executed and completed.

Foundation Modifications and Repairs

An assessment of the current condition (Sect. 5.3.2.2) frequently leads to the conclusion that the existing foundation (independent of the building technique) meets neither the requirements of the current nor intended use. A solution to the problem can be found in structural modifications to the existing foundation. Because such work always poses a high risk to workers, the affected building, as well as any

Fig. 5.49 Foundation repair work: underpinning of the foundation of an earth block wall [85]

adjacent buildings, appropriate safety measures must be carefully planned and implemented. The survey of the current condition forms the basis of an evaluation of the performance of the renovated building during and after execution of the foundation work as well as any damage which might occur. In addition, this survey can be used to evaluate potential damage claims by a third party.

Foundation modifications and repairs can generally be divided into four work stages [56]:

- Bracing and shoring (stabilization) of the building
- Exposing the foundation and section-by-section load reduction (shoring) of the existing foundation, for example, by using crossbeams with lateral supports (Fig. 5.49)
- Execution of the foundation work (widening, deepening, underpinning using posts or pillars, building ground stabilization)
- Frictional transfer of structure loads onto the modified foundation

By widening the foundation, existing bearing pressure can be distributed across a larger surface which helps to remove possible load transgressions at the edges. A suitable reinforcement configuration guarantees a frictional connection between the foundation and the added components (which are initially tension-free). This connection is only established when settling occurs.

Older foundations often consist of quarry stones which were laid without joint mortar. They are generally shallow and do not reach below the frost line. A repair or modification of quarry stone foundations typically involves the construction of a concrete footing. Through a deepening of the foundation, the bottom of the footing reaches below the frost line and onto firm ground. This method requires underpinning of the existing foundation section by section.

If the building ground near the surface is not sufficiently firm, underpinning can be carried out using piles. The building ground can also be stabilized by injecting cement slurry.

Exterior Moisture Sealing

As protection against hydrostatic water pressure, clay seals where often installed in older buildings with basements (as long as suitable soil material was available locally) [57]. These seals were installed as rings around the exterior of the foundation. The clay was permanently wet and its swelling properties protected the foundation, stem wall, and rising walls from groundwater and rising damp. This sealing function was maintained as long as the ring around the entire foundation remained closed. Subsequent remodeling, however, typically involved the installation of new underground utility connections which broke through the clay seal and rendered it ineffective. Insufficient or missing horizontal moisture barriers led to moisture in the rising walls after having been dry for decades [58].

Throughout the decades of the building's use, the function of clay seals was gradually forgotten. Today, bitumen seals in connection with horizontal moisture barriers and circumferential drainage systems are the state-of-the-art method of groundwater sealing.

5.3.3.2 Walls

While most earthen wall repairs in Germany occur in the field of half-timber construction, rammed earth and cob as well as earth block construction also play a role. In this context, archaeological ruins of earthen structures require very specific restoration strategies (Sects. 1.1, 1.2, and 5.3.3.5).

Half-Timber Construction Using Earth Building Materials

Planning Criteria

In Germany, half-timber construction has been firmly rooted in traditional architecture for centuries. Half-timber houses still shape the character of many towns and rural regions and are part of Germany's cultural identity. Their current numbers are estimated to be approx. two million [47]. Compared to structures using other earth building techniques, half-timber construction comprises the largest group of buildings by far. It is also the oldest composite construction technique: wood is used as the structural framework in connection with earth serving a room-enclosing function (Sects. 1.1 and 4.3.3.2). Owners, architects, and planners therefore have a special obligation to preserve this cultural heritage for future generations.

German trade organizations and institutions have published building regulations, Technical Information Sheets, and guidelines which define the field's current state of the art and can be used for the planning of restoration and renovation measures of half-timber structures. Such institutions include:

- The German Association for Building with Earth (DVL) [15, 59, 60]
- The Scientific and Technological Association for the Restoration of Buildings and Preservation of Monuments (WTA) [61]
- The Timber Information Service [19]
- The German Center for Craftsmanship and Historic Preservation (ZHD) [62]
- The Farmhouse Association (IGB) [63]
- The Ministry for Construction and Housing of the German state of North Rhine-Westphalia [64]

Whereas the DVL's Lehmbau Regeln [15] represent official regulations introduced by the building authorities (Sect. 4.2.1.2), the Technical Information Sheets and the guidelines published by other organizations are recommendations which might deviate from one another.

When planning restoration and renovation work on half-timber structures and determining a workflow, two different structural components must be distinguished between which differ in terms of their function:

- The load-bearing timber frame
- The non-load-bearing infill according to Sect. 4.3.3.2 which must meet room-enclosing and building physics requirements

The sequence of operations always begins with the structural framework. The non-load-bearing panel infill can only be repaired or replaced after all work on the timber frame, the foundation, and the roof has been completed. In addition to earth building materials, panel infills also consisted of other locally available materials, such as bricks, tufaceous limestone blocks [65], and stones, which were laid in earth masonry mortar.

Thermal Insulation

Overall heat transfer coefficient U: In terms of thermal insulation, the values for historical half-timber construction with earth infill are considerably below the required minimum values. The respective values for the overall heat transfer coefficient U are:

- According to DIN 4108-2:

  $U=0.73\ \text{W/m}^2\ \text{K}$ for the infill panel only
  $U=0.85\ \text{W/m}^2\ \text{K}$ as an average value for the entire building element

- According to EnEV [3]:

  $U=0.45\ \text{W/m}^2\ \text{K}$

Table 5.10 Thermal retrofitting of historical half-timber construction, calculation example of *U*-value

No.	Wall building material	Dry bulk density ρ_d [kg/dm³]	Thermal conductivity λ [W/mK][a]	Layer thickness *d* [m]	d/λ [m²K/W]	Overall heat transfer coefficient *U* [W/m²K]
1	Interior earth plaster	1.50	0.65	0.02	0.031	
2	Earth blocks	1.40	0.60	0.12	0.200	
3	Exterior earth plaster	1.50	0.65	*0.02*	0.031	
4	d_{total}			0.16		
5	R_{se+si}				*0.170*	
6	$R_T = 1/U = \Sigma d/\lambda + R_{se+si}$				0.432	
7	$U = 1/R_T$					2.315

[a]λ values according to the Lehmbau Regeln [15]

Table 5.11 Thermal retrofitting of historical half-timber construction, calculation example of *U*-value

No.	Wall building material	Dry bulk density ρ_d [kg/dm³]	Thermal conductivity λ [W/m K][a]	Layer thickness *d* [m]	d/λ [m²K/W]	Overall heat transfer coefficient *U* [W/m² K]
1	Interior earth plaster	1.50	0.65	0.020	0.031	
2	Light-clay blocks 2DF	0.80	0.25	0.115	0.460	
3	Wood-chip light-clay infill	0.80	0.25	0.060	0.240	
4	Earth blocks	1.40	0.60	0.120	0.200	
5	Exterior lime plaster	1.50	0.65	0.020	0.031	
6	d_{total}			0.335		
7	R_{se+si}				0.170	
8	$R_T = 1/U = \Sigma d/\lambda + R_{se+si}$				1.132	
9	$U = 1/R_T$					0.883

[a]λ values according to the Lehmbau Regeln [15]

Therefore, thermal retrofitting (Sect. 5.3.2.2) in the form of additional thermal insulation for residential buildings or buildings which are being converted into residual buildings is one of the key areas of restoration and renovation work carried out on half-timber structures with earth infill.

Based on the following assumptions, traditional infill using earth blocks of $d = 12$ cm with 2 cm each of interior and exterior plaster attains the following overall heat transfer coefficient $U = 2.315$ W/m² K (Table 5.10):

Using a wall lining of light-clay blocks and a leveling fill of wood-chip light clay significantly improves the thermal insulation when compared to the initial situation (Table 5.11, Figure 5.50 [86], see Table 4.9 new construction).

When using an interior thermal insulation layer, [19] recommends a minimum value of $R \leq 0.8$ m² K/W ($U \geq 1.0$ W/m² K) as the overall heat transfer resistance.

a

b

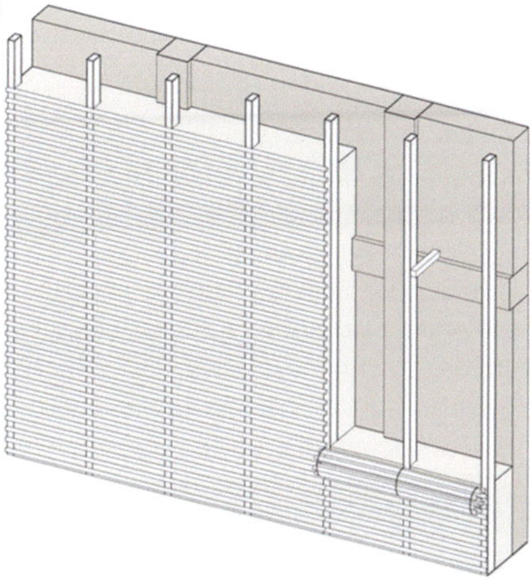

Fig. 5.50 Thermal insulation of half-timber construction: wall lining made of a light-clay mix which is installed wet. (**a**) Temporary formwork [86] and (**b**) permanent formwork, reed mat [59]

For the entire wall, an average value of $R \geq 1.0$ m² K/W ($U \leq 0.85$ W/m² K) should be attained with a diffusion equivalent air layer thickness of $0.5 < s_d < 2.0$ m.

Location of Thermal Insulation Layer. When planning thermal insulation for exterior walls of half-timber structures, the question of whether the insulation layer should be attached to the interior or exterior side of the wall is an important aspect.

With regard to building physics, *exterior insulation* is generally recommended because the temperature amplitude attenuation is most effective on the exterior surface of the building elements (Sect. 5.1.1.2). In this context, it is also easier to prevent thermal bridges because the insulation layer can be installed without gaps around the entire building envelope. The exterior insulation layer can be protected from the elements with the help of a plaster or a rear-ventilated facade system made of weatherproof material. If the insulation layer is to be plastered, a diffusion-open plaster should be selected which matches the earthen substrate.

Historic preservation regulations often restrict the use of exterior insulation on exposed half-timber construction. In such cases, *interior insulation* must be installed. This means that in winter, the exterior walls cool down all the way to the insulation layer on the interior, leading to an increase in capillary moisture in the earth building materials and possible undercutting of the dew point temperature along the layer boundaries between infill panel—interior insulation—wood. Condensation can then lead to problems along the timber frame (Sect. 5.1.2.3).

When exposed to strong sunlight, exterior wall surfaces in sunny and sheltered locations can also heat up considerably in spring and fall. This can lead to an increase in indoor temperature if the walls are not insulated. Thermal insulation (especially on the exterior) could prevent this effect. "Thermal insulation" might have to be seen in a new light as a result of a proven increase in the average air temperature over the next decades, also in Central Europe.

Building Materials. In traditional half-timber construction, the water-sensitive earthen material in weather-exposed exterior walls was especially protected, for example, by cladding made from locally available weatherproof materials such as slate or wooden shingles. Weather-resistant plasters in connection with water-repellent, vapor permeable paints, for example, lime, were also used.

In the past decades, still intact earth infill in exterior walls was often replaced by insulation material or building materials with better insulation properties in order to improve thermal insulation. These measures did not pay attention to a special characteristic of historical half-timber construction: the wooden framework continuously "moves" depending on the climatic conditions, and the earth in the timber-frame panels was able to adjust to these deformations. Masonry or ready-mix mortars with better insulation properties, on the other hand, were often rigid and restricted the wooden frame in its "natural" movements. After moisture penetration, building materials with a closed-pore structure dried more slowly and caused damage to the timber frame, particularly around the edges of the panels. The use of sealants accelerated this process. Many half-timber houses which were renovated in this manner needed to be repaired again after only a few years.

To improve thermal insulation in modern renovation work, the building element surfaces are completely covered with thermal insulation panels made of reed or wood fibers (e.g., soft wood fiber boards, wood wool panels) [15]. Used as interior insulation, they can be attached using sufficiently cohesive earth mortar. On the exterior, standard fasteners are used for attaching the panels which are then finished with a lime plaster to improve weather resistance.

In addition, light-clay building materials can be installed as continuous wall linings placed on the inside directly in front of exterior walls, either in the form of masonry using earth masonry mortar or as earth mixes which are installed wet using an appropriate formwork system.

Panel Infill Repair

Moisture-penetrated foundations or stem walls of half-timber buildings frequently lead to serious damage of the ground sill of the load-bearing frame. The sill can only be replaced after the causes of the moisture penetration have been eliminated (Sect. 5.2.1.2). After the new ground sill has been installed, the earth infill can be repaired. In this context, a differentiation is made between a repair and a complete replacement with new earth building materials. Today, both types are generally tied to additional thermal insulation measures in the form of light-clay linings or insulation panels which are attached with mortar.

Replacement of the Ground Sill. The ground sill lies on top of a stem wall that is made of masonry or natural stone with a minimum height of 40 cm. A moisture barrier against rising damp is installed into the masonry of the stem wall. The ground sill is laid on top of the stem wall in a bed of lime or lime–earth mortar and protrudes a few millimeters over the edge of the wall to allow draining rainwater to drip off. The medullary rays of the ground sill should point downward. Vertical holes should be drilled into the mortises for the posts to facilitate the draining of collected rainwater.

Repair of the Panels. For the repair of *straw-clay panel infill over stakes*, sections which are still intact are repaired with a straw-clay mix which largely matches the original earthen material in terms of its composition. Another possibility is to reuse salvaged earth modified with sand as recycled earthen material (Sect. 2.2.1.3). Earth building materials taken from salt-loaded wall sections cannot be reused.

As a first step in the repair process, all loose and hollow-sounding sections are knocked or brushed off. Additionally, all repair work which was done at an earlier point in time using inappropriate materials, such as fired brick or cement mortar, is removed. The wooden members of the load-bearing frame and the stake work which are exposed in the process are checked for their proper functionality and repaired or replaced if necessary. Infill which is still intact but loose is stabilized with the help of shims made of dry wood and/or wood screws.

In the next step, the cohesive strength of the clay minerals in the existing material needs to be activated in order to form a "bonding course" with the applied repair material. For this, the existing earthen material is thoroughly wetted the evening before as well as immediately before commencement of the repair work, particularly in the corners and along the edges of the panels. These are also the sections where the new earthen material should be applied first, with the help of a trowel or by hand. Then, the wet material is compressed using a float and the surface is roughened before it dries in order to promote good adhesion for subsequent layers or the plaster. The panels should always be worked across their entire surface and the thickness of each layer should not exceed 3 cm.

Repair work of exposed half-timber construction on weathered exterior walls is typically finished with a single- or double-layer lime plaster with a maximum thickness of 1.5 cm. The plaster is applied flush with the outer edge of the half-timber frame. The thickness of the applied lime plaster therefore needs to correspond to the depth of the recess between the flush edge and the restored straw-clay surface. It is important to create a plaster layer with an even thickness across the entire panel surface. If necessary, the layer can become slightly thinner toward the panel edges.

Exterior walls which are not exposed to the weather, or only to a limited extent, can also be covered with an earth plaster. In order to improve their weather resistance, exterior plasters were typically finished with a limewash (Sect. 4.3.6.4).

The repair of *earth block* infill is carried out in the same manner. The earth blocks and the earth masonry mortar should match the original building materials in terms of their composition whenever possible.

Complete Replacement of the Panels. Often, the panel infill is damaged to such an extent that parts are missing or repair is no longer an option. In such situations, a complete replacement with new earth building materials becomes necessary. This might also be the case when load-bearing timbers need to be replaced and the panel has to be newly filled. In addition, historic preservation stipulations for the restoration of half-timber structures might require that an infill be executed using the original historical building technique (Sect. 4.3.3.2). "New" earth building materials which can be used for panel infills are wet straw-clay and light straw-clay mixes or earth blocks and light-clay blocks.

The *wet light clay with organic or mineral lightweight aggregates* is poured into temporary or permanent formwork in layers and compacted (Sect. 4.3.3.2).

It is also possible to install *light-clay panel infill using the spray method* [66]. Suitable materials for this method are earth or light-clay mortars which are easily pumpable. This method requires an adhesive surface (e.g., lightweight panels) serving as permanent formwork as well as a load-bearing frame or cavity frame which is used as a lateral boundary and guide for plumbing the wall.

The material is prepared to the required consistency and applied in multiple layers at a maximum thickness of 5 cm each with the help of plastering machines. The individual layers must be completely dry before a new layer can be applied. This needs to be considered when scheduling the work. After the surface has been smoothed and has dried, the sprayed infill can be finished with a plaster, paint, or cladding.

Recently, new spraying methods have become available which make it possible to use nearly dry soil mixes. With these techniques, the material can be applied under very high pressure as a single layer up to the panel thickness resulting in minimal shrinkage deformation [67].

For panel infill using *earth blocks and clay panels made of straw clay or light straw clay*, the earth blocks should be laid into a full bed of earth masonry mortar using standard masonry bonding practices (Fig. 4.41). The panel infill can be secured (to keep it from falling out) with the help of triangular or trapezoidal slats attached to the center of the posts or sills and/or stainless steel nails driven into the posts at 25-cm intervals. Some manufactures offer grooved straw-clay or light-clay blocks which can be used in connection with slats in the bedding joints for additional bracing of the panels. To improve plaster adhesion, cutting 1-cm-deep channels into the masonry joints while they are still wet is recommended.

Light-Clay Linings. The construction of wall linings and the replacement of panels which can no longer be repaired can be combined into one process using the same earth building material. The following conditions must be met for the installation of wall linings:

- The timber frame must be intact and strong enough to absorb and structurally transfer the additional load of the lining on every floor.
- Existing panel infill in the exterior walls needs to be checked and repaired as needed. Water-vapor-impermeable paints on the wall surfaces should be removed.
- Existing room sizes or room sizes altered by the conversion must permit the respective decrease in size.
- According to the Lehmbau Regeln [15], the ceiling height of wall linings in the form of non-load-bearing light-clay panels is limited to 4 m.
- The effects of splash back and standing water, e.g., in the event of accidental water damage, must be prevented according to Sect. 5.2.1.2.

The lining is "placed" in front of the existing exterior walls on the inside. To facilitate manual installation of the *wet light-clay mix* and create a firm bond between the lining and the existing exterior wall, a cavity frame is used which is attached to the existing load-bearing frame and serves as a formwork guide (Sect. 4.3.3.2, Fig. 5.50). Interior light-clay linings can also be installed using the spray method.

For existing exterior walls, the thickness of light-clay linings which are installed wet should not exceed 15 cm. If the exterior walls are made of diffusion-open building materials promoting capillary action (earth building materials, bricks), interior light-clay linings with a maximum thickness of 20 cm can be installed. For drying, the instructions given in Sect. 3.3 should be referred to.

When installing linings made of *light-clay blocks*, the blocks can be laid completely flush against the inside of the exterior wall. A separate cavity frame is not required up to a maximum slenderness ratio of $h/d=15$.

The interior lining can also be divided into two partial linings (Fig. 5.51): a lining of light-clay blocks is laid as permanent formwork at the appropriate distance from the exterior wall and anchored to it. This forms a cavity which is filled with a wet light-clay mix or insulation material. The light-clay block work and installation of the wet light-clay mix are carried out simultaneously and in gradual steps. The thickness of the masonry lining and the complete light-clay fill needs to be a minimum of 11.5 cm. The side of the masonry wall which faces the room can be left exposed or finished with a (two-layer) earth plaster.

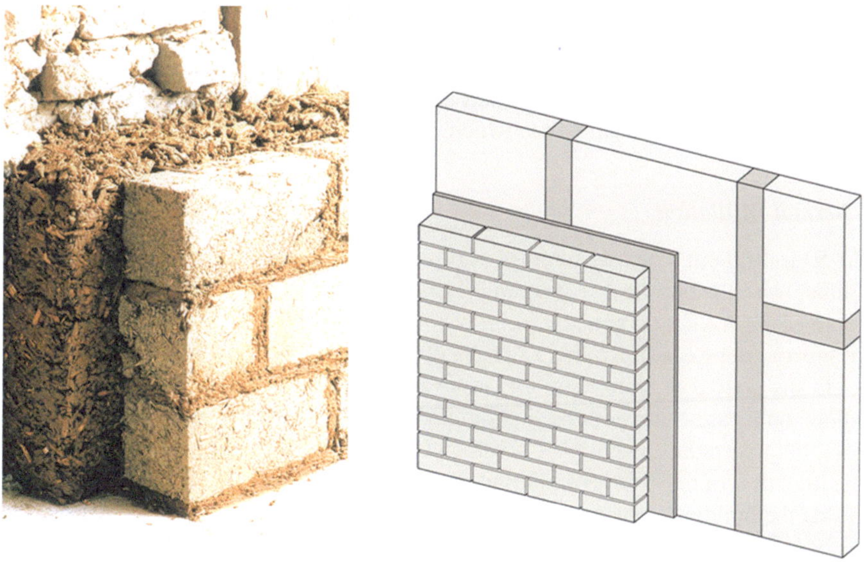

Fig. 5.51 Thermal insulation of half-timber construction: wall lining made of light-clay blocks with leveling fill of wood-chip light clay [59, 60]

Cob and Rammed Earth Structures

Planning Criteria

Only rough estimates can be made regarding the number of existing older cob and rammed earth buildings in Germany. According to [48], the number of "pure earthen buildings" in Germany is approx. 200,000. Güntzel [29] specifies a total of 2500. Based on an inventory analysis in the German states of Saxony, Saxony-Anhalt, and Thuringia [30, 33, 68], the number of earth buildings is assumed to be at least ten times greater. This is only about one hundredth the amount of half-timber buildings. Accordingly, the number of current normative documents which can assist in the planning of restoration and renovation work in this field is also lower. In addition to the Lehmbau Regeln [15], useful information for renovation strategies can be found in these sources: [68–71].

Historical cob and rammed earth buildings are often confused with each other. The location of the technique-related construction joints in cob and rammed earth construction is very similar: the horizontal bedding joints are created by continuous cob or rammed earth lifts with heights of approx. 0.8–1 m. The "butt joints," which are vertical in rammed earth construction and slightly slanted in cob construction, correspond to the length of the rammed earth formwork or the length of a cob section. These joints are "pre-installed" weak zones which are often a source of future damage.

A relatively sure indication of rammed earth can be found in the horizontal layers of brick chunks which were incorporated into the wall to improve erosion resistance and plaster adhesion. On the other hand, a sign of cob construction is found in the straw fibers wrapped in earthen material which can be seen in breaks or cracks in the walls. Based on their similarities, the following renovation strategies apply to both building techniques. The different strength properties of both building materials, however, are one main difference which needs to be taken into consideration.

Thermal Insulation

At 50 to >60 cm, cob and rammed earth walls are significantly thicker than half-timber walls which have a maximum thickness of 15 cm. In addition, cob and rammed earth walls do not have joints between a timber frame and earth infill which, in timber-frame construction, is a problem zone which is difficult to control.

In the course of restoration and renovation work, it should therefore be determined on a case-by-case basis if additional thermal insulation measures are necessary. In this context, it should be noted that additional insulation layers will increase the thickness of the exterior walls and could negatively affect the lighting conditions inside the building.

Foundations, stem walls, rising walls, and wall copings of cob and rammed earth structures are frequently penetrated with moisture. They will be able to dry out thoroughly after the causes of damage have been removed which, as a result, also improves thermal insulation properties.

Repair

Foundation, Stem Wall, and Wall Base. Moisture-penetrated foundation and stem wall areas are among the most common causes of damage. Restoration work should therefore begin with an identification of the sources of moisture and their removal. This task can be very complex because different causes often overlap (Sect. 5.2.1.2).

First, a decision should be made if and how vertical moisture transport in the cross section of the wall can be prevented. Available methods for the retrofit installation of a horizontal moisture barrier include a number of techniques which are common in standard masonry construction ([28, 70–72]). The practicality and advantage of each measure should be determined on a case-by-case basis.

Repair work on the base of the wall is connected to a weakening of the wall cross section. Such work involves certain risks and requires that appropriate safety measures be taken to protect the workers, the building itself, as well as any adjacent buildings. Bracing should be used to ensure that the damaged wall remains stable during the entire renovation period (Fig. 5.49).

For the *mechanical* method of installing a horizontal moisture barrier in the form of a layer which acts as a capillary break (e.g., plastic membrane or tar paper), the wall is opened at the base by cutting, chiseling, and/or replacing the damaged wall building material in sections. The lengths of the wall sections which need to be replaced depend on the individual case but should not exceed 1 m (Fig. 5.55).

Mechanical methods require a high degree of technical craftsmanship which is reflected in their costs. They should be considered if a significant reduction of the wall cross section has already occurred.

Instead of "opening" the wall cross section, stainless steel sheets can be driven horizontally into the joint between the top of the foundation or stem wall and the rising earth wall. This method is more cost-effective but also involves risks connected to a loss in structural stability and additional crack formation caused by the vibrations which occur when inserting the sheets. Additionally, the sheets can bend during installation. The risk of the metal being attacked by salts contained in the particular earth wall also needs to be considered.

For the *chemical* method, a liquid seal is injected into the wall cross section (with or without pressure) through holes which have been drilled above the stem wall at appropriate intervals.

In the past, horizontal "barrier layers" made of various materials were used as protection against rising damp in walls. These layers included sheets of lead or slate, layers of clinker bricks laid in asphalt or cement mortar, asphalt–sand mixes, birch bark, or reed and bamboo. It was not until the early 1900s that horizontal moisture barriers made of tar paper became the state of the art [57].

Before the destabilized base of the wall can be repaired, the extent of the salt damage and concentration of harmful substances in the earth building material must be determined. With the help of chemical analysis methods, moisture and salt profiles in the damaged wall sections are identified (Fig. 5.12 [6]). Working with respective specialists (e.g., structural engineers and chemists), the determined contaminant profiles can then be used to decide how much of the salt-loaded section

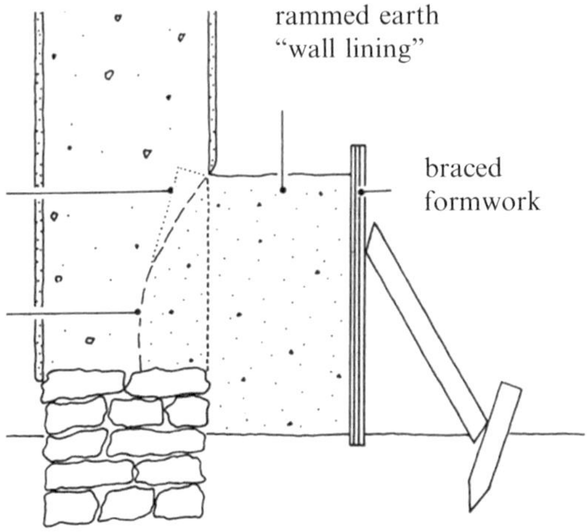

Fig. 5.52 Repair of rammed earth and cob walls: repair of the wall base using a rammed earth mix [71]

needs to be replaced (Fig. 5.13). As an alternative, the damaged wall section can be removed to a maximum of depth approx. 10 cm into the intact and "healthy" section which is not visibly loose. The salt-loaded material cannot be reused for earth building purposes and needs to be disposed of. It should be replaced by earth building materials which match the characteristics of the original building materials.

Suitable materials for the repair are prepared mixes which are installed in the same manner as rammed earth or earth blocks which are laid as masonry. It is important to establish a frictional connection between the weakened wall cross section and the replacement earth building material. First, a wedge is cut at a right angle into the undamaged wall section at the upper edge of the weakened zone (Fig. 5.52 [71]). Next, vertical formwork is set up in front of the exposed and weakened base of the wall. The distance from the formwork to the wall should be 2.5–3 times the depth of the weakened cross section, but a minimum of 20 cm. This "lining" can be used to effectively compact the replacement material without obstructions.

The prepared rammed earth mix is installed at a semisolid consistency. It is poured in layers with a maximum height of 10 cm each and compacted to a height of approx. 6–7 cm using 4–5 compaction passes. When testing the materials by hand, the soil should just barely form a ball which easily falls apart. When the soil is mixed to the correct consistency, settling of the replacement building material can be kept to a minimum after drying.

The wedge at the upper edge of the weakened zone should be compacted with particular care. In this section, the mix should be installed from the side. When loads are applied to the wall, the wedge creates a frictional connection to the replacement building material which guarantees that the load transfer to the foundation is restored across the entire wall cross section.

After completion of the rammed earthwork, the "lining" is trimmed off in vertical layers with the help of a spade until the repaired section is flushed with the existing wall. The surfaces and transition to the existing material can then be smoothed and evened out.

Wall sections which have been repaired in this manner maintain the character of the building technique used in the existing wall and are barely distinguishable in terms of their surface texture. The repair can also be carried out using earth blocks. The disadvantage of applying the earth block technique to unplastered cob or rammed earth walls is that the repaired sections remain visible and might spoil the aesthetics of the wall.

Rising Wall and Wall Coping. Vertical cracks which extend across the entire wall cross section are typical damage patterns in cob and rammed earth walls (Figs. 5.43, 5.44, and 5.45). This type of damage can also be found in earth block masonry (Fig. 5.19). Crack repair involves two different aspects: the actual closing of the crack with the help of tensile-resistant elements and earth building fillers and strengthening of the load-bearing system.

The following tensile-resistant elements were traditionally used for closing cracks and are still in use today: anchors, straps, grouts, and pins made of traditional building materials [70, 71, 73, 74]. In addition, modern anchoring, pinning, and injection methods from the field of masonry construction have also become available for the restoration of historical earth buildings.

Figure 5.53 [71] shows the stabilization of a vertical crack at the corner of a rammed earth building with the help of *straps*. This technique employs horizontal steel L-brackets which wrap around the corners. They are attached to the exterior surface of the wall using staggered bolts which are inserted perpendicular to the wall axis using predrilled openings and anchored to the interior wall surface with the help of a vertical strip of steel.

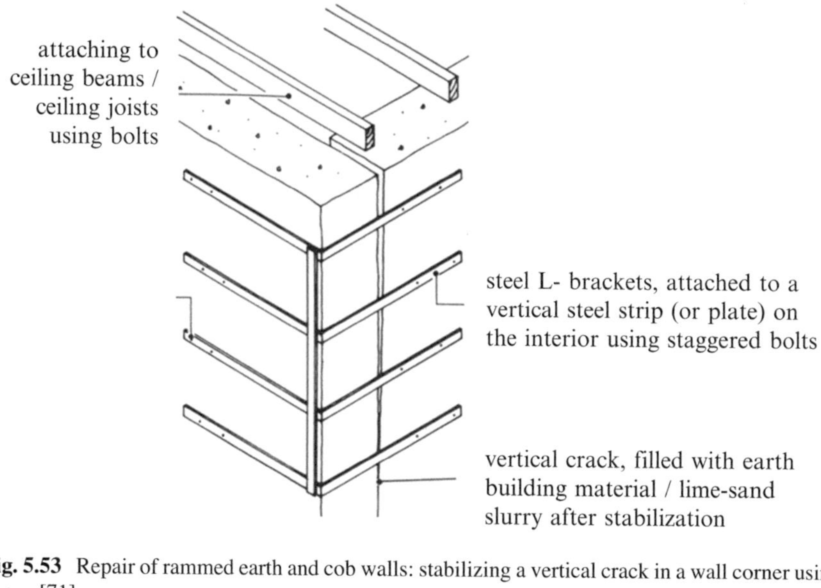

Fig. 5.53 Repair of rammed earth and cob walls: stabilizing a vertical crack in a wall corner using straps [71]

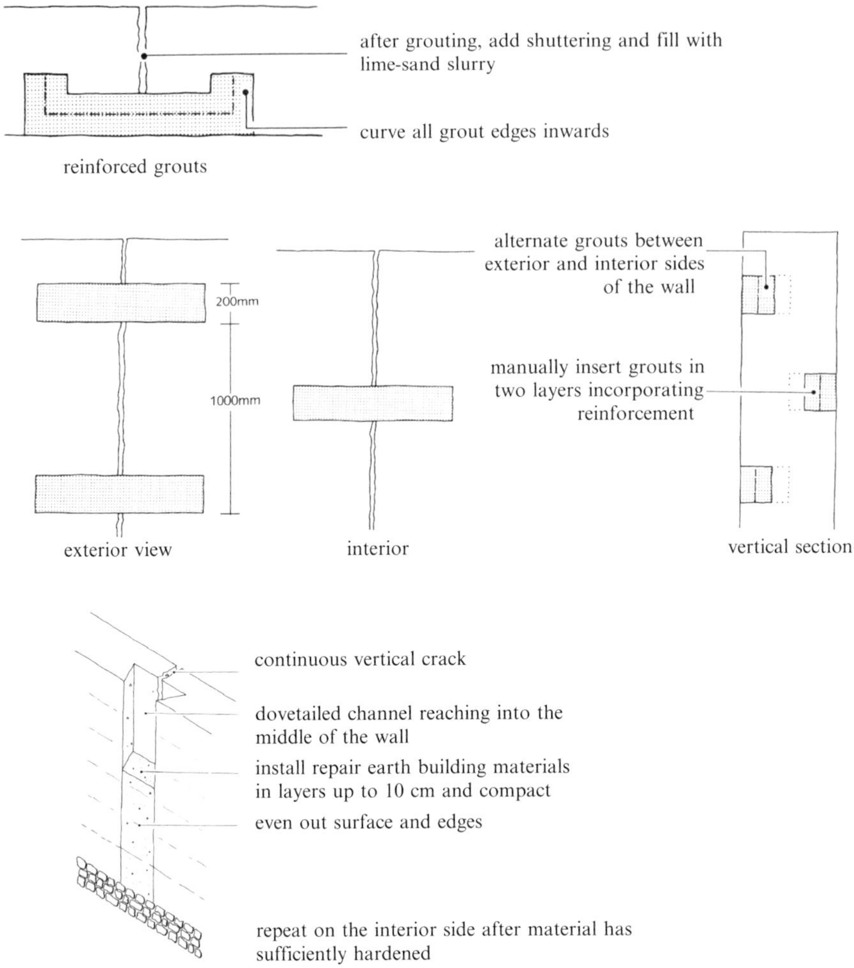

Fig. 5.54 Repair of rammed earth and cob walls: crack repair using grouts made of lime-stabilized rammed earth and lime-sand slurry [71]

Rammed earth *grouts* can be inserted (flush with the wall surface) into vertical or horizontal channels which have been cut into the damaged wall (Fig. 5.54 [71]). Depending on the individual case, the approx. 1-m-long horizontal grouts are installed across the entire height of the crack at intervals of 1 m, alternating between the interior and exterior sides. Rammed earth, straw-clay block masonry (approx. four layers, reinforced with geogrid if necessary), or straw-clay reels can be used as grout material [73, 74]. The grout should be placed so that its center bridges the opening of the crack. The strength of the cracked section can be increased if the grouts curve inward (hook in) at both ends and are reinforced with stainless steel mesh or geogrid.

For vertical cracks extending the entire height of the wall, a vertical channel is first dovetailed into the exterior wall surface (which typically shows a higher level of damage). The channel should be centered over the entire length of the crack and reach into the middle of the wall cross section. The vertical edges of the channel are pre-wetted and the channel is filled with approx. 10-cm-thick layers of cob and compacted. After settling is completed, the procedure is repeated on the interior side of the wall.

When restoring wall copings, the same methods are applied as for the base of the wall. The repair work should only begin after all causes of damage (typically holes in the outer skin of the roof) are eliminated.

The decision if bond beams or other supporting elements should or need to be used to reinforce the damaged building depends on a number of aspects: the individual damage pattern, existing structural conditions, as well as a planned conversion of the building which is generally connected to a change in working loads. In modern renovation work, bond beams are typically made of reinforced concrete. The load-bearing effect is achieved through a sufficient shear- and tensile-resistant anchoring of the load-bearing elements to the existing structure.

Surface Erosion and Plaster Damage. With average wall thicknesses of 60 cm, damage caused by surface erosion of historical cob and rammed earth structures has already been accounted for in the design. Investigations carried out on cob buildings in the German state of Thuringia have shown that after a lifetime of approx. 100 years, only a few centimeters of cob material had been lost due to weathering of the exterior wall surfaces. Wall surface erosion in "common" structures is generally deemed harmless (Fig. 5.21 [30]) unless it occurs in connection with the damage patterns described above.

In the past, only residential buildings were plastered which is why these buildings are difficult to recognize as cob or rammed earth structures. Barns and other agricultural buildings were generally left unplastered or they were regularly whitewashed. It is therefore not easy to determine (and primarily the owner's decision) if renovation measures for historical cob and rammed earth structures should include exterior plasters. Unplastered historical cob and rammed earth structures add unique character to the landscape. Newly plastered cob and rammed earth buildings, on the other hand, appear too perfect, sterile, and replaceable.

Damage to the original plaster on cob and rammed earth buildings should be repaired professionally. Holes in the earth plaster should be wetted and can be repaired with the help of earth mortar. It is generally recommended to rework the entire surface (Sect. 4.3.6).

Earth Block Structures

Planning Criteria

There is no exact data about the number of load-bearing earth block structures in Germany today, but their total is estimated to be several tens of thousands. German-language normative documents therefore contain very little information specific to the planning of restoration and renovation work for load-bearing earth block walls [15].

On an international level, various national regulations focus particularly on the issue of seismic retrofitting of load-bearing earth block walls (Sect. 4.2.1.3).

Thermal Insulation

In order to meet structural requirements, load-bearing earth block walls should have a minimum wall thickness of 24 cm. Exterior walls are typically 36 cm thick. These dimensions are significantly greater than the wall thicknesses found in half-timber construction with earth block infill. It should therefore be determined on a case-by-case basis (and possibly verified with calculations) if and how much thermal insulation needs to be added in conjunction with renovation measures.

Moisture-penetrated foundations, stem walls, rising walls, and wall copings will be able to dry out completely after the causes of damage have been removed. As a result, this also improves thermal insulation properties.

As is the case with half-timber structures with earth infills, the thermal insulation properties of earth block walls can also be improved by adding light-clay wall linings.

Repair

The restoration of earth block walls focuses on the same areas as cob and rammed earth construction and employs similar repair measures.

Foundation, Stem Wall, and Wall Base. To install a horizontal moisture barrier against rising damp, the damaged, loosened, and salt-loaded material is removed along the joints by cutting channels into the undamaged section to a depth of approx. 10 cm (but not further than to the middle of the wall). The earth block layers which are still intact must be braced with the help of blocks. This forms a cavity in the wall which is filled with courses of earth blocks which are laid into a full bed of mortar. The earth blocks must correspond to the existing masonry in terms of their composition and dimensions. As the work progresses, the height of the blocking should be reduced to ensure that the existing earth block courses are always braced (Fig. 5.55 [75]).

In order to ensure that the repair is a success, a frictional connection between the replacement material and the existing, undamaged wall section must be attained. The final horizontal joint which connects the repair work to the undamaged wall section needs to be filled with mortar particularly carefully, if necessary with the help of a pointing trowel. The horizontal and vertical joints should not be thicker than 10 mm.

If the section in need of repair extends the entire length of the wall, the use of earth blocks stabilized with lime or cement and lime masonry mortar might be necessary. In this case, only full courses of stabilized earth blocks should be used to replace courses in the rising masonry wall above the foundation. In other sections

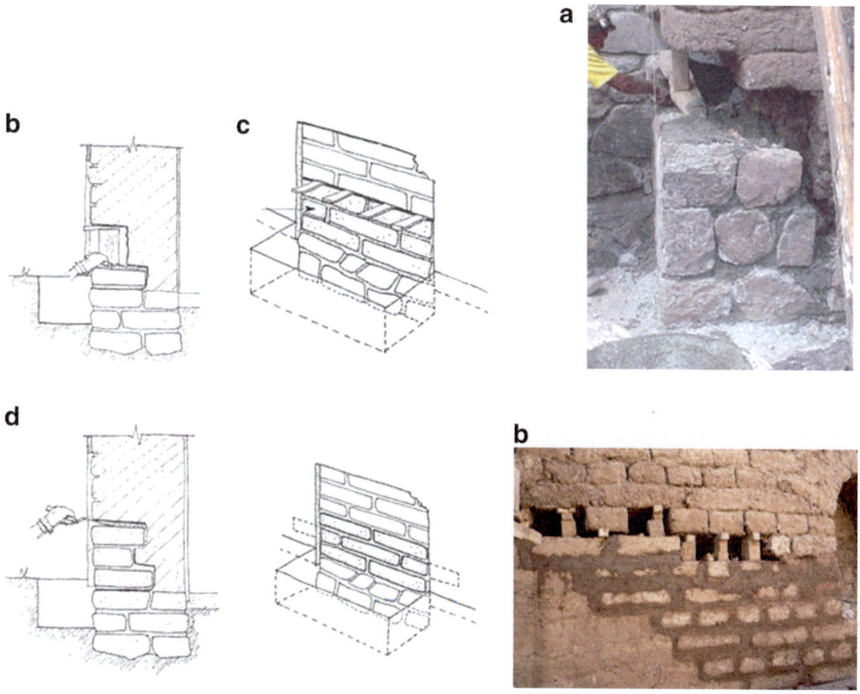

Fig. 5.55 Repair of earth block walls [75, 85]. (**a**) Repair of stem wall/wall base using natural stone and earth blocks, (**b**) removal of the weathered earth block masonry to depth of structurally stable section, bracing of cavity, (**c**) course-by-course filling with fully bedded masonry, allowing joint mortar to dry sufficiently, and (**d**) forming a friction connection, packing mortar using a pointing trowel

which do not extend the entire length of the wall, earth blocks which correspond to the quality of the original blocks should be used.

The masonry mortar should also match the original mortar in order to prevent the formation of rigid "plates" which exhibit a higher strength than the existing material.

The construction soil of the replacement materials used in the restoration of architectural earth block monuments should correspond to the original building materials in terms of its origin and composition. This also applies to the particular method used to produce the earth blocks.

Rising Wall and Wall Coping. Continuous vertical cracks in earth block walls can be repaired by following the recommendations for cob and rammed earth construction. The following case study is used to illustrate the application of modern techniques (which are applied in standard masonry construction) for repairing cracks in load-bearing earth block masonry [76]:

At the location of the structure in Kasbah Ait el Caid, Asslim, Agdz, Morocco, the building ground changes from rock in the north to river sediments of the Draa valley in the south. After an earthquake, a continuous vertical crack had formed in the southwestern tower of the kasbah ("fortress" residence), starting at the exterior wall coping. At the tower coping, the crack was approx. 15 cm wide and 6 m long.

A staircase tower which was integrated into the east wall asymmetrically next to the northeast tower had separated from the continuous wall structure. The main crack opened to a width of 15 cm at the wall coping and ran almost vertically down the wall for approx. 10 m, extended by a window opening.

The cracked wall sections had already begun to bulge out and were at risk of collapsing. The crack repair employed two restoration techniques which are common in modern masonry construction: anchoring and pinning.

The *anchoring* method was used in the restoration of the cracked southwest tower (Fig. 5.56). For this process, two loose steel cables ($d = 12$ mm) were placed into precut channels located on the floor of the viewing platform of the tower. The two channels were positioned at the one third points of the surface. The ends of the cables were pushed through holes which had been drilled into the exterior walls and guided through the eyes of exterior anchor plates. Using turnbuckles, the cables

Fig. 5.56 Repair of earth block walls, crack repair: anchoring [76]. (**a**) Before repair, view of crack from exterior; (**b**) before repair, view of crack from interior; (**c**) guiding the cables through the eyes of the anchor plates; (**d**) running of cables through ceiling; (**e**) turnbuckle; and (**f**) after repair, view of crack closed with a cement slurry

were slightly pretensioned which led to a stabilization of the crack. The crack had been cleaned and stitched in advance with earth block masonry on the exterior and filled with rammed earth on the interior. To finish, rammed earth was used to close up the channels on the floor of the platform and the crack was plastered on both sides.

For the *pinning* process, round steel rods ($d=40$ mm) with a length of approx. 80 cm were used to secure the crack (Fig. 5.57). Starting at the wall coping, four pairs of pins were spaced at intervals of approx. 60 cm over a length of approx. 3.5 m. The pins were prepared by inserting them into narrow tubes of gauze

Fig. 5.57 Repair of earth block walls, crack repair: pinning [76]. (**a**) Reinforced steel frame installed as emergency shoring before crack repair, (**b**) filling of cleaned crack with earth blocks, (**c**) steel pins and plastic hose inserted in gauze tube, (**d**) steel pin inserted into drill hole, and (**e**) view of wall after the crack repair

together with plastic hoses for connecting to a mortar pump. Next, the pins were inserted into the drilled holes to cover the crack roughly in the center and injected with a cement slurry. Before injection, the crack was opened up, cleaned, and stitched with earth blocks.

Before the pinning process could start, a reinforced concrete frame was erected to provide emergency shoring to stabilize the bulging wall section.

Pinning does not apply any added pretensioning force to the damaged wall. It only structurally stabilizes the existing condition. The pins absorb any additional tensile forces and restrict further movement.

The described anchoring and pinning measures were executed in March 2004 and 2005, respectively. During an on-site visit in March 2007, the repaired sections proved to be stable.

Surface Erosion and Plaster Damage. For large areas of surface erosion of the earth block masonry (up to a depth of 10 cm), the loose particles were brushed off and a lime plaster was applied. Flat pieces of broken bricks were then pushed into the plaster spaced tightly next to each other, while the plaster was still wet (Fig. 5.58 [74]). After setting, a two-layer lime plaster (base coat with coarse sand, finish coat with fine sand) was applied and finished with a coat of limewash.

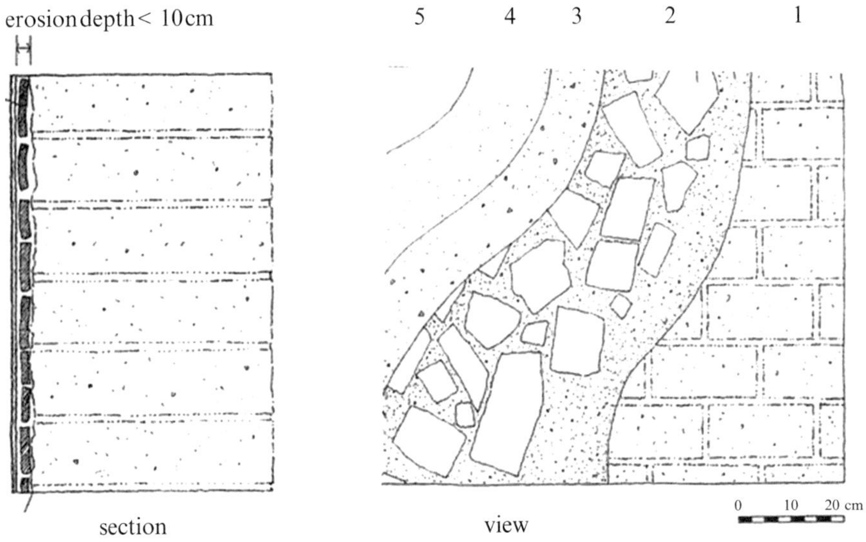

1 earth block masonry, undamaged section
2 lime base coat with embedded brick chunks
3 lime base coat, with coarse sand
4 lime finish coat, with fine sand
5 limewash

Fig. 5.58 Repair of earth block walls: repair of surface erosion [74]

5.3.3.3 Ceilings and Flat Roofs

As is the case with other building elements, damage to ceilings and flat roofs can only be repaired (Sect. 4.3.5.2) after the causes of damage have been analyzed. They lie primarily in:

- The effects of the elements (precipitation, extreme temperature fluctuations caused by high temperature amplitudes of the air, material deterioration caused by extreme solar radiation)
- Mechanical wear caused by the occupants
- Damage to the wooden support structure as a result of various impacts (moisture penetration, earthquake, termites, etc.)

If flat roofs are not regularly maintained, particularly after rainfall, serious damage can occur after only a short period of time. Maintenance should especially focus on the following measures [73, 77]:

- Sealing and crack repair of roof surfaces which occupants walk on
- Function of rainwater pipes which penetrate the parapet wall (Fig. 4.62)
- Stability of the wall coping covers (parapet)
- Support structure

Figure 5.59 shows the repair of a support structure of a flat roof made of peeled round timbers in Ait Benhaddou, Morocco. A plastic membrane has been integrated into the roof system. However, if this roof is not restored correctly, it will soon need to be repaired again (Fig. 5.60).

Fig. 5.59 Traditional flat roofs made of earth building materials: restoration of the supporting structure

Fig. 5.60 Traditional flat roofs made of earth building materials: repair of the roof seal (plastic membrane)

5.3.3.4 Earth Plaster

There are two different strategies for maintaining and restoring earth plasters based on their initial functions:

- Plasters as flat, thin, and continuous coats used to protect the surfaces of building elements (Sect. 4.3.6)
- Plasters as bases for ornaments, reliefs, and paintings in the preservation of architectural monuments

Historical Earth Plasters

Existing historical earth plasters which are still strong can be repaired in individual sections or across their entire surface using the methods described in Sect. 4.3.6.

Stabilizers, which are applied to the plaster and permeate it, are used to secure brittle or pulverized sections.

Separated plaster sections are "reattached" to the substrate by injecting suitable chemicals or glues (Fig. 5.61 [78]). Holes need to be filled with new plaster.

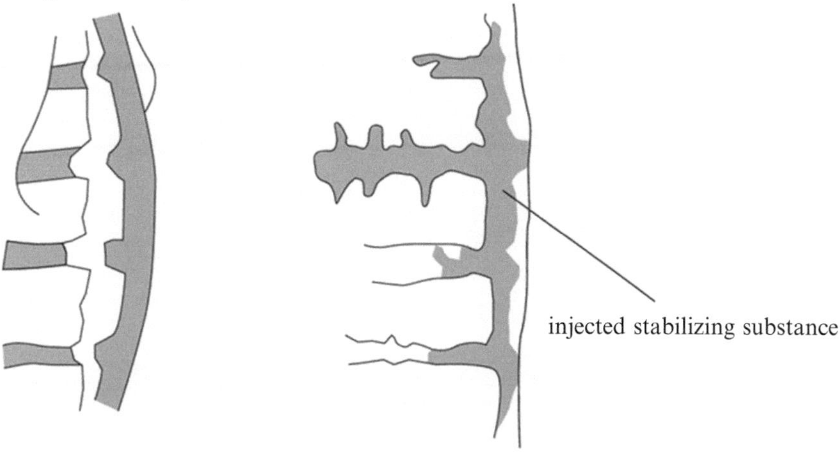

Fig. 5.61 Repair of historical plasters on earth block substrates [78]

Often, the plasters are also dirty, penetrated by salts and infested with mold. Therefore, measures for chemically neutralizing and cleaning the plasters also need to be included in the planning.

These types of damage are frequently caused by moisture penetration of the foundation and stem wall which needs to be eliminated before plaster restoration work can begin.

Plasters as a Picture Base

The preservation and restoration of historical earth and lime plasters which serve as the base for ornaments and wall paintings represents a complex field for restorers and archaeologists. Various types of impacts might have caused the plasters to become brittle, to separate from the substrate or to bulge or fall off.

Figure 5.62 shows the successful restoration of an earth plaster used as a plaster base. The painted earth plaster served as a finish for a historical cob wall (pachsah) (Sect. 5.3.2.1) which is part of the ruins of Afrasiab (Samarkand), Uzbekistan (Fig. 1.12). The earthen wall belonged to the local ruler's palace reception hall in the seventh century AD. In 1965, a wall with the depiction of a court reception of foreign ambassadors was discovered by coincidence during excavation work. The earthen wall, with adhering earth plaster and ornamentation, was stabilized, cut into blocks, and restored. In 1985, it was installed in the central hall of the Museum of History of Samarkand and has become one of the main attractions of the permanent collection.

Fig. 5.62 Earth plaster as picture base: "Afrasiab painting"—depiction of the procession of ambassadors to the reception of the local ruler of Afrasiab (Samarkand, Uzbekistan), seventh century AD

Sustainable Restoration

From 2009 to 2012, the former tower building for members of the German Bundestag in Bonn, the so-called Langer Eugen (built in 1953) was converted into the new headquarters of the secretariat of the United Nations Framework Convention on Climate Change (UNFCCC). Construction was commissioned by the Federal Office for Building and Regional Planning of Germany. The principles of sustainable building (Sect. 1.4.1) were strictly implemented into this "modern" renovation task by applying the evaluation system "Sustainable Building for Federal Buildings (BNB)" published by the German Federal Ministry of Transport and Digital Infrastructure [79, 80].

After assessing various materials based on this evaluation system, earth building materials were used in the area of interior construction and plaster. Clay panels made by a well-known German manufacturer ($w = 150$ cm, $l = 62.5$ cm, $d = 20$ and 25 cm) were used in non-load-bearing partition walls. They were attached to a framework of light steel framing profiles (Fig. 5.63 [81]) and finished with a fine-finish earth plaster.

5.3.3.5 Archaeological Ruins

In many archaeological ruins of earthen structures, only the remains of walls can be found due to the original functions of the buildings being lost long ago. There is archaeological evidence that some of the structures are more than 10,000 years old (Sect. 1.1), while others were built in the modern era.

Fig. 5.63 Sustainable renovation using clay plaster boards, UN Framework Convention on Climate Change, Bonn, 2011 [81]

For the contemporary earth builder, archaeological ruins of earthen structures provide a glance into the past. They reveal a technological "fingerprint" of earth building techniques which were common at the time of construction hundreds or thousands of years ago (Fig. 1.19 [82], straw-clay infill, 1289). Figure 5.64 shows the right-hand fingerprints left by a worker while forcefully throwing an earth clump onto the coping of a rising wall. This wall was part of the Bam Citadel whose core is more than 2000 years old. The fingerprints tragically resurfaced during the devastating earthquake on December 26, 2003 (Figs. 5.33 and 5.41). The findings clearly illustrate the "tschineh" technique, a regional version of cob construction which is still common, under various local names, in Iran, Central Asia, and Western China (Fig. 4.36).

Preservation measures of archaeological ruins of earthen structures include the conservation and structural stabilization of the remains. The planning of the preservation measures can be carried out according to the steps described in Sect. 5.3.2.2. Initial protection from the elements and from trespassers can be provided through transparent roofs and enclosures (Fig. 1.2). Structural measures include the application of "sacrificial layers" of stabilized earth to the wall copings (Fig. 1.13) and the construction of support walls to protect the base of the walls. These steps can help delay deterioration but are not permanent solutions.

In addition to the repair measures already mentioned, earthen structures can also be stabilized and waterproofed. Chemically modifying, natural organic, or synthetic

Fig. 5.64 Historical technological "fingerprint" of the "tschineh" building method, Bam Citadel, Iran, after 2003 earthquake

substances (Fig. 3.31b) are used to reduce capillary water absorption in the earth building material. The stabilizing or waterproofing substance should thoroughly penetrate the earthen material, not leave a sticky film on its surface and not seal the capillaries. The substances used must be alkali and weather resistant. Stabilization and waterproofing measures are irreversible.

The controlled backfill of archaeological sites is another measure which is used on structures which, due to their specific surfaces, would rapidly deteriorate if left exposed to the elements.

At best, the results of the described measures can only be seen as satisfactory in the medium term. The lack of permanent conservation methods [83] makes the need for close and interdisciplinary cooperation between specialists working with "earth" (Fig. 1.28) all the more urgent.

5.3.3.6 Earthquake-Resistant Restoration and Retrofitting

In regions with expected earthquake intensities of $I \geq 5$ (Sect. 5.2.4.2), earthquake-resistant restoration of buildings in general, and buildings made of earth building materials in particular, poses a very specific problem. According to EMS 98 (Fig. 5.34a), earthen structures are considered particularly vulnerable to earthquake damage.

Earthquake-resistant restoration of existing earthen structures can be divided into two action strategies:

- Preventive structural reinforcement or seismic retrofitting of existing buildings
- Restoration of damaged buildings after an earthquake

Structural Reinforcement

The extent of preventive structural reinforcement in buildings made of earth building materials depends on the expected earthquake intensity (which can be found in geological service maps of the respective region or country), the construction technique, the floor plan, the number of stories, but also on the protection value of the affected building. Schools, daycare facilities for children, public buildings with pedestrian traffic, and residential buildings have a higher protection value than agricultural buildings or workshops.

Seismic reinforcement of existing earthen structures begins with a thorough analysis of the buildings regarding their earthquake resistance deficits. This primarily includes a professional assessment of the condition of load-bearing and non-load-bearing building elements in accordance with valid national standards on earthquake-resistant construction.

After a comprehensive analysis of earth buildings in Central Asia, Chakimov [45] determined the following main structural deficits which, in the event of an earthquake, would lead to damage:

- Lack of an appropriate foundation based on the building's intended use
- Lack of a continuous bond beam at the level of the wall copings and bearing surfaces for ceilings
- Lack of corner reinforcement which would prevent a falling over of individual walls
- Lack of collapse reinforcement for earth block infill used in half-timber construction
- Insufficient and unreinforced pillar cross sections between wall openings

Foundations can be reinforced according to the methods described in Sect. 5.3.3.1.

Traditional half-timber construction using timber framing and earth infill requires a shear-resistant connection between ground sill plates and the foundation (Fig. 5.65 [41]). For this, anchors are inserted into the foundation at the necessary intervals and attached to the ground sill plate. The vertical posts and diagonal braces in the corner sections also need to form a shear-resistant connection with the sill. This can be achieved with the help of tenon joints, L-brackets, or straps and ties.

For rising walls made of earth block masonry or half-timber walls with earth infill (earth block masonry, earth clumps), reinforcement needs to be applied across the entire exterior and interior surfaces to function as collapse reinforcement. The reinforcement is anchored to the masonry on both sides with the help of a web of evenly distributed fasteners to form tensile- and shear-resistant connections.

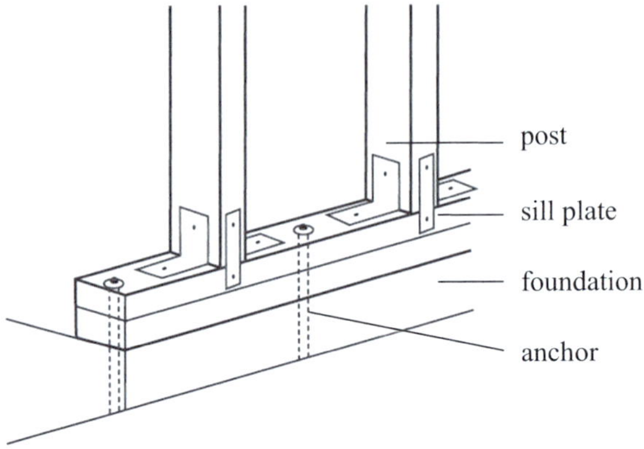

Fig. 5.65 Seismic retrofitting of half-timber construction: shear-resistant anchoring between foundation and ground sill [41]

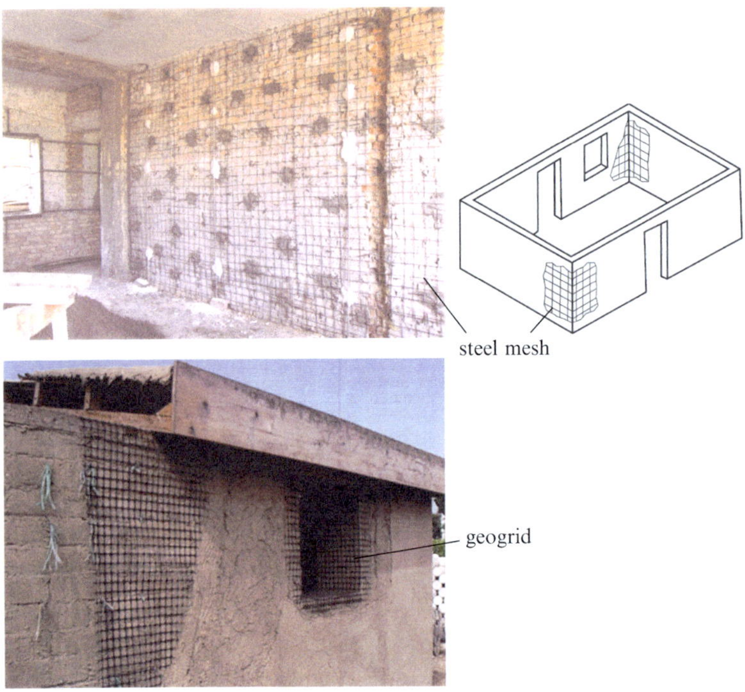

Fig. 5.66 Seismic retrofitting of earth block structures: wall containment using continuous reinforcement with tensile-resistant steel netting and geogrids [41, 45, 84]

In half-timber construction, the surface reinforcement is attached to the structural framework.

Weblike steel mesh or tensile-resistant geogrid material (Fig. 5.66 [41, 45, 84]) can be used as suitable surface reinforcement. To finish, a cement mortar is used to cover the anchored or attached surface reinforcement material.

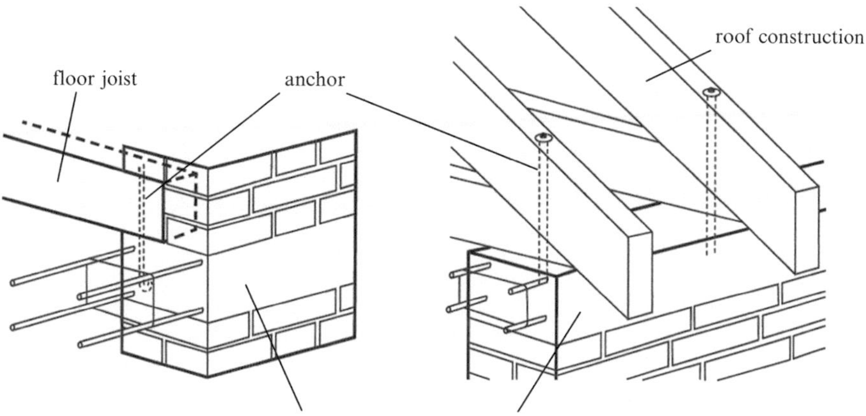

Fig. 5.67 Seismic retrofitting of earth buildings: shear-resistant anchoring of ceiling and roof systems using a continuous bond beam [41]

If there is no bond beam at the level of the ceiling joists or the wall coping, one should be installed retroactively. This also applies to all load-bearing earth building techniques such as rammed earth and cob and their regional versions.

The joist ends at the ceiling level as well as the roof structure at the level of the wall copings must form shear-resistant connections with the bond beam (or top plate in half-timber construction) using appropriate fasteners (steel tie rods, L-brackets) (Fig. 5.67 [41]). This also applies to connecting the bond beam to the wall copings to prevent shearing of the plane-like ceiling and roof structure (Fig. 5.37).

Reinforced concrete and steel (U- or I-profiles) are suitable materials for bond beams. In half-timber construction with a wooden structural framework, the bond beam can also be made of wood as long as it is of the same quality as all other building elements.

In order to prevent the damage in half-timber construction with timber framing and earth infill shown in Fig. 5.39, the vertical posts and diagonal braces need to form a shear-resistant connection with the continuous top plate which, in turn, must be securely connected to the roof system.

The strength of walls made of earth block masonry can be increased by integrating steel frames into window and door openings (Fig. 5.68 [45]). The earthquake resistance of walls can also be improved by adding tapered buttresses. They are placed at regular intervals along the load-bearing exterior longitudinal walls forming shear- and tensile-resistant connections. Buttresses reduce deflection in the longitudinal walls. Figure 5.69 shows a school building which lacks a supporting buttress in front of the corner seen in the foreground. This type of buttress could have prevented the longitudinal wall from tearing away from the end wall.

These recommended preventive measures involve significant costs. A cost–benefit analysis might lead to the realization that an earthquake-resistant new construction might be preferred over the retrofitting of an existing earthen structure.

Fig. 5.68 Seismic retrofitting of earth buildings: reinforcement of pillars using steel frames in window openings and continuous steel netting [45]

Fig. 5.69 Seismic retrofitting of earth buildings, reinforcement using buttresses: school building constructed in "pachsah" method in the Kamashi Region, Uzbekistan, after earthquake in the year 2000 [45]

Restoration of Damaged Buildings

After earth buildings have been damaged by an earthquake, the first decision to be made is if the structures can be repaired or if they should be demolished. In general, damage at damage levels 1 and 2 (Fig. 5.34b) is worth repairing. The decision if a building should be demolished or not primarily depends on the damage level of the building but also on a number of other factors, for example, the building's protection value. Owners have a special connection to their homes and will try to save and repair them within their means. In this context, independent specialists often need to make difficult decisions in the interest of the occupants.

For school buildings in Central Asia, Chakimov [45] has outlined some general recommendations for the assessment of the stability of damaged structures with regard to possible restoration work. The following displacement values and crack widths which occurred in building elements after an earthquake have been classified as impermissible or dangerous:

Dangerous displacement values:

- Ceiling system displaced from one of the three or four bearing surfaces with deflection exceeding 1/50 of the span
- Tilting of vertical load-bearing/non-load-bearing walls of more than 1/6 of the wall thickness in relation to the ceiling height

Impermissible displacement values:

- Displacement of ceiling systems exceeding 25 % of the projected bearing length with respective deflection of 1/100 of the span with no danger of collapse
- Tilting δ of vertical members in the range of $10<\delta<h/6$ [mm], with h being the smallest dimension in relation to the cross section of the structure
- Deformations of individual load-bearing building elements to an extent at which their load resistance appears completely exhausted

Impermissible crack widths:

- Crack widths in building elements >10 mm
- Crack widths in structural connections between building elements >15 mm

The recommended displacement values apply to school buildings which belong to the building category with the highest protection value. They can therefore only be used a guide. Different values can be applied to buildings of lower protection value (e.g., agricultural buildings).

Cracks can be repaired with the help of the methods described in Sect. 5.3.3.2 (Figs. 5.56 and 5.57).

Today, earthen structures can still be found in many earthquake-prone areas of the world. In the affected countries, regulations concerning earthquake-resistant construction, seismic retrofitting of existing earthen structures, as well as adequate restoration of earthen structures damaged by earthquakes are urgently needed. Appropriate regulations and their enforcement by the building authorities could save lives and reduce structural damage to buildings made of earth building materials (Sect. 4.2.1.3).

References

1. Reincke, D.: Untersuchungen zum Emissionsverhalten von Lehmoberflächen. Unpublished diploma thesis, FB Maschinenbau, Fachhochschule, Stralsund (2000)
2. BV der Deutschen Ziegelindustrie e.V., Mauerziegel, A.G. (Hrsg.): Baulicher Wärmeschutz. Ziegel Information GmbH, Bonn (1994)
3. Zweite Verordnung zur Änderung der Energieeinsparverordnung (EnEV 2009) v. 18.11.2013. Bundesgesetzblatt I, Nr. 61, Berlin (2013)
4. Gut, P., Ackerknecht, D.: Climate Responsive Building—Appropriate Building Construction in Tropical and Subtropical Regions. BASIN/SKAT, St. Gallen (1993)
5. Stark, J., Krug, H.: Baustoffkenngrößen. Schriften der Bauhaus-Universität Nr. 102, F.A.-Finger—Institut f. Baustoffkunde, Weimar (1996)
6. Dettmering, T., Kollmann, H.: Putze in Bausanierung und Denkmalpflege. Verlag für Bauwesen, Berlin (2001)
7. Niemeyer, R.: Der Lehmbau und seine praktische Anwendung (Reprint der Originalausg. von 1946). Ökobuch Verlag, Grebenstein Staufen (1982/1990)
8. Minke, G.: Handbuch Lehmbau. Baustoffkunde, Techniken, Lehmarchitektur, 7nd erw. Aufl. Ökobuch, Staufen (2009)
9. Eckermann, W., Ziegert, C.: Auswirkung von Lehmbaustoffen auf die Raumluftfeuchte. Unpublished manuscript, Berlin (2006)
10. Edwards, R.: Basic Rammed Earth—An Alternative Method to Mud Brick Building, 7th edn. The Rams Skill Press, Kuranda (2004)
11. CSIRO Australia, Division of Building, Construction and Engineering: Earth-Wall Construction, Bull. 5, 4th ed. CSIRO Australia, Division of Building, Construction and Engineering, North Ryde (1995)
12. Standards New Zealand: NZS 4297: 1998 engineering design of earth buildings, NZS 4298: 1998 materials and workmanship for Earth buildings, NZS 4299: 1998 earth building not requiring specific design. Standards New Zealand, Wellington (1998)
13. Lippsmeier, G.: Tropenbau/Building in the Tropics, 2nd Aufl. Callwey, München (1980)
14. Europäisches Parlament, Rat der Europäischen Union: Verordnung Nr. 305/2011 vom 9.März 2011 zur Festlegung harmonisierter Bedingungen für die Vermarktung von Bauprodukten und zur Aufhebung der Richtlinie 89/106/EWG des Rates. Amtsblatt der Europäischen Union L 88/5 v, Brüssel (2011)
15. Dachverband Lehm e.V. (Hrsg.): Lehmbau Regeln—Begriffe, Baustoffe, Bauteile, 3rd überarbeitete Aufl. Vieweg+Teubner/GWV Fachverlage, Wiesbaden (2009)
16. Schmidt, P.: Urteil über Lehmhäuser—eine Umfrage. In: das bauzentrum 8/95, pp. 70–74
17. Röhlen, U.: Schallschutz-Werte von Lehmbau-Konstruktionen. In: LEHM 2000 Berlin, Beiträge zur 3. Int. Fachtagung für Lehmbau des Dachverbandes Lehm e.V., pp. 64–68. Overall, Berlin (2000)
18. Bundesverband der Deutschen Ziegelindustrie e.V. (Hrsg.): Baulicher Schallschutz—Schallschutz mit Ziegeln. AG Mauerziegel e.V., Bonn (1996)
19. Holzabsatzfonds, Deutsche Gesellschaft f. Holzforschung (Hrsg.): Erneuerung von Fachwerkbauten. In: Holzbau Handbuch, Reihe 7, Teil 3, Folge 1. Informationsdienst Holz, Bonn (2004)
20. Stark, J., Wicht, B.: Umweltverträglichkeit von Baustoffen. Schriften der Bauhaus-Universität Nr. 104, Weimar (1996)
21. European Commission (ed.): Radiological protection principles concerning the natural radioactivity of building materials, Radiation protection 112. Directorate-General Environment, Nuclear Safety and Civil Protection, Brussels (1999)
22. Bundesamt für Strahlenschutz (Hrsg.): Strahlenthemen—Radon in Häusern. Informationsblatt BfS, Salzgitter (2007)
23. Dachverband Lehm e.V. (Hrsg.): Qualitätsüberwachung von Baulehm als Ausgangsstoff für industriell hergestellte Lehmbaustoffe—Richtlinie. Technische Merkblätter Lehmbau, TM 05. Dachverband Lehm e.V., Weimar (2011)

24. Lindenmann, M., Leimer, H.P., Rusteberg, C.: Möglichkeiten und Grenzen der Abschirmwirkung von Gebäuden gegen elektromagnetische Wellen. In: Ganzheitliche Bauwerkssanierung und Bauwerkserhaltung nach WTA, WTA-Schriftenreihe H. 28, pp. 101–116. WTA, München (2006)
25. Pauli, P., Moldan, D.: Reduzierung hochfrequenter Strahlung im Bauwesen—Baustoffe und Abschirmmaterialien. Univ. d Bundeswehr, München (2000)
26. Stahr, M. (Hrsg.): Bausanierung—Erkennen und Beheben von Bauschäden, 2nd Aufl. Vieweg, Braunschweig (2002)
27. Hughes, R.: Material and structural behaviour of soil constructed walls. Monumentum **26**(3), 175–188 (1983)
28. Goretzki, L.: Instandsetzung von feucht- und salzbelastetem Mauerwerk. Workshop "Erdbebengerechtes Bauen mit lokal verfügbaren Materialien in Zentralasien" Fergana, Usbekistan 1996. In: Thesis, Wiss. Zeitschr. d. 45(6), pp. 49–53. Bauhaus-Universität, Weimar (1999)
29. Güntzel, J.: Zur Geschichte des Lehmbaus in Deutschland, Bd. 1. Dissertation, Gesamthochschule, Kassel (1986)
30. Köhler, T.: Analyse von Bauschäden und Maßnahmen der Instandsetzung an Konstruktionen aus Lehmbaustoffen. Unpublished diploma thesis, Fak. Bauingenieurwesen, Bauhaus-Universität, Weimar (2003)
31. Künzel, H.: Regenbeanspruchung und Regenschutz von Holzfachwerk-Außenwänden. In: Statusseminar "Erhaltung von Fachwerkbauten," Sonderheft Verbundforschungsprojekt Fachwerkbautenverfall und Fachwerkbautenerhaltung, pp. 32–39, ZHD, Fulda (1991)
32. Pollack, E., Richter, E.: Technik des Lehmbaus. Verlag Technik, Berlin (1952)
33. Maiwald,Y, Hein, B.: Lehmwellerbauten in Thüringen und Sachsen-Anhalt. Unpublished student research paper, Bauhaus-Universität, Weimar (1998)
34. Bielenberg, H.: Der Einfluss des Stalles auf die Schweinemast. Dissertation, Fak. f. Bauwesen, Technische Hochschule, Braunschweig (1963)
35. Burger, H.: Sanfter Baustoff Lehm in der Landwirtschaft—Auswirkungen auf das Stallklima. Unpublished diploma thesis, FB Landwirtschaft/Agrartechnik, Univ.-GH, Kassel Witzenhausen (1995)
36. Pacific Science Association: Information Bulletin (Honolulu), 42(1) (1990)
37. Manu, F.W., Amoah-Mensah, P.D., Bai-den-Amissah, Nana K. Nsiah-Achampong: Die Überschwemmung von Sandema (Ghana) im September 2007—Auswirkungen auf Lehmbauten. In: LEHM 2008 Koblenz, Beiträge zur 5. Int. Fachtagung für Lehmbau, pp. 80–89. Dachverband Lehm e.V., Weimar (2008)
38. Rauch, P.: Hochwasserschäden an Lehmbauten. www.ib-rauch.de (2003)
39. Cornerstones Community Partnerships (ed.): How to save your adobe home in the event of a flood disaster—An emergency flood mitigation manual for historic and traditional earthen architecture, Santa Fe (2008)
40. Arnold, W.: Eroberung der Tiefe. VEB Verlag f Grundstoffindustrie, Leipzig (1974)
41. Coburn, A., Hughes, R., Pomonis, A., Spence, R.: Technical Principles of Building for Safety. Intermediate Technology, London (1995)
42. Minke, G.: Construction Manual for Earthquake-Resistant Houses Built of Earth. GATE-BASIN/GTZ, Eschborn (2001)
43. Grünthal, G., Musson, R.M.W., Schwarz, J., Stucchi, M.: European Macroseismic Scale 1992 (up-dated MSK-Scale). Centre Europeén de Géodynamique et de Séismologie, Luxembourg (1998)
44. Schroeder, H., Schwarz, J., Chakimov, S.A., Tulaganov, B.A.: Traditional earthen architecture in Uzbekistan—evaluation of earthquake resistance and strategies for improvement. In: Proceedings of Terra 2003—9th international conference on the study and conservation of earthen architecture, Yazd, Iran, pp. 513–530 (2003)
45. Хакимов, Ш. А. и др.: Технологические приёмы антисейсмического усиления школных зданий—Пособие для строителей (Technological procedures for seismic reinforcement of school buildings—guideline for building professionals). Extremum, Ташкент (2009)

46. When the earth quakes—earthquake resistant construction with local materials in the Iran. Unpublished project thesis, Bauhaus-Universität, Fak. Architektur, Weimar (2005)
47. Lamers, R., Rosenzweig, D., Abel, R.: Bewährung innen wärmegedämmter Fachwerkbauten—Problemstellung und daraus abgeleitete Konstruktionsempfehlungen. In: Bauforschung für die Praxis, Band 54. Fraunhofer IRB, Stuttgart (2000)
48. Schneider, U., Schwimann, M., Bruckner, H.: Lehmbau für Architekten und Ingenieure—Konstruktion, Baustoffe und Bauverfahren, Prüfungen und Normen, Rechenwerte. Werner, Düsseldorf (1996)
49. Petzet, M.: Grundsätze der Denkmalpflege. In: Sana`a—die Restaurierung der Samsarat al-Mansurah, pp. 92–98. ICOMOS Hefte des Deutschen Nationalkomitees XV, München (1995)
50. Wirth, H.: Denkmalpflege in erdbebengefährdeten Regionen. Workshop "Erdbebengerechtes Bauen mit lokal verfügbaren Materialien in Zentralasien" Fergana, Usbekistan 1996. In: Thesis, Wiss. Zeitschr. d. 45(6), pp. 54–57. Bauhaus-Universität, Weimar (1999)
51. Hähnel, E.: Fachwerk-Instandsetzung—ein Praxishandbuch. Bauwesen/Fraunhofer IRB, Berlin Stuttgart (2003)
52. Happe, M.: Gebäudetranslozierungen des Thüringer Freilichtmuseums Hohenfelden—Projekte und Technologien unter den Bedingungen der DDR und der Zeit nach 1990. Fachtagung "Vorfahrt mit Blaulicht für Museumshäuser," Tagungsband, pp. 43–50. AG der regionalen ländlichen Freilichtmuseen in Baden-Württemberg, Bad Waldsee (2005)
53. Kienzle, P., Ziegert, C.: Die Rekonstruktion römischer Stampflehmbauten im Archäologischen Park Xanten. In: LEHM 2008 Koblenz, Beiträge zur 5. Int. Fachtagung für Lehmbau, pp. 148–161. Dachverband Lehm e.V., Weimar (2008)
54. Dahlhaus, U., Kortlepel, U., et al: Lehmbau 2004—aktuelles Planungshandbuch für den Lehmbau, Ergänzungen 2001 u. 2004. manudom, Aachen (1997)
55. Schmitz, H., Krings, E., Dahlhaus, U., Meisel, U.: Baukosten 2012/13, Band 1: Instandsetzung, Sanierung, Modernisierung, Umnutzung, Band 2: Preiswerter Neubau von Ein- u. Mehrfamilienhäusern, 21st Aufl. Verlag Hubert Wingen, Essen (2013)
56. Kinze, W., Franke, D.: Grundbau. VEB Verlag f Bauwesen, Berlin (1981)
57. Ahnert, R., Krause, K.H.: Typische Baukonstruktionen von 1860 bis 1960 zur Beurteilung der vorhandenen Bausubstanz; Bd. I: Gründungen, Wände, Decken, Dachtragwerke; Bd. II: Stützen, Treppen, Erker und Balkone, Bogen, Fußböden, Dachdeckungen. VEB Verlag f. Bauwesen, Berlin 1985 (I); 1988 (II)
58. Schroeder, H.: Gutachterliche Stellungnahme zu einem Gründungsschaden in Kornwestheim. Unpublished, Weimar (2000)
59. Dachverband Lehm e.V. (Hrsg.): Kurslehrbuch Fachkraft Lehmbau. Dachverband Lehm e.V., Weimar (2005)
60. Dachverband Lehm e.V. (Hrsg.): Lehmbau Verbraucherinformation. Dachverband Lehm e.V., Weimar (2004)
61. Wissenschaftlich-Technischen Arbeitsgemeinschaft für Bauwerkserhaltung und Denkmalpflege e.V. WTA: Merkblätter zur Fachwerkinstandsetzung 2000-2013. 8-1: Bauphysikalische Anforderungen an Fachwerkgebäude (2012); 8-2: Checkliste zur Sanierungsplanung und—durchführung (2007); 8-3: Ausfachungen von Sichtfachwerken (2010) (Kap. 5.4 Lehmausfachungen); 8-4: Außenbekleidungen (2008); 8-5: Innendämmungen (2008); 8-6: Beschichtungen auf Fachwerkwänden—Ausfachungen/Putze (2009); 8-7: Beschichtungen auf Fachwerkwänden—Holz (2010); 8-8: Tragverhalten von Fachwerkbauten (2007); 8-9: Gebrauchsanweisung für historische Fachwerkhäuser (2013); 8-10: EnEV: Möglichkeiten und Grenzen (2011); 8-11: Schallschutz bei Fachwerkgebäuden (2008); 8-12: Brandschutz bei Fachwerkgebäuden (2011)
62. Deutsches Zentrum für Handwerk u. Denkmalpflege Fulda ZHD (Hrsg.): Arbeitsblätter, AB Fachwerksanierung, AB Wärmedämmung bei bestehendem Fachwerk. ZHD, Fulda (1990)
63. Kraft, J.: Was wie machen? Instandsetzen u. Erhalten alter Bausubstanz. Interessengemeinschaft Bauernhaus IGB, Weyhe (1992)
64. Ministerium für Bauen u. Wohnen des Landes Nordrhein-Westfalen (Hrsg.): Fachwerkgebäude erhalten und instandsetzen. MBW-Ratgeber Nr. 4, Düsseldorf (1991)

65. Schüler, G., Schroeder, H: Naturbelassene Baustoffe—Beispiel Kalktuff Magdala, pp. 337–341. In: Wiss. Z. Hochsch. Archit. Bauwes. -A/B- 39(4), Weimar (1993)
66. Kortlepel, U.: Lehmspritzverfahren. Landesinstitut f. Bauwesen u. angewandte Bauschadensforschung NRW, Aachen (1994)
67. Dingeldein, T.: Konzeption einer neuen Lehmbauweise mit maschineller Einblastechnik. In: KirchBauhof Berlin (Hrsg.) "Moderner Lehmbau 2002," pp. 9–13. Fraunhofer IRB, Stuttgart (2002)
68. Ziegert, C: Lehmwellerbau—Konstruktion, Schäden und Sanierung. Berichte aus dem konstruktiven Ingenieurbau, Heft 37. Fraunhofer IRB, Stuttgart (2003)
69. Morris, W.: 1—History, building methods and conservation; Keefe, L.: 2—Repair and maintenance. Devon Historic Building Trust, Exeter (1993)
70. Keefe, L.: Earth Building—Methods and Materials, Repair and Conservation. Taylor & Francis, London (2005)
71. Pearson, G.T.: Conservation of Clay and Chalk Buildings. Donhead, Shaftesbury (1997)
72. Henes-Klaiber, U.: Ursachen und Behandlungsmethoden von Feuchteschäden an historischen Bauwerken. In: Klimatisierung und bauphysikalische Konzepte—Wege zur Nachhaltigkeit bei der Pflege des Weltkulturerbes, pp. 129–138. ICOMOS Hefte des Deutschen Nationalkomitees XLII, und Deutscher Kunstverlag, München Berlin (2005)
73. CERKAS, CRATerre-EAG: Manuel de conservation du patrimoine architectural en terre des vallées présahariennes du Maroc. Ouarzazate, CERKAS Centre du Patrimoine Mondial de l'UNESCO (2005)
74. Fodde, E.: Architetture di terra in Sardegna—Archeometria e conservazione. Cagliari, Aipsa edizione (2004)
75. Cornerstones Community Partnerships (ed.): Adobe Conservation—A Preservation Handbook. Sunstone, Santa Fe (2006)
76. Fahnert, M., Schroeder, H.: Sanierung traditioneller Stampflehmbauten in Südmarokko. In: LEHM 2008 Koblenz, Beiträge zur 5. Int. Fachtagung für Lehmbau, pp. 250–251. Dachverband Lehm e.V., Weimar (2008)
77. Klessing, J. M.: Planung und Ausführung der Restaurierung. In: Sana`a—die Restaurierung der Samsarat al-Mansurah, pp. 49–91. ICOMOS Hefte des Deutschen Nationalkomitees XV, München (1995)
78. Crosby, A.: Conservation of painted lime plaster on mud brick walls at Tumacacori National Monument, U.S.A. In: Proceedings of 3rd international symposium on mud brick (adobe) preservation, pp. 59–78. ICOMOS, Ankara (1980)
79. Bundesministerium für Verkehr, Bau und Stadtentwicklung (BMVBS): Bekanntmachung über die Nutzung und die Anerkennung von Bewertungssystemen für das nachhaltige Bauen. Bundesanzeiger Nr. 70, v. 07, p. 1642 (2010)
80. Bundesministerium für Verkehr, Bau und Stadtentwicklung (BMVBS) (Hrsg.): Leitfaden Nachhaltiges Bauen. BMVBS, Berlin (2011)
81. Röhlen, U., Mai, D.: Trockenbau mit Lehmbau im "Klimareferat der Vereinten Nationen," Bonn. In: LEHM 2012 Weimar, Beiträge zur 6. Int. Fachtagung für Lehmbau, pp. 50–53. Dachverband Lehm e.V., Weimar (2012)
82. Volhard, F.: Historische Lehmausfachungen und Putze. In: LEHM '94 Internationales Forum für Kunst u. Bauen mit Lehm, Beiträge, Aachen, pp. 25–27 (1994)
83. Simon, S.: Kulturelles Erbe Erdarchitektur—Materialien, Forschung, Konservierung. In: Naturwissenschaft & Denkmalpflege, pp. 263–273. Innsbruck University Press, Innsbruck (2007)
84. Mantovani, A.: Architetture e tipologie della terra cruda in Perù. In: Terra—Ragionamenti e Progetti (a cura di Stefan Pollak), pp. 27–40. ARACNE editrice S.r.l., Roma (2012)
85. Torrealva, D., Esquivel, Y.: Structural Intervention in Adobe Monuments in Cusco. Catholic University Peru CUP, Lima, Peru (2008)
86. Naasner, C., Paulick, R., Stendel, G., Wunderlich, A.: Lehmbaupraktikum Lindig 2004. Unpublished internship report, Fak. Architektur, Bauhaus-Universität, Weimar (2004)
87. Volhard, F.: Leichtlehmbau, alter Baustoff—neue Technik, 5th Aufl. C. F. Müller, Heidelberg (1995)

6

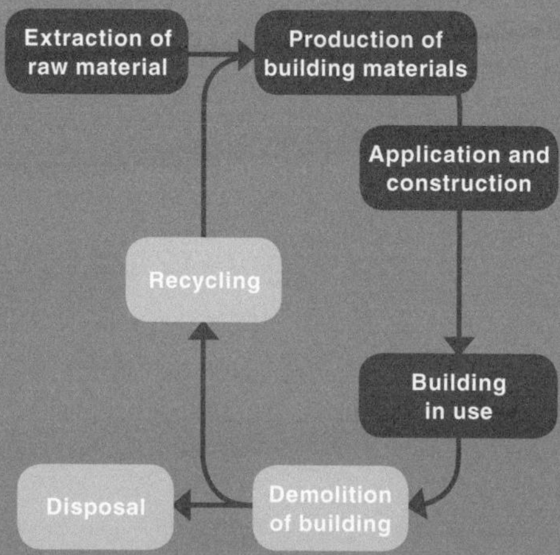

Demolition, Recycling and Disposal of Earth Building Materials

The German Closed Substance Cycle Waste Management Act of 1996 (Kreislaufwirtschafts -und Abfallgesetz / KrW-/AbfG) describes the mandatory procedures for the demolition of buildings, recycling and, if necessary, the disposal of demolition materials. In this context, the law gives priority to waste prevention over recycling.

This is one of the principles which must be adhered to if unspoiled environmental conditions are to remain accessible to future generations. As a consequence, the quality of today's buildings cannot only be judged with regard to meeting specific requirements in terms of design, structure, material technology and building practice. They also need to be measured by their fulfillment of the following stipulation: buildings need to be designed for recyclability.

© Springer International Publishing Switzerland 2016
H. Schroeder, *Sustainable Building with Earth*,
DOI 10.1007/978-3-319-19491-2_6

6.1 Demolition of Buildings

When a building has reached the end of its useful life, it is demolished. The point in time of demolition is determined by a number of scenarios:

- Wear and tear or a lack of upkeep has led to severe damage and the repair costs outweigh the expected benefits.
- The building has been seriously impacted by accidental damage or exceptional natural disasters.
- New user requirements cannot be met by the existing building design.
- Decisions based on urban and land-use planning.

6.1.1 Legal Framework

Based on the German Construction Contract Procedures VOB, Part C (Sect. 4.2.1.1), the German Demolition Association (Deutscher Abbruchverband e.V.) has published technical regulations for demolition work and corresponding procedures [1]. These regulations have been further supplemented by a guideline [2] which refines the specifications of the contract documents for demolition work.

According to these regulations, it is the building owner's responsibility to plan and supervise the demolition work and the disposal of the demolition materials. This responsibility entails:

- Drawing up a work description (by a qualified professional) which includes "Additional Services (protection and stabilization measures, safety and health measures, supervision, disposal)"
- Obtaining a demolition permit (which includes informing the building inspection authorities and obtaining permits from the traffic, business, and environmental authorities)
- Contracting the demolition work (restricted calls for bids) and supplying the demolition company with the obtained permits in order to comply with the specified requirements

The work description needs to define the buildings which are to be demolished in terms of their mass (material), their dimensions (enclosed space), as well as their structural properties (e.g., wall thickness). These specifications form the basis for pricing and the separation of materials.

It is particularly important to supply information about possible harmful substances specific to the use of the building, related to production, or intrinsic to the building materials. For health and safety reasons, this also applies to work involving building elements containing asbestos and synthetic mineral fibers. Work involving these materials must be reported to the responsible authorities in advance.

6.1.2 Disassembly Steps

Generally, the term *demolition* refers to the removal of structures or building components through disassembly or wrecking. The demolition of buildings or building elements can be partial or complete.

According to the German Closed Substance Cycle Waste Management Act (KrW-/AbfG) [3], a controlled disassembly should be given preference over an "uncontrolled" total demolition, because it can be used specifically to obtain pure demolition material. Modern equipment and the professional expertise of demolition companies as well as an overall increased awareness of environmental and safety issues facilitate the controlled disassembly of buildings.

The controlled dismantling of buildings follows a specific sequence of disassembly steps:

Step 1: Nondestructive removal of components which can be immediately reused (technical devices, doors, windows, fixtures, etc.)
Step 2: Removal of accessible, usable components (wall paneling, window glass, pipes, flooring, etc.)
Step 3: Removal of usable components which are part of the building (structural steelwork, synthetic materials, plumbing, etc.)
Step 4: Removal of materials which cannot be reused (insulation panels, filling foams, tar paper, glued waterproofing membranes, etc.)
Step 5: Dismantling of the building's primary building materials (earthen material, wood and other building materials)
Step 6: Removal of underground structures

Time restrictions often lead to an "uncontrolled" total demolition followed by the sorting of the demolition materials on site or in a sorting facility.

6.1.3 Demolition Methods

Structures made from earth building materials are demolished using mechanical methods [4]. The methods should allow for a targeted recovery of correctly sorted demolition materials for recycling. Which particular method is used depends on a number of factors: the available space on site, the technical equipment and qualifications of the demolition company, time restrictions, and, lastly, the receiving conditions and fees of the recycling plant or, if applicable, the waste disposal site. The "fee" factor, in particular, can be used as an instrument to encourage companies and owners to recycle more and send less waste to landfills.

Demolition work on earthen structures typically generates a high amount of dust which poses a threat to the health of the workers. In order to bind the dust, the building elements scheduled for demolition can be sprayed down with water. This needs to be done carefully due to the water solubility of the earth building materials and the possible risk of mixing the earthen materials with other demolition materials.

6.1.3.1 Mechanical Striking and Hammering

Handheld tools (chisels, hammers, crowbars, pickaxes) and *demolition hammers* (pneumatic, electric, hydraulic) are used to remove building elements from a safe work position.

This method is mainly applied during small-scale demolition work, such as the demolition, dismantling, or remodeling of one- to two-story residential buildings (disassembly Steps 1–4), as well as in disassembly Step 5 of half-timber or earth block buildings. It can also be used in preparation for other demolition methods, for example, to break off and salvage earth plasters. Other demolition methods are subsequently used to remove the support structure. This method is not recommended, or only to a very limited extent, for cob and rammed earth structures due to the thickness of the building elements. It can, however, be used for breaking up larger chunks at a later stage.

The advantage of this method is found in the relatively low damage risk to adjacent buildings or traffic infrastructure. It also allows for a high degree of pure sorting of the demolition material which is a precondition for subsequent recycling. The disadvantages lie in the high degree of physical labor and the exceptionally high risk for the workers. This is compounded by a comparatively low demolition rate of return.

In order to ease the physical labor, demolition hammers can be attached to carrier equipment (e.g., hydraulic excavators, small mobile machines). Such carriers require sufficient space as well as a stable work surface.

6.1.3.2 Mechanical Breaking

A steel wrecking ball is used to *knock down* building elements. The wrecking ball is suspended from the arm of a carrier, typically a cable-operated excavator. Depending on the purpose, the weight of the wrecking ball ranges from 500 (for masonry) to 5000 kg (for reinforced concrete).

Exposure to noise and dust place a high degree of physical and mental stress on the machine operator who must be experienced in carrying out this type of work.

This method is used in disassembly Step 5. Theoretically, it can also be used for cob and rammed earth structures, but so far there has not been enough practical experience or written documentation regarding this type of application.

Hydraulic machines (e.g., bulldozers or loaders) are used to demolish building elements with the help of the "pushing in and pulling down" method. For this method, the demolition equipment must be tall enough to reach the highest point of the building.

The method is suitable for the demolition of earth block and half-timber structures (disassembly Step 5).

Cables are used to *tear down* relatively large building parts. This method requires a large enough space to allow for the towing equipment to be placed a sufficient distance away from the building (roughly three times the height of one story of the building). In addition, the equipment must be placed on stable ground, and the building elements scheduled for demolition need to be able to safely absorb the tensile force of the cable.

This method is suitable for earth block and half-timber structures (disassembly Step 5).

The *grabbing* method is used to demolish building parts mechanically from the top down using grabbing devices (excavators with grabbers). This method also requires sufficient space. In addition, the building must be accessible from several sides. Unstable building elements need to be removed beforehand.

This method is suitable for the demolition of earth block structures (disassembly Step 5).

6.1.3.3 Mechanical Cutting and Drilling

Diamond saw blades or *diamond wire saws* are used to apply relatively clean cuts to existing structures without causing extensive vibration. This method is used primarily for renovating and modernizing older buildings, particularly for the removal of load-bearing building elements. It is also suitable for repair work around the base of cob and rammed earth walls (Sect. 5.3.3.2) as well as for preparation work for further demolition.

This method can theoretically be used in the preparation work of disassembly Step 5 for cob and rammed earth structures, but so far there has not been sufficient documentation regarding the practical application of this procedure.

The carrying out of *core hole drilling* or slitting is another field of application. It is used to create openings in the course of the technical retrofitting of historical cob and rammed earth structures during remodeling and renovation work (Sect. 4.3.7).

6.2 Reuse of Earth Building Materials

For centuries, construction work did not require more than a few, locally available building materials used by generations of builders to construct fully functional structures. Using and reusing building materials, even multiple times, was standard practice for our ancestors. Ruins of medieval fortresses and monasteries typically served as popular building material sources for the surrounding population. Probably the most well-known example of the reuse of mass-produced conventional building materials in the recent past is the work of the "Trümmerfrauen," women who helped clear the rubble and reconstruct Germany's destroyed cities after World War II.

It has been common practice to reuse earth building materials for thousands of years. Archaeological findings have shown that earth building materials were reused as fill material or for the production of earth blocks as early as 8000 years ago in Çatal Höyük, the oldest earthen house structures presently known. Afrasiab, the ancient predecessor of today's Samarkand, rose to a height of more than 40 m before it was destroyed by Genghis Khan in the middle of the thirteenth century (Fig. 1.12). This height was possible because buildings which were no longer needed, or which were dilapidated or destroyed, were demolished, and the salvaged materials were reused as fill material or for making earth blocks. This ancient idea of recycling could also be observed in the Iranian city of Bam which was almost completely destroyed by an earthquake in 2003 (Fig. 5.41). The owners of the destroyed houses used the rubble to make new earth blocks for the repair or new construction of their homes.

Today's "throwaway society" stands in stark contrast: currently, the total volume of waste in Germany is around 400 million tons per year with construction waste accounting for about three quarters of the amount, or 300 million tons per year. Sixty percent, around 180 million tons per year, goes into landfills [5] [6]. This construction waste includes soil which is generated in all areas of subgrade construction in the form of non-contaminated, undisturbed "excavation soil."

Landfill space is becoming scarce and more expensive. People living in the vicinity of potential new landfill sites increasingly oppose arguments for the development of such sites. Their fear of a rise in pollution and falling property values is justified. This stresses the urgent need for measures to reduce the overall volume of waste. Construction waste represents the largest share of the total waste volume by far. Reducing the yearly volume of construction waste could contribute considerably to a decrease in demand for landfill space, particularly by finding alternative uses for "excavation soil" waste.

This is where the opportunities for modern earth building can be found: The industrial processing of suitable "excavation soil" into earth building materials with properties comparable to conventional building materials could result in a lower demand for landfill space. This is already being successfully carried out by a number of producers of earth building materials in Germany where the current total annual sales of earth building materials amount to approx. €40 million with an upward trend [7].

Furthermore, non-contaminated earth building materials can easily be returned to geogenic and biogenic cycles by using them as backfill material in road construction and landscaping.

Recently, the problem of overflowing landfills described above has led to a rethinking process resulting in the idea of building material recycling. What started with windows, doors, stoves, floor boards, and timber beams has expanded to wall bricks and roof tiles as well as earth blocks and plasters. Today, this field basically covers any building material which can be salvaged. In Germany, this development has created an independent industry, the "recycling of historical building materials," which is represented by the Business Association of Historical Building Materials e.V (Unternehmensverband Historische Baustoffe e.V., www.historische-baustoffe.de).

6.2.1 *Planning Criteria*

The situation described above is also reflected in legal regulations. Whereas a safe disposal of waste was the goal of the Waste Disposal Law (Abfallbeseitigungsgesetz) of 1972, the 1986 amendment to the law aimed for as much waste reduction and reuse as possible. Raw materials taken from nature should be kept within the material cycle whenever possible. For the construction field, this means that the recyclability or the reuse of the building materials at the end of a building's "lifetime" have to be taken into consideration as early as during the planning phase.

The German Closed Substance Cycle Waste Management Act (KrW-/AbfG) [3] dictates resource-conserving, low-waste recycling management and the environmentally safe disposal of unavoidable waste. Avoiding waste is preferred over the recovery of materials and energy. This puts the "polluter-pays" principle into practice by establishing the producers' product stewardship.

Based on this law, the German Construction Contract Procedures (VOB) (Sect. 4.2.1.1) stipulate equal treatment of recycling products and primary building materials as long as they meet the quality requirements of the individual use.

6.2.2 *Recycling*

Today, the recycling of building materials has become a part of sustainable building. It forms the last step in the life cycle of a building (Sect. 1.4.2): consumed materials and production waste are reprocessed and turned into raw material for new products, thus returning them to the material cycle.

True recycling allows for the reuse of the recycling product in the same place as the primary product. This closes the life cycle of the material. *Downcycling*, on the other hand, produces a product which is reused at a lower level of quality, leading the product down the path to where it finally becomes waste. Depending on the building material, the methods required for recycling can range from simple to energy-intensive processes.

6.2.2.1 Prerequisites for the Use of Recycled Earth Building Materials

The general prerequisites for the use of recycled building materials are:

- Technical suitability
- Environmental safety
- Cost-efficient usage
- User acceptance

If these prerequisites are met, the salvaged earthen material obtained during demolition becomes a reusable building material called *recycled earthen material* (Sect. 2.2.1.3). This completes the life cycle of the material (Fig. 1.27).

Technical Suitability

The technical suitability of recycled earth building materials depends largely on the purity of the material obtained during demolition. Frequently, earthen materials are combined with gypsum, lime, paints, etc., as a result of structural changes such as renovation and repair work carried out during the lifetime of the building. These material can usually not be separated from each other and therefore need to be disposed of. The effort required to sort the materials would lower the cost-effectiveness of using recycled earthen materials.

As a result of their clay mineral structure, one characteristic property of earth building materials used in the production of building elements is their ability to replasticize. This means that they regain their plasticity when water is added (Sect. 2.2.3.2). In outdoor areas exposed to the weather, earth building materials can therefore generally only be used to a limited extent or by adding "stabilization." From a structural point of view, this characteristic property presents a disadvantage when compared to other mineral building materials (fired brick, concrete). From a building ecology perspective, however, the earth's ability to replasticize offers nearly unlimited recycling possibilities without additional energy demands. For this reason, earth building materials meet one of the most important requirements of sustainable building (Sect. 1.4.1).

The addition of stabilizers during production, in the form of aggregates and additives, decreases the replastification properties of earth building materials (Sect. 3.4.2). If synthetic binders are added, the decrease in plasticity becomes apparent when the earth building materials are in a wet state. Earth building materials "stabilized" with the most common mineral binders lime and cement can easily be recycled at a lower quality level.

In terms of usability, no technical requirements with quantified criteria for recycled earthen materials have been formulated thus far.

Environmental Safety

During the lifetime of the building or structure, the earth building elements might be exposed to a number of different substances. This can limit or rule out the reuse of these earth building materials. In some cases this exposure can even lead to problems with the disposal of the materials in an environmentally safe manner. This includes:

- Salts (Sect. 5.2.1.2)
- Air pollutants (Sect. 5.2.2)

- Dry rot and mold (Sect. 5.2.3)
- Hygiene-related concerns, such as the binding of odors and germs in dismantled livestock barns (Sect. 5.2.3)

Salts

Readily soluble salts (sulfates, chlorides, nitrates) with ions of the alkali and alkaline groups are the chemical compounds which dissolve in groundwater and get transported via capillary action into the finished building elements where they then crystallize (Sect. 5.2.1.2). Depending on their concentration, they break down in the soil after a certain amount of time. Generally, they do not pose any direct health risks and are therefore not considered harmful substances.

Permissible salt contents of earth blocks and clay mortars are defined in DIN 18945–47 (Sects. 3.5.6 and 3.5.7).

Harmful Substances

Discussions about harmful substances in building materials have led to the question if earth building materials could also contain harmful substances with potential health risks.

Any evaluation of a (chemical) substance in terms of its hazardousness or risk to humans consists of two aspects which are independent of each other: *exposure* and *toxicity*. Looking at one aspect alone does not allow for an adequate assessment of a substance's harmful effects. Put very simply, the "most toxic" substance at an exposure level of "zero" is completely safe, whereas a "less toxic" substance at a very high level of exposure is connected to significant risks. In other words, according to Paracelsus, the *dose* makes the poison. It is therefore important to first know the dose of a (chemical) substance in order to assess the effects which might actually arise. Thus, it is necessary to agree on *limits* which, when exceeded, can result in harmful effects on humans. It is at this point that a substance becomes a *harmful substance*. Specifying these limits has become particularly important today because modern measuring methods can detect potential "harmful substances" in building materials at a nanoscale.

Whereas the toxicity of a substance is a "material-inherent" quality, exposure is only partially based on material properties. Humans spend the greatest part of their lives indoors. When assessing a situation, it is therefore also important to examine in which manner and to what extent a substance can be absorbed and distributed by the environment—which is the indoor air in the case of buildings. Additional factors which need to be considered are the substance's impact duration and possible combinations with other materials which could increase the toxicity.

Building biology, as a subarea of building ecology, examines the health effects of building materials and buildings on people. From the outset, uncontaminated earth building materials which do not contain synthetically altered raw materials are "recommended based on the principles of building biology" because their properties in terms of life cycle assessment, hygroscopicity, diffusivity, heat storage, toxicity, and recycling receive positive ratings compared to other building materials. Today, an assessment which recommends a building material "based on the principles of building biology" is an important criterion for clients when selecting materials. According to the Institute of Building Biology + Ecology, Neubeuern (2003), no limits for harmful substances in earth building materials, which could affect their positive ratings, are known thus far, or the limits have not been agreed upon. Radioactivity (Sect. 5.1.6.1) and mold growth (Sect. 4.3.6.3), however, could be problematic from a building biology perspective.

For issuing its quality seal, the organization natureplus e.V. has defined limits for substances (Table 6.1) and emissions (Table 6.2) which could be potentially harmful in the production of earth plaster mortars (LPM), clay paint, and clay thin-layer finishes (LDB) as well as clay panels (LP) [8]. Permissible limits have been specified for heavy metals and their compounds, pesticides, as well as harmful organic substances and provide a test criterion for radioactivity. In addition to volatile organic compounds (VOCs, TVOCs), the emissions also include odors:

Additional requirements are a pH level of ≤8 (ISO 10390) and testing for possible asbestos fibers. The following information is given for testing for natural radioactivity: measuring is conducted according to ÖNORM S 5200 as the cumulative value of the partial activity of the radioactive nuclides K-40 and Cs-137 as well as the Th series, the U series, and the Ac series using gamma-ray spectroscopy (Sect. 5.1.6.1). A value of 0.5 Bq/kg is given as the detection limit, while ≤0.75 Bq/kg is the limit which must not be exceeded.

The substances listed in Table 6.1 refer to:

Metals and metalloids: Heavy metals and their compounds. They do not break down in the soil, can accumulate in the food chain through different channels, and are toxic at certain concentration levels.

Pesticides: Organochlorine pesticides, e.g., DDT, hexachlorobenzene, lindane, pentachlorophenol, and pyrethroids. Product-specific pesticides need to be determined for individual cases.

Adsorbable organic halogen compounds (AOXs): The sum parameter for adsorbable organic *halogen* compounds when determining the quality of waste water. "X" refers to the halogens fluorine, chlorine, bromine, and iodine. Among the products in this group are toxic pesticides.

When it comes to the recycling of earth building materials, possible pathways for contamination with harmful substances need to be considered. They can differ greatly depending on the type of exposure during the building's lifetime. For example, earth plasters on exterior walls which have been exposed to the exhaust fumes of heavy traffic are unsuitable for recycling. However, no detailed specifications have been drawn up thus far.

Harmful substances according to Table 6.1 can also find their way into building elements as additives during the production phase of the earth building materials. Earth building standards in various countries allow for the addition of asphalt

Table 6.1 Limits for harmful substances in earth building materials [8]

No.	Test parameter/substances	Limit			Test method
		LPM	LP	LDB	
1	*Metals and metalloids* (mg/kg)				Digestion nitric acid/hydrofluoric acid
1.1	As arsenic	≤5	20	20	DIN EN ISO 11885 or DIN EN ISO 17294-2
1.2	Cd cadmium	≤1	1	1	Same
1.3	Co cobalt	≤20	20	20	Same
1.4	Cr chromium	≤20	200 (Cr total)	200 (Cr total)	Same
1.5	Cu copper	≤35	–	–	Same
1.6	Hg mercury	≤0.5	0.5	0.5	DIN EN ISO 12846
1.7	Ni nickel	≤20	100	100	DIN EN ISO 11885 or DIN EN ISO 17294-2
1.8	Pb lead	≤15	20	20	Same
1.9	Sb antimony	≤5	–	–	Same
1.10	Sn tin	≤5	–	–	Same
1.11	Zn zinc	≤150	–	–	Same
1.12	Cr VI (mg/L)			2	Eluate analysis TRGS 613
2	*Pesticides* (mg/kg)				Same as DFG S19
2.1	Sum		1	1	
3	*Organic substances* (mg/kg)				
3.1	AOXs	≤1	1	1	According to natureplus—execution standard "AOX/EOX"
4	*Asbestos fibers in talc*	–	–	Asbestos-free according to the German Pharmacopoeia (Deutsches Arzneibuch—DAB)	REM
5	*Radioactivity* (Bq/kg)				
5.1	Natural radioactivity: cumulative value I according to ÖNORM S 5200	0.75	0.75	0.75	Determination of activity of the radioactive nuclides K-40 and Cs-137 as well as the Th series, U series, and the Ac series using gamma-ray spectroscopy, detection limit: 0.5 Bq/kg (Sect. 5.1.6.1)

Table 6.2 Limits for emissions from earth building materials [8]

		Limit			Test method (test chamber method according to natureplus—execution standard)
No.	Test parameters/emissions (after conditioning)	LPM	LP	LDB	
1	*Volatile organic compounds (VOCs)* (μg/m^3)				DIN ISO 16000-6; -9; -11
1.1	VOCs classified into: Regulation EC No. 1272/2008; TGRS 905: K1, K2, M1, M2, R1, R2 in MAK List III.1 and MAK III.2	No significant trace	No significant trace	No significant trace	3 days after test chamber loading
1.2	TVOCs (total), classified into: Regulation EC No. 1272/2008; TGRS 907	≤3,000	≤3,000	3,000	3 days after test chamber loading
		≤300	≤300	300	28 days after test chamber loading
2	*Formaldehyde* (μg/m^3)	≤24	≤24	≤24	DIN EN 717-1, DIN ISO 16000-3, 28 days after test chamber loading
3	*Acetaldehyde* (μg/m^3)	≤24	≤24	≤24	DIN ISO 16000-3, after 28 days PKB
4	*Odor/odor grade*	≤3	3	3	VDA 270; 23 °C; natureplus—execution standard "odor test," 6-step odor grading scale, 24 h after test chamber loading

emulsion for increasing weather resistance (Sect. 4.2.1.3). Possible indoor air pollution in the finished building or the question of future building material recycling have so far not been addressed or tested in connection with this additive. Other harmful substances include pesticides which can enter the soil mixture through the addition of organic fiber aggregates.

In developing countries in particular, earth blocks are stabilized by adding locally available waste products as an alternative to expensive cement-based binders (Sect. 3.4.2). In many cases these additives include industrial waste with heavy metal content. Cement, as well, can contain ground industrial slag with heavy metal content added in the grinding process.

An "uncontrolled" total demolition of buildings (Sect. 6.1.2) can furthermore lead to the mixing of residual earth building materials with asbestos fibers. If the asbestos content is too high, this residual material mix should not be recycled but instead needs to be disposed of.

Asbestos is a naturally formed rock. Its weathered products include fibrous clay-rich soils which have been used in traditional earth building for a long time.

In [9] the health hazards of using soils or buildings containing asbestos in traditional Turkish construction are described. So far, no solutions on how to restore or dispose of these contaminated buildings have been found.

Using the limits listed in Table 6.1, it has become possible for the first time to answer the question of permissible levels of harmful substances in recycled earth building products using standardized test methods. It remains to be seen to what extent these limits can be established in actual building practices.

The emissions listed in Table 6.2 refer to:

Total volatile organic compounds (TVOCs): The sum of all volatile organic compounds. This term combines chemically different organic compounds with boiling ranges between 50–100 °C as the lower and 240–260 °C as the upper limit. They can be separated and detected using the gas chromatography method. Typical sources of VOCs in private homes and offices are:

- Cleaning agents and care products
- Paints, finishes, paint thinners
- Glues, adhesives
- Fragrances, scented oils
- Solid wood furniture made of pine and spruce

Very volatile organic compounds (VVOCs) with low boiling ranges up to around 50 °C include, for example, methanol CH_3OH and formaldehyde CH_2O. They can lead to headaches when inhaled indoors. According to the Technical Regulations for Hazardous Materials (Technischen Regeln für Gefahrstoffe—TRGS), these substances can also be referred to as volatile organic solvents and are listed separately in Table 6.2.

Polycyclic aromatic hydrocarbons (PAHs): Umbrella term for a group of chemical substances with a molecular structure derived from *Benzene* C_6H_6. They escape from tar products or form when organic materials (mineral tar oil, diesel fumes, tobacco smoke, barbecue products, etc.) are only partially burnt. A large number of substances in the PAH group is highly carcinogenic.

Phenol index: The sum parameter for a group of chemical substances with a molecular structure derived from *phenol* (hydroxy derivative of hydrocarbons). Typical products containing phenol include disinfectants and preservatives, intermediate products of dyes, synthetic resins, plastics, pesticides, plasticizers, and detergents. Chlorophenols, in particular, are toxic and have a strong odor and taste.

Threshold limit value (Maximale Arbeitsplatz-Konzentration—MAK): The maximum permissible concentration of a substance at the workplace in the form of gas, steam, or particles suspended in the air which generally does not affect or harm the workers' health during continuous exposure. The MAK cannot be applied to assess the indoor air quality where people are exposed to chemicals from building materials or materials used in interior furnishings. There is no prescribed evaluation procedure in Germany for this field. Common assessment values for air pollutants can be found in the different rules and regulations of the Federal Immission Control Act (see [10]).

Odor: For the odor test according to VDA 270 (German Association of the Automotive Industry), the odor of a material (in this case: an earth building material) is determined by a team of specialists which then issues corresponding grades. The grading scale ranges from 1 "imperceptible" to 6 "unbearable." A typical limit for the odor test is a grade of <3.

At certain concentrations in the indoor air, all organic substances listed in Table 6.2 pose health hazards to varying degrees. All of the harmful substances mentioned above are either non- or only very poorly degradable in the ground.

Fungal Spores

Salvaged earth building materials which are contaminated with dry rot spores pose a risk even after the building has been demolished (Sect. 5.2.3). They should therefore not be recycled and reused in earthen structures.

Mold spores in salvaged earth building materials do not pose a risk for the occupants (indoor air) or the structures as long as the materials have been installed correctly. Mold requires a certain moisture level in order to survive, and this level is not found in the finished state of dry building elements.

Odors

Odors are volatile chemical compounds in the indoor air which (also) affect building elements in their finished state and can be adsorbed by them. Earth building element surfaces with open pores have high adsorption capacity. Earth building materials integrate these compounds into their clay mineral structure, thereby neutralizing them and making the odors imperceptible. However, there are limits to the absorbing capacity, such as in agricultural buildings (Sect. 5.2.3 [11, 12]).

When recycled earthen materials are prepared in a wet state, the "odors" are released again (e.g., cigarette smoke). They disappear quickly, which means that a wall which has been finished with recycled earth plaster is odorless after drying. The released odors do not pose any health hazard. They can, however, create a psychological barrier against using recycled earthen materials. For hygienic reasons, recycled earthen materials salvaged from agricultural buildings should generally not be used for residential structures.

The first odor test applied to earth building materials was a test based on the grading scale developed by the German Association of the Automotive Industry (VDA 270). It was part of the certification process for the quality seal issued by the organization natureplus e.V. for the production of earth panels (Table 6.2).

Cost-Effectiveness

According to the Business Association of Historical Building Materials e.V. (2006), there is currently no data on the commercial use of recycled earth building materials. It is therefore not possible to make a statement about the cost-effectiveness of using these materials from an economic point of view.

On the other hand, recycled earth building materials are widely used by owner-builders for repair and renovation work which typically only require small quantities. The "cost-effectiveness" of work carried out by owner-builders cannot be calculated based on economic criteria or only to a very limited extent.

For owner-builders, the advantage of using recycled earth building materials lies in the fact that the materials are typically of a composition which was suitable for their initial use. If the recycled materials are intended for the same use, the search for the "correct mix" becomes relatively easy. In the end, this also represents a form of cost-effectiveness.

User Acceptance

According to data provided by the Federal Union of Recycled Building Materials e.V. [13], 52 million tons of recycled building materials was produced in Germany in 1997, representing a mere 7.4 % of the total production of construction rock. Eighty percent of this material is used in earthwork and road construction. The remaining volume is used for asphalt plannings, aggregates for concrete, and other applications. The area referred to as "other applications" apparently includes "aboveground construction" and is thus the field which applies to recycled earth building materials. In this area in particular, there should be more use of recycled earthen materials. It is very possible that the use of recycled building materials will increase to 90 million tons by 2012. Reliable information on recycled earth building materials is not available.

These numbers show that the acceptance of recycled building materials for the construction of buildings, including recycled earth building materials, currently appears to be at a very low level despite existing legal frameworks and discussions about the need and significance of building material recycling. There appears to be a general skepticism among owners and builders in connection with the actual characteristics of recycled building materials when compared to "new and modern" materials. The field of historical preservation represents the only exception.

Additional reasons for the current reluctance to use recycled building materials, including recycled earth building materials, for the construction of buildings, might be found in the limited availability of pure materials and differing opinions about pricing. Uncertainties in connection with possible warranties as well as psychological barriers might also play a role. If nothing else, a low acceptance of recycled building materials, including recycled earth building materials, is caused by a lack of objective information by all parties involved in construction.

6.2.2.2 Possible Applications of Recycled Earthen Materials

After the abovementioned prerequisites have been met, recycled earthen materials (Sect. 2.2.1.3) can be kept in the material's life cycle after a building's demolition as true recycling products or through downcycling. The following applications are possible (Table 6.3) (•):

Table 6.3 Possible applications of recycled earthen materials

No.	Earth building material [18]	True recycling, reuse	True recycling, preparation + shaping	Downcycling	Comment
1	Rammed earth		●	●	Milled construction soil
2	Cob		●	●	Aggregate, milled
3	Straw clay		●	●	Aggregate, milled
4	Light clay			●	
5	Loose fill	●			Aggregate, milled if necessary
6	Earth blocks/"green" unfired bricks	●	●	●	
7	Clay panels			●	Take surface coating into account
8	Earth mortar		●	●	Take surface coating into account, aggregate or milled

True Recycling

Reusing earth blocks for earth block masonry, which means using them for the same purpose, constitutes the highest form of recycling and is most desirable. This is due to the fact that the incorporated work (entropy) which was applied to the earth building material is preserved in the "shaped" building material itself (Fig. 6.1 [14]).

Particularly in the field of renovation and retrofitting of old buildings by owner-builders, it is common to use salvaged earth plaster mortar with the addition of sand (if necessary). The salvaged material can also be used as an additive for a "new" earth plaster mix which is prepared using the method described in Sect. 3.1.2 (Fig. 6.2 [14]).

Downcycling

The value of an earth building material lies in the already existing optimal mix of construction soil and additives. The term "downcycling" can therefore only be applied to earth building materials to a limited extent.

It is possible, however, to mechanically process earth blocks which can no longer be used or are broken. The same is true for chipped off, finely milled cob or rammed earth building elements for use as construction soil or loose fill material in building construction.

Recycled earth building materials (including salt-damaged materials) can also be used as fill material in subgrade construction. The same applies to earth building materials contaminated with dry rot spores or salvaged from agricultural buildings. Explicit criteria for such applications have not yet been drawn up, but the materials

Fig. 6.1 "True recycling": repair work on an existing earth block masonry wall using salvaged earth blocks [14]

Fig. 6.2 Reuse of salvaged soil: soaking and adding sand [14]

must conform to the limits of environmentally harmful substances according to LAGA guidelines (Sect. 6.3.2).

The German Closed Substance Cycle Waste Management Act defines "waste" as a residue which cannot be used as material or through energy recovery and therefore needs to be disposed of in landfills. This type of waste does not occur during the recycling of earth building materials, with the exception of non-separable material compounds or building materials containing harmful substances. Residue, such as earth block breakage, can be returned to a wet state and, if needed, undergo a new shaping process (Sects. 3.1.2 and 3.2).

6.3 Disposal of Earth Building Materials

The German Closed Substance Cycle Waste Management Act of 1996 (KrW-/AbfG) [3] requires waste prevention be placed above recycling. Nonetheless, even if all recycling measures are intensified, it is not always possible to return residual building materials to the material cycle. In this case, their disposal as "construction waste" cannot be avoided.

6.3.1 *Construction Waste*

Germany's Joint Waste Commission of the Federal States (LAGA—Landesarbeitsgemeinschaft "Abfall") divides the term "construction waste" into four material groups [15]: excavation soil, road construction waste, building rubble, and building site waste. Table 6.4 describes the origins and components of these four material groups and shows the total volume for the year 2000 [16]. The total quantities are further divided into backfilling/direct utilization, recycling, and residues (waste).

The classification "waste" is applied to (contaminated) building material residues which are no longer needed and are thus taken out of the material cycle and deposited into landfills in an environmentally safe manner.

After the preparation and quality control of building waste materials intended for direct reuse, backfill, or recycling, these materials regain defined properties as building products and become equal to primary materials. This completes the material's life cycle.

Except for the material group "road construction waste," all remaining categories also include clay-rich soil of varying degrees of quality as building waste material. However, no quantitative information is available on this subject.

6.3.2 *Concentration of Harmful Substances*

The concentration of harmful substances in construction waste can be assessed with the help of the technical regulations for the "Requirements for the material recycling of mineral waste material/waste" (LAGA List) [17]. This guideline defines

Table 6.4 Construction waste: material groups according to LAGA guidelines and their total generated volume for the year 2000 [16]

No.	Material group according to LAGA	Origin	Components	Total volume (in millions of tons)	Backfill, direct utilization	Recycling	Residue (waste)
1	Excavated soil	Solid mineral waste material generated by the construction of buildings, below-grade construction, and road construction	Non-contaminated, undisturbed, or previously used soil or rock material (topsoil, gravel, sand, *clay-rich soil*, *clay*, stones, and rock)	163.6	126.5	11.2	25.9 (15.8 %)
2	Road construction waste	Solid mineral waste material generated by demolition, improvement and expansion of traffic infrastructure	Hydraulically bound building materials, paving and curb stones, paving slabs, sand, gravel, crushed rock, chippings, etc.	54.5	40.6	8.6	5.3 (9.7 %)
3	Construction waste	Solid mineral waste material mainly generated by the demolition of all types of buildings as well as through renovation and reconstruction	Concrete and/or fired-brick masonry, *clay-rich soil*, lime-sand bricks, natural stone, concrete and lightweight concrete blocks, aerated concrete, mortar, plaster, tiles, mineral wool, etc.	22.3	2	19.1	1.2 (5.4 %)
4	Construction site waste	Mixed mineral and organic waste material generated by new construction, renovation, and demolition of buildings	Concrete, masonry, mortar (earth) plaster, wood, synthetic materials, glass, ceramics, metals, cardboard, paper, etc., but also cables, paints, varnishes, adhesives, sealants	11.8	–	1.7	10.1 (85.6 %)

Table 6.5 Usage classes for soils and mineral waste based on concentrations of environmentally harmful substances (assignment criteria)

No.	Assignment criterion Z	Description
1	Z 0	Concentration of substances up to assignment criterion Z 0 identifies natural soils. Values lower than assignment criterion Z 0 allow for unrestricted use. Soil excavation is not required for these values. The soil is suitable for ecologically sensitive use such as in playgrounds or soccer fields for children, school yards, kitchen gardens
2	Z 1	The Z 1 assignment criteria define the upper limit of unrestricted use taking certain limitations into consideration. Applicable regulations for preserving the groundwater quality play a decisive role in determining these values. If the assignment criteria are exceeded, an unrestricted use on nonsensitive land is possible. This includes industrial sites, commercial land, and storage areas. The abovementioned ecologically sensitive areas and uses are excluded
3	Z 1.1	If these values are not exceeded, no negative changes in groundwater quality are to be expected, even under unfavorable hydrological conditions
4	Z 1.2	If these values reach the upper limit Z 1.2, erosion protection such as full ground cover is required for unrestricted use
5	Z 2	The Z 2 assignment criteria represent the upper limit for the use of soil with defined technical protection measures. Soils of this class can only be used as a base layer under water-impermeable top layers or as a leveling layer between the waste and surface sealing. It must be ensured that harmful substances are not released into the groundwater. These soils are also not allowed to be used on ecologically sensitive land

limits for substances in mineral waste (including soil) which are harmful to the environment.

Different usage classes determine if the soil can be used in earthworks, road construction, landscaping, and the construction of landfill sites as well as for backfilling and land re-cultivation measures (Table 6.5). For these usage classes, assignment criteria Z are defined for the upper limits of harmful substances, namely, for the total content (Table 6.6) of substances and for the existing dissolved portion of harmful substances (eluate analysis) (Table 6.7) in the original material.

In addition to the national guidelines for the treatment of residual materials and waste materials published by LAGA, "The decision of the EU Council of December 19, 2002 on establishing criteria and procedures for the acceptance of waste at landfills pursuant to Article 16 and Annex II of Directive 1999/31/EC" needs to be taken into consideration [19]. The ruling took effect on July 16, 2004, giving all member states the option of implementing it into their national legislation within 1 year.

Germany's strategy is to make sure that waste material which is inevitably headed to the landfill is suitable for depositing and no longer chemically reactive. This can typically only be achieved through thermal pretreatment.

Table 6.6 Permissible concentrations contained in solid substances for the assignment criteria according to LAGA guidelines

Parameter (mg/kg TS)	Z 0	Z 1.1	Z 1.2	Z 2
Hydrocarbons	100	300	500	1.000
PCBs	0.02	0.1	0.5	1
EOX	1	3	10	15
PHA, total according to EPA	1	5	15	20
LHKW, total	<1	1	3	5
BTEX aromatic compounds	<1	1	3	5
Cyanide, total	1	10	30	100
Cadmium	0.6	1	3	10
Nickel	40	100	200	600
Lead	100	200	300	1000
Arsenic	20	30	50	150
Chromium, total	50	100	200	600
Copper	40	100	200	600
Zinc	120	300	500	1.500
Mercury	0.3	1	3	10
Thallium	0.5	1	3	10

Table 6.7 Permissible values of existing dissolved content of harmful substances in the eluate analysis for the assignment criteria according to LAGA guidelines

Parameter (mg/L)	Z 0	Z 1.1	Z 1.2	Z 2
pH value	6.5–9.0	6.5–9.0	6.0–12.0	5.5–12.0
Conductivity (μS/cm)	500	500	1,000	1,500
Phenol index	<0.01	0.01	0.05	0.10
Cyanide, total	<0.01	0.01	0.05	0.10
Cadmium	0.002	0.002	0.005	0.01
Nickel	0.040	0.050	0.150	0.20
Lead	0.020	0.040	0.100	0.20
Arsenic	0.010	0.01	0.040	0.06
Chromium, total	0.015	0.03	0.075	0.15
Copper	0.050	0.05	0.150	0.20
Zinc	0.100	0.10	0.300	0.60
Mercury	0.0002	0.0002	0.001	0.002
Thallium	<0.001	0.001	0.003	0.005
Chlorite	10	10	20	30
Sulfate	50	50	100	150

References

1. Deutscher Abbruchverband e. v. (Hrsg.): Technische Vorschriften für Abbrucharbeiten. Düsseldorf, 3. Aufl (1997)
2. Deutscher Abbruchverband e. v. (Hrsg.): Anforderungen an Verdingungsunterlagen bei Abbrucharbeit en und Rückbauprojekten—Handlungshilfe, Düsseldorf (2004)

3. Gesetz zur Vermeidung, Verwertung und Beseitigung von Abfällen. Artikel 1: Gesetz zur Förderung der Kreislaufwirtschaft und zur Sicherung der umweltverträglichen, Beseitigung von Abfällen (Kreislaufwirtschafts- und Abfallgesetz—KrW-/AbfG), Bundesgesetzblatt Teil I, 6.10.1994. Letzte Neufassung/Änderung BGBl. I S.1324, 1346 v. 22.05, pp. 2705–2732 (2013)
4. Röbenack, K.-D.; Müller, A.: Baustoffrecycling—Einführung Abbruchverfahren. Weimar, Bauhaus-Universität, Fak. Bauingenieurwesen, Lehrunterlagen Vertiefung. Bauwerkserhaltung und Baustoffrecycling (1996)
5. Löfflad, H.: Das globalrecyclingfähige Haus—Fallstudie über die Möglichkeiten der Wiedereingliederung von Baurückständen in den Naturkreislauf am Beispiel eines globalrecyclingfähigen Hauses mit Klassifizierung von Baustoffen und Planerkatalog sowie Öko- und Energiebilanz. Technische Universität, Diss., Eindhoven, NL (2002)
6. Katalyse GmbH/Löfflad, H.; Justen, M.; Ebert, L.: Das recyclingfähige Haus. Studie über die Notwendigkeit und die Möglichkeiten der Wiedereingliederung von Rückständen in den Naturkreislauf, am Beispiel des globalrecyclingfähigen Hauses, Köln (1993)
7. Schroeder, H.: Die Produktion von Lehmbaustoffen in Deutschland—aktuelle Situation und Tendenzen. Bauhaus-Universität, Weimar, unpublished study (2001)
8. natureplus e.V. (Hrsg.) Richtlinien zur Vergabe des Qualitätszeichens "natureplus". RL 0607 Lehmanstriche und Lehmdünnlagenbeschichtungen (September 2010); RL 0803 Lehmputzmörtel (September 2010); RL 0804 Stabilisierte Lehmputzmörtel (proposed); RL 1006 Lehmbauplatten (September 2010); RL 1101 Lehmsteine (proposed); RL 0000 Basiskriterien (May 2011); Neckargemünd (2010)
9. Vural, S.M.: Health effects of earthen building products. In: "Living in Earthen Cities—Kerpic 05", Proc. 1st International Conference, Istanbul Technical University 2005, pp. 204-211. Istanbul Technical University ITU, Istanbul (2005)
10. Zwiener, G.; Mötzl, H.: Ökologisches Baustoff-Lexikon: Bauprodukte, Chemikalien, Schadstoffe, Ökologie, Innenraum., 3., erw. Aufl. C. F. Müller, Heidelberg (2006)
11. Bielenberg, H.: Der Einfluss des Stalles auf die Schweinemast. Braunschweig, Technische Hochschule, Fak. f. Bauwesen, Diss. (1963)
12. Burger, H.: Sanfter Baustoff Lehm in der Landwirtschaft—Auswirkungen auf das Stallklima. Kassel/Witzenhausen, Univ.-GH, FB Landwirtschaft/Agrartechnik, Unpublished diploma thesis (1995)
13. Rese, F.; Strauß, H. (Hrsg.): Die Steine- und Erden-Industrie—Baustoffe 2000 (2001/2002, 2003). Stein-Verlag GmbH, Baden-Baden 2000 (2002, 2003)
14. Kurz, J.; Steinbichl. S.: Dokumentation Lehmbauarbeiten Haus Kurzbichl, Weimar. Weimar, Bauhaus-Universität, Fak. Architektur, unpublished student research paper (2005)
15. Landesarbeitsgemeinschaft Abfall LAGA (Hrsg.): Abfallarten. Berlin: In: Abfallwirtschaft in Forschung u. Praxis, Bd. 41 (1992)
16. Glücklich, D. (Hrsg.): Ökologisches Bauen—von Grundlagen zu Gesamtkonzepten. München: Deutsche Verlags-Anstalt (2005)
17. Landesarbeitsgemeinschaft Abfall LAGA (Hrsg.): Anforderungen an die stoffliche Verwertung von mineralischen Reststoffen/Abfällen, Heft 20, 5. Aufl. LAGA Mitteilung, Berlin (2004)
18. Dachverband Lehm e.V. (Hrsg.): Lehmbau Regeln—Begriffe, Baustoffe, Bauteile., 3., überarbeitete Aufl. Vieweg + Teubner | GWV Fachverlage, Wiesbaden (2009)
19. Entscheidung des Rates vom 19.12.2002 zur Festlegung von Kriterien und Verfahren für die Annahme von Abfällen auf Abfalldeponien gem. Art. 16 u. Anhang II der Richtlinie 1999/31/EG

7

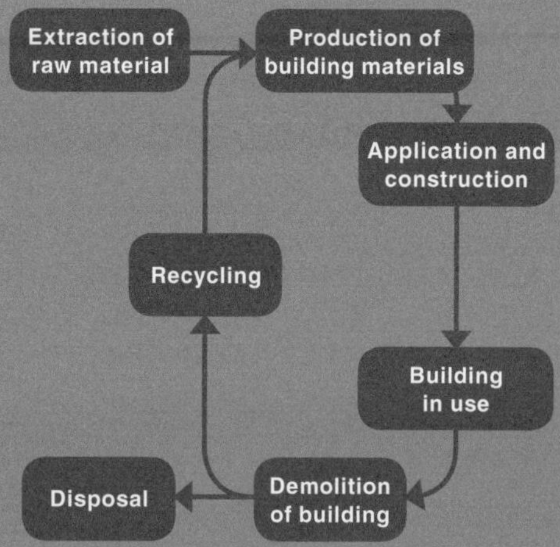

The Future of Earth Building

As a last point, the question of the future of earth building will be examined. Can the approaches described in the preceding chapters guarantee that earth as a building material can be successful in the market place or even be able to expand its position? Is the "supremacy" of the building material giants not too overwhelming for earthen materials? Maybe the new Shanghai – Pudong airport, the Burj al Khalifa – Hotel in Dubai or the Taipeh-101 skyscraper in Taiwan's capital are the trend-setting architectural goals? So then what role does earth as a building material still play?

© Springer International Publishing Switzerland 2016
H. Schroeder, *Sustainable Building with Earth*,
DOI 10.1007/978-3-319-19491-2_7

When talking to people from the new "booming" regions of the world, one thing stands out: many of them, whether in the ultramodern skyscrapers or at the busy airports constructed at record-breaking speeds, are afflicted by a strange longing for their own roots. The rapidly growing economic powers have lost sight of their own cultural identity, and people have now started to look for and "re"discover their traditions. Time and again, this search also leads them to earth as a building material and presents a great opportunity for the field.

In developing countries, earth is still the only available building material for millions of people who are on the lowest rung of the social ladder. Here, earth building is ever present. Even though people aspire to building solid houses out of concrete or fired brick, for most of them it remains a dream. This is where most earth building activities take place today and rather inadequately at that. Earth is equated with poverty in these countries, similar to the situation in Germany after WWII. Particularly in seismic regions, people generally do not possess adequate knowledge of how to construct earthquake-resistant homes out of earth. A high number of casualties during the next earthquake is therefore inevitable. This lack of information and inadequate education poses a big challenge for the world of earth building.

People in industrial nations live in a society largely influenced by the media. Information about stock and energy prices is at the center of attention with politicians presenting themselves in front of melting glaciers in order to justify further energy-saving measures. These measures always pose new difficulties when it comes to purchasing and using people's favorite consumer goods: cars and homes. Wellness, health, and fitness are of at least equal importance.

This very same media could be utilized to inform the public about the health benefits of earth building materials in interior spaces, their virtually nonexistent energy costs during production when compared to cement, as well as their generally uncomplicated recycling process. It would be an important milestone for the future of earth building if a large number of people arrived at this insight, and their realization, in turn, had a positive effect on the price of oil.

In addition, the building process in the European Union is governed by a sheer unmanageable number of rules and regulations. Within this process, earth is a "non-regulated" building material in most EU countries, and, although highly valued in terms of building ecology and building economy, this official classification makes it unmarketable in the long run. The classification of earth as a "regulated" building product requires the development of appropriate product and application standards based on comprehensive research on material behavior. Earth building has been neglected for decades in Germany, and systematic research has therefore not been carried out. In this respect, as well, changes need to occur.

This leads to three areas which are important for the future of earth building:

- Education
- Networking
- Research and standardization

Above all, however, lies the most important question: does earth building make economic sense?

7.1 Education

The following highly simplified matrix can be used to apply the structure of the German education system to prospective target groups for earth building education (Table 7.1):

Educational earth building offerings for the individual target groups vary widely:

7.1.1 Advanced Training for the Building Trades

The "tradespeople" target group is directly involved in the building process. In order to remain competitive, this group needs to fill information gaps in earth building most urgently. Educational activities for this group are therefore essential. In a pilot project undertaken from 1999 to 2001, the DVL developed a "Specialist for Building with Earth" continuing vocational training [1–4] in cooperation with the Environmental Center for Crafts and Trade in Rudolstadt, Thuringia, at the Eastern Thuringian Chamber of Crafts and Trade in Gera. This project was funded by the Thuringian Ministry for Economic Affairs and Infrastructure. The main target group consists of tradespeople who have finished their vocational training in the building trades. A participant course book is included in the course [5].

This course has been successful and economically self-sustaining since 2005. It has been integrated into a German-wide continuing education structure for trades and is the first continuing education earth building training in Germany to be recognized and implemented by the trades authorities [6, 7]. The legal basis of the training is a "Special Legal Provision" which is issued by each participating chamber of crafts and trades (HWK) for their respective chamber district.

Graduates of the course receive a certificate issued by the respective HWK. With this certificate, they may list their business in Trades Register A for the professional trade of "mason and concrete worker." Within this professional trade, the listing is restricted to the special field of "earth building" because earth building does not currently qualify as an independent trade. The legal basis for the business listing is the German Trade and Crafts Code, Section 8 "Exemptions."

In addition to the "Specialist for Building with Earth" certificate issued by the HWK, the graduate's company is entitled to carry the "company specialized in building with earth DVL" seal which is registered with the DVL. The seal identifies the company as a competent specialist in the field of earth building. It can be used

Table 7.1 Target groups for earth building education in Germany

	Professional	Academic
Initial vocational and undergraduate training	Apprentices in building trades	Students of architecture/civil engineering
Advanced, continuing education	Tradespeople in building trades	Architects/civil engineers/planners

for advertising purposes and assists clients, architects, and planners searching for earth building specialists.

As part of a European Union LEONARDO project, the German organization FAL e.V. in Ganzlin, Mecklenburg, Western Pomerania, developed a "Clay Plaster" pilot course from 2002 to 2005 in cooperation with the HWK Schwerin. This course has also been recognized and implemented by the trades authorities. The course material is available on CD-ROM [8].

7.1.2 *Initial Vocational Training for the Building Trades*

In Germany, Austria, and Switzerland, apprentices for the trades divide their initial vocational education between practical training in a company and theoretical training in a vocational school (= dual vocational education). Germany also has a third pillar of vocational training for specialized professions in the form of intercompany vocational training centers (= three-part system).

In Germany, vocational education for federally recognized occupations requiring formal training is based on the Vocational Training Act (Berufsbildungsgesetzes—BBiG) [9]. All federally recognized occupations are subject to their respective training regulations. Company training is the responsibility of instructors or masters who are required to demonstrate their qualifications in a trainer aptitude examination. The vocational school provides specialized theoretical, practical, and general education.

The task of vocational training is "to teach the vocational skills, competence and knowledge required for a changing work environment in the form of an organized and regulated course" [9, Section 1.3]. The process for testing new occupations requiring formal training is defined in [9, Section 6]. Currently, earth building is not a recognized independent trade.

The "Specialist for Building with Earth" course officially introduced earth building as a "special field" within the trades which represents a major improvement in the acceptance of earth building.

In order to obtain the status of a "professional trade," a corresponding federally recognized occupation requiring formal training according to [9, Sections 4 and 6] needs to be developed, tested, and implemented for earth building. This means that within the structure of the initial vocational training, the following, highly simplified, steps need to be taken:

1. An internal "test run" of an "earth building" module (approx. 40 h) at one (or multiple) vocational school(s).
2. If the test run is successful, an "official school trial" must be applied for with the responsible Ministry of Education and Cultural Affairs and an "earth building" required (elective) module (qualification module) needs to be developed for the second and third year of the vocational training.
3. Long-term goal: to develop a stand-alone occupation and apply for its federal accreditation with the German Federal Institute for Vocational Education and Training (BIBB).

Several vocational schools in Germany have already integrated earth building modules into their curricula (step 1). However, strategic coordination and exchange of information are currently missing.

In 2008, the DVL supported a project to develop a national qualification module called "Construction of Earth Building Structures" based on BBiG Sections 68 ff. at the Knobelsdorff Vocational School in Berlin. In 2009, this qualification module was tested and certified by the HWK Berlin according to the national criteria of the Central Agency for Continuing Vocational Training in the Skilled Crafts (Zentralstelle für Weiterbildung im Handwerk—ZDH). This means that the module can now be integrated into vocational training for masons on a national level.

Currently, first steps are being taken to make vocational training within the European Union more "mobile" and transparent in order to promote the mobility of apprentices and to ensure that the learning outcomes achieved abroad are mutually recognized in the apprentices' home countries. For this purpose, a credit point system is being developed for the initial vocational training phase—the *European Credit System for Vocational Education and Training ECVET*.

The basis of this system is the common "European Qualifications Framework for Lifelong Learning (EQR)" which is expected to link the qualification systems of different countries and serve as a "translation device" [10]. This framework consists of eight levels with defined descriptors. They are used to qualify the learning outcomes required for each level in order to attain the knowledge, skills, and competences in all qualification systems. Levels 1 to 5 cover the area of vocational training. From a German perspective, levels 4 and 5 more or less correspond to the professional qualifications of "journeyman" and "master."

With the help of respective descriptors, the EQR facilitates compatibility with the Framework for Qualifications of the European Higher Education Area—the so-called Bologna Process. Levels 5 to 8 of the EQR cover the area of academic education with bachelor's and master's degrees up to and including a doctoral thesis (PhD).

Looking ahead to the European harmonization of the technical (building) standards (Sect. 4.2.1.3), it seems sensible to take steps toward the creation of a "harmonized" European education area as a long-term objective. In this regard, the main problem lies in the vast differences between the national vocational education systems which also results in a lack of comparability.

In the field of earth building, a first step toward transborder, mutual recognition of learning outcomes achieved through national educational systems is the EU project (LEONRDO) ECVET Lehmbau/Lern•Lehm (LearnWithClay) initiated by the organization FAL e.V. and carried out between 2007 and 2009. The project resulted in the development of learning modules from Level 1 (beginner) to Level 4 (self-employed building tradesman/journeyman) in the field of earth plaster [11]. In an EU follow-up project entitled "Provide Instructions and Resources for Assessment and Training in Earth Building (PIRATE)," learning modules are being developed for additional earth building techniques as well as for levels covering academic education through 2015.

7.1.3 *Undergraduate Academic Training for Architects, Civil Engineers, and Planners*

After WWII, there were attempts in the German state of Thuringia to combine practical earth building training with the academic education of students at the University of Building and Fine Arts in Weimar (which was reestablished in 1946 and is the present-day Bauhaus University). On December 1, 1947, the Earth Building School of the State of Thuringia opened as an independent organizational structure within the university. The participants (students of the university) were able to put their theoretical knowledge of typical earth building methods (earth block and rammed earth construction) into practice at experimental agricultural stations (model projects) (Fig. 7.1). After only a few years in operation, this educational institution closed due to the decline of earth building.

Currently, only a few German universities and colleges include ongoing earth building education in their curricula. The University of Kassel (Prof. Minke) was a pioneer in this field. At the Bauhaus University as well, "earth building" was offered as a required elective for diploma/master's courses in architecture and civil engineering from 1993 to 2012. This course established an interesting link to the field of vocational training through the degree program "Teaching Qualification in the Building Trades": a number of students in this program took "earth building" as a required elective making them potential trainers in the field of initial earth building training at vocational schools.

Fig. 7.1 Practical training at the State Earth Building School Weimar (1948)

Courses for academic training can be divided into levels similar to those of vocational training:

1. Earth (building) as part of an interdisciplinary course, such as building material science, building restoration, or ecological building, as well as design for architects and planners, often in connection with non-European traditional earth building architecture.
2. Earth (building) as a separate required elective within the curriculum followed by an exam, generally over the course of one semester.
3. Earth (building) as a separate master's degree, typically over the course of four semesters ending with the respective degree (e.g., Master of Arts MA, Master of Science MSc, Master of Engineering M.Eng.) without the addition of specialist.

The few university-level courses on earth building offered in Germany are courses at levels 1 and 2 in bachelor's and master's degree programs. They are integrated into *consecutive* programs which means that the bachelor's and master's courses are related in terms of their subject matter. The individual colleges or university departments can decide if earth building courses are implemented in their respective curricula. A level 3 course requires the approval of the responsible department of education. Before approval, the course must successfully pass an extensive evaluation process conducted by a specialist commission made up of external experts. Currently, there is only one level 3 earth building program, offered by the University of Grenoble (ENSAG), France [12].

Within the framework of the Bologna Process, a system of comparable and (internationally) transferable credit points has been developed for the European Higher Education Area—the *European Credit Transfer and Accumulation System ECTS*. Its main objective is to facilitate students' mobility within the EU. After the academic system in the EU has fully completed its shift from the present-day diploma programs to comparable bachelor's and master's programs, the ECT system will show its full potential.

Experiences with this system at the Bauhaus University Weimar have shown that there is great interest in "earth building" as a required elective among foreign students, not only from EU countries but especially from Asia and Latin America. Worldwide, students can now research their specific subject interests on university websites and make their selections accordingly. To facilitate this process, the DVL has developed its own website www.uni-terra.org which can be used by universities to advertise their earth building courses to a global audience [13].

7.1.4 Continuing Education for Architects, Civil Engineers, and Planners

In order to be able to meet the demands of a "changing work environment," architects, civil engineers, and planners are required to continuously expand and refresh their professional skills and knowledge. They are required by their professional associations, the so-called chambers, to earn continuing education credits through regular participation in workshops.

Currently, architects, civil engineers, and planners have several options to fill potential gaps in their knowledge of earth building. As the official organizers of such activities, the Chambers of Architects or Engineers can act as state associations, universities, or colleges.

Continuing education offered by the chambers on the topic of earth building generally consists of seminars lasting one to several days, organized in cooperation with earth building specialists (associations, companies, individuals, etc.).

In addition to academic education in the traditional sense, continuing education offerings for architects, civil engineers, and planners are increasingly gaining importance at universities and colleges. These courses are based on the master's degree defined by the Bologna Process. They are *nonconsecutive* or *continuing* master's degree programs which typically take place over the course of four semesters. As in the case of undergraduate courses, earth building is again embedded as a partial subject in interdisciplinary programs such as building restoration or ecological building.

Universities also offer continuing education programs at a lower level than the master's programs. These programs include earth building as one or two-semester certificate courses. Earth building courses at universities can also be audited by students who then receive a certificate of participation but are usually not allowed to take exams.

In addition, there have been examples of "exporting" earth building courses into non-EU countries. In 2011 and 2014, such courses were developed and taught on location by the DVL on behalf of the Abu Dhabi Authority for Culture and Heritage (ADACH) and the Egyptian Earth Construction Association (EECA) [14, 15].

7.2 Networking

The term *network* basically refers to a system of individual parts which are linked with each other in numerous ways based on cause-and-effect relationships and general or specific system properties.

A well-linked system is able to quickly access important information through a tight network of relationships. By linking specific information, required action strategies can be developed, for example, to prevent a crisis situation or to improve one's public image. In order to improve the public perception of the "earth building system" as part of ecological or sustainable building, relevant sectors of society must identify with it. These sectors include the trades, the building industry, education, standardization, research, finances, and law. The structure of these sectors needs to be analyzed and suitable action strategies must be defined. The better the "network" between the individual sectors, the more successful the "earth building system" can be, for example, with regard to its public perception.

This cannot be accomplished by one individual. As Germany's national organization for the promotion of earth building, the Association for Building with Earth (Dachverband Lehm e.V.) has set itself the task of linking regional activities in the areas of construction, planning, building materials, and information with national activities in the areas of education, standardization, research, and international cooperation. In this context, the association carries out its own projects and serves as a partner in larger undertakings.

A crucial element is a persistent, targeted, and technically sound engagement with the media. In addition to print media, the Internet is the most common and most effective form of information exchange. Since 1999, the DVL has provided information about earth building on its website www.dachverband-lehm.de. A "discussion forum" on the site serves as a platform for the exchange of general information and experiences in the field of earth building. The website www.uni-terra.org serves as a platform for the global exchange of information on earth building in the academic field (Sect. 7.1). Every four years the DVL organizes international conferences and trade fairs in different "earth building regions" of Germany to facilitate the direct exchange of experiences in all areas of earth building. LEHM 2012 in Weimar was the sixth; LEHM 2016 will be the seventh conference based on this concept. All presentations are published in conference proceedings [16–19].

In Germany, a national earth building association serving as an "umbrella organization" has proven to be a successful model. Earth building specialists in a number of countries have followed similar paths, for example, in Switzerland, Australia, and New Zealand. More or less inspired by Germany's DVL, earth builders in other European countries have recently formed national earth building associations, for example, in the Czech Republic, France, Portugal, and Great Britain. It is becoming increasingly clear that a European or international earth building network needs to be formed in order to collectively tackle important tasks, such as the European standardization for earth building

A global network for the preservation of historic earthen architecture has formed under the umbrella of UNESCO-ICOMOS. Under the patronage of ICOMOS, international conferences on the preservation of earthen monuments have taken place in different regions of the world approximately every four years since 1972. These conferences have used the title "terra" since 1993. "Terra 2012" was the eleventh conference based on this concept and took place in Lima/Peru. "Terra 2016," the twelfth conference, will be held in Lyon/France. The presentations were published in conference proceedings. A system of international specialist committees based on the rules of UNESCO-ICOMOS was created to coordinate the work. The committee responsible for the field of the preservation of earthen architecture is the *International Scientific Committee on Earthen Architectural Heritage ISCEAH* (http://isceah.icomos.org).

7.3 Research and Standardization

In addition to education, standardization is another crucial area for the general acceptance of earth as a building material. In the future, earth can only be successful on national and international markets if the material is embedded in the system of technical building regulations and standards.

In Germany, the use of earth as a building material was officially regulated by the building authorities in 1999 in the form of Technical Regulations (Sect. 4.2.1.2). These regulations were updated in 2007 as part of a DVL project funded by the German Federal Environmental Foundation (DBU) and published as the third revised edition in 2008 [20, 32].

Today, most earth building materials in Germany are manufactured industrially. This made it necessary to develop product standards for earth building materials based on DIN 18200. In October 2012, the DIN work committee for "earth building," initiated by the DVL, presented drafts for three DIN product standards which were introduced in April 2013. Prospects for further development in the field of earth building standardization are described in Sect. 4.2.1.2.

In this context, research mainly focused on determining material parameters for earth building materials and finding suitable test methods. The research findings were included in the DIN drafts for earth building and are described in Sect. 3.6. Research needs to continue and expand to include building element testing.

The international development of earth building standards [21] shows that this topic is also becoming increasingly important in developing countries, especially within the context of earthquakes.

The parameter matrix in Table 1.1 serves as a guide for planning parameter studies. This matrix shows the relevant parameter groups and individual parameters for the processing stages of "construction soil," "earth building material," and "earth building structures" (Sect. 1.4.2).

At first, however, the suitability of the test methods currently used for earth building must be verified and, if necessary, modified. Test intervals must also be defined. These test methods have been largely borrowed from other building material sectors and more or less adapted to earth building materials. This matrix can be used to derive and define the relevant parameters for a specific earth building material based on its intended application.

Research and standardization is still a "vast and complex field." But regardless of the many questions which remain unanswered, it has become apparent that earth as a building material is returning to the standard references on building material science and on architectural and structural design [22–27].

7.4 Economic Development

After looking at the future prospects of earth building, the all-important question regarding the possibilities of economic development must be answered. The demand for earth building materials is growing. But is this demand stable enough for earth building to permanently assert itself as an (small) independent sector within the building industry? There has been no considerable increase in annual construction volume in Germany in recent years. This means that an increase in sales can only occur through a redistribution of market shares.

The development of the pit and quarry industry in Germany (which includes the production of earth building materials) shows that 6,500 companies with approx. 156,000 employees generated a revenue of approx. €23.5 billion in 2001 [28]. In 2008, the German Building Materials Association (BV Steine u. Erden e.V.) recorded a revenue of approx. €28 billion which was generated by approx. 6,000 companies employing approx. 140,000 workers (www.bvbaustoffe.de). The total construction investment volume for the same year was €251.3 billion (€140.8 billion for housing construction). How does earth building and the production of earth building materials relate to these numbers?

A study conducted at the Bauhaus University Weimar [29] analyzed the production of earth building materials in Germany from 1995 to 2000. In questionnaires, producers answered questions about the following topics:

- Internal company structures
- Production volumes/product groups/revenue
- Market segments/sales
- Qualification needs

Although the data is no longer current, it nonetheless illustrates a crucial period in the market development of earth building in Germany: showing the transition from its very beginnings to a small but stable market position as a result of the continually rising demand.

In those 5 years, the following trends were observed:

Many manufacturers started as "all-in-one" companies which means they combined all aspects of the production, marketing, and processing of building materials under one roof. Over time, however, a development toward a concentration on individual business segments became apparent.

This development was also true for the number of companies. At the beginning, a high fluctuation was observed. By the end of the study, the absolute number of companies increased at a much slower rate than the generated revenue per company. This indicates that production had become more stable and profitable. Today, some of the companies from that period are still active in the marketplace, and their business operations remain very strong and stable.

During the abovementioned study [29], the manufactured earth building materials were classified into four product groups: ready mixes/loose fill, ready-to-use mortars/plasters, earth blocks, and earth panels. The development of the production volumes for the individual product segments was then examined for the duration of the study. It was found that developments in the abovementioned segments differed considerably.

In 1995, the product groups "earth blocks" and "earth mortars" clearly dominated the earth building market with approx. 64 % and 29 %, respectively. In 1999, however, the market segment "earth panels" rose from practically 0 to 40 % of total sales. In the same year, the products "earth blocks" and "earth mortars" reached only 55 % combined, according to the collected data.

All of this changed starting in 2000 when the first major manufacturers of ready-mixed mortars "discovered" earth for themselves and expanded their product ranges to include earth building materials. There is no reliable data for the time period after the year 2000. The "perceived" development of the earth building material market shows that the product group "earth mortars" (especially earth plasters) is currently in the leading position.

It should also be mentioned that the willingness to provide information differed greatly from company to company. Revenue data should therefore only be seen as a rough estimate. According to the collected data, the generated annual sales tripled from approx. €3.4 million (converted from German marks) in 1995 to almost €9.3 million in 1999. These numbers represent 30 % average annual increases in revenue, with the years 1998 and 1999 clearly exceeding this amount.

When projecting these numbers onto the year 2014 based on "perceived" development, a steady annual growth in the lower double-digit range can be assumed. A conservative estimate of a 10 % annual growth rate would result in approx. €40 million in annual sales for 2014. Based on this, the market share of the production of earth building materials is in the range of 0.1 % of the total building material market. It is important to keep in mind that this number was as low as 0.01 % in 2001.

Although earth building has seen remarkable development over the past 15 years, the generated revenue (still) remains negligible when compared to the total volume of building material production. On the other hand, a slow but steady increase, contrary to the overall trend, also brings stability. Therefore, a "significant shift" or "major breakthrough" is not a reasonable expectation.

A "perceived" consumer open-mindedness regarding questions of environmental protection and, in particular, ecological building can be observed. However, when it comes to actual implementation, for example, through the use of earth building materials, skepticism is still widespread. If consumers' reservations can be successfully addressed, earth building will continue to develop in a positive direction. Potential strategies for achieving this were formulated in Sects. 7.1–7.31. In this context, major public buildings which adhere to politically enforced sustainable building principles [30] can serve as important role models for earth building [31].

In the field of manufacturing, investments in modern production facilities and the development of new products are limited primarily by insufficient financial resources. In addition, many manufactures adhere to the traditional way of doing things, feel reluctant to take risks, and do not see the necessity of good marketing. It is also becoming apparent that the increased amount of testing required for the production of earth building materials according to the DIN standards is a financial burden, especially for smaller companies.

Looking back on the past 20 years, earth building has developed in a very favorable way. It has transformed itself from a field made up of individual activities, which were belittled by many, into a credible, small but stable, independent sector of the building industry. Building with earth is becoming a common practice again.

References

1. Hohle, F.: Qualifizierung für den Lehmbau. In: LEHM 2000, Beiträge zur 3. Int. Fachtagung des Dachverbandes Lehm e.V., pp. 85–89. Overall, Berlin (2000)
2. Pitzing, U.-D.: Fachkraft im Lehmbau—Die erste handwerksrechtlich anerkannte deutsche Lehmbauausbildung. In: Zukunft Lehmbau 2002—Fachtagung 10 Jahre Dachverband Lehm e.V., Beiträge pp. 123–132. Dachverband Lehm e.V./Bauhaus-Universität Weimar, Weimar (2002)
3. Beuchel, E; Röhlen, U.: Modellhafte Qualifizierung zur "Fachkraft im Lehmbau". In: Zukunft Lehmbau 2002—Fachtagung 10 Jahre Dachverband Lehm e.V., Beiträge, pp. 111–122. Dachverband Lehm e.V./Bauhaus-Universität, Weimar (2002)
4. Beuchel, E.: Weiterbildung zur Fachkraft für Lehmbau—Inhalte und bisherige Erfahrungen. In: Moderner Lehmbau 2003, pp. 101–104. Fraunhofer IRB, Stuttgart (2003)
5. Dachverband Lehm e.V. (Hrsg.): Kurslehrbuch Fachkraft Lehmbau. Dachverband Lehm e.V., Weimar (2005)
6. Röhlen, U: Überführung der Qualifikation "Fachkraft Lehmbau" in eine bundesweit anwendbare Struktur. In: Moderner Lehmbau 2005, pp. 99–102, Umbra Umwelt- u. Unternehmensberatung GmbH (Hrsg.), Berlin (2005)
7. Schroeder, H.; Röhlen, U.; Jörchel, S.: Professional Training and Academic Teaching in Earth Architecture: The Activities and Experiences of the Dachverband Lehm e.V., Germany. In: Mediterra 2009; 1st Mediterranean Conference on Earth Architecture, Proc, Monfalcone, pp. 657–663. Edicom Edizioni, Italia (2009)
8. FAL e.v.: Lehmputze und Gestaltung/Clay Plaster. MVP, Ganzlin, CD (2005)
9. Bundesministerium für Bildung u. Forschung: Berufsbildungsgesetz (BBiG) v. 23 March 2005 BGBl. I, p. 931
10. Europäische Kommission (Hrsg.): Der Europäische Qualifikationsrahmen für lebenslanges Lernen (EQR). Amt für amtliche Veröffentlichungen der Europäischen Gemeinschaften, Luxemburg (2008)
11. Fal e.v. (Hrsg.): Handbuch ECVET Lehmbau, Ganzlin MVP: 2009 (www.earthbuilding.eu)
12. École Nationale Supérieure d'Architecture de Grenoble: Diplôme de spécialisation et d'approfondissement en architecture DSA—Architecture de Terre. Laboratoire CRATerre—ENSAG, Grenoble (2005)
13. Schreckenbach, H.: *Vernetzen im Lehmbau.* In: LEHM 2008 Koblenz, Beiträge zur 5. Internationalen Fachtagung für Lehmbau., pp. 74–77. Dachverband Lehm e.V., Weimar (2008)
14. Dachverband Lehm E.V./Abu Dhabi Authority for Culture & Heritage (ed.): Building with Earth—Course handbook for building trades: masons, carpenters, plasterers, Weimar: unpublished (2009)
15. Dachverband Lehm e.v./Egyptian Earth Construction Association (ed.): Planning and Building with Earth—Course Handbook for Architects, Civil Engineers And Urban Planners. Weimar: unpublished (2013)
16. KirchBauhof gGmbH (Hrsg. im Auftrag des DVL): LEHM 2000 Berlin—Beiträge zur 3. Internationalen Fachtagung Lehmbau des Dachverbandes Lehm e.V. Overall, Berlin (2000)
17. Dachverband Lehm e.V. (Hrsg.): LEHM 2004 Leipzig—Tagungsbeiträge der 4. Internationalen Fachtagung für Lehmbau. Eigenverlag Dachverband Lehm e.V., Weimar (2004)
18. Dachverband Lehm e.V. (Hrsg.): LEHM 2008 Koblenz—Tagungsbeiträge der 5. Internationalen Fachtagung für Lehmbau. Eigenverlag Dachverband Lehm e.V., Weimar (2008)
19. Dachverband Lehm e.V. (Hrsg.): LEHM 2012 Weimar—Tagungsbeiträge der 6. Internationalen Fachtagung für Lehmbau. Eigenverlag Dachverband Lehm e.V., Weimar (2012)
20. Dachverband Lehm e.V. (Hrsg.): Lehmbau Regeln—Begriffe, Baustoffe, Bauteile, 3., überarbeitete Aufl. Vieweg + Teubner I GWV Fachverlage, Wiesbaden (2009)
21. Schroeder, H.: Modern earth building codes, standards and normative development. In: Hall, M.R., Lindsay, R., Krayenhoff, M. (eds.) Modern earth buildings—Materials, engineering, construction and applications, pp. 72–109. Woodhead Publishing Series in Energy: Nr 33. Woodhead, Oxford (2012)

22. Backe, H.; Hiese, W.: Baustoffkunde für Berufs- und Technikerschulen und zum Selbstunterricht, 8. Aufl. Werner, Düsseldorf (1997)
23. Dierks, K.; Wormuth, R. (Beitrag Ziegert): Baukonstruktion, 6. Aufl. Werner, Düsseldorf (2007)
24. Hegger, M.; Auch-Schwelk, V.; Fuchs, M., Rosenkranz, T.: Baustoff Atlas. Architektur/Edition Detail. Birkhäuser-Verlag f., Basel/München (2005)
25. Achtziger, J.; Pfeifer, G.; Ramcke, R.; Zilch, K.: .: Mauerwerk Atlas. Architektur/Edition Detail. Birkhäuser, Basel (2001)
26. Schroeder, H.: Konstruktion und Ausführung von Mauerwerk aus Lehmsteinen. In: Mauerwerkkalender 2009, pp. 271–290. W. Ernst & Sohn, Berlin (2009)
27. Minke, G.; Schroeder, H.; Ziegert, C.: Das Technische Merkblatt "Anforderungen an Lehmputze" des Dachverbandes Lehm e.V. In: Europäischer Sanierungskalender 2009, pp. 105–111. Beuth, Berlin (2009)
28. Rese, F.; Strauß, H. (Hrsg.): Die Steine- und Erden-Industrie—Baustoffe 2000 (2001/2002, 2003). Stein-Verlag GmbH, Baden-Baden, 2000 (2002, 2003)
29. Schroeder, H.: Die Produktion von Lehmbaustoffen in Deutschland—aktuelle Situation und Tendenzen. Weimar, Bauhaus-Universität, unpublished study (2001)
30. Röhlen, U.; Mai, D.: Trockenbau mit Lehmbau im "Klimareferat der Vereinten Nationen", Bonn. In: LEHM 2012 Weimar, Beiträge zur 6. Int. Fachtagung für Lehmbau, pp. 50–53. Dachverband Lehm e.V., Weimar (2012)
31. Bundesministerium für Verkehr, Bau und Stadtentwicklung (BMVBS) (Hrsg.): Leitfaden Nachhaltiges Bauen. BMVBS, Weimar (2011)
32. Schroeder, H.; Volhard, F.; Röhlen, U.: Die Lehmbau Regeln 2008–10 Jahre Erfahrungen mit der praktischen Anwendung. In: LEHM 2008 Koblenz, Beiträge zur 5. Internationalen Fachtagung für Lehmbau, pp. 12–21. Dachverband Lehm e.V., Weimar (2008)

Cited Standards as of March 2015

Number	Edition	Title
DIN 105-100	2012-01	Clay masonry units—Part 100: Clay masonry units with specific properties
DIN 276-1	2008-12	Building costs—Part 1: Building construction
DIN 276-4	2009-08	Building costs—Part 4: Civil constructions
DIN 1048-2	1991-06	Testing concrete; testing of hardened concrete (specimens taken in situ)
DIN 1048-5	1991-06	Testing concrete; testing of hardened concrete (specimens prepared in mould)
DIN 1053-1	1996-11	Masonry—Part 1: Design and construction
DIN 1055-2	2010-11	Actions on structures—Part 2: Soil properties
DIN 4102-1	1998-05	Fire behaviour of building materials and building components—Part 1: Building materials; concepts, requirements and tests
DIN 4102-2	1977-09	Fire behaviour of building materials and building components; Building Components, Definitions, Requirements and Tests
DIN 4102-4 (D)	2014-06	Fire behaviour of building materials and building components—Part 4: Synopsis and application of classified building materials, components and special components
DIN 4103-1	2014-03	Internal non-loadbearing partitions
DIN 4108-2	2013-02	Thermal protection and energy economy in buildings—Part 2: Minimum requirements to thermal insulation
DIN 4108-3	2014-11	Thermal protection and energy economy in buildings—Part 3: Protection against moisture subject to climate conditions—requirements and directions for design and construction
DIN 4108-4	2013-02	Thermal insulation and energy economy in buildings—Part 4: Hygrothermal design values

(continued)

© Springer International Publishing Switzerland 2016

H. Schroeder, *Sustainable Building with Earth*,
DOI 10.1007/978-3-319-19491-2

(continued)

Number	Edition	Title
DIN V 4108-6	2004-03	Thermal protection and energy economy in buildings—Part 6: Calculation of annual heat and energy use; Corrigendum 1
DIN 4108-7	2011-01	Thermal insulation and energy economy in buildings—Part 7: Air tightness of buildings—requirements, recommendations and examples for planning and performance
DIN 4109-1 (D)	2013-06	Sound insulation in buildings—Part 1: Requirements to sound insulation
DIN 18121-1	1998-04	Soil, investigation and testing—water content—Part 1: Determination by drying in oven
DIN 18122-1	1997-07	Soil—investigation and testing—consistency limits—Part 1: Determination of liquid limit and plastic limit
DIN 18122-2	2000-09	Soil—investigation and testing—consistency limits—Part 2: Determination of the shrinkage limit
DIN 18123	2011-04	Soil—investigation and testing—determination of grain-size distribution
DIN 18124	2011-04	Soil—investigation and testing—determination of density of solid particles—capillary pycnometer, wide mouth pycnometer, gas pycnometer
DIN 18125-1	2010-07	Soil—Investigation and testing—determination of density of soil—Part 1: Laboratory tests
DIN 18127	2012-09	Soil—investigation and testing—Proctor-test
DIN 18128	2002-12	Soil—investigation and testing—determination of ignition loss
DIN 18129	2011-07	Soil—investigation and testing—determination of lime content
DIN 18132	2012-04	Soil, testing procedures and testing equipment—determination of water absorption
DIN 18134	2011-07	Soil—testing procedures and testing equipment—plate load test
DIN 18135	2012-04	Soil—investigation and testing—oedometer consolidation test
DIN 18136	2003-11	Soil—investigation and testing—unconfined compression test
DIN 18137-1	2010-07	Soil, Investigation and testing—determination of shear strength—Part 1: Concepts and general testing conditions
DIN 18137-2	2011-04	Soil, Investigation and testing—determination of shear strength—Part 2: Triaxial test
DIN 18137-3	2002-09	Soil, Investigation and testing—determination of shear strength—Part 3: Direct shear test

(continued)

(continued)

Number	Edition	Title
DIN 18196	2011-05	Earthworks and foundations—soil classification for civil engineering purposes
DIN 18200	2000-05	Assessment of conformity for construction products—certification of construction products by certification body
DIN 18300	2012-09	German construction contract procedures (VOB)—Part C: General technical specifications in construction contracts (ATV)—earthworks
DIN 18319	2012-09	German construction contract procedures (VOB)—Part C: General technical specifications in construction contracts (ATV)—trenchless pipelaying
DIN 18330	2012-09	German construction contract procedures (VOB)—Part C: General technical specifications in construction contracts (ATV)—masonry work
DIN 18331	2012-09	German construction contract procedures (VOB)—Part C: General technical specifications in construction contracts (ATV)—concrete works
DIN 18340	2012-09	German construction contract procedures (VOB)—Part C: General technical specifications in construction contracts (ATV)—dry lining and partitioning work
DIN 18350	2012-09	German construction contract procedures (VOB)—Part C: General technical specifications in construction contracts (ATV)—plastering and rendering
DIN 18353	2012-09	German construction contract procedures (VOB)—Part C: General technical specifications in construction contracts (ATV)—laying of floor screed
DIN 18550-2	2015-06	Design, preparation and application of external rendering and internal plastering–Part 2: Supplementary provisions for DIN EN 13914-2 for internal plastering
DIN 18555-6	1987-11	Testing mortars containing mineral binders; determination of bond strength of hardened mortar
DIN V 18599-1	2013-05	Energy efficiency of buildings—calculation of the net, final and primary energy demand for heating, cooling, ventilation, domestic hot water and lighting—Part 1: General balancing procedures, terms and definitions, zoning and evaluation of energy sources; Corrigendum 1
DIN 18945	2013-08	Earth blocks—terms and definitions, requirements, test methods

(continued)

(continued)

Number	Edition	Title
DIN 18946	2013-08	Earth masonry mortar—terms and definitions, requirements, test methods
DIN 18947/1	2015-03	Earth plasters—terms and definitions, requirements, test methods/Amendment 1
DIN 52108	2010-05	Testing of inorganic non-metallic materials—wear test using the grinding wheel according to Böhme—grinding wheel method
DIN 52612-3	1979-09	Testing of thermal insulating materials; determination of thermal conductivity by the guarded hot plate apparatus; thermal resistance of laminated materials for use in building practice
DIN 20000-401	2012-11	Application of building products in structures—Part 401: Rules for the application of clay masonry units according to DIN EN 771-1:2011-07
DIN EN 717-1	2005-01	Wood-based panels—determination of formaldehyde release—Part 1: Formaldehyde emission by the chamber method
DIN EN 771-1	2011-07	Specification for masonry units—Part 1: Clay masonry units
DIN EN 772-1	2011-07	Methods of test for masonry units—Part 1: Determination of compressive strength
DIN EN 772-9	2005-05	Methods of test for masonry units—Part 9: Determination of volume and percentage of voids and net volume of clay and calcium silicate masonry units by sand filling
DIN EN 998-1	2010-12	Specification for mortar for masonry—Part 1: Rendering and plastering mortar
DIN EN 998-2	2010-12	Specification for mortar for masonry—Part 2: Masonry mortar
DIN EN 1015-1	2007-05	Methods of test for mortar for masonry—Part 1: Determination of particle size distribution (by sieve analysis)
DIN EN 1015-2	2007-05	Methods of test for mortar for masonry—Part 2: Bulk sampling of mortars and preparation of test mortars
DIN EN 1015-3	2007-05	Methods of test for mortar for masonry—Part 3: Determination of consistence of fresh mortar (by flow table)
DIN EN 1015-10	2007-05	Methods of test for mortar for masonry—Part 10: Determination of dry bulk density of hardened mortar
DIN EN 1015-11	2007-05	Methods of test for mortar for masonry—Part 11: Determination of flexural and compressive strength of hardened mortar

(continued)

(continued)

Number	Edition	Title
DIN EN 1015-12	2000-06	Methods of test for mortar for masonry—Part 12: Determination of adhesive strength of hardened rendering and plastering mortars on substrates
DIN EN 1015-19	2005-01	Methods of test for mortar for masonry—Part 19: Determination of water vapour permeability of hardened rendering and plastering mortars
DIN EN 1052-3	2007-06	Methods of test for masonry—Part 3: Determination of initial shear strength
DIN EN 1990 /NA/A1	2012-08	Eurocode: Basis of structural design/National annex/Amendment 1
DIN EN 1991-1-1 /NA/A1	2014-07	Eurocode 1: Actions on structures—Part 1-1: General actions—densities, self-weight, imposed loads for buildings/national annex—nationally determined parameters/Amendment 1
DIN EN 1995-1-1 /A2	2014-07	Eurocode 5: Design of timber structures—Part 1-1: General—common rules and rules for buildings/Amendment 2
DIN EN 1996-1-1 /NA/A2	2015-01	Eurocode 6: Design of masonry structures—Part 1-1: General rules for reinforced and unreinforced masonry structures/national annex—nationally determined parameters/Amendment 2
DIN EN 1996-2	2012-01	Eurocode 6: Design of masonry structures—Part 2: Design considerations, selection of materials and execution of masonry
DIN EN 1998/A1	2013-05	Eurocode 8: Design of structures for earthquake resistance—Part 1: General rules, seismic actions and rules for buildings/Amendment 1
DIN EN 12354-1	2000-12	Building acoustics—estimation of acoustic performance of buildings from the performance of products—Part 1: Airborne sound insulation between rooms
DIN EN 12354-2	2000-09	Building acoustics—estimation of acoustic performance of buildings from the performance of elements—Part 2: Impact sound insulation between rooms
DIN EN 12390-6	2010-09	Testing hardened concrete—Part 6: Tensile splitting strength of test specimens
DIN EN 12620	2008-07	Aggregates for concrete
DIN EN 12664	2001-05	Thermal performance of building materials and products—determination of thermal resistance by means of guarded hot plate and heat flow meter methods—dry and moist products with medium and low thermal resistance

(continued)

(continued)

Number	Edition	Title
DIN EN 12878	2014-07	Pigments for the colouring of building materials based on cement and/or lime—specifications and methods of test
DIN EN 13055 (D)	2012-04	Lightweight aggregates for concrete, mortar, grout, bituminous mixtures, surface treatments and for unbound and bound applications
DIN EN 13139	2004-12	Aggregates for mortar/Corrigenda 1
DIN EN 13300	2002-11	Paints and varnishes—water-borne coating materials and coating systems for interior walls and ceilings—classification
DIN EN 13501-1	2010-01	Fire classification of construction products and building elements—Part 1: Classification using data from reaction to fire tests
DIN EN 13501-2	2010-02	Fire classification of construction products and building elements—Part 2: Classification using data from fire resistance tests, excluding ventilation services
DIN EN 13914-2 (D)	2013-09	Design, preparation and application of external rendering and internal plastering—Part 2: Design considerations and essential principles for internal plastering
DIN EN 15643-1	2010-12	Sustainability of construction works—sustainability assessment of buildings—Part 1: General framework
DIN EN 15643-2	2011-05	Sustainability of construction works—sustainability assessment of buildings—Part 2: Framework for the assessment of environmental performance
DIN EN 15643-3	2012-04	Sustainability of construction works—sustainability assessment of buildings—Part 3: Framework for the assessment of social performance
DIN EN 15643-4	2012-04	Sustainability of construction works—sustainability assessment of buildings—Part 4: Framework for the assessment of economic performance
DIN EN 15804 /A1	2014-07	Sustainability of construction works—environmental product declarations—core rules for the product category of construction products/Amendment 1
DIN EN 15933	2012-11	Sludge, treated biowaste and soil—determination of ph
DIN EN 15942	2012-01	Sustainability of construction works—environmental product declarations—communication format business-to-business
DIN EN 15978	2012-10	Sustainability of construction works—assessment of environmental performance—calculation method

(continued)

(continued)

Number	Edition	Title
DIN EN 16309 /A1	2014-12	Sustainability of construction works—assessment of social performance—calculation methodology/Amendment 1
DIN EN 16627	2013-07	Sustainability of construction works—assessment of economical performance—calculation method
DIN EN ISO 6946	2008-04	Building components and building elements—thermal resistance and thermal transmittance—calculation method
DIN EN ISO 7345	1996-01	Thermal insulation—physical quantities and definitions
DIN EN ISO 7500-1 (D)	2014-05	Metallic materials—verification of static uniaxial testing machines—Part 1: Tension/compression testing machines—verification and calibration of the force-measuring system
DIN EN ISO 7730	2006-05	Ergonomics of the thermal environment—analytical determination and interpretation of thermal comfort using calculation of the PMV and PPD indices and local thermal comfort criteria
DIN EN ISO 9346	2008-02	Hygrothermal performance of buildings and building materials—physical quantities for mass transfer—vocabulary
DIN EN ISO 11885	2009-09	Water quality—determination of selected elements by inductively coupled plasma optical emission spectrometry
DIN EN ISO 12571	2013-12	Hygrothermal performance of building materials and products—determination of hygroscopic sorption properties
DIN EN ISO 12572 (D)	2015-01	Hygrothermal performance of building materials and products—determination of water vapour transmission properties
DIN EN ISO 12846	2012-08	Water quality—determination of mercury—method using atomic absorption spectrometry (AAS) with and without enrichment
DIN EN ISO 14021	2012-04	Environmental labels and declarations—self-declared environmental claims (Type II environmental declaration)
DIN EN ISO 14024	2001-02	Environmental labels and declarations—type I environmental labeling—principles and procedures
DIN EN ISO 14025	2011-10	Environmental labels and declarations—type III environmental declarations—principles and procedures
DIN EN ISO 14040	2009-11	Environmental management—life cycle assessment—principles and framework
DIN EN ISO 14044	2006-10	Environmental management—life cycle assessment—requirements and guidelines

(continued)

(continued)

Number	Edition	Title
DIN EN ISO 14688-1	2013-12	Geotechnical investigation and testing—identification and classification of soil—Part 1: Identification and description
DIN EN ISO 14688-2	2013-12	Geotechnical investigation and testing—identification and classification of soil—Part 2: Principles for a classification
DIN EN ISO 15148	2003-03	Hygrothermal performance of building materials and products—determination of water absorption coefficient by partial immersion
DIN (EN) ISO 16000-3	2013-01	Indoor air—Part 3: Determination of formaldehyde and other carbonyl compounds in indoor air and test chamber air—active sampling method
DIN (EN) ISO 16000-6	2012-11	Indoor air—Part 6: Determination of volatile organic compounds in indoor and test chamber air by active sampling on Tenax TA sorbent, thermal desorption and gas chromatography using MS or MS-FID
DIN EN ISO 16000-9	2008-04	Indoor air—Part 9: Determination of the emission of volatile organic compounds from building products and furnishing—emission test chamber method
DIN EN ISO 16000-11	2006-06	Indoor air—Part 11: Determination of the emission of volatile organic compounds from building products and furnishing—sampling, storage of samples and preparation of test specimens
DIN EN ISO 17294-2 (D)	2014-12	Water quality—application of inductively coupled plasma mass spectrometry (ICP-MS)—Part 2: Determination of 62 elements
DIN EN ISO 22475-1	2007-01	Geotechnical investigation and testing—sampling methods and groundwater measurements—Part 1: Technical principles for execution
ISO 15392	2008-05	Sustainable building in construction—general principles
ISO 21929-1	2011-11	Sustainability in building construction—sustainability indicators—Part 1: Framework for the development of indicators and a core set of indicators for buildings
ISO 21930	2007-10	Sustainability in building construction—environmental declaration of building products
ISO 21931-1	2010-06	Sustainability in building construction—framework for methods of assessment of the environmental performance of construction works—Part 1: Buildings

(D) Draft

List of Illustrations

The author and publishers wish to thank everyone who contributed to the production of this book by providing their images, granting reproduction authorization or making information available. Despite intensive efforts it was not possible to identify all copyright holders. Their copyrights are nevertheless protected. Please contact us if you have not been identified as the owner of an image. All photos and illustrations not listed here are by the author.

Chapter 1: The Development of Earth Building

Figure	Bibliography no.	Author
1.1, 1.16	[1]	
1.3		Susanne Schroeder, Weimar, Germany
1.4	[2]	
1.5	[3]	
1.6	[4]	
1.7	[5]	
1.8	[6]	
1.9		Hannah Schreckenbach, Magdeburg, Germany
1.10	[7]	
1.11, 1.13	[8]	Prof. Dr. Stefan Simon, Berlin, Germany
1.14a	[12]	
1.15b		b: www.wmf.org
1.17	[21]	Bianca Isensee, Weimar, Germany
1.18, 4.39	[25]	
1.19	[26]	Franz Volhard, Darmstadt, Germany
1.20	[27]	Stadtmuseum Nürnberg
1.21, 1.22	[29]	
1.24, 1.25	[33]	

© Springer International Publishing Switzerland 2016

H. Schroeder, *Sustainable Building with Earth*,
DOI 10.1007/978-3-319-19491-2

Chapter 2: Construction Soil—Sourcing, Extraction and Classification

Figure	Bibliography no.	Author
2.1, 2.4, 2.5, 2.6, 2.7	[1]	
2.2	[2]	
2.3, 2.33	[3]	
2.8	[5]	
2.9	[15]	
2.15	[9, 37]	
2.11, 2.12, 2.13, 2.22, 2.23, 2.24, 2.25, 2.29	[10]	Thomas Schnellert, Weimar, Germany
2.14, 2.16, 2.31	[11]	2.31 REM: Wikipedia
2.17, 2.18	[14]	
2.19 top, 2.27, 2.28, 2.30	[9]	
2.19 bottom	[15]	
2.26	[27]	
2.32	[20, 27]	
2.33	[3, 23]	
2.34	[20, 27]	
2.35	[34, 35]	
2.15	[38]	

Chapter 3: Earth Building Materials—Production, Requirements and Testing

Figure	Bibliography no.	Author
3.1	[3]	
3.2	[2]	
3.3	[4]	
3.6	[7]	
3.7, 3.28	[6]	
3.8	[8]	Claytec, Viersen, Germany
3.9	[9, 10, 74]	
3.10, 3.20, 3.35	[7]	Dachverband Lehm e. V. (DVL)
3.11, 3.38, 3.39	[11]	
3.13	[2, 14]	
3.24, 3.21	[16]	Bachrom A. Tulaganow, Weimar, Germany
3.14	[10, 15]	
3.15	[8]	
3.17	[75]	Diagram
3.22	[3, 76]	

(continued)

(continued)

Figure	Bibliography no.	Author
3.23	[17]	Photo: Wikipedia
3.24	[2, 16, 18]	
3.25	[14, 19, 20]	
3.26		System Putzmeister
3.27	[14, 25]	(b) Quentin Wilson, El Rito, NM, USA
		(c) Claytec, Viersen, Germany
3.29, 3.47	[26]	
3.31	[28]	
3.32		According to DIBt
3.33	[10]	
3.34	[77]	Tom Hiersemann, Weimar, Germany
3.36	[81]	
3.37	[11, 43]	Thomas Schnellert, Weimar, Germany
3.40	[37]	
3.41, 3.42	[42]	
3.43		According to DIN 18945
3.44	[40]	
3.45	[38, 78]	
3.46	[43]	Photo: Thomas Schnellert, Weimar, Germany
3.48	[51]	
3.49	[40, 78]	
3.50, 3.54	[49, 50]	
3.51	[40]	
3.52	[55]	
3.54	[50]	
3.55	[61]	
3.56	[62]	
3.57, 3.58	[63]	
3.59	[64]	
3.60		According to DIN 18952-2
3.61, 3.62, 3.63	[66]	
3.4, 3.12	[5]	
3.5 diagram	[74]	

Chapter 4: Earthen Structures—Planning, Building and Construction Supervision

Figure	Bibliography no.	Author
4.2, 4.5, 4.7, 4.8, 4.10, 4.11, 4.12, 4.77	[1]	
4.3	[2]	
4.9		www.santafeproperties.com
4.13	[4]	
4.16	[107]	Drawings
4.17	[6]	
4.18		According to www.dibt.de
4.20, 4.38, 4.49, 4.50, 4.55, 4.56, 4.57, 4.78, 4.79	[52]	DVL/Julian Reisenberger (JR), Weimar and Claytec, Viersen, Germany
4.20	[55]	Julian Reisenberger, Weimar, Germany
4.21, 4.43, 4.68		Hannah Schreckenbach, Magdeburg, Germany
4.24, 4.29	[61]	
4.26	[54]	
4.30	[66]	
4.31	[68]	
4.32	[69]	David Easton, Napa, CA, USA
4.33	[72]	
4.34	[73]	Ivana Zabičkova, Brno, Czech Republic
4.36, 4.38, 4.42	[76]	Bachrom A. Tulaganow, Weimar, Germany
4.35, 4.60, 4.68	[74, 98]	Hannah Schreckenbach, Magdeburg, Germany
4.38 (JR), 4.41, 4.45, 4.69, 4.70, 4.71, 4.73, 4.74,	[51]	Dachverband Lehm e.V. (DVL)
4.39	[77]	
4.40	[79, 80]	
4.44	[84]	Anke Richter, Weimar, Germany
4.46		Claytec, Viersen, Germany
4.47	[85]	
4.48	[87, 88]	(b) Tom Morton, Auchtermuchty, UK, (c) Diagram: claytec
4.51	[89]	
4.53		André Sudmann, Weimar, Germany
4.54	[53]	
4.56		Anna Migliaccio, Weimar, Germany
4.64	[91, 112]	
4.65	[94]	Richard Orgel, Weimar, Germany
4.67	[97]	
4.72	[100]	
4.75	[103]	
4.76	[104]	

(continued)

(continued)

Figure	Bibliography no.	Author
4.81	[105]	
4.82	[106]	
4.59	[83] (a)	(b) Claytec

Chapter 5: Structures Built of Earth Building Materials—Impacts, Structural Damages and Preservation

Figure	Bibliography no.	Author
5.3	[2]	
5.4	[4]	
5.5	[9]	
5.6	[5]	
5.8	[11]	
5.9	[13]	
5.10	[25]	
5.11	[28]	
5.12	[6]	
5.14, 5.15, 5.17, 5.18, 5.19, 5.21, 5.23, 5.25, 5.28, 5.45	[30]	Tino Köhler, Weimar, Germany
5.20	[32]	
5.22, 5.24, 5.26, 5.44	[33]	Yvonne Maiwald; Bettina Hein, Weimar, Germany
5.27		Schader, Weimar, Germany
5.30		UNDP
5.31	[40]	
5.34	[43]	
5.35	[44]	.
5.37, 5.38, 5.39, 5.40, 5.66, 5.68, 5.69	[45]	Shamil A. Chakimov, Tashkent, Uzbekistan
5.46	[47]	
5.48	[52]	
5.49, 5.55	[85]	Daniel Torrealva, Lima, Peru
5.50	[86]	
5.50 (JR), 5.51 (JR/Claytec)	[59]	
5..52, 5.53, 5.54	[71]	
5.55	[75]	
5.56, 5.57	[76]	Manfred Fahnert, Flammersfeld, Germany
5.58	[74]	
5.61	[78]	

(continued)

(continued)

Figure	Bibliography no.	Author
5.62		Archaeological Museum Samarkand, Uzbekistan
5.63	[81]	
5.65, 5.66, 5.67	[41]	
5.66	[84]	
5.42	[46]	

Chapter 6: Demolition, Recycling and Disposal of Earth Building Materials

Figure	Bibliography no.	Author
6.1, 6.2	[14]	Jürgen Kurz, Weimar, Germany

Chapter 7: The Future of Earth Building

Figure	Bibliography no.	Author
7.1		Archive City Museum Weimar, Germany

Index

© Springer International Publishing Switzerland 2016

H. Schroeder, *Sustainable Building with Earth*,
DOI 10.1007/978-3-319-19491-2